Springer Tracts in Modern Physics

Volume 256

For further volumes:
http://www.springer.com/series/426

Springer Tracts in Modern Physics

Springer Tracts in Modern Physics provides comprehensive and critical reviews of topics of current interest in physics. The following fields are emphasized: Elementary Particle Physics, Condensed Matter Physics, Light Matter Interaction, Atomic and Molecular Physics, Complex Systems, Fundamental Astrophysics.

Suitable reviews of other fields can also be accepted. The Editors encourage prospective authors to correspond with them in advance of submitting a manuscript. For reviews of topics belonging to the above mentioned fields, they should address the responsible Editor as listed below.

Special offer: For all clients with a print standing order we offer free access to the electronic volumes of the Series published in the current year.

Wolfgang Quint · Manuel Vogel
Editors

Fundamental Physics in Particle Traps

Springer

Editors
Wolfgang Quint
Manuel Vogel
GSI Helmholtz-Zentrum für
Schwerionenforschung
Darmstadt
Germany

ISSN 0081-3869 ISSN 1615-0430 (electronic)
ISBN 978-3-642-45200-0 ISBN 978-3-642-45201-7 (eBook)
DOI 10.1007/978-3-642-45201-7
Springer Heidelberg New York Dordrecht London

Library of Congress Control Number: 2013957878

Printed on acid-free paper

Springer is part of Springer Science+Business Media (www.springer.com)

Foreword

It is surprising—even ironic—that this volume on precision measurements in ion traps exists. For it is simple to prove that it is impossible to make a good trap for ions—that is a static trap wherein the force on an ion is always toward the center. This follows from Earnshaw's theorem: in a field-free region, the lines of electric field are continuous. Thus if the field along the z-axis is inward so the z-force is inward, causing the particle to oscillate stably along z, then those field lines must diverge radially outwards in the equatorial plane, i.e. we have an electrostatic quadrupole field configuration. Hence, the ion will escape (with exponential motion) out from the sides of the trap.

Fortunately, there are two different ways to make a useful ion trap. The first way to defeat Earnshaw's theorem is by varying the quadrupole field with time. Imagine that an ion is created in the field described above. The ion might start to accelerate outwards in the equatorial plane. After a short time, it will be further from the center in the equatorial plane than where it started. If the quadrupole field is now reversed, then the ion will receive an inwards kick that is stronger than the outward kick (remember that the force is proportional to the distance from the center), resulting in a net inward impulse and hence confinement. If the ion were initially at positive z, it would receive an inward impulse, and later a weaker outward one, again resulting in net confinement. However, if the time of the field reversal is too long, the ions may have moved beyond the center of the trap to negative z, in which case the reversal may give an outward impulse that is greater than the first inward one, hence expelling the ion. Thus there are only some combinations of frequencies and quadrupole field strengths for which the trap is stable. Under these conditions, the ion undergoes a micro-motion at the oscillation frequency of the field reversals, and the average force gives a harmonic secular oscillation at a lower frequency. The regions of stability and instability have been turned to advantage in making the quadrupole mass filters so useful as mass spectrometers for large molecules [1].

The second method for making a useful ion trap is to frustrate the ions' escape due to the radially outward force in the equatorial plane of the quadrupole field. By adding a magnetic field along z, the initial radial outward motion gives the ion a tangential force that causes it to circle the trap center—the so-called magnetron orbital motion. The Penning trap is thereby dynamically stabilized—however, if residual gas causes a slight damping, the ion's magnetron motion will be slowed and the ion will consequently make its way radially outward—toward the lower electrostatic potential. The combination of a quadrupole electric field plus an axial magnetic field is called a "Penning Trap" to acknowledge Penning's use of it to make better glow discharges at low pressure [2].

Ion traps offer the possibility of high precision due to their combination of long measurement times on isolated particles. I long admired the precise measurements on single electrons by the Dehmelt group, and when I met him I was curious about what made him commit to tackling such an ambitious experiment. His response was "when I realized that the magnetron frequency was independent of radius"—presumably because this made him think that mode coupling might work. Thus his group became a pioneer in making precision measurements of both electrons and ions.

This brief Foreword gives some background for the current volume—a rich collection of reports of breathtaking precision and important physical measurements done on confined electrons, protons, ions, antiparticles, and antimatter. It is important to emphasize that this volume describes only a moment in time: every result reported here can be improved by applying the techniques that underlie these measurements—imagination and persistent hard work. I certainly hope that some readers of this volume will take up the implicit challenge of going far beyond the measurements and scientific results presented here.

Cambridge, October 2013 David E. Pritchard

References

1. W. Paul, Electromagnetic traps for charged and neutral particles. Rev. Mod. Phys. **62**, 531 (1990)
2. F.M. Penning, Die Glimmentladung bei niedrigem Druck zwischen koaxialen Zylindern in einem axialen Magnetfeld. Physica. **3,** 873 (1936)

Preface

Particle traps are successful tools for a broad range of high-precision spectroscopy experiments on all frequency scales. Single particles or defined particle ensembles can be well-isolated from external influences, prepared with respect to their internal and external states and confined for times up to months. Traps provide a multitude of manipulation and measurement techniques including nondestructive detection methods down to the single-particle level. They are hence nearly ideal surroundings for fundamental physics at low energies, particularly for high-precision measurements of fundamental physical quantities which employ spectroscopic methods.

In this book, we intend to give detailed insight into fundamental physics with particles confined in traps. This comprises experiments and theoretical predictions at the highest precisions reached so far. Such investigations allow stringent tests of quantum electrodynamics and the Standard Model, antiparticle and antimatter research, study of fundamental symmetries, constants, and their possible variations with time and space. They are key to fundamental aspects in metrology such as mass measurements and time standards, as well as promising to further developments in quantum information processing. We focus on experiment and theory of magnetic moments, both for unbound and bound electrons, as well as on the particle–antiparticle cases of the electron and the positron and of the proton and the antiproton. This is directly connected with antimatter research on the antihydrogen system, the bound state of the positron and the antiproton. We treat precision mass measurements of radionuclides using traps and quantum information processing with confined particles. A second focus is on fundamental physics with highly charged ions, which feature a number of unique properties that make them interesting under various aspects to be discussed.

Darmstadt, October 2013

Manuel Vogel
Wolfgang Quint

Contents

Contributors

Julian C. Berengut School of Physics, University of New South Wales, Sydney, NSW 2052, Australia, e-mail: julianberengut@gmail.com

Michael Block GSI Helmholtzzentrum für Schwerionenforschung, Planckstrasse 1, 64291 Darmstadt, Germany, e-mail: m.block@gsi.de

José R. Crespo López-Urrutia Max-Planck-Institut für Kernphysik, Saupfercheckweg 1, 69117 Heidelberg, Germany, e-mail: jose.crespo@mpi-hd.mpg. de

Joshua Dorr Department of Physics, Harvard University, 17 Oxford Street, Cambridge, MA 02138, USA, e-mail: z.harman@mpi-hd.mpg.de

Victor V. Flambaum School of Physics, University of New South Wales, Sydney, NSW 2052, Australia, e-mail: v.flambaum@unsw.edu.au

Gerald Gabrielse Department of Physics, Harvard University, 17 Oxford Street, Cambridge, MA 02138, USA, e-mail: gabrielse@physics.harvard.edu

Dmitry A. Glazov Department of Physics, St. Petersburg State University, Oulianovskaya 1, Petrodvorets, St. Petersburg, Russia 198504; Institut für Theoretische Physik, Technische Universität Dresden, Mommsenstraße 13, 01062 Dresden, Germany

Jeffrey S. Hangst Department of Physics and Astronomy, Ny Munkegade 120, 8000 Aarhus C, Denmark, e-mail: jeffrey.hangst@cern.ch

Zoltan Hárman Max-Planck-Institut für Kernphysik, Saupfercheckweg 1, 69117 Heidelberg, Germany, e-mail: z.harman@mpi-hd.mpg.de

Masashi Hayakawa Department of Physics, Nagoya University, Nagoya 464-8602, Japan, e-mail: masashi.hayakawa@gmail.com

Shannon Fogwell Hoogerheide Department of Physics, Harvard University, 17 Oxford Street, Cambridge, MA 02138, USA

Peter J. Mohr Atomic Physics Division, 100 Bureau Drive, Gaithersburg, MD 20899, USA, e-mail: mohr@nist.gov

Elise Novitski Department of Physics, Harvard University, 17 Oxford Street, Cambridge, MA 02138, USA

Andrew Ong School of Physics, University of New South Wales, Sydney, NSW 2052, Australia, e-mail: ong.andrew@gmail.com

Günter Plunien Institut für Theoretische Physik, Technische Universität Dresden, Mommsenstraße 13, 01062 Dresden, Germany, e-mail: Guenter.Plunien@tudresden.de

Wolfgang Quint GSI Helmholtzzentrum für Schwerionenforschung, Planckstrasse 1, 64291 Darmstadt, Germany, e-mail: w.quint@gsi.de

Christian F. Roos Institut für Experimentalphysik, Universtität Innsbruck, Technikerstr. 25, Innsbruck 6020, Austria, e-mail: Christian.Roos@uibk.ac.at

Vladimir M. Shabaev Division of Quantum Mechanics, Department of Physics, Saint-Petersburg State University, St. Petersburg, Russia, e-mail: shabaev@pcqnt1.phys.spbu.ru

Christian Smorra RIKEN Advanced Science Institute, 2-1 Hirosawa, Wako, Saitama 351-0198, Japan, e-mail: christian.smorra@cern.ch

Joseph N. Tan Atomic Physics Division, 100 Bureau Drive, Gaithersburg, MD 20899, USA, e-mail: joseph.tan@nist.gov

Stefan Ulmer RIKEN Advanced Science Institute, 2-1 Hirosawa, Wako, Saitama 351-0198, Japan, e-mail: stefan.ulmer@cern.ch

Manuel Vogel Institut für Angewandte Physik, Technische Universität Darmstadt, Schlossgartenstrasse 7, 64289 Darmstadt, Germany, e-mail: m.vogel@gsi.de

Andrej V. Volotka Department of Physics, St. Petersburg State University, Oulianovskaya 1, Petrodvorets, St. Petersburg, Russia 198504; Institut für Theoretische Physik, Technische Universität Dresden, Mommsenstraße 13, 01062 Dresden, Germany

Acronyms

AD	Antiproton Decelerator (CERN)
AEgIS	Antihydrogen Experiment: Gravity, Interferometry, Spectroscopy (CERN)
AFOSR	Air Force Office of Scientific Research, USA
AI	Auto-Ionization
ALPHA	Antihydrogen Laser PHysics Apparatus (CERN)
AME	Atomic Mass Evaluation
ASACUSA	Atomic Spectroscopy And Collisions Using Antiprotons (collaboration)
ATRAP	Antihydrogen Trap (collaboration)
BASE	Baryon Antibaryon Symmetry Experiment (CERN)
BNL	Berkeley National Laboratory, USA
BS-QED	Bound State Quantum Electrodynamics
CCD	Charge Coupled Device
CERN	Conseil Européen pour la Recherche Nucléaire
CI	Configuration Interaction (method)
CI-DFS	Configuration Interaction Dirac-Fock-Sturmian (method)
CMB	Cosmic Microwave Background
CNOT	Controlled NOT (gate)
CODATA	The Committee on Data for Science and Technology
CPT	Charge, Parity and Time (transformation)
CTMC	Classical Trajectory Monte Carlo (method)
DF	Dirac-Fock (approach)
DFS	Dirac-Fock-Slater (approach)
DHF	Dirac-Hartree-Fock (approximation)
DPI	Direct photoionization
DR	Dielectronic recombination
EAMM	Electron Anomalous Magnetic Moment
EBIS	Electron Beam Ion Source
EBIT	Electron Beam Ion Trap
EIE	Electron Impact Excitation
EII	Electron Impact Ionization
ELENA	Extra Low Energy Antiproton Ring (CERN)

ESR	Experiment Storage Ring (GSI)
FEL	Free-Electron Laser
FET	Field-Effect Transistor
FFT	Fast Fourier Transform
FLASH	Free-Electron LASer in Hamburg, Germany
FS	Fine Structure
GANIL	Grand Acclélérateur National d'Ions Lourds, France
GBAR	Gravitational Behaviour of Antihydrogen at Rest (CERN)
GHZ	Greenberger-Horne-Zeilinger (state)
GS-HFS	Ground-State Hyperfine Structure
GSI	Gesellschaft für Schwerionenforschung, Germany
GUT	Grand Unified Theory
HCI	Highly Charged Ion
HF	Hartree-Fock (approximation)
HFS	Hyperfine Splitting
ICSU	International Council for Science
IGBT	Isolated-Gate Bipolar Transistor
IGISOL	Ion Guide Isotope Separator On-Line
IMME	Isobaric Multiplet Mass Equation
IMPATT	Impact Ionization Avalanche Transit-Time (diode)
IR	Infrared
ISOL	Isotope Separation On-Line
ISOLDE	Isotope Separator On-Line Detector (CERN)
JLAB	Thomas Jefferson National Accelerator Facility, USA
LBNL	Lawrence Berkeley National Laboratory, USA
LCLS	Linac Coherent Light Source, SLAC, USA
LEAR	Low-Energy Antiproton Ring
LEP	Large Electron–Positron (collider), CERN
LHC	Large Hadron Collider (CERN)
LLNL	Lawrence Livermore National Laboratory, USA
LSA	Least-Square Adjustment
MAMI	Mainz Microtron (electron accelerator), Germany
MBPT	Many-Body Perturbation Theory
MCDF	Multi-Configuration Dirac-Fock (method)
MCP	Multi-Channel Plate (detector)
MeVVA	Metal Vapour Vacuum Arc (ion source)
MIT	Massachusetts Institute of Technology
NIST	National Institute for Standards and Technology
NMR	Nuclear Magnetic Resonance
NMS	Nuclear Mass Shift
NSCL	National Superconducting Cyclotron Laboratory, USA
NSF	National Science Foundation, USA

PnA	Pulse and Amplify (method)
PnP	Pulse and Phase (method)
PSI	Paul Scherrer Institute, Switzerland
PTMS	Penning Trap Mass Spectrometry
QCD	Quantum Chromodynamics
QED	Quantum Electrodynamics
QIP	Quantum Information Processing
QND	Quantum Non-Demolition (measurement)
QR	Quadruelectronic Recombination
RATIP	Relativistic calculations of Atomic Transition, Ionization and Recombination Properties
RCI	Relativistic Configuration Interaction (method)
RF	Radio-Frequency
RFQ	Radio-Frequency Quadrupole
RIBF	Radioactive Isotope Beam Factory (RIKEN)
RIKEN	Rikagaku Kenkyujo, The Institute of Physical and Chemical Research, Japan
RMBPT	Relativistic Many-Body Perturbation Theory
RNMS	Relativistic Nuclear Mass Shift
RPI	Resonant Photoionization
RR	Radiative Recombination
RSMS	Relativistic Specific Mass Shift
SE	Self Energy (Feynman diagrams)
SM	Standard Model
SMS	Specific Mass Shift
SLAC	Stanford Linear Accelerator Center, USA
TOF	Time-Of-Flight
TOF-ICR	Time-Of-Flight Ion-Cyclotron Resonance
TOF-MS	Time-Of-Flight Mass Spectrometry
TR	Trielectronic Recombination
UV	Ultra-Violet
UW	University of Washington
VP	Vacuum Polarization (Feynman diagrams)
WHIM	Warm-Hot Intergalactic Medium
WTI	Ward-Takahashi Identity
XUV	Extreme Ultra-Violet

Chapter 1
Precise Matter and Antimatter Tests of the Standard Model with e^-, e^+, p, $\overline{p}$ and $\overline{H}$

G. Gabrielse, S. Fogwell Hoogerheide, J. Dorr and E. Novitski

Abstract Extremely precise tests of the Standard Model of particle physics and its CPT theorem come from low energy measurements of the electron, positron, proton and antiproton magnetic moments and charge-to-mass ratios. Ground state antihydrogen atoms are now available for measurements that could eventually reach a higher precision, though no precise $\overline{H}$ measurements have yet been carried out. After a brief summary of the results and status of such measurements, the focus is upon the most precisely measured and precisely calculated property of an elementary particle. The electron magnetic moment, measured to 3 parts in 10^{13}, is a probe of the interaction of the electron with the fluctuating vacuum described by quantum electrodynamics (QED). It is also a probe for electron substructure not predicted by the Standard Model. The measured magnetic moment, together with fine structure constant determined by a different method, is the most stringent test of QED and the Standard Model of particle physics. The measured magnetic moment and QED theory together yield the most precise measured value of the fine structure constant. The summary includes the antiproton magnetic moment that was recently measured precisely for the first time. The 4 parts in 10^6 precision is much less than the electron precision or the 9 parts in 10^{11} at which the antiproton and proton charge-to-mass ratios have been compared, but very large increases in precision seem possible as quantum methods are incorporated.

1.1 Overview Summary

Electron Magnetic Moment

The most precisely measured property of an elementary particle is the electron magnetic moment, measured in Bohr magnetons,

---

G. Gabrielse (✉) · S. Fogwell Hoogerheide · J. Dorr · E. Novitski
Department of Physics, Harvard University, 17 Oxford Street, Cambridge, MA 02138, USA
e-mail: gabrielse@physics.harvard.edu

W. Quint and M. Vogel (eds.), *Fundamental Physics in Particle Traps*,
Springer Tracts in Modern Physics 256, DOI: 10.1007/978-3-642-45201-7_1,

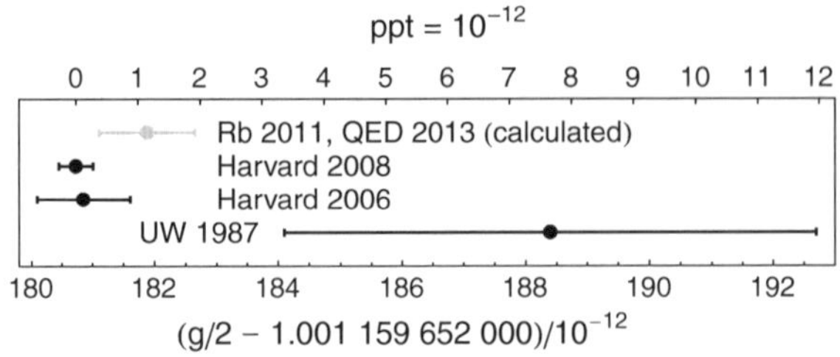

Fig. 1.1 The electron magnetic moment in Bohr magnetons, $\mu/\mu_B = -g/2$, is the most precisely measured property of an elementary particle. The good agreement of the measurement and the prediction of the Standard Model of particle physics is a great triumph of the Standard Model and QED

$$\mu/\mu_B = -g/2 = -1.001\,159\,652\,180\,73\,(28) \quad [0.28 \text{ ppt}]. \tag{1.1}$$

A negative μ indicates that the magnetic moment $\boldsymbol{\mu}$ points opposite to the spin $\mathbf{S}$, with $\boldsymbol{\mu} = \mu\hat{\mathbf{S}}$ and $\hat{\mathbf{S}} = \mathbf{S}/(\hbar/2)$. The distinguishing feature of our Harvard measurements was using quantum jump spectroscopy of the lowest cyclotron states and spin states of a single electron bound to a cylindrical Penning trap apparatus [1–3]. Cooling to the lowest quantum states makes possible a much higher accuracy than realized in a long history of applying new methods to measuring the electron magnetic moment [4]. Our quantum measurements superseded the UW result that had stood for about 20 years [5], with an uncertainty 15 times lower and a measured value shifted by 1.7 standard deviations (Fig. 1.1). The new methods that made possible this large increase in precision are summarized in following sections.

Triumph of the Standard Model

Calculations based upon the Standard Model of particle physics relate the electron magnetic moment in Bohr magnetons to the fine structure constant α. Since the electron magnetic moment was measured, there has been impressive progress in the calculations [6] (reviewed in this volume by M. Hayakawa) and an improved determination of α that is independent of our electron magnetic moment measurement [7]. Together these predict a magnetic moment in Bohr magnetons,

$$\mu/\mu_B = -g/2 = -1.001\,159\,652\,181\,78\,(77) \quad [0.77 \text{ ppt}], \tag{1.2}$$

though the uncertainty in this "calculated" value is mostly from the uncertainty with which the independent value of α is measured.

The more precisely measured electron moment, the improved standard model calculations, and the more precise independently measured value of α present the Standard Model with its most stringent test. The difference between the predicted magnetic moment $\mu(SM)$ and the measured electron magnetic moment μ is

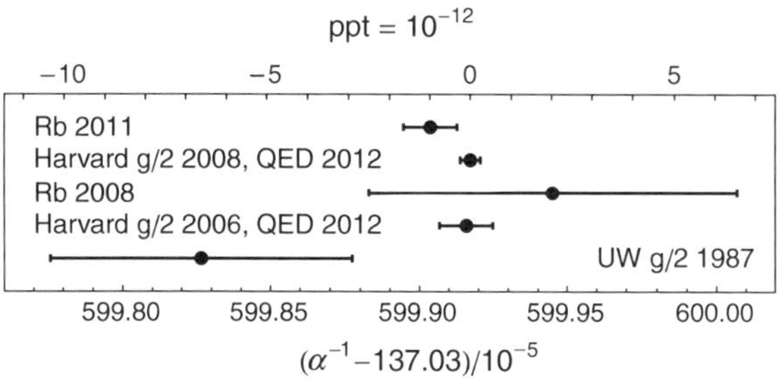

Fig. 1.2 The most accurate determinations of α are determined from the measured electron $g/2$ and standard model theory. The best independently measured value relies on measurements of a Rb atom recoil, the Rydberg constant, and several mass ratios

$$\frac{\mu - \mu(SM)}{\mu} = 0.000\,000\,000\,001\,05\,(82)\;[0.82\;\text{ppt}], \tag{1.3}$$

$$= 1.1\,(0.8) \times 10^{-12}\;[0.8\;\text{ppt}]. \tag{1.4}$$

The measurement and the prediction agree to 1.4 standard deviations (Fig. 1.1). The strikingly precise agreement to better than 12 significant figures is arguably the greatest triumph of the Standard Model and QED.

Fine Structure Constant

The electron magnetic moment is measured much more precisely than the fine structure constant. The result is that the fine structure constant is determined most precisely from the measurement of the electron moment and the Standard Model calculation (QED plus hadronic contributions) to be

$$\alpha^{-1} = 137.035\,999\,173\,(33)\,(8)\;\;[0.24\;\text{ppb}]\,[0.06\;\text{ppb}],$$
$$= 137.035\,999\,173\,(34)\qquad[0.25\;\text{ppb}], \tag{1.5}$$

and is compared with all other precise measurements in Fig. 1.2. The total 0.25 ppb uncertainty is now mostly from the measurement (0.24 ppb) rather than from the Standard Model calculations (0.06 ppb). (Until last year the theoretical uncertainty was slightly larger than the experimental uncertainty.)

Antiproton Charge-to-Mass Ratio

The most precise test of the CPT theorem of the Standard Model made with a baryon and an antibaryon is a comparison of the charge-to-mass ratios of the antiproton and proton [8, 9],

$$\frac{\frac{q}{m}(\overline{\text{p}})}{\frac{q}{m}(p)} = -0.999\,999\,999\,839\,(90)\quad[90\;\text{ppt}]. \tag{1.6}$$

It is intriguing that this ratio is greater than the -1 predicted by the CPT theorem by 161 (90) ppt, which is about 1.8 standard deviations. Does this suggest that the

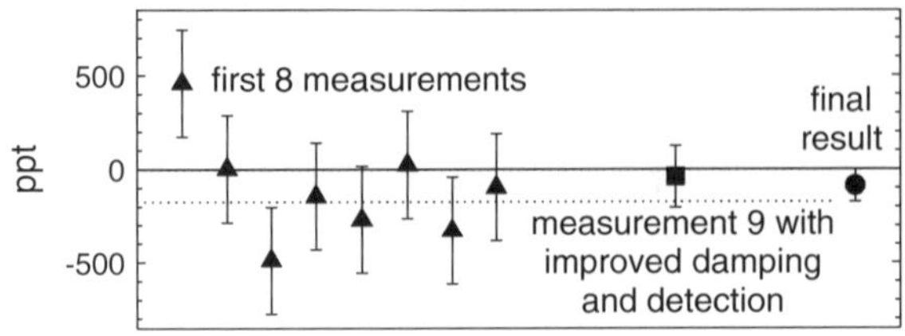

Fig. 1.3 Nine measurements of fractional differences in $|q/m|$ for the $\bar{p}$ and p, and their weighted average

magnitude of the charge-to-mass ratios of the antiproton and proton differ? It would be premature to make this conclusion insofar as the uncertainties were assigned to represent a standard deviation, and thus there is a reasonable probability that the true ratio could lie slightly outside the quoted uncertainty (Fig. 1.3).

A much more accurate measurement of q/m for the antiproton could certainly be carried out. This is most evident in the series of measurements that went into the final determination of q/m for the antiproton. An apparatus and technique improvement made the last 1-day measurement to be much more accurate than the earlier 8 measurements. Before this new level of accuracy could be exploited, unfortunately, LEAR closed down. Our ATRAP collaboration intends to make a more precise measurement.

Antiproton Magnetic Moment

The first one-particle measurement of the antiproton magnetic moment in nuclear magnetons,

$$\mu_{\bar{\mathrm{p}}}/\mu_N = -2.792\,845 \pm 0.000\,012 \quad [4.4 \text{ ppm}]. \tag{1.7}$$

has just been reported [10]. Again the negative μ indicates that the magnetic moment $\boldsymbol{\mu}$ points opposite to the spin $\mathbf{S}$, with $\boldsymbol{\mu} = \mu\hat{\mathbf{S}}$ and $\hat{\mathbf{S}} = \mathbf{S}/(\hbar/2)$. The uncertainty is 680 times smaller than for any previous measurement (Fig. 1.4). To this precision, the measurement is consistent with the prediction of the Standard Model and its CPT theorem that the antiproton and proton have exactly opposite magnetic moments—equal in magnitude but opposite in sign.

This first precise antiproton magnetic moment measurement did not resolve individual spin flips for quantum jump spectroscopy as was done for the electron measurement. With a trapped proton, however, the possibility to resolve one-proton spin flips was reported even more recently by our Harvard group [11] and by a group from Mainz [12]. These demonstrations bode well for future increases in precision that could eventually improve by a factor as large as 10^4. A participant in this latter effort, S. Ulmer, discusses their proton measurements and their antiproton aspirations in a chapter in this volume.

Better Electron and Positron Magnetic Moment Measurements

An entirely new apparatus has been constructed at Harvard for making a much more precise comparison of the positron and electron magnetic moments, and for

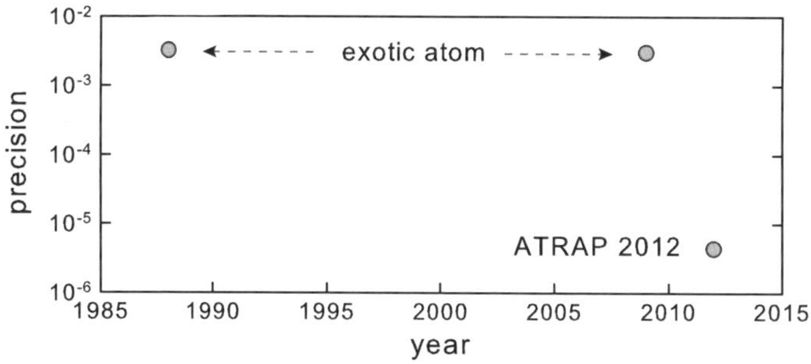

Fig. 1.4 Uncertainties in measurements of the $\overline{p}$ magnetic moment measured in nuclear magnetons. From [10]

more precise electron magnetic moment measurements. Positrons have recently been captured in this apparatus as preparation for measuring the positron magnetic moment at the electron precision. It should be possible to thereby make a lepton CPT test that compares positron and electron magnetic moments 15–20 times more precisely than a previous comparison [5]. The positron magnetic measurement will be followed with an attempt to measure the electron and positron magnetic moments at a much higher precision since no uncertainty that would prevent a substantial increase in precision has yet been identified.

Aspirations to Compare of Antihydrogen and Hydrogen

The possibility to use low energy antiprotons to make low energy antihydrogen cold enough to be trapped for precise measurements was proposed long ago [13] just after antiprotons were first captured in an ion trap [14]. The proposed goal was precise comparisons of antihydrogen and hydrogen, made at a precision that exceeds other precise tests of CPT invariance and the Standard Model with leptons and baryons. Now three collaborations are using cold antiprotons to produce cold antihydrogen to pursue this goal, and a fourth hopes to do so soon.

The nested Penning trap was invented to make possible the interaction of oppositely charged antiprotons and positrons that are trapped simultaneously during the positron cooling of the antiprotons. Essentially all the antihydrogen produced so far was produced using this device and method, by ATRAP [15, 16] and ATHENA [17] in 2002, and then by ASACUSA in 2010 [18].

ATRAP has also demonstrated a second method for producing antihydrogen, though much less antihydrogen has been produced so far. Lasers were used to control resonant charge exchange [19, 20]. ATRAP recently greatly increased the number of $\overline{H}$ atoms produced by this second method, and plans to attempt to trap atoms so produced when CERN restarts antiproton production.

The proposal to trap cold antihydrogen in its ground state in a neutral particle trap [13] has so far been realized by ALPHA [21] and ATRAP [22]. However, the largest number of ground state $\overline{H}$ atoms trapped per trial so far (Fig. 1.5) is ATRAP's 5 ± 1 atoms per trial [22]. More simultaneously trapped $\overline{H}$ atoms are clearly needed, and no precise comparisons of antihydrogen and hydrogen have been made to date.

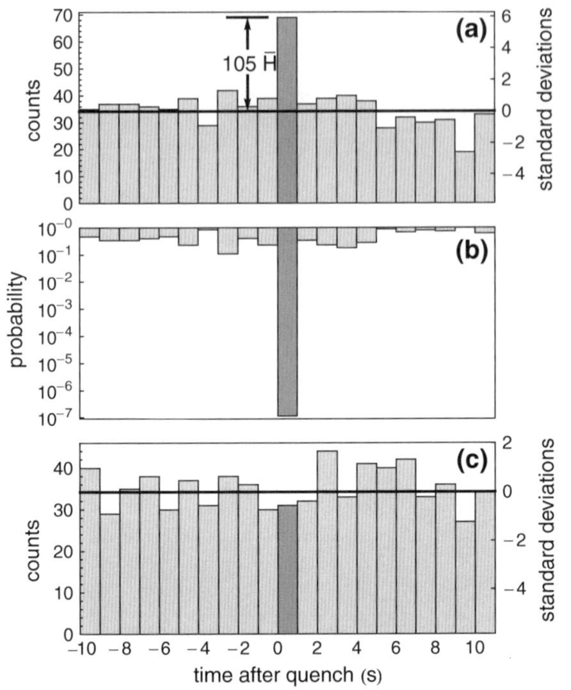

Fig. 1.5 The largest number of simultaneously trapped, ground state $\overline{\mathrm{H}}$ atoms is ATRAP's 5 ± 1 atoms per trial. (**a**) Detector counts in 1 s intervals for 20 trials. The radial Ioffe trap field turns off and releases trapped $\overline{\mathrm{H}}$ between $t = 0$ and 1 s. The counts in this interval above the average cosmic ray counts (*solid line*) correspond to 105 trapped $\overline{\mathrm{p}}$ for our detection efficiency. (**b**) Probability that cosmic rays produce the observed counts or more. (**c**) Quenching the Ioffe trap generates no false signals in 20 control trials

Antihydrogen experiments are mentioned here because of their promise for the future. It is hoped that precise comparisons of the antihydrogen and hydrogen will eventually reach a precision higher than the precise tests summarized above. J. Hangst of the ALPHA collaboration has provided a chapter in this volume on antihydrogen experiments.

1.2 Magnetic Moments

An electron with charge $-e$ with mass m in a magnetic field $B\hat{\mathbf{z}}$ circles in a cyclotron orbit with angular frequency $eB/m\hat{\mathbf{z}}$ and a magnetic moment

$$\boldsymbol{\mu} = -\frac{e}{2m}\mathbf{L} = -\mu_B\frac{\mathbf{L}}{\hbar}. \tag{1.8}$$

The Bohr magneton $\mu_B = e\hbar/(2m)$ gives the scale of the magnetic moment insofar as the orbital angular momentum $\mathbf{L}$ comes in units of $\hbar$ for a quantum system. The magnetic moment for an antiproton cyclotron orbit is instead scaled by the nuclear magneton, μ_N, which is about 2,000 times smaller than μ_B, by the ratio of the electron and proton masses.

The magnetic moment from the electron spin $\mathbf{S} = \frac{1}{2}\hbar\boldsymbol{\sigma}$ is often written in terms of the dimensionless Pauli operator $\boldsymbol{\sigma}$ for a spin 1/2 particle. It is also often written equivalently in terms of a dimensionless g value.

$$\boldsymbol{\mu} = \mu\boldsymbol{\sigma} = -g\frac{e}{2m}\mathbf{S} = -\frac{g}{2}\mu_B\boldsymbol{\sigma}. \tag{1.9}$$

The electron magnetic moment in Bohr magnetons is thus given by $\mu/\mu_B = -g/2$ which is the dimensionless size of the magnetic moment that must be measured and calculated. For an antiproton, the analogous dimensionless constant to be measured and calculated instead is the antiproton magnetic moment in nuclear magnetons, $\mu_{\bar{p}}/\mu_N \equiv -g_{\bar{p}}/2$.

The electron magnetic moment in Bohr magnetons can be very precisely calculated within the Standard Model of particle physics. It has the form

$$g/2 = 1 + a_{QED}(\alpha) + a_{hadronic} + a_{weak}. \tag{1.10}$$

The leading term $g/2 = 1$ is a prediction of the Standard Model insofar as it is the Dirac equation prediction for a point particle. Quantum electrodynamics (QED) gives the Standard Model prediction that vacuum fluctuations and polarization slightly increase $g/2$ by the small "anomaly" $a_{QED}(\alpha) \approx 10^{-3}$ that is a function of the fine structure constant α. The hadronic addition calculated within the Standard Model is much smaller, and the weak interaction addition is negligible at the current level of precision. One intriguing question is whether electron substructure (or other deviations from the Standard Model) could make $g/2$ deviate by the addition of some a_{new} from the Standard Model prediction to Eq. 1.10, as quark-gluon substructure does for an antiproton and proton.

Why measure the electron magnetic moment in Bohr magnetons, $g/2$? The motivations include:

1. The magnetic moment in Bohr magnetons is the property that can be most accurately measured for an electron—the important component of our universe that is unusual in that no internal structure has been predicted or detected.
2. The most stringent test of QED comes from measuring $g/2$ and comparing to the value $g(\alpha)$ calculated using an independently determined α in Eq. 1.10.
3. The most accurate determination of the fine structure constant comes from solving Eq. 1.10 for α in terms of the measured $g/2$. (No physics beyond the Standard Model, i.e. $a_{new} = 0$, is assumed.)
4. A search for physics beyond the Standard Model (e.g. electron substructure) comes from using the best measurement of $g/2$ and the best independent α (with

calculated values of $a_{hadronic}$ and a_{weak}) in Eq. 1.10 to set a limit on a possible a_{new} addition.
5. Comparing $g/2$ for an electron and a positron is the most stringent test of CPT invariance with leptons.

Owing to the great importance of the dimensionless magnetic moment, there have been many measurements of the electron $g/2$. A long list of measurements of this fundamental quantity has been compiled [4]. Worthy of special mention is a long series of measurements at the University of Michigan [23]. The spin precession relative to the cyclotron rotation of keV electrons was measured. Also worthy of special mention is the series of measurements at the University of Washington [5, 24].

Our measurements, like the UW measurements, made use of a single electron in a Penning trap. We were able to measure the electron magnetic moment about 15 times more precisely, and show that the measured value was different by about 1.7 standard deviations (Fig. 1.1).

The unifying idea for the new methods is that of a one-electron quantum cyclotron—with fully resolved cyclotron and spin energy levels, and a detection sensitivity sufficient to detect one quantum transitions is achieved in our fully quantum measurements [1, 2].

The substantially higher accuracy of the new measurements was the result of new experimental methods, developed and demonstrated one thesis at a time over 20 years by a string of excellent Ph.D. students—C.H. Tseng, D. Enzer, J. Tan, S. Peil, B. D'Urso, B. Odom and D. Hanneke. The new methods included:

1. A cylindrical Penning trap was used to suspend the electron. The cylindrical trap was invented to form a microwave cavity that could inhibit spontaneous emission. The calculable cavity shape made it possible to understand and correct for cavity shifts of the measured cyclotron frequency.
2. Cavity-inhibited spontaneous emission (by a factor of up to 250) narrowed measured linewidths and gave us the crucial averaging time that we needed to resolve one-quantum changes in the electron's cyclotron state.
3. The cavity was cooled to 100 mK rather than to 4.2 K so that in thermal equilibrium the electron's cyclotron motion would be in its ground state.
4. Detection with good signal-to-noise ratio came from feeding back a signal derived from the electron's motion along the magnetic field to the electron to cancel the damping due to the detection impedance. The "classical measurement system" for the quantum cyclotron motion was this large self-excited motion of the electron, with a quantum nondemolition coupling between the classical and quantum systems.
5. A silver trap cavity avoided the magnetic field variations due to temperature fluctuations of the paramagnetism of conventional copper trap electrodes.
6. The measurement was entirely automated so that the best data could be taken at night, when the electrical, magnetic and mechanical disturbances were lowest, with no person present.
7. A parametric excitation of electrons suspended in the trap was used to measure the radiation modes of the radiation field in the trap cavity.

8. The damping rate of a single trapped electron was used as a second probe of the radiation fields within the trap cavity.

1.3 One-Electron Quantum Cyclotron

1.3.1 A Homemade Atom

A one-electron quantum cyclotron is a single electron suspended within a magnetic field, with the quantum structure in its cyclotron motion fully resolved. Accurate measurements of the resonant frequencies of driven transitions between the energy levels of this homemade atom—an electron bound to our trap—reveals the electron magnetic moment in units of Bohr magnetons, $g/2$. The energy levels and what must be measured to determine $g/2$ are presented in this section. The experimental devices and methods needed to realize the one-electron quantum cyclotron are discussed in following sections.

A nonrelativistic electron in a magnetic field has energy levels

$$E(n, m_s) = \frac{g}{2} h\nu_c m_s + (n + \tfrac{1}{2}) h\nu_c. \tag{1.11}$$

These depend in a familiar way upon the electron's cyclotron frequency ν_c and its spin frequency $\nu_s \equiv (g/2)\nu_c$. The electron $g/2$ is thus specified by the two frequencies,

$$\frac{g}{2} = \frac{\nu_s}{\nu_c} = 1 + \frac{\nu_s - \nu_c}{\nu_c} = 1 + \frac{\nu_a}{\nu_c}, \tag{1.12}$$

or equivalently by their difference (the anomaly frequency $\nu_a \equiv \nu_s - \nu_c$) and ν_c. Because ν_s and ν_c differ by only a part-per-thousand, measuring ν_a and ν_c to a precision of 1 part in 10^{10} gives $g/2$ to 1 part in 10^{13}.

Although one electron suspended in a magnetic field will not remain in one place long enough for a measurement, two features of determining $g/2$ by measuring ν_a and ν_c are apparent in Eq. 1.12.

1. Nothing in physics can be measured more accurately than a frequency (the art of time keeping being so highly developed) except for a ratio of frequencies.
2. Although both of these frequencies depend upon the magnetic field, the field dependence drops out of the ratio. The magnetic field thus needs to be stable only on the time scale on which both frequencies can be measured, and no absolute calibration of the magnetic field is required.

To confine the electron for precise measurements, an ideal Penning trap includes an electrostatic quadrupole potential $V \sim z^2 - \frac{1}{2}\rho^2$ with a magnetic field $B\hat{\mathbf{z}}$ [25]. This potential shifts the cyclotron frequency from the free-space value ν_c to $\bar{\nu}_c$. The latter frequency is also slightly shifted by the unavoidable leading imperfections of a real

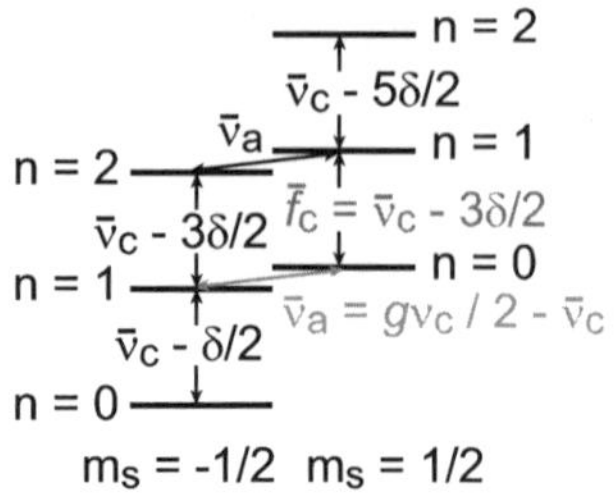

Fig. 1.6 Lowest cyclotron and spin levels of an electron in a Penning trap

laboratory trap—a misalignment of the symmetry axis of the electrostatic quadrupole and the magnetic field, and quadratic distortions of the electrostatic potential.

The lowest cyclotron energy levels (with quantum numbers $n = 0, 1, \ldots$) and the spin energy levels (with quantum numbers $m_s = \pm 1/2$) are given by

$$E(n, m_s) = \tfrac{g}{2} h\nu_c m_s + (n + \tfrac{1}{2}) h\bar{\nu}_c - \tfrac{1}{2} h\delta (n + \tfrac{1}{2} + m_s)^2. \tag{1.13}$$

The lowest cyclotron and spin energy levels are represented in Fig. 1.6.

Special relativity is important for even the lowest quantum levels. The third term in Eq. 1.13 is the leading relativistic correction [25] to the energy levels. Special relativity makes the transition frequency between two cyclotron levels $|n, m_s\rangle \leftrightarrow |n+1, m_s\rangle$ decrease from $\bar{\nu}_c$ to $\bar{\nu}_c + \Delta\bar{\nu}_c$, with the shift

$$\Delta\bar{\nu}_c = -\delta(n + 1 + m_s) \tag{1.14}$$

depending upon the spin state and cyclotron state. This very small shift, with

$$\delta/\nu_c \equiv h\nu_c/(mc^2) \approx 10^{-9}, \tag{1.15}$$

is nonetheless significant at our precision. An important new feature of our measurement is that special relativity adds no uncertainty to our measurements. Quantum transitions between identified energy levels with a precisely known relativistic contribution to the energy levels are resolved. When the average cyclotron frequency of an unknown distribution of cyclotron states was all that could be measured [5], figuring out the size of the relativistic frequency shift was difficult.

We have seen how $g/2$ is determined by the anomaly frequency ν_a and the free-space cyclotron frequency $\nu_c = eB/(2\pi m)$. However, neither of these frequencies is an eigenfrequency of the trapped electron. We actually measure the transition frequencies

$$\bar{f}_c \equiv \bar{\nu}_c - \frac{3}{2}\delta \tag{1.16}$$

$$\bar{\nu}_a \equiv \frac{g}{2}\nu_c - \bar{\nu}_c \tag{1.17}$$

represented by the arrows in Fig. 1.6 for an electron initially prepared in the state $|n = 0, m_s = 1/2\rangle$.

The needed $\nu_c = eB/(2\pi m)$ is deduced from the three observable eigenfrequencies of an electron bound in the trap by the Brown-Gabrielse invariance theorem [26],

$$(\nu_c)^2 = (\bar{\nu}_c)^2 + (\bar{\nu}_z)^2 + (\bar{\nu}_m)^2. \tag{1.18}$$

The three measurable eigenfrequencies on the right include the cyclotron frequency $\bar{\nu}_c$ for the quantum cyclotron motion we have been discussing. The second measurable eigenfrequency is the axial oscillation frequency $\bar{\nu}_z$ for the nearly-harmonic, classical electron motion along the direction of the magnetic field. The third measurable eigenfrequency is the magnetron oscillation frequency for the classical magnetron motion along the circular orbit for which the electric field of the trap and the motional electric field exactly cancel.

The invariance theorem applies for a perfect Penning trap, but also in the presence of the mentioned imperfection shifts of the eigenfrequencies for an electron in a trap. This theorem, together with the well-defined hierarchy of trap eigenfrequencies, $\bar{\nu}_c \gg \bar{\nu}_z \gg \bar{\nu}_m \gg \delta$, yields an approximate expression that is sufficient at our accuracy. We thus determine the electron $g/2$ using

$$\frac{g}{2} = \frac{\bar{\nu}_c + \bar{\nu}_a}{\nu_c} \simeq 1 + \frac{\bar{\nu}_a - \bar{\nu}_z^2/(2\bar{f}_c)}{\bar{f}_c + 3\delta/2 + \bar{\nu}_z^2/(2\bar{f}_c)} + \frac{\Delta g_{cav}}{2}. \tag{1.19}$$

The cavity shift $\Delta g_{cav}/2$ that arises from the interaction of the cyclotron motion and the trap cavity is presently discussed in detail.

1.3.2 Cylindrical Penning Trap Cavity

A cylindrical Penning trap (Fig. 1.7) is the key device that makes these measurements possible. It was invented [27] and demonstrated [28] to provide boundary conditions that produce a controllable and understandable radiation field within the trap cavity, along with the needed electrostatic quadrupole potential. Spontaneous emission can be significantly inhibited at the same time as corresponding shifts of the electron's oscillation frequencies are avoided. We shall see that this is critical to the new Harvard measurements in several ways (Table 1.1).

A necessary function of the trap electrodes is to produce a very good approximation to an electrostatic quadrupole potential. This is possible with cylindrical electrodes but only if the relative geometry of the electrodes is carefully chosen [27]. The electrodes of the cylindrical trap are symmetric under rotations about the center axis ($\hat{\mathbf{z}}$), which is parallel to the spatially uniform magnetic field ($B\hat{\mathbf{z}}$). The potential (about 100 V) applied between the endcap electrodes and the ring electrode provides the basic trapping potential and sets the axial frequency $\bar{\nu}_z$ of the nearly harmonic oscillation of the electron parallel to the magnetic field. The potential applied to the

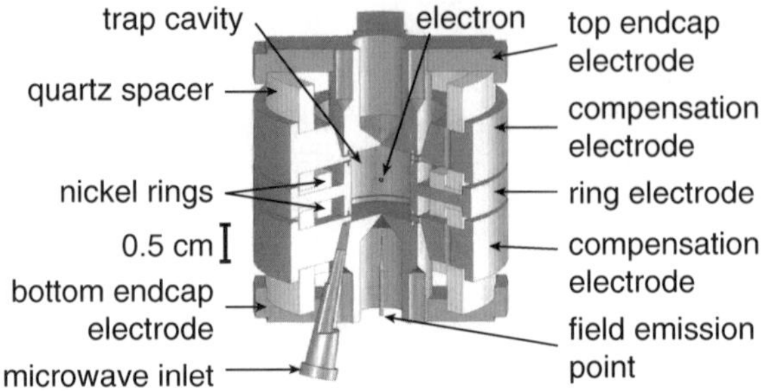

Fig. 1.7 Cylindrical Penning trap cavity used to confine a single electron and inhibit spontaneous emission

Table 1.1 Properties of the trapped electron

Cyclotron frequency	$\omega_c/(2\pi)$	150 GHz
Trap-modified cyc. freq.	$\omega_+/(2\pi)$	150 GHz
Axial frequency	$\omega_z/(2\pi)$	200 MHz
Magnetron frequency	$\omega_-/(2\pi)$	133 kHz
Cyclotron damping (free space)	τ_+	0.09 s
Axial damping	τ_z	200 ms
Magnetron damping	τ_-	10^9 yr

compensation electrodes is adjusted to tune the shape of the potential, to make the oscillation as harmonic as possible. The tuning does not change $\bar{\nu}_z$ very much owing to an orthogonalization [27, 29] that arises from the geometry choice. What we found was that one electron could be observed within a cylindrical Penning trap with as good or better signal-to-noise ratio than was realized in hyperbolic Penning traps.

The principle motivation for the cylindrical Penning trap is to form a microwave cavity whose radiation properties are well understood and controlled—the best possible approximation to a perfect cylindrical trap cavity. (Our calculation attempts with a hyperbolic trap cavity were much less successful [30].) The modes of the electromagnetic radiation field that are consistent with this boundary condition are the well known transverse electric TE_{mnp} and transverse magnetic TE_{mnp} modes (see e.g. [31, Sect. 8.7]).

For a right circular cylinder of diameter $2\rho_0$ and height $2z_0$ the TE and TM modes have characteristic resonance frequencies,

$$TE: \quad \omega_{mnp} = c\sqrt{\left(\frac{x'_{mn}}{\rho_0}\right)^2 + \left(\frac{p\pi}{2z_0}\right)^2} \tag{1.20a}$$

$$TM: \quad \omega_{mnp} = c\sqrt{\left(\frac{x_{mn}}{\rho_0}\right)^2 + \left(\frac{p\pi}{2z_0}\right)^2}. \tag{1.20b}$$

They are indexed with integers

$$m = 0, 1, 2, \ldots \tag{1.21}$$

$$n = 1, 2, 3, \ldots \tag{1.22}$$

$$p = 1, 2, 3, \ldots, \tag{1.23}$$

and are functions of the nth zeros of Bessel functions and their derivatives

$$J_m(x_{mn}) = 0 \tag{1.24}$$

$$J'_m(x'_{mn}) = 0 \tag{1.25}$$

The zeros force the boundary conditions at the cylindrical wall. All but the $m = 0$ modes are doubly degenerate.

Of primary interest is the magnitude of the cavity electric fields that couple to the cyclotron motion of an electron suspended in the center of the trap. For both TE and TM modes, the transverse components of **E** are proportional to

$$\sin\left(\tfrac{p\pi}{2}\left(\tfrac{z}{z_0}+1\right)\right) = \begin{cases} (-1)^{p/2}\sin\left(\frac{p\pi z}{2z_0}\right) & \text{for even } p, \\ (-1)^{(p-1)/2}\cos\left(\frac{p\pi z}{2z_0}\right) & \text{for odd } p. \end{cases} \tag{1.26}$$

For an electron close to the cavity center, ($z \approx 0$), only modes with odd p thus have any appreciable coupling.

The transverse components of the electric fields are also proportional to either the order-m Bessel function times m/ρ for the TE modes, or to the derivative of the order-m Bessel function for the TM modes. Close to the cavity center ($\rho \approx 0$),

$$\frac{m}{\rho} J_m\left(x^{(\prime)}_{mn}\frac{\rho}{\rho_0}\right) \sim \begin{cases} \dfrac{\rho^{m-1}}{(m-1)!}\left(\dfrac{x^{(\prime)}_{mn}}{2\rho_0}\right)^m & \text{for } m > 0 \\ 0 & \text{for } m = 0 \end{cases} \tag{1.27a}$$

$$\frac{x^{(\prime)}_{mn}}{\rho_0} J'_m\left(x^{(\prime)}_{mn}\frac{\rho}{\rho_0}\right) \sim \begin{cases} \dfrac{\rho^{m-1}}{(m-1)!}\left(\dfrac{x^{(\prime)}_{mn}}{2\rho_0}\right)^m & \text{for } m > 0 \\ -\dfrac{x^{(\prime)2}_{0n}}{2\rho_0^2}\rho & \text{for } m = 0. \end{cases} \tag{1.27b}$$

In the limit $\rho \to 0$, all but the $m = 1$ modes vanish.

For a perfect cylindrical cavity the only radiation modes that couple to an electron perfectly centered in the cavity are $\mathrm{TE}_{1n(\mathrm{odd})}$ and $\mathrm{TE}_{1n(\mathrm{odd})}$. If the electron is moved slightly off center axially it will begin to couple to radiation modes with $mnp = 1n(\mathrm{even})$. If the electron is moved slightly off-center radially it similarly begins to couple to modes with $m \neq 1$.

In the real trap cavity, the perturbation caused by the small space between the electrodes is minimized by the use of "choke flanges"—small channels that tend to reflect the radiation leaking out of the trap back to cancel itself, and thus to minimize the losses from the trap. The measured radiation modes, discussed later, are close enough to the calculated frequencies for a perfect cylindrical cavity that we have been able to identify more than 100 different radiation modes for such trap cavities [32–34]. The spatial properties of the electric and magnetic field for the radiation that builds up within the cavity are thus quite well understood. Some of the modes couple to cyclotron motion of an electron centered in the cavity, others couple to the spin of a centered electron, and still others have the symmetry that we hope will one day allow us to sideband-cool the axial motion.

1.3.3 *100* mK *and 5 T*

Detecting transitions between energy levels of the quantum cyclotron requires that the electron-bound-to-the-trap system be prepared in a definite quantum state. Two key elements are a high magnetic field, and a low temperature for the trap cavity. A high field makes the spacing of the cyclotron energy levels to be large. A high field and low temperature make a very large Boltzmann probability to be in the lowest cyclotron state, $P \propto \exp[-h\bar{\nu}_c/(kT)]$, which is negligibly different from unity.

The trap cavity is cooled to 0.1 K or below via a thermal contact with the mixing chamber of an Oxford Instruments Kelvinox 300 dilution refrigerator (Fig. 1.8). The electrodes of this trap cavity are housed within a separate vacuum enclosure that is entirely at the base temperature. Measurements on an apparatus with a similar design but at 4.2 K found the vacuum in the enclosure to be better than 5×10^{-17} torr [35]. Our much lower temperature make our background gas pressure much lower. We are able to keep one electron suspended in our apparatus for as long as desired—regularly months at a time. Substantial reservoirs for liquid helium and liquid nitrogen make it possible to keep the trap cold for five to seven days before the disruption of adding more liquid helium or nitrogen is required.

The trap and its vacuum container are located within a superconducting solenoid (Fig. 1.8) that makes a very homogeneous magnetic field over the interior volume of the trap cavity. A large dewar sitting on top of the solenoid dewar provides the helium needed around the dilution refrigerator below. The superconducting solenoid is entirely self-contained, with a bore that can operate from room temperature down to 77 K. It possesses shim coils capable of creating a field homogeneity better than a part in 10^8 over a 1 cm diameter sphere and has a passive "shield" coil that reduces fluctuations in the ambient magnetic field [36, 37]. When properly energized (and after the steps described in the next section have been taken) it achieves field stability better than a part in 10^9 per hour. We regularly observe drifts below 10^{-9} per night.

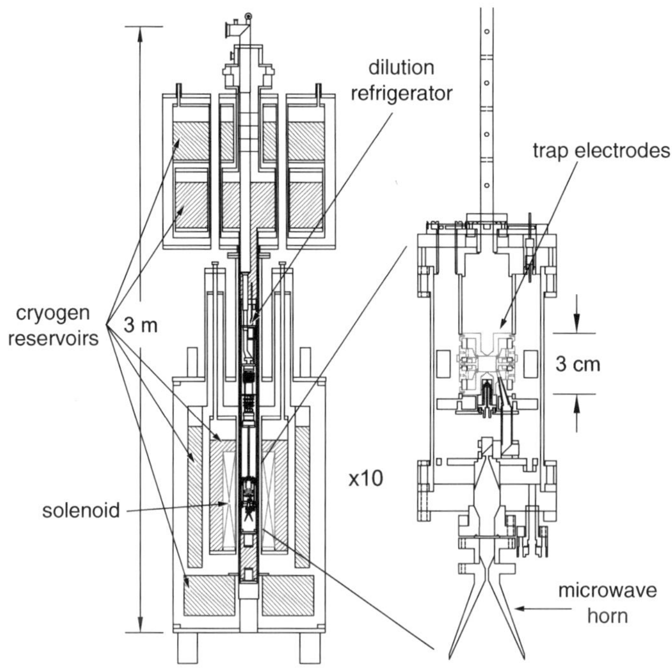

Fig. 1.8 The apparatus includes a trap electrodes near the central axis, surrounded by a superconducting solenoid. The trap is suspended from a dilution refrigerator

1.3.4 Stabilizing the Energy Levels

Measuring the electron $g/2$ with a precision of parts in 10^{13} requires that the energy levels of our homemade atom, an electron bound to a Penning trap, be exceptionally stable. The energy levels depend upon the magnetic field and upon the the potential that we apply to the trap electrodes. The magnetic field must be stable at least on the time scale that is required to measure the two frequencies, $\bar{f}_c$ and $\bar{\nu}_a$, that are both proportional to the magnetic field.

One defense against external field fluctuations is a high magnetic field. This makes it so that field fluctuations due to outside sources are relatively smaller. The largest source of ambient magnetic noise is a subway that produces 50 nT (0.5 mG, 10 ppb) fluctuations in our lab and that would limit us to four hours of data taking per day (when the subway stops running) if we did not shield the electron from them. Eddy currents in the high-conductivity aluminum and copper cylinders of the dewars and the magnet bore shield high-frequency fluctuations [38]. For slower fluctuations, the aforementioned self-shielding solenoid [36] has the correct geometry to make the central field always equal to the average field over the solenoid cross-section. This translates flux conservation into central-field conservation, shielding external fluctuations by more than a factor of 150 [37].

Stabilizing the field produced by the solenoid requires that care is taken when the field value is changed, since changing the current in the solenoid alters the forces between windings. Resulting stresses can take months to stabilize if the coil is not pre-stressed by "over-currenting" the magnet. Our recipe is to overshoot the target value by a few percent of the change, undershoot by a similar amount, and then move to the desired field.

The apparatus in Fig. 1.8 evolved historically rather than being designed for maximum magnetic field stability in the final configuration. Because the solenoid and the trap electrodes are suspended from widely separated support points, temperature and pressure changes can cause the electrodes to move relative to the solenoid. Apparatus vibrations can do the same insofar as the magnetic field is not perfectly homogeneous, despite careful adjusting of the persistent currents in ten superconducting shim coils. Any relative motion of the electron and solenoid changes the field seen by the electron.

To counteract this, we regulate the five He and N_2 pressures in the cryostats to maintain the temperature of both the bath and the solenoid itself [39, 40]. Recently we also relocated the dilution refrigerator vacuum pumps to an isolated room at the end of a 12 m pipe run. This reduced vibration by more than an order of magnitude at frequencies related to the pump motion and reduced the noise level for the experimenters but did not obviously improve the $g/2$ data.

Because some of the structure establishing the relative location of the trap electrodes and the solenoid is at room temperature, changes in room temperature can move the electron in the magnetic field. The lab temperature routinely cycles 1–2 K daily, so we house the apparatus in a large, insulated enclosure within which we actively regulate the air temperature to 0.1 K. A refrigerated circulating bath (ThermoNeslab RTE-17) pumps water into the regulated zone and through an automobile transmission fluid radiator, heating and cooling the water to maintain constant air temperature. Fans couple the water and air temperatures and keep a uniform air temperature throughout.

The choice of materials for the trap electrodes and its vacuum container is also crucial to attaining high field stability [1, 41]. Copper trap electrodes, for example, have a nuclear paramagnetism at 100 mK that makes the electron see a magnetic field that changes at an unacceptable level with very small changes in trap temperature. We thus use only low-Curie-constant materials such as silver, quartz, titanium, and molybdenum at the refrigerator base temperature and we regulate the mixing chamber temperature to 1 mK or better.

A stable axial frequency is also extremely important since small changes in the measured axial frequency reveal one-quantum transitions of the cyclotron and spin energy (as will be discussed in Sect. 1.4.1). A trapping potential without thermal fluctuations is provided by a charged capacitor (10 μF) that has a very low leakage resistance at low temperature. We add to or subtract from the charge on the capacitor using 50 ms current pulses sent to the capacitor through a 100 MΩ resistor as needed to keep the measured axial frequency constant. Because of the orthogonalized trap design [27] already discussed, the potential applied to the compensation electrodes (to make the electron see as close to a pure electrostatic quadrupole potential as possible) has little effect upon the axial frequency.

1.3.5 Motions and Damping of the Suspended Electron

We load a single electron using an electron beam from a sharp tungsten field emission tip. A hole in the bottom endcap electrode admits the beam, which hits the top endcap electrode and releases gas atoms cryopumped on the surface. Collisions between the beam and gas atoms eventually cause an electron to fall into the trap. Adjusting the beam energy and the time it is on determines the number of electrons loaded.

The electron has three motions in the Penning trap formed by the $B = 5.4$ T magnetic field, and the electrostatic quadrupole potential. The cyclotron motion in the trap has a cyclotron frequency $\bar{\nu}_c \approx 150\,\text{GHz}$. The axial frequency, for the harmonic oscillator parallel to the magnetic field direction, is $\bar{\nu}_z \approx 200\,\text{MHz}$. A circular magnetron motion, perpendicular to **B**, has an oscillation frequency, $\bar{\nu}_m \approx 133\,\text{kHz}$. The spin precession frequency, which we do not measure directly, is slightly higher than the cyclotron frequency. The frequency difference is the anomaly frequency, $\bar{\nu}_a \approx 174\,\text{MHz}$, which we do measure directly.

The undamped spin motion is essentially uncoupled from its environment [25]. The cyclotron motion is only weakly damped. By controlling the cyclotron frequency relative to that of the cavity radiation modes, we alter the density of radiation states and inhibit the spontaneous emission of synchrotron radiation [25, 42] by 10–50 times the $(90\,\text{ms})^{-1}$ free-space rate. Blackbody photons that could excite from the cyclotron ground state are eliminated because the trap cavity is cooled by the dilution refrigerator to 100 mK [43]. The axial motion is cooled by a resonant circuit at a rate $\gamma_z \approx (0.2\ \text{s})^{-1}$ to as low as 230 mK (from 5 K) when the detection amplifier is off. The magnetron radius is minimized with axial sideband cooling [25].

1.4 Non-destructive Detection of One-Quantum Transitions

1.4.1 QND Detection

Quantum nondemolition (QND) detection has the property that repeated measurements of the energy eigenstate of the quantum system do not change the state of the quantum system [44, 45]. This is crucial for detecting one-quantum transitions in the cyclotron motion insofar as it avoids transitions produced by the detection system. In this section we discuss the QND coupling, and in the next section the self-excited oscillator readout system.

Detecting a single 150 GHz photon from the decay of one cyclotron energy level to the level below would be very difficult—because the frequency is so high and because it is difficult to cover the solid angle into which the photon could be emitted. Instead we get the one-quantum sensitivity by coupling the cyclotron motion to the orthogonal axial motion at 200 MHz, a frequency at which we are able to make sensitive detection electronics [46]. The QND detection keeps the thermally driven axial motion of the electron from changing the state of the cyclotron motion.

We use a magnetic bottle gradient that is familiar from plasma physics and from earlier electron measurements [5, 47],

$$\boldsymbol{\Delta B} = B_2 \left[\left(z^2 - \rho^2/2 \right) \hat{\mathbf{z}} - z\rho \hat{\boldsymbol{\rho}} \right] \tag{1.28}$$

with $B_2 = 1540\,\mathrm{T/m^2}$. This is the lowest order gradient that is symmetric under reflections $z \to -z$ and is cylindrically symmetric about $\hat{\mathbf{z}}$. The gradient arises from a pair of thin nickel rings (Fig. 1.7) that are completely saturated in the strong field from the superconducting solenoid. To lowest order the rings modify B by $\approx -0.7\,\%$—merely changing the magnetic field that the electron experiences without affecting our measurement.

The formal requirement for a QND measurement is that the Hamiltonian of the quantum system (i.e. the cyclotron Hamiltonian) and the Hamiltonian describing the interaction of the quantum system and the classical measurement system must commute. The Hamiltonian that couples the quantum cyclotron and spin motions to the axial motion does so. It has the form $-\mu B$, where μ is the magnetic moment associated with the cyclotron motion or the spin. The coupling Hamiltonian thus has a term that goes as μz^2. This term has the same spatial symmetry as does the axial Hamiltonian, $H = \frac{1}{2} m (2\pi \bar{\nu}_z)^2 z^2$. A change in the magnetic moment that takes place from a one-quantum change in the cyclotron or spin magnetic moment thus changes the observed axial frequency of the suspended electron.

The result is that the frequency of the axial motion $\bar{\nu}_z$ shifts by

$$\Delta \bar{\nu}_z = \delta_B (n + m_s), \tag{1.29}$$

in proportion to the cyclotron quantum number n and the spin quantum number m_s. Figure 1.9 shows the $\Delta \bar{\nu}_z = 4\,\mathrm{Hz}$ shift in the 200 MHz axial frequency that takes

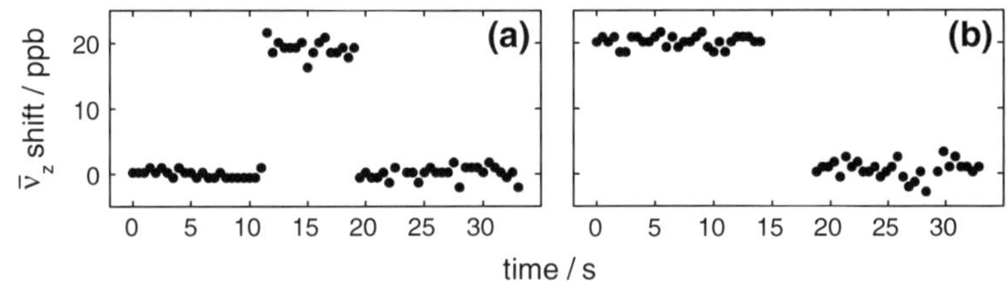

Fig. 1.9 Two quantum jumps: A cyclotron jump (**a**) and spin flip (**b**) measured via a QND coupling to shifts in the axial frequency

place for one-quantum changes in cyclotron (Fig. 1.9a) and spin energy (Fig. 1.9b). The 20 ppb shift is easy to observe with an averaging time of only 0.5 s. We typically measure with an averaging time that is half this value.

1.4.2 One-Electron Self-Excited Oscillator

The QND coupling makes small changes in the electron's axial oscillation frequency, the signature of one-quantum cyclotron transitions and spin flips. Measuring these small frequency changes is facilitated by a large axial oscillation amplitude. To this end we use electrical feedback which we demonstrated could be used effectively to either cool the axial motion [48] or to make a large self-excited axial oscillation [49]. Cyclotron excitations and spin flips are generally induced while the detection system is off, as will be discussed. After an attempt to excite the cyclotron motion or to flip the spin has been made, the detection system is then turned on. The self-excited oscillator rapidly reaches steady state, and its oscillation frequency is then measured by fourier transforming the signal.

The 200 MHz axial frequency lies in the radio-frequency (rf) range which is more experimentally accessible than the microwave range of the 150 GHz cyclotron and spin frequencies, as mentioned. Nevertheless, standard rf techniques must be carefully tailored for our low-noise, cryogenic experiment. The electron axial oscillation induces image currents in the trap electrodes that are proportional to the axial velocity of the electron [25, 50]. An inductor (actually the inductance of a cryogenic feedthrough) is placed in parallel with the capacitance between two trap electrodes to cancel the reactance of the capacitor which would otherwise short out the induced signal. The rf loss in the tuned circuit that is formed is an effective resistance that damps the axial motion.

The voltage that the electron motion induces across this effective resistance is amplified with two cryogenic detection amplifiers. The heart of each amplifier is a single-gate high electron mobility transistor (Fujitsu FHX13LG).

The first amplifier is at the 100 mK dilution refrigerator base temperature. Operating this amplifier without crashing the dilution refrigerator requires operating with a power dissipation in the FET that is three orders of magnitude below the transistor's 10 mW design dissipation. The effective axial temperature for the electron while current is flowing through the FET is about 5 K, well above the ambient temperature. Very careful heat sinking makes it possible for the effective axial temperature of the electron to cool to below 230 mK in several seconds after the amplifier is turned off, taking the electron axial motion to this temperature. Cyclotron excitations and spin flips are induced only when the axial motion is so cooled, with the detection amplifiers off, since the electron is then making the smallest possible excursion in the magnetic bottle gradient.

The second cryogenic amplifier is mounted on the nominally 600 mK still of the dilution refrigerator. This amplifier counteracts the attenuation of a thermally-isolating but lossy stainless steel transmission line that carries the amplified signal out of the refrigerator. The second amplifier boosts the signal above the noise floor of the first room-temperature amplifier.

Because the induced image-current signal is proportional to the electron's axial velocity, feeding this signal back to drive the electron alters the axial damping force, a force that is also proportional to the electron velocity. Changing the feedback gain thus changes the damping rate. As the gain increases, the damping rate decreases as does the effective axial temperature of the electron, in accord with the fluctuation dissipation theorem [51]. Feedback cooling of the one-electron oscillator from 5.2 to 850 mK was demonstrated [48]. The invariant ratio of the separately measured damping rate and the effective temperature was also demonstrated, showing that the amplifier adds very little noise to the feedback.

Setting the feedback gain to make the feedback drive exactly cancel the damping in the attached circuit could sustain a large axial oscillation amplitude, in principle. However, since the gain cannot be perfectly adjusted, noise fluctuations will always drive the axial oscillation exponentially away from equilibrium. We thus stabilize the oscillation amplitude using a digital signal processor (DSP) that Fourier transforms the signal in real time, and adjusts the feedback gain to keep the signal size at a fixed value.

The one-particle self-excited oscillator is turned on after an attempt has been made to excite the cyclotron energy up one level, or to flip the spin. The frequency of the axial oscillation that rapidly stabilizes at a large and easily detected amplitude is then measured. Small shifts in this frequency reveal whether the cyclotron motion has been excited by one quantum or whether the spin has flipped, as illustrated in Fig. 1.9.

1.4.3 Inhibited Spontaneous Emission

The spontaneous emission of synchrotron radiation in free space would make the damping time for an electron's cyclotron motion to be less than 0.1 s. This is not enough time to average down the noise in our detection system to the level that would allow the resolution of one-quantum transitions between cyclotron states. Also, to drive cyclotron transitions "in the dark", with the detection system off, requires that the cyclotron excitations persist long enough for the detection electronics to be turned on. Cavity-inhibition of the spontaneous emission gives us the averaging time that we need.

One of the early papers in what has come to be know as cavity QED was an observation of inhibited spontaneous emission within a Penning trap [42]—the first time that inhibited spontaneous emission was observed within a cavity and with only one particle—as anticipated earlier [52, 53]. As already mentioned, the cylindrical Penning trap [27] was invented to provide understandable boundary conditions to control the spontaneous emission rate with only predictable cavity shifts of the electron's cyclotron frequency.

The spontaneous emission rate is measured directly, by making a histogram of the time the electron spends in the first excited state after being excited by a microwave

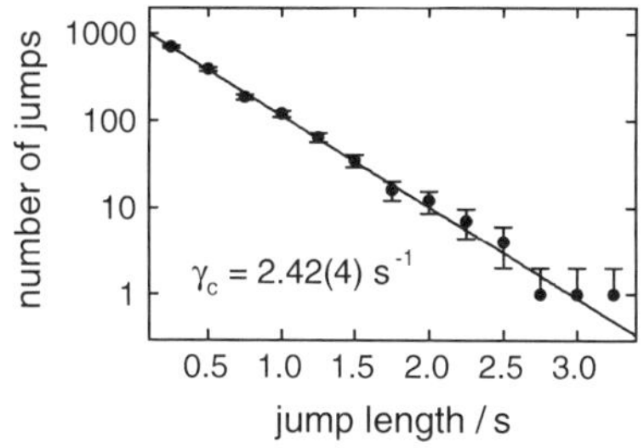

Fig. 1.10 A histogram of the time that the electron spends in the first excited state that is fit to an exponential reveals the substantial inhibition of the spontaneous emission of synchrotron radiation. The decay time, 0.41 s in this example, depends on how close the cyclotron frequency is to neighboring radiation modes of the trap cavity. Lifetimes as long as 16 s have been observed

drive injected into the trap cavity with the detector left on. Figure 1.10 shows a sample histogram which fits well to an exponential (solid curve) with a lifetime of 0.41 s in this example.

Stimulated emission is avoided by making these observations only when the cavity is at low temperature so that effectively no blackbody photons are present. The detector causes thermal fluctuations of the axial oscillation amplitude, and these in turn make the cyclotron frequency fluctuate. For measuring the cyclotron decay time, however, this does not matter as long as the fluctuations in axial amplitude are small compared to the 2 mm wavelength of the radiation that excites the cyclotron motion.

The spontaneous emission rate into free space is [25]

$$\gamma_+ = \frac{4}{3}\frac{r_e}{c}(\omega_c)^2 \approx \frac{1}{0.89\ \text{ms}}\,. \tag{1.30}$$

The measured rate in this example is thus suppressed by a factor of 4.5. The density of states within the cylindrical trap cavity is not that of free space. Instead the density of states for the radiation is peaked at the resonance frequencies of the radiation modes of the cavity, and falls to very low values between the radiation modes. We attain the inhibited spontaneous emission by tuning the magnetic field so that the cyclotron frequency is as far as possible from resonance with the cavity radiation modes. With the right choice of magnetic field we have increased the lifetime to 16 s, which is a cavity suppression of spontaneous emission by a factor of 180.

In a following section we report on using the direct measurements of the radiation rate for electron cyclotron motion to probe the radiation modes of the cavity, with the radiation rate increasing sharply at frequencies that approach a resonant mode of the cavity.

1.5 Elements of a Electron $g/2$ Measurement

1.5.1 Quantum Jump Spectroscopy

We determine the cyclotron and anomaly frequencies using quantum jump spectroscopy, in which a near resonance drive attempts to either excite the cyclotron motion or flip the spin. After each attempt we check whether a one-quantum transition has taken place, and build up a histogram of transitions per attempt. Figure 1.11 shows the observed quantum jump lineshapes upon which our 2008 measurement is based.

A typical data run consists of alternating scans of the cyclotron and anomaly lines. The runs occur at night, with daytime runs only possible on Sundays and holidays when the ambient magnetic field noise is lower. Interleaved every three hours among these scans are periods of magnetic field monitoring to track long-term drifts using

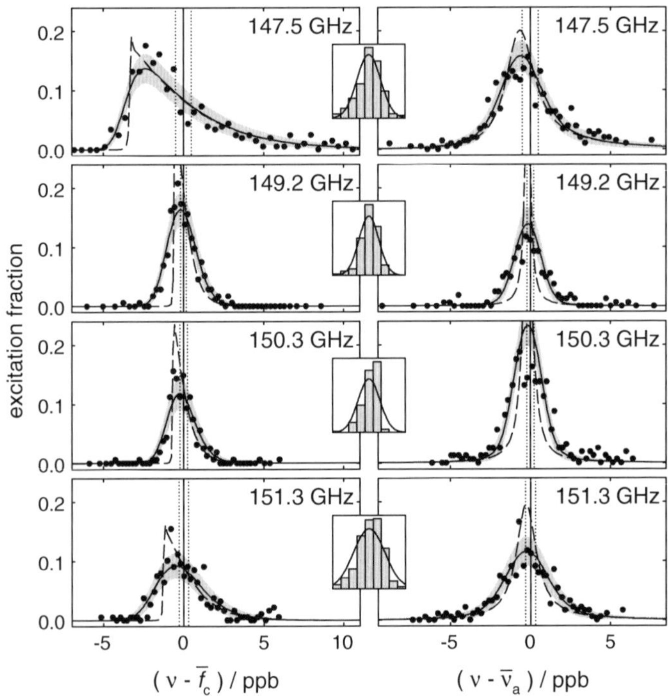

Fig. 1.11 Quantum-jump spectroscopy lineshapes for cyclotron (*left*) and anomaly (*right*) transitions with maximum-likelihood fits to broadened lineshape models (*solid*) and inset resolution functions (*solid*) and edge-tracking data (histogram). *Vertical lines* show the $1-\sigma$ uncertainties for extracted resonance frequencies. Corresponding unbroadened lineshapes are dashed. Gray bands indicate $1-\sigma$ confidence limits for distributions about broadened fits. All plots share the same relative frequency scale

the electron itself as the magnetometer. In addition, we continuously monitor over fifty environmental parameters such as refrigerator temperatures, cryogen pressures and flows, and the ambient magnetic field in the lab so that we may screen data for abnormal conditions and troubleshoot problems.

Cyclotron transitions are driven by injecting microwaves into the trap cavity. The microwaves originate as a 15 GHz drive from a signal generator (Agilent E8251A) whose low-phase-noise, 10 MHz oven-controlled crystal oscillator serves as the time-base for all frequencies in the experiment. After passing through a waveguide that removes all subharmonics, the signal enters a microwave circuit that includes an impact ionization avalanche transit-time (IMPATT) diode, which multiplies the frequency by ten and outputs the $\bar{f}_c$ drive at a power of 2 mW. Voltage-controlled attenuators reduce the strength of the drive, which is broadcast from a room temperature horn through teflon lenses to a horn at 100 mK (Fig. 1.8) and enters the trap cavity through an inlet waveguide (Fig. 1.7).

Anomaly transitions are driven by potentials, oscillating near $\bar{\nu}_a$, applied to electrodes to drive off-resonant axial motion through the magnetic bottle gradient (Eq. 1.28). The gradient's $z\rho\hat{\boldsymbol{\rho}}$ term mixes the driven oscillation of z at $\bar{\nu}_a$ with that of ρ at $\bar{f}_c$ to produce an oscillating magnetic field perpendicular to **B** as needed to flip the spin. The axial amplitude required to produce the desired transition probability is too small to affect the lineshape (Sect. 1.5.4); nevertheless, we apply a detuned drive of the same strength during cyclotron attempts so the electron samples the same magnetic gradient.

Quantum jump spectroscopy of each resonance follows the same procedure. With the electron prepared in the spin-up ground state $|0, \frac{1}{2}\rangle$, the magnetron radius is reduced with 1.5 s of strong sideband cooling at $\bar{\nu}_z + \bar{\nu}_m$ with the SEO turned off immediately and the detection amplifiers turned off after 0.5 s. After an additional 1 s to allow the axial motion to thermalize with the tuned circuit, we apply a 2 s pulse of either a cyclotron drive near $\bar{f}_c$ or an anomaly drive near $\bar{\nu}_z$ with the other drive applied simultaneously but detuned far from resonance. The detection electronics and SEO are turned back on; after waiting 1 s to build a steady-state axial amplitude, we measure $\bar{\nu}_z$ and look for a 20 ppb shift up (from a cyclotron transition) or down (from an anomaly transition followed by a spontaneous decay to $|0, -\frac{1}{2}\rangle$) in frequency. Cavity-inhibited spontaneous emission provides the time needed to observe cyclotron transitions before decay. The several-cyclotron-lifetimes wait for a spontaneous decay after an anomaly attempt is the rate-limiting step in the spectroscopy. After a successful anomaly transition and decay, simultaneous cyclotron and anomaly drives pump the electron back to $|0, \frac{1}{2}\rangle$. All timing is done in hardware. We probe each resonance line with discrete excitation attempts spaced in frequency by approximately 10 % of the linewidth. We step through each drive frequency on the $\bar{f}_c$ line, then each on the $\bar{\nu}_a$ line, and repeat.

1.5.2 The Electron as Magnetometer

Slow drifts of the magnetic field are corrected using the electron itself as a magnetometer. Accounting for these drifts allows the combination of data taken over many days, giving a lineshape signal-to-noise that allows the systematic investigation of lineshape uncertainties at each field. For a half-hour at the beginning and end of a run and again every three hours throughout, we alter our cyclotron spectroscopy routine by applying a stronger drive at a frequency below $\bar{f}_c$. Using the same timing as above but a ten-times-finer frequency step, we increase the drive frequency until observing a successful transition. We then jump back 60 steps and begin again.

We model the magnetic field drift by fitting a polynomial to these "edge" points (so-called because the ideal cyclotron lineshape has a sharp low-frequency edge). Since we time-stamp every cyclotron and anomaly attempt, we use the smooth curve to remove any field drift. This edge-tracking adds a 20 % overhead, but allows the use of data from nights with a larger than usual field drift, and the combination of data from different nights.

1.5.3 Measuring the Axial Frequency

In addition to $\bar{f}_c$ and $\bar{\nu}_a$, measuring $g/2$ requires a determination of the axial frequency $\bar{\nu}_z$ (Eq. 1.19). To keep the relative uncertainty in $g/2$ from $\bar{\nu}_z$ below 0.1 ppt, we must know $\bar{\nu}_z$ to better than 50 ppb (10 Hz). This is easily accomplished. We routinely measure $\bar{\nu}_z$ when determining the cyclotron and spin states. However, the large self-excited oscillation amplitude in the slightly anharmonic axial potential typically results in a 10 ppb shift, compared to the $\bar{\nu}_z$ for the thermally-excited amplitude during the cyclotron and anomaly pulses. We cannot directly measure the axial frequency under the pulse conditions because the amplifiers are off. We come close when measuring $\bar{\nu}_z$ with the amplifiers on and all axial drives off. This thermal axial resonance appears as a dip on the amplifier noise resonance [50], and we use it as our measurement. The difference in $\bar{\nu}_z$ with the amplifiers on and off is negligible. A second shift comes from the interaction between the axial motion and the amplifier, which both damps the motion and shifts $\bar{\nu}_z$. The maximum shift of $\bar{\nu}_z$ is 1/4 of the damping rate, which at $\approx$1 ppb is negligible at our current precision. A third shift of $\bar{\nu}_z$ comes from the anomaly drive, which induces both a frequency-pulling from the off-resonant axial force and a Paul-trap shift from the change in effective trapping potential [54]; based on extrapolation from measured shifts at higher powers, we estimate these shifts combine to 1 ppb at the highest anomaly power used for the measurement—too small to affect $g/2$.

1.5.4 Frequencies from Lineshapes

The cyclotron frequency $\bar{f}_c$ and anomaly frequency $\bar{\nu}_a$ (Fig. 1.6) must be determined from their respective quantum jump spectroscopy lineshapes (Fig. 1.11). The observed lineshapes are much broader than the natural linewidth that arises because the excited cyclotron state decays by the cavity-inhibited spontaneous emission of synchrotron radiation. The shape arises because the electron experiences a magnetic field that varies during the course of a measurement. Variations arise because of the electron's thermal axial motion within the magnetic bottle gradient, for example. Other possible variations could arise because the magnetic field for the Penning trap fluctuates in time, or because of a distribution of magnetron orbit sizes for the quantum jump trials.

Once the slow drift of the magnetic field (p. 24) has been removed, there is no reason for the electron to sample a different distribution of magnetic field values while the anomaly frequency is being measured compared to when the trap-modified cyclotron frequency is being measured. Each resonance shape converts the distribution of sampled magnetic fields values into the corresponding distribution of frequency values. Dividing the quantum jump lineshapes into frequency bins, we obtain average cyclotron and anomaly frequencies by weighting the frequency of each bin by the number of quantum jumps in the bin, and use these average frequencies in Eq. 1.19.

Using the weighted average frequencies will remove shifts to $g/2$ caused by the thermal axial motion of the electron within the magnetic bottle gradient, the largest source of the observed linewidth. The use of weighted average frequencies should also account for temporal fluctuations in the magnetic field of the Penning trap on the measurement time scale for the frequencies. If there is a distribution of magnetron radii for the quantum jump trials, the weighted average method should account for the resulting distribution of magnetic field values as well.

To verify the weighted averages method, and to assign safe uncertainties to the average frequencies that we deduce using it, we also analyze our measured lineshapes in a very different way. We start with an analytic calculation of the lineshape for thermal Brownian motion of the axial motion for a given axial temperature T_z [55]. We then fit the measured cyclotron and anomaly lineshapes (Fig. 1.11) to the ideal lineshape convolved with a Gaussian broadening function to take into account other sources of the magnetic field distribution. The analytically calculated lineshapes are the dashed curves in Fig. 1.11, the maximum-likelihood fits to the broadened lineshapes are solid curves, and the gray bands indicate where we would expect 68 % of the measured points to lie. The insets to Fig. 1.11 show the best-fit resolution functions. We assign a lineshape uncertainty that is the size of the differences between the $g/2$ value determined from the fitting and our preferred weighted averages method.

The linewidths are wider for two of the four measurements in Fig. 1.11, and they remained reproducible over the weeks required to take each data point. A wider cyclotron linewidth indicates a higher axial temperature. We know of no reason that the axial temperatures should be different for different values of the Penning trap field;

this is one reason that we assign the larger uncertainties that reflect the difference between the two methods. The narrower lineshapes have better agreement between the weighted average method and the fit method, and hence the assigned lineshape uncertainties are smaller. Not surprisingly, the narrower lines better determine the corresponding frequencies.

For the 2008 measurement the lineshape uncertainty is larger than any other. Future efforts will focus upon understanding and reducing the lineshape broadening and uncertainty.

1.5.5 Cavity Shifts

Despite the precision reached in this measurement, only one correction to the directly measured $g/2$ value is required, the $\Delta g_{cav}/2$ included in Eq. 1.19. The correction is a cavity shift correction that depends upon interaction of the electron with nearby cavity radiation modes. The trap cavity modifies the density of states of the radiation modes of free space, though not enough to significantly affect QED calculations of g [56]. Since the cavity shift correction depends upon the electron cyclotron frequency, we measure $g/2$ at four different cyclotron frequencies to make sure that the same $g/2$ is deduced when cavity shifts of different sizes are applied.

The cavity-inhibited spontaneous emission narrows the cyclotron resonance line, giving the time in the excited state that is needed to turn on the self-excited oscillator, and to average its signal long enough determine the cyclotron state. Cavity shifts are the unfortunate downside of the cavity, arising because the cyclotron oscillator has its frequency pulled by its coupling to nearby radiation modes of the cavity.

The cylindrical Penning trap was invented to make a microwave cavity with a calculable geometry. Section 1.3.2 describes a perfect cylindrical trap cavity and the radiation fields that it can support. However, the trap is not perfectly machined, it changes its size as it cools from room temperature down to 0.1 K, and it has small slits that make it possible to bias sections to form a Penning trap.

The shape of the radiation fields near the center of the trap cavity are not greatly altered for the real cavity, but the resonant frequencies of the modes are slightly shifted. The frequency shifts are not enough to keep us from identifying most modes by comparing to calculated frequencies, but are large enough that we must measure the mode frequencies if we are to characterize the interaction of the cavity and an electron. The mode quality factors (resonant frequencies divided by energy damping rates) must also be determined. The decay of the radiation field within the cavity depends upon power dissipated by currents (induced in the electrodes and modified by the slits), and upon the loss of microwave power that escapes the trap despite the choke flanges in the slits.

We developed two methods to learn the resonant frequencies of the radiation modes of real trap cavity.

1. A cloud of electrons near the center of the trap is heated using a parametric driving force. The electrons cool via synchrotron radiation with a rate that is highest when their cyclotron frequency is resonant with a cavity radiation mode, and that is very small far from resonance [32–34, 57]. Figure 1.12a shows the peaks in the signal from the electrons that correspond to resonance with cavity radiation modes that are labeled as described earlier.
2. The measured spontaneous emission rate for a single electron near the center of the trap cavity (Fig. 1.12b), and the dependence of this rate upon the amplitude of the axial oscillation of the electron (Fig. 1.12c), both depend upon the proximity of the electron cyclotron frequency to cavity radiation modes that couple to a nearly centered electron. Figure 1.13 illustrates how the one-electron damping rate and dependence upon axial oscillation amplitude are measured.

From the cavity spectra in Fig. 1.12a–c we deduce the mode frequencies and uncertainties represented by the gray bands in these figures. Our identification of the modes is aided by several features of the spectra. Modes that are strongly coupled to the electrons (the coupling increases with electron number) can split into two normal modes. A large axial oscillation during measurements of the cavity spectrum produces sidebands at the axial frequency for modes with a node at the trap center, and at twice the sidband frequency for radiation modes with an antinode at the center. Modes which would not couple to a perfectly centered electron will couple more strongly to the electrons as their number is increased so that they occupy a larger volume. From 2006 to 2008 our understanding of the cavity improved when we became aware of and were able to measure a small displacement of the electrostatic center of the trap (where the electron resides), and the center for the cavity radiation modes.

So far we have used the calculable cylindrical trap geometry to know which radiation modes can couple to an electron near the center of the trap, and we have recognized these modes in measured cavity spectra by comparing their measured frequencies to what is calculated for a perfect cavity. Next we use the measured radiation mode frequencies and quality factors as input to a calculation of the cavity shift of the electron cyclotron frequency as a function of the electron cyclotron frequency (Fig. 1.12d).

A calculation of the shifts [55, 58] must carefully distinguish and remove the electron self-energy from the electron-cavity interaction. The uncertainty in the measured inputs give a cavity shift uncertainty (Fig. 1.12e) that is small between the resonance frequencies of modes that couple strongly to a centered electron, and then increases strongly closer to the resonant frequencies of these modes. The diamonds at the top of the figure show how in our four measurements of $g/2$ we avoid the electron cyclotron frequencies for which the uncertainty is the largest. Figure 1.14 shows the good agreement attained between the four measurements when the cavity shifts are applied.

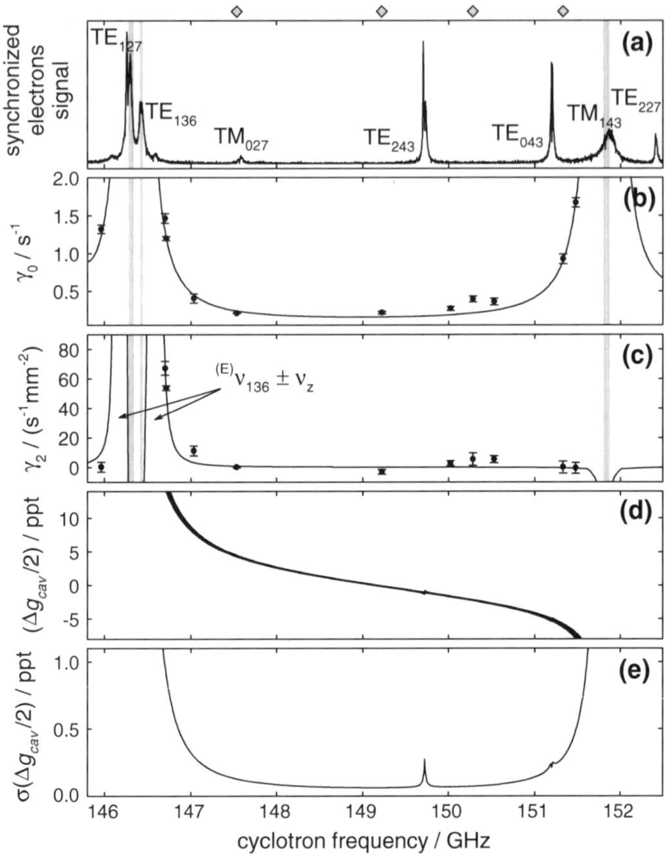

Fig. 1.12 Cavity shift results come from synchronized electrons (**a**) and from direct measurements with one electron of γ_c (**b**) and its dependence on axial amplitude (**c**). Together, they provide uncertainties in the frequencies of coupled cavity radiation modes (*gray*) that translate into an uncertainty band of cavity shifts $\Delta g_{cav}/2$ (**d**) whose half-width, i.e., the cavity shift uncertainty, is plotted in (**e**). The diamonds at the top indicate the cyclotron frequencies of the four $g/2$ measurements

1.6 Results and Applications

1.6.1 Most Accurate Electron $g/2$

The measured values, shifts, and uncertainties for the four separate measurements of $g/2$ are in Table 1.2. The uncertainties are lower for measurements with smaller cavity shifts and smaller linewidths, as might be expected. Uncertainties for variations of the power of the $\bar{\nu}_a$ and $\bar{f}_c$ drives are estimated to be too small to show up in the table. A weighted average of the four measurements, with uncorrelated and correlated errors combined appropriately, gives the electron magnetic moment in Bohr magnetons,

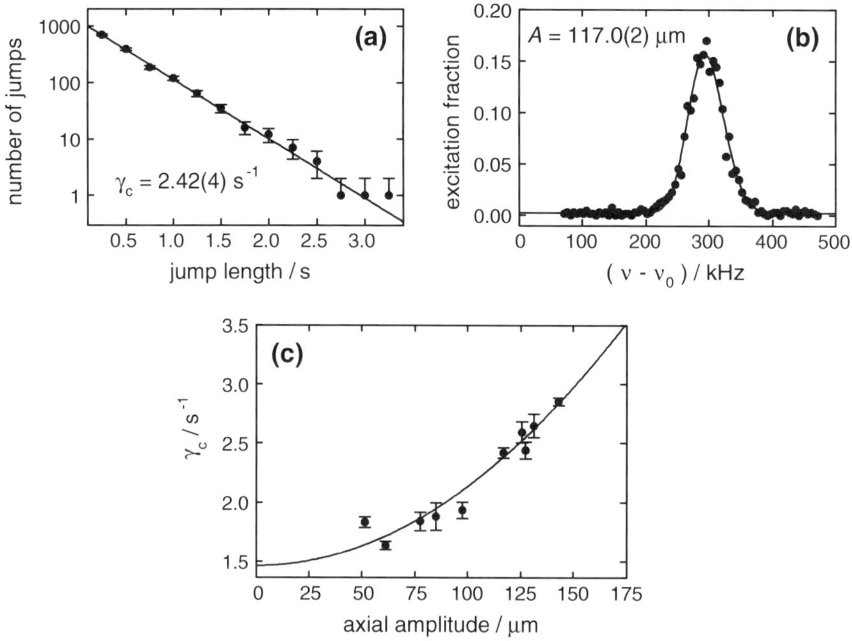

Fig. 1.13 Measurement of the cyclotron damping rate at 146.70 GHz, near the upper sideband of TE_{136}. The cyclotron damping rate as a function of axial amplitude (**c**) extrapolates to the desired lifetime. Each point in (**c**) consists of a damping rate measured from a fit to a histogram of cyclotron jump lengths (**a**) as well as an axial amplitude measured from a driven *cyclotron line* (**b**)

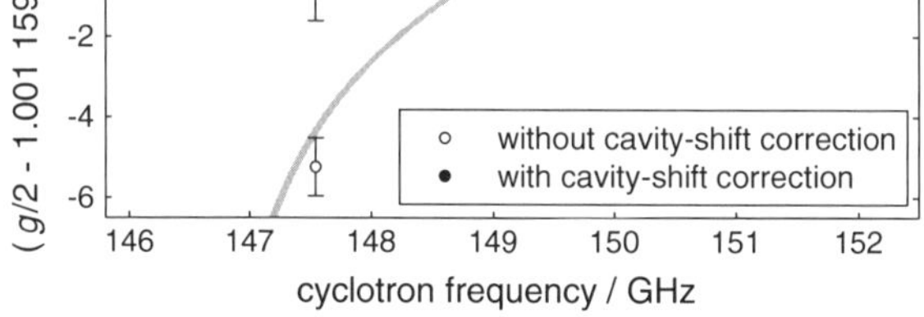

Fig. 1.14 Four measurements of $g/2$ without (*open*) and with (*filled*) cavity-shift corrections. The *light gray* uncertainty band shows the average of the corrected data. The *dark gray* band indicates the expected location of the uncorrected data given our result in Eq. 1.1 and including only the cavity-shift uncertainty

Table 1.2 Measurements and shifts with uncertainties multiplied by 10^{12}

$\bar{f}_c$	147.5 GHz	149.2 GHz	150.3 GHz	151.3 GHz
$g/2$ raw	−5.24(0.39)	0.31 (0.17)	2.17 (0.17)	5.70 (0.24)
Cav. shift	4.36(0.13)	−0.16 (0.06)	−2.25 (0.07)	−6.02 (0.28)
Lineshape				
correlated	(0.24)	(0.24)	(0.24)	(0.24)
uncorrelated	(0.56)	(0.00)	(0.15)	(0.30)
$g/2$	−0.88(0.73)	0.15 (0.30)	−0.08 (0.34)	−0.32 (0.53)

The cavity-shifted "$g/2$ raw" and corrected "$g/2$" are offset from our result in Eq. 1.1

$$\mu/\mu_B = -g/2 = -1.001\,159\,652\,180\,73\,(28) \quad [0.28\text{ ppt}]. \tag{1.1}$$

The uncertainty is 2.7 and 15 times smaller than the 2006 and 1987 measurements, and 2,300 times smaller than has been achieved for the heavier muon lepton [59].

1.6.2 Most Accurate Determination of α

The new measurement determines the fine structure constant, $\alpha = e^2/(4\pi\varepsilon_0\hbar c)$, much more accurately than does any other method. The fine structure constant is the fundamental measure of the strength of the electromagnetic interaction in the low energy limit, and it is also a crucial ingredient of our system of fundamental constants [60]. Only the bare essentials of what is needed to determine α from $g/2$ are summarized here.

The standard model relates g and α by

$$\begin{aligned}\frac{g}{2} &= 1 + C_2\left(\frac{\alpha}{\pi}\right) + C_4\left(\frac{\alpha}{\pi}\right)^2 + C_6\left(\frac{\alpha}{\pi}\right)^3 + C_8\left(\frac{\alpha}{\pi}\right)^4 \\ &\quad + C_{10}\left(\frac{\alpha}{\pi}\right)^5 + \cdots + a_{\text{hadronic}} + a_{\text{weak}},\end{aligned} \tag{1.31}$$

with the asymptotic series and the values of the C_k coming from QED. Very small hadronic and weak contributions are included, along with the assumption that there is no significant modification from electron substructure or other physics beyond the standard model (Fig. 1.15).

QED calculations give the constants C_k,

$$C_2 = 0.500\,000\,000\,000\,00 \text{ (exact)} \tag{1.32}$$

$$C_4 = -0.328\,478\,444\,002\,55\,(33) \tag{1.33}$$

$$C_6 = 1.181\,234\,016\,816\,(10) \tag{1.34}$$

$$C_8 = -1.909\,7\,(20) \tag{1.35}$$

$$C_{10} = 9.16\,(0.57). \tag{1.36}$$

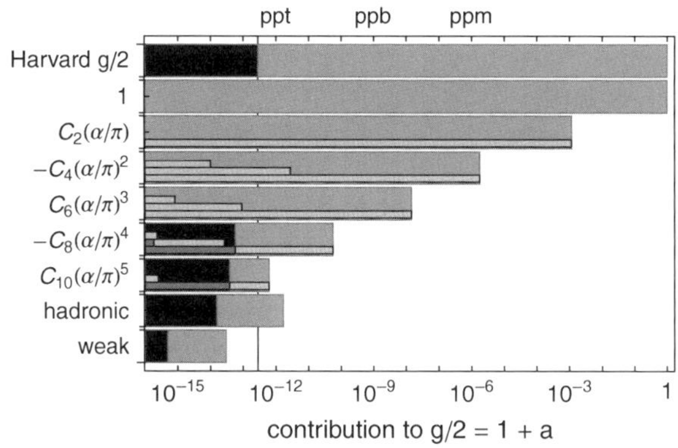

Fig. 1.15 Contributions to $g/2$ for the experiment (*top bar*), terms in the QED series (*below*), and from small distance physics (*below*). Uncertainties are *black*. The inset *light gray bars* represent the magnitude of the larger mass-independent terms (A_1) and the smaller A_2 terms that depend upon either m_e/m_μ or m_e/m_τ. The even smaller A_3 terms, functions of both mass ratios, are not visible on this scale

The QED theory for C_2 [61], C_4 [62–64], and C_6 [65] is exact, with no uncertainty, except for an essentially negligible uncertainty in C_4 and C_6 that comes from a weak functional dependence upon the lepton mass ratios, m_μ/m_e and m_τ/m_e. Numerical QED calculations give the value and uncertainty for C_8 and C_{10} [6]. The first evaluation of C_{10} is the most significant theoretical advance since the electron magnetic moment was measured. A remarkable 12,672 Feynman diagrams are involved.

The hadronic anomaly a_{hadronic}, calculated within the context of the Standard Model,

$$a_e^{\text{hadronic}} = 1.678(15) \times 10^{-12}\,, \tag{1.37}$$

contributes at the level of several times the current experimental uncertainty, but the calculation uncertainty in the hadronic anomaly is not important [66–69]. The weak anomaly is negligible for the current experimental precision.

The most accurately determined fine structure constant is given by

$$\begin{aligned}\alpha^{-1} &= 137.035\,999\,173\,(33)\,(8) \quad [0.24\text{ ppb}]\,[0.06\text{ ppb}],\\ &= 137.035\,999\,173\,(34) \quad [0.25\text{ ppb}]. \end{aligned} \tag{1.38}$$

The first line shows that the 0.24 ppb experimental uncertainty (33) is larger than the 0.06 ppb theoretical uncertainty (8) now that C_{10} has been calculated. The theory uncertainty contribution to α is divided as (7) and (5) for C_8 and C_{10}. The uncertainties from the experiment and theory are represented in Fig. 1.16. The dashes show the uncertainties before the recent theoretical advances.

The total 0.25 ppb uncertainty in α is nearly three times smaller than the 0.66 ppm uncertainty for the next most precise determination [7] (Fig. 1.2). The so-called atom recoil determination that is used actually requires measurements of the Rydberg

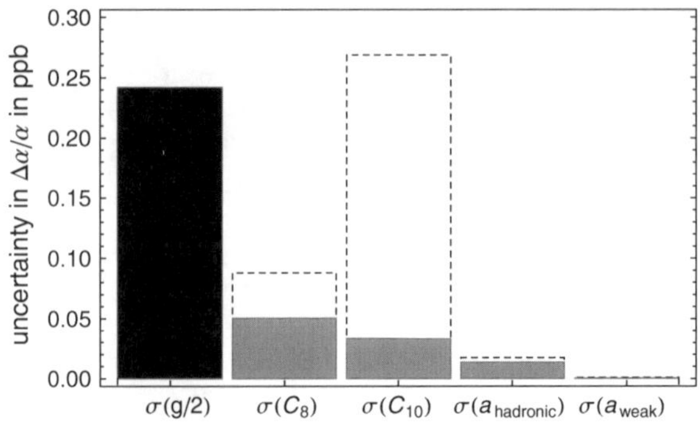

Fig. 1.16 Experimental uncertainty (*black*) and theoretical uncertainties (*gray*) that determine the uncertainty in the α that is determined from the measured electron $g/2$. The dashes present the uncertainties before the very recent theoretical advances

constant [70, 71], transition frequencies [7, 72], mass ratios [73–75], and either a Rb [7] or Cs [76] recoil velocity measured in an atom interferometer.

1.6.3 Testing the Standard Model and QED

The dimensionless electron magnetic moment g that is measured can be compared to the $g(\alpha)$ that is predicted by the Standard Model of particle physics. The input needed to calculate $g(\alpha)$ is the measured fine structure constant α (that is determined without the use of the electron magnetic moment). The most accurately measured and calculated values of $g/2$ are currently given by

$$g/2 = 1.001\ 159\ 652\ 180\ 73\ (28)\ [0.28\ \text{ppt}], \tag{1.39}$$

$$g(\alpha)/2 = 1.001\ 159\ 652\ 181\ 78\ (77)\ [0.77\ \text{ppt}]. \tag{1.40}$$

The measurement is our one-electron quantum cyclotron measurement [2]. The calculated value $g(\alpha)/2$ comes from using the most precise Rb value of α [7] in Eq. 1.31. The uncertainty in this "calculated" value actually comes almost entirely from the uncertainty in the Rb α. The Standard Model prediction is thus tested and verified to 0.8 ppt. The smaller 0.3 ppt uncertainty in the measured $g/2$, along with the comparable uncertainty in the QED calculation, would allow a better test of QED if a more precise, independent value of α could be measured.

About 1 part per thousand of the electron $g/2$ comes from the unavoidable interaction of the electron with the virtual particles of "empty space", as described by quantum electrodynamics (QED) and represented in Fig. 1.17. Where testing QED is the primary focus, measured and calculated values of the so-called anomalous magnetic moment of the electron (defined by $a = g/2 - 1$ so that the Dirac contribution

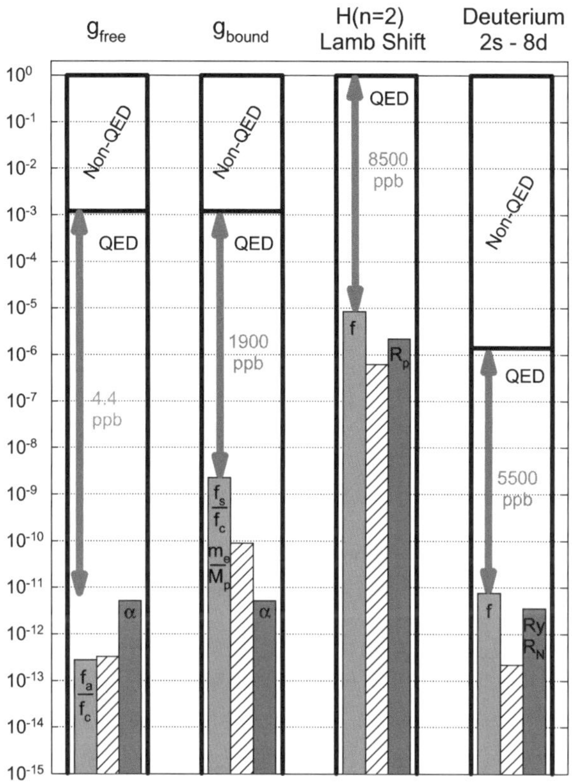

Fig. 1.17 Comparisons of precise tests of QED. The arrows represent the fractional accuracy to which the QED contribution to the measured g values and frequencies that are measured

is subtracted out) are often compared. The measured and calculated values of a that correspond to the $g/2$ values above are

$$a = 0.001\,159\,652\,180\,73\,(\,28)\,[0.24\text{ ppb}], \tag{1.41}$$

$$a(\alpha) = 0.001\,159\,652\,181\,78\,(\,77)\,[0.66\text{ ppb}], \tag{1.42}$$

At the one standard deviation level, the difference of the measured and calculated values is

$$\delta a = a - a(\alpha) \tag{1.43}$$

$$= g/2 - g(\alpha)/2 \tag{1.44}$$

$$= 1.05(0.82) \times 10^{-12}. \tag{1.45}$$

The possible difference between the measurement and calculation is thus bounded by

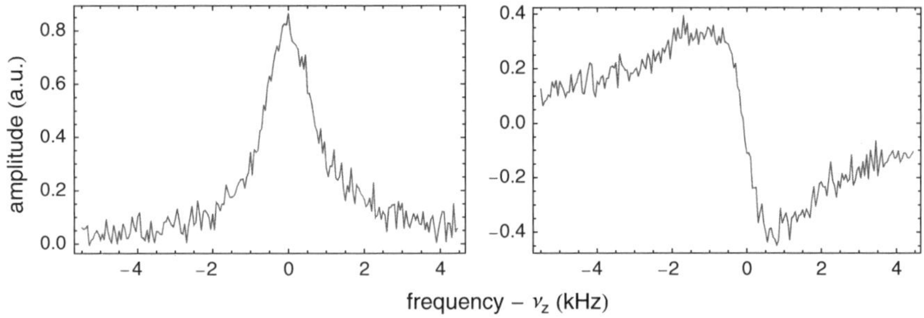

Fig. 1.18 A driven resonance for positrons captured and stored next to a Penning trap designed for a new generation of electron and posiron magnetic moment measurements

$$|\delta a| < 1.9 \times 10^{-12}, \tag{1.46}$$

at the one standard deviation level, with this bound arising almost entirely from the uncertainty in the measurement of α from Rb.

Some of the most precise tests of bound-state QED are compared in Fig. 1.18 with the electron $g/2$. The QED test based upon the measurements [74] and calculation [77] of $g/2$ for an electron bound in an ion tests QED less precisely. In fact, the calculated value of the bound g values depends upon the mass of the electron strongly enough that this measurement is now being used to determine the electron mass in amu, much as we determine α from our measurements of the magnetic moment of the free electron. The n $=$ 2 Lamb shift in hydrogen is essentially entirely due to QED. However, the measurements are much less precise so that QED is again tested less precisely.

The last example in the figure is a QED test based upon a number of measurements of hydrogen and deuterium transition frequencies—the QED contribution to which are typically at the ppm level. Theoretical calculations that depend upon the Rydberg constant, the fine structure constant, the ratio of the electron and proton masses and the size of the nucleus are fit to a number of accurately measured transition frequencies for hydrogen and deuterium. The fit determines values for the mentioned constants. The QED test come from removing one of the measured lines from the fit, and using the best fit to predict the value of the transition frequency that was omitted. This process tests the Standard Model prediction at a comparable precision to that provided using the magnetic moment of the electron. However, it tests the size of the QED contribution to a much lower fractional precision.

The QED tests described so far test QED predictions to the highest precision and the highest order in α. There are many other tests of QED with a much lower precision. Although these tests are outside of the scope of this work it is worth mentioning that it is interesting to probe QED in other ways. For example, it seems interesting to test QED for systems whose binding energy is very large, even comparable to the electron rest mass energy as can be done in high Z systems. Another example is probing the QED of positronium, the bound state of an electron and a positron, insofar as annihilation and exchange effects are quite different than what must be calculated for normal atoms.

1.6.4 Probe for Electron Substructure

Comparing experiment and theory probes for possible electron substructure at an energy scale one might only expect from a large accelerator. An electron whose constituents would have mass $m^* \gg m$ has a natural size scale, $R = \hbar/(m^* c)$. The simplest analysis of the resulting magnetic moment [78] gives $\delta a \sim m/m^*$, suggesting that $m^* > 260{,}000\,\mathrm{TeV}/c^2$ and $R < 9 \times 10^{-25}\,\mathrm{m}$. This would be an incredible limit, since the largest e^+e^- collider (LEP) probed for a contact interaction at an $E = 10.3\,\mathrm{TeV}$ [79], with $R < (\hbar c)/E = 2 \times 10^{-20}\,\mathrm{m}$.

However, the simplest argument also implies that the first-order contribution to the electron self-energy goes as m^* [78]. Without heroic fine tuning (e.g., the bare mass cancelling this contribution to produce the small electron mass) some internal symmetry of the electron model must suppress both mass and moment. For example, a chirally invariant model [78], leads to $\delta a \sim (m/m^*)^2$. In this case, $m^* > 360\,\mathrm{GeV}/c^2$ and $R < 5 \times 10^{-19}\,\mathrm{m}$. These are stringent limits to be set with an experiment carried out at 100 mK, although they are not yet at the LEP limits. With a more precise measurement of α, so this was limited only by the experimental uncertainty in a, then we could set a limit $m^* > 1\,\mathrm{TeV}$ and $R < 2 \times 10^{-19}$.

1.6.5 Comparison to the Muon $g/2$

The electron $g/2$ is measured about 2300 times more accurately than is $g/2$ for its heavier muon sibling [2, 59]. Because the electron is stable there is time to isolate one electron, cool it so that it occupies a very small volume within a magnetic field, and to resolve the quantum structure in its cyclotron and spin motions. The short-lived muon must be studied before it decays in a very small fraction of a second, during which times it orbits in a very large orbit over which the same magnetic field homogeneity realized with a nearly motionless electron cannot be maintained.

Why then measure the muon $g/2$? The compelling reason is that the muon $g/2$ is expected to be more sensitive to physics beyond the standard model by about a factor of 4×10^4, which is the square of the ratio of the muon to the electron mass. In terms of Eq. 1.10 this means that a_{new} is expected to be bigger for the muon than for the electron by this large factor, making the muon a more attractive probe for new physics.

Unfortunately, the other Standard Model contribution, $a_{\mathrm{hadronic}} + a_{\mathrm{weak}}$, is also bigger by approximately the same large factor, rather than being essentially negligible as in the electron case. Correctly calculating these terms is a significant challenge to detecting new physics. These large terms, and the much lower measurement precision, also make the muon an unattractive candidate (compared to the electron) for determining the fine structure constant and for testing QED.

The measured electron $g/2$ makes two contributions to using the muon system for probing for physics beyond the Standard Model. Both relate to determining the muon QED anomaly $a_{\mathrm{QED}}(\alpha)$

1. The electron measurement of $g/2$ makes possible the most accurate determination of the fine structure constant (discussed in the previous section) as is needed to calculate $a_{\mathrm{QED}}(\alpha)$.
2. The electron measurement of $g/2$ and an independently measured value of α test QED calculations of the very similar $a_{\mathrm{QED}}(\alpha)$ terms in the electron system.

The QED contribution must be accurately subtracted from the measured muon $g-2$ if the much smaller possible contribution from new physics is to be observed.

1.7 Prospects and Conclusion

In conclusion, our 2008 measurement of the electron $g/2$ is 15 times more accurate than the 1987 measurement that provided $g/2$ and α for nearly 20 years, and 2.7 times more accurate than our 2006 measurement that superseded it.

With no uncertainty show-stoppers yet presenting themselves, an entirely new solenoid, dewar, and trap apparatus has been designed and constructed. Figure 1.18 shows the signals from positrons now being stored in the new apparatus. Achieving the reported electron $g/2$ uncertainty with a positron to make the most stringent lepton CPT test seems feasible.

We also intend to seek a much more precise measurement of the electron magnetic moment. Several experimental items warrant further study. First is the broadening of the expected lineshapes which limits the splitting of the resonance lines. Second, the variation in axial temperatures in the observed resonance lineshapes is not understood, and a larger uncertainty comes from the wider lineshapes. Third, cavity sideband cooling could cool the axial motion to near its quantum ground state for a more controlled measurement. Fourth, the new apparatus should be much less sensitive to vibration and other variations in the laboratory environment.

With QED and the assumption of no new physics beyond the standard model of particle physics, the new measurement determines α nearly three times more accurately than any independent method. Given the recent theoretical advance, a better measurement of the electron $g/2$ will produce a more precise value of α.

The measured $g/2$ and an independently measured α test QED and probe for electron size at an unprecedented precision. Prospects are good that better measurements will enable even more sensitive tests of the Standard Model.

Acknowledgments The electron magnetic moment measurements were supported by the atomic physics program of the NSF. The antiproton magnetic moment measurements and antihydrogen experiments were supported by the NSF and the AFOSR.

References

1. B. Odom, D. Hanneke, B. D'Urso, G. Gabrielse, New measurement of the electron magnetic moment using a one-electron quantum cyclotron. Phys. Rev. Lett. **97**, 030801 (2006)
2. D. Hanneke, S. Fogwell, G. Gabrielse, New measurement of the electron magnetic moment and the fine structure constant. Phys. Rev. Lett. **100**, 120801 (2008)
3. D. Hanneke, S. Fogwell Hoogerheide, G. Gabrielse, Cavity control of a single-electron quantum cyclotron: measuring the electron magnetic moment. Phys. Rev. A **83**, 052122 (2011)
4. B. Lautrup, H. Zinkernagel, g-2 and the trust in experimental results. Stud. Hist. Phil. Mod. Phys. **30**, 85–110 (1999)
5. R.S. Van Dyck, Jr., P.B. Schwinberg, H.G. Dehmelt, New high-precision comparison of electron and positron g factors. Phys. Rev. Lett. **59**, 26–29 (1987)
6. T. Aoyama, M. Hayakawa, T. Kinoshita, M. Nio, Quantum electrodynamics calculation of lepton anomalous magnetic moments: numerical approach to the perturbation theory of QED. Prog. Theor. Exp. Phys. **01A**, 107 (2012)
7. R. Bouchendira, P. Clade, S. da Guellati-Khe'lifa, F. Nez, F. Biraben, New determination of the fine structure constant and test of the quantum electrodynamics. Phys. Rev. Lett. **106**, 080801 (2011)
8. G. Gabrielse, A. Khabbaz, D.S. Hall, C. Heimann, H. Kalinowsky, W. Jhe, Precision mass spectroscopy of the antiproton and proton using simultaneously trapped particles. Phys. Rev. Lett. **82**, 3198–3201 (1999)
9. G. Gabrielse, Antiproton mass measurements. Int. J. Mass Spectrom. **251**, 273 (2006)
10. J. DiSciacca, M. Marshall, K. Marable, G. Gabrielse, and the ATRAP Collaboration, One-particle measurement of the antiproton magnetic moment. Phys. Rev. Lett. **110**, 130801 (2013)
11. J. DiSciacca, M. Marshall, K. Marable, G. Gabrielse, Resolving an individual one-proton spin flip to determine a proton spin state. Phys. Rev. Lett. **110**, 140406 (2013)
12. A. Mooser et al., Resolution of single spin flips of a single proton. Phys. Rev. Lett. **110**, 140405 (2013)
13. G. Gabrielse, Penning traps, masses and antiprotons, in *Fundamental Symmetries*, ed. by P. Bloch, P. Pavlopoulos, R. Klapisch (Plenum, New York, 1987), pp. 59–75
14. G. Gabrielse, X. Fei, K. Helmerson, S.L. Rolston, R.L. Tjoelker, T.A. Trainor, H. Kalinowsky, J. Haas, W. Kells, First capture of antiprotons in a penning trap: a keV source. Phys. Rev. Lett. **57**, 2504–2507 (1986)
15. G. Gabrielse, N.S. Bowden, P. Oxley, A. Speck, C.H. Storry, J.N. Tan, M. Wessels, D. Grzonka, W. Oelert, G. Schepers, T. Sefzick, J. Walz, H. Pittner, T.W. Hänsch, E.A. Hessels, Background-free observation of cold antihydrogen with field-ionization analysis of its states. Phys. Rev. Lett. **89**, 213401 (2002)
16. G. Gabrielse, N.S. Bowden, P. Oxley, A. Speck, C.H. Storry, J.N. Tan, M. Wessels, D. Grzonka, W. Oelert, G. Schepers, T. Sefzick, J. Walz, H. Pittner, T.W. Hänsch, E.A. Hessels, Driven production of cold antihydrogen and the first measured distribution of antihydrogen states. Phys. Rev. Lett. **89**, 233401 (2002)
17. M. Amoretti et al., Production and detection of cold antihydrogen atoms. Nature **419**, 456–459 (2002)
18. Y. Enomoto et al., Phys. Rev. Lett. **105**, 243401 (2010)
19. E.A. Hessels, D.M. Homan, M.J. Cavagnero, Two-stage Rydberg charge exchange: an efficient method for production of antihydrogen. Phys. Rev. A **57**, 1668–1671 (1998)
20. A. Speck, C.H. Storry, E.A. Hessels, G. Gabrielse, Laser-controlled production of Rydberg positronium via charge exchange collisions. Phys. Lett. B **597**, 257–262 (2004)
21. G.B. Andresen et al., Confinement of antihydrogen for 1000 seconds. Nature Phys. **7**, 558–564 (2011).
22. G. Gabrielse, R. Kalra, W.S. Kolthammer, R. McConnell, P. Richerme, D. Grzonka, W. Oelert, T. Sefzick, M. Zielinski, D.W. Fitzakerley, M.C. George, E.A. Hessels, C.H. Storry, M. Weel, A. Müllers, J. Walz, Trapped antihydrogen in its ground state. Phys. Rev. Lett. **108**, 113002 (2012)

23. A. Rich, J.C. Wesley, Current status of the lepton *g* factors. Rev. Mod. Phys. **44**, 250–283 (1972)
24. R.S. Van Dyck, Jr., P.B. Schwinberg, H.G. Dehmelt, The Electron (Kluwer Academic Publishers, The Netherlands, 1991), pp. 239–293
25. L.S. Brown, G. Gabrielse, Geonium theory: physics of a single electron or ion in a penning trap. Rev. Mod. Phys. **58**, 233–311 (1986)
26. L.S. Brown, G. Gabrielse, Precision spectroscopy of a charged particle in an imperfect penning trap. Phys. Rev. A **25**, 2423–2425 (1982)
27. G. Gabrielse, F. Colin MacKintosh, Cylindrical penning traps with orthogonalized anharmonicity compensation. Int. J. Mass Spec. Ion Proc. **57**, 1–17 (1984)
28. J.N. Tan, G. Gabrielse, One electron in an orthogonalized cylindrical penning trap. Appl. Phys. Lett. **55**, 2144–2146 (1989)
29. G. Gabrielse, Relaxation calculation of the electrostatic properties of compensated penning traps with hyperbolic electrodes. Phys. Rev. A **27**, 2277–2290 (1983)
30. L.S. Brown, G. Gabrielse, J.N. Tan, K.C.D. Chan, Cyclotron motion in a penning trap microwave cavity. Phys. Rev. A **37**, 4163–4171 (1988)
31. J.D. Jackson, *Classical Electrodynamics*, 2nd edn. (Wiley, New York, 1975)
32. J. Tan, G. Gabrielse, Synchronization of parametrically pumped electron oscillators with phase bistability. Phys. Rev. Lett. **67**, 3090–3093 (1991)
33. J.N. Tan, G. Gabrielse, Parametrically-pumped electron oscillators. Phys. Rev. A **48**, 3105 (1993)
34. G. Gabrielse, J.N. Tan, L.S. Brown, Cavity shifts of measured electron magnetic moments (World Scientific, Singapore, 1990), pp. 389–418
35. G. Gabrielse, X. Fei, L.A. Orozco, R.L. Tjoelker, J. Haas, H. Kalinowsky, T.A. Trainor, W. Kells, Thousand-fold improvement in the measured antiproton mass. Phys. Rev. Lett. **65**, 1317–1320 (1990)
36. G. Gabrielse, J. Tan, Self-shielding superconducting solenoid systems. J. Appl. Phys. **63**, 5143–5148 (1988)
37. G. Gabrielse, J.N. Tan, L.A. Orozco, S.L. Rolston, C.H. Tseng, R.L. Tjoelker, A superconducting solenoid system which cancels fluctuations in the ambient magnetic field. J. Mag. Res. **91**, 564–572 (1991)
38. E.S. Meyer, I.F. Silvera, B.L. Brandt, Eddy current shielding and heating: reduction of dissipation for very low—temperature experiments in the presence of magnetic field ripple. Rev. Sci. Instrum. **60**, 2964–2968 (1989)
39. D.F. Phillips, A precision comparison of the $\bar{p} - p$ charge-to-mass ratios, Ph.D. thesis (Harvard University, Cambridge, 1996)
40. R.S. Van Dyck, Jr., D.L. Farnham, S.L. Zafonte, P.B. Schwinberg, Ultrastable superconducting magnet system for a Penning trap mass spectrometer. Rev. Sci. Instrum. **70**, 1665–1671 (1999)
41. B. Odom, Fully quantum measurement of the electron magnetic moment, Ph.D. thesis (Harvard University, Cambridge, 2004)
42. G. Gabrielse, H. Dehmelt, Observation of inhibited spontaneous emission. Phys. Rev. Lett. **55**, 67–70 (1985)
43. S. Peil, G. Gabrielse, Observing the quantum limit of an electron cyclotron: QND measurements of quantum jumps between fock states. Phys. Rev. Lett. **83**, 1287–1290 (1999)
44. K.S. Thorne, R.W.P. Drever, C.M. Caves, Quantum nondemolition measurement of harmonic oscillators. Phys. Rev. Lett. **40**(11), 667–670 (1978)
45. V.B. Braginsky, F. Ya, Khalili, Quantum nondemolition measurements: the route from toys to tools. Rev. Mod. Phys. **68**(1–11), 1 (1996)
46. B. D'Urso, Cooling and self-excitation of a one-electron oscillator, Ph.D. thesis (Harvard University, Cambridge, 2003)
47. R. Van Dyck, Jr., P. Ekstrom, H. Dehmelt, Axial, magnetron, cyclotron and spin-cyclotron-beat frequencies measured on single electron almost at rest in free space (geonium). Nature **262**, 776 (1976)

48. B. D'Urso, B. Odom, G. Gabrielse, Feedback cooling of a one-electron oscillator. Phys. Rev. Lett. **90**(4), 043001 (2003)
49. B. D'Urso, R. Van Handel, B. Odom, D. Hanneke, G. Gabrielse, Single-particle self-excited oscillator. Phys. Rev. Lett. **94**, 113002 (2005)
50. D.J. Wineland, H.G. Dehmelt, Principles of the stored ion calorimeter. J. Appl. Phys. **46**, 919–930 (1975)
51. R. Kubo, The fluctuation-dissipation theorem. Rep. Prog. Phys. **29**(1), 255–284 (1966)
52. E.M. Purcell, Spontaneous emission probabilities at radio frequencies. Phys. Rev. **69**, 681 (1946)
53. D. Kleppner, Inhibited spontaneous emission. Phys. Rev. Lett. **47**, 233 (1981)
54. F.L. Palmer, Excitation of spin flips in geonium at small cyclotron quantum numbers: transition rates and frequency shifts. Phys. Rev. A **47**, 2610 (1993)
55. L.S. Brown, Geonium lineshape. Ann. Phys. (N.Y.), 159:62–98 (1985)
56. D.G. Boulware, L.S. Brown, T. Lee, Apparatus-dependent contributions to $g-2$?. Phys. Rev. D **32**, 729–735 (1985)
57. G. Gabrielse, J.N. Tan, One electron in a cavity (Academic Press, New York, 1994), pp. 267–299
58. L.S. Brown, Line shape for a precise measurement of the electron's magnetic moment. Phys. Rev. Lett. **52**, 2013–2015 (1984)
59. G.W. Bennett et al., Final report of the e821 muon anomalous magnetic moment measurement at BNL. Phys. Rev. D **73**, 072003 (2006)
60. P.J. Mohr, B.N. Taylor, D.B. Newell, Codata recommended values of the fundamental physical constants: 2010. Rev. Mod. Phys. **84**, 1527 (2012)
61. J. Schwinger, On quantum-electrodynamics and the magnetic moment of the electron. Phys. Rev. **73**, 416–417 (1948)
62. A. Petermann, Fourth order magnetic moment of the electron. Helv. Phys. Acta **30**, 407–408 (1957)
63. C.M Sommerfield, Magnetic dipole moment of the electron. Phys. Rev. **107**, 328–329 (1957)
64. C.M. Sommerfield, The magnetic moment of the electron. Ann. Phys. (N.Y.) **5**, 26–57 (1958)
65. S. Laporta, E. Remiddi, Analytical value of the electron $(g-2)$ at order a^3 in QED. Phys. Lett. B **379**, 283–291 (1996)
66. P.J. Mohr, B.N. Taylor, CODATA recommended values of the fundamental physical constants: 2002. Rev. Mod. Phys. **77**, 1–107 (2005)
67. A. Czarnecki, B. Krause, W.J. Marciano, Electroweak corrections to the muon anomalous magnetic moment. Phys. Rev. Lett. **76**, 3267–3270 (1996)
68. J. Prades, E. de Rafael, A. Vainshtein, *Hadronic Light-by-Light Scattering Contribution to the Muon Anomalous Magnetic Moment, chapter Hadronic Light-by-Light Scattering Contribution to the Muon Anomalous Magnetic Moment* (World Scientific, Singapore, 2009)
69. D. Nomura, T. Teubner, Hadronic contributions to the anomalous magnetic moment of the electron and the hyperfine splitting of muonium. Nucl. Phys. B **867**, 236 (2013)
70. Th Udem, A. Huber, B. Gross, J. Reichert, M. Prevedelli, M. Weitz, T.W. Hänsch, Phase-coherent measurement of the hydrogen $S-2S$ transition frequency with an optical frequency interval divider chain. Phys. Rev. Lett. **79**, 2646–2649 (1997)
71. C. Schwob, L. Jozefowski, B. de Beauvoir, L. Hilico, F. Nez, L. Julien, F. Biraben, O. Acef, J.J. Zondy, A. Clairon, Optical frequency measurement of the 2S–12D transitions in hydrogen and deuterium: rydberg constant and lamb shift determinations. Phys. Rev. Lett. **82**, 4960–4963 (1999)
72. V. Gerginov, K. Calkins, C.E. Tanner, J.J. McFerran, S. Diddams, A. Bartels, L. Hollberg, Optical frequency measurements of $6s^2S_{1/2} - 6p^2P_{1/2}(D_1)$ transitions in ^{133}Cs and their impact on the fine structure constant. Phys. Rev. A **73**, 032504 (2006)
73. M.P. Bradley, J.V. Porto, S. Rainville, J.K. Thompson, D.E. Pritchard, Penning trap measurements of the masses of ^{133}Cs, ^{87,85}Rb, and ^{23}Na with uncertainties ≤ 0.2 ppb. Phys. Rev. Lett. **83**, 4510–4513 (1999)
74. G. Werth, J. Alonso, T. Beier, K. Blaum, S. Djekic, H. Häffner, N. Hermanspahn, W. Quint, S. Stahl, J. Verdú, T. Valenzuela, M. Vogel, Highly charged ions, quantum electrodynamics, and the electron mass. Int. J. Mass Spectrom. **251**, 152 (2006)

75. B. Mount, M. Redshaw, E. Myers, Atomics Masses of ^{6}Li, ^{23}Na, ^{38,41}K, ^{85,86}Rb, and ^{133}Cs. Phys. Rev. A **82**, 042513 (2010)
76. A. Wicht, J.M. Hensley, E. Sarajlic, S. Chu, A preliminary measurement of the fine structure constant based on atom interferometry. Phys. Scr. **T102**, 82–88 (2002)
77. K. Pachucki, A. Czarnecki, U. Jentschura, V.A. Yerokhin, Complete two-loop correction to the bound-electron g factor. Phys. Rev. A **72**, 022108 (2005)
78. S.J. Brodsky, S.D. Drell, Anomalous magnetic moment and limits on fermion substructure. Phys. Rev. D **22**, 2236–2243 (1980)
79. D. Bourilkov, Hint for axial-vector contact interactions in the data on $e^+e^- \rightarrow e^+e^-(\gamma)$ at center-of-mass energies 192–208 GeV. Phys. Rev. D **64**, 071701R (2001)

Chapter 2
Theory of Anomalous Magnetic Dipole Moments of the Electron

Masashi Hayakawa

Abstract The anomalous magnetic dipole moment of the electron, the so-called electron $g-2$, provides us with a high-precision test of quantum electrodynamics (QED), which is the relativistic and quantum-mechanical generalization of electromagnetism, and helps to determine the value of the fine structure constant α, one of the fundamental physical constants. This article intends to give a pedagogical introduction to the theory of $g-2$, in particular, the computation of high-order quantum electrodynamics in $g-2$.

2.1 Introduction

A single electron is known to become magnetized due to its intrinsic charge and spin. Its magnetic dipole moment is given by

$$\boldsymbol{\mu} = -g_e \frac{e\hbar}{2m_e c} \mathbf{s}\,, \tag{2.1}$$

where m_e and $\mathbf{s}$ denote the mass and spin of an electron, respectively. The constant g_e represents the strength of the magnetic dipole moment in units of the Bohr magneton, and is called the g-factor of the electron.

At the zeroth order of perturbation, QED predicts that g_ψ is equal to 2 for every massive particle ψ with spin $\frac{1}{2}$. The quantum correction in general shifts g_ψ from 2, depending on the particle species ψ. It is thus convenient to focus on this shift, called the 'anomalous magnetic dipole moment', by introducing a symbol

M. Hayakawa (✉)
Department of Physics, Nagoya University, 464-8602 Nagoya, Japan
e-mail: hayakawa@eken.phys.nagoya-u.ac.jp

W. Quint and M. Vogel (eds.), *Fundamental Physics in Particle Traps*,
Springer Tracts in Modern Physics 256, DOI: 10.1007/978-3-642-45201-7_2,

$$a_\psi \equiv \frac{g_\psi - 2}{2}. \tag{2.2}$$

We shall call this quantity $g - 2$ of ψ hereafter.

An important point to notice is the fact that the electron $g-2$, a_e, can be measured very accurately. The most accurately measured value of the electron $g - 2$

$$a_e(\mathrm{HV08}) = 1\ 159\ 652\ 180.73\ (28) \times 10^{-12}, \tag{2.3}$$

where the numerals in the parenthesis represent the uncertainty in the final few digits, was obtained by the Harvard group using a Penning trap with cylindrical cavity [1, 2]. The uncertainty (2.3) is smaller than the one obtained by Washington university group in 1987 [3] by a factor 15. See the previous chapter by Gabrielse et al. for a detailed explanation about how such a large reduction of uncertainty has been achieved.

Even if $g-2$ is measured very accurately, one may wonder what physical implication it possesses. Worthy of note is the fact that a_ψ is a predictable quantity in so far as the theory is renormalizable in the framework of quantum field theory. (Section 2.2 introduces the notion and the foundation of the quantum field theory for the readers not specialized in particle physics.) We are thus inclined to question the validity of a renormalizable model of elementary particles by asking the compatibility of its theoretical prediction $a_\psi(\mathrm{th})$ with the experimentally measured value $a_\psi(\mathrm{exp})$.

The *standard model* of elementary particles has endured most of tests for these forty years. It is a renormalizable theory and thus gives a prediction to $g-2$, referred hereafter as $a_\psi(\mathrm{SM})$. The measured value $a_\psi(\mathrm{exp})$ in the experiment may contain the potential contribution $a_\psi(\mathrm{new})$ from the structures which are not built in the standard model

$$a_\psi(\mathrm{exp}) = a_\psi(\mathrm{SM}) + a_\psi(\mathrm{new}). \tag{2.4}$$

One's actual interest is the existence of such new structures, and its effect on $g - 2$ of ψ. Nevertheless, in order to explore its existence through the study of $g - 2$, it is indispensable to compute all relevant dynamics of the standard model contributing to $a_\psi(\mathrm{SM})$, and ask if the difference, $a_\psi(\mathrm{exp}) - a_\psi(\mathrm{SM})$, is not zero within available precision.

As is elucidated in full detail in Table 2.1, all the contributions relevant to $a_\mu(\mathrm{SM})$ can be classified unambiguously and exclusively into three components;

Table 2.1 Classification into QED, QCD and weak contributions

Type	Definition
$a_\psi(\mathrm{QED})$	QED with charged leptons[a] only
$a_\psi(\mathrm{QCD})$	(QED + QCD) \(QED with charged leptons only)[b]
$a_\psi(\mathrm{weak})$	All others

[a] All known charged leptons are electron (e), muon (μ) and tau-lepton (τ)

[b] $A\backslash B$ for the sets A and B denotes the intersection of A and the complement of B

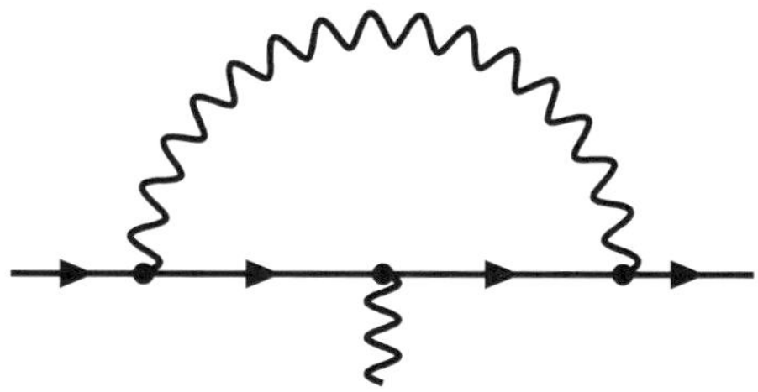

Fig. 2.1 The Feynman diagram giving the lowest-order (second-order) contribution to the anomalous magnetic moment of the lepton

$$a_\psi(\mathrm{SM}) = a_\psi(\mathrm{QED}) + a_\psi(\mathrm{QCD}) + a_\psi(\mathrm{weak})\,. \tag{2.5}$$

The QED contribution $a_\psi(\mathrm{QED})$ is the most dominant among these. For instance, the contribution of the leading order, $O(\alpha)$,[1] of perturbation is induced from the Feynman diagram shown in Fig. 2.1 [4]. (Brief explanation on Feynman diagrams can be found in Sect. 2.2.1.) Figure 2.1, where time flows in the right direction, describes a process in which a single virtual photon is emitted before the lepton ψ couples to the external magnetic field, and is absorbed after that. The theory of $a_\psi(\mathrm{QED})$ is our primary concern in this article.

In Eq. (2.5), $a_\psi(\mathrm{QCD})$ represents the contribution from *QCD*. QCD is the gauge theory of *strong interaction* which binds quarks to form nucleons. Such a contribution arises because every quark has a non-zero electric charge. Figure 2.2 illustrates the QCD contribution to the $g-2$. There, a virtual photon turns into a pair of quark and anti-quark, which receive dynamics of strong interaction and then annihilate into a virtual photon again. $a_\psi(\mathrm{QCD})$ requires evaluation of non-perturbative dynamics of QCD in some manner, and is thus the most difficult to compute. The weak contribution $a_\psi(\mathrm{weak})$ comes from the Feynman diagrams, each of which contains at least one

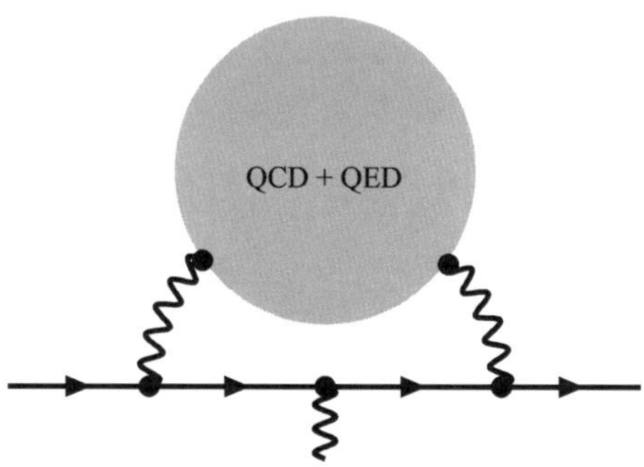

Fig. 2.2 The $O(\alpha^2)$ QCD contribution to the anomalous magnetic moment of the lepton, called the 'leading-order hadronic vacuum polarization contribution', $a_\psi(\mathrm{had.v.p.})$. The blob part denotes the renormalized correlation function $\langle 0|\, T j^{\mathrm{em}}_\mu(x)\, j^{\mathrm{em}}_\nu(y)\, |0\rangle$ of the *hadronic* electromagnetic current j^{em}_μ in QCD + QED

[1] The fine structure constant α is given in terms of the elementary charge e and vacuum permittivity ε_0 by

$$\alpha = \frac{e^2}{4\pi\epsilon_0\hbar c}\,. \tag{2.6}$$

virtual W boson, Z boson or Higgs particle. Section 2.3 explains the basic materials on these non-QED contributions.

Importantly, the sizes of individual contributions on the right hand side of Eq. (2.5) depend sensitively on the mass m_ψ of the species ψ. For larger m_ψ, a_ψ(QCD) and a_ψ(weak) are enhanced relative to a_ψ(QED). Since the muon is heavier than the electron by a factor of about 200, a_μ(QCD), for instance, is about 40,000 times larger than a_e(QCD). Likewise, a_μ(new), the contribution from the as-yet-unknown structures, may also be much larger than a_e(new). This motivates us to study the muon $g-2$ in a serious manner on both sides of experiment and theory, although it is not discussed fully here.

As advocated above, this article intends to give a pedagogical introduction to the QED contribution, a_e(QED), to the electron $g-2$. Section 2.2 provides the generic readers with the introduction of QED, quantum field theory, perturbation theory, and the anomalous magnetic dipole moment in terms of the field theory. Section 2.3 describes the current status of the non-QED contributions. The basic materials of the numerical approach to the computation of high-order QED contributions is explained in Sect. 2.4. Section 2.5 summarizes the latest result for a_e(QED).

2.2 QED and Anomalous Magnetic Dipole Moment

2.2.1 Perturbation Theory of QED

Throughout this article, we work with the natural unit in which the speed of light c and the reduced Planck constant $\hbar$ are set to unity, and with the signature of the metric $\eta^{\mu\nu} = \mathrm{diag}\,(1,\ -1,\ -1,\ -1) = \eta_{\mu\nu}$ with respect to the coordinates $x^\mu = (ct, \mathbf{x})$ in four-dimensional Minkowski space. Einstein convention is employed; the sum over 0, 1, 2, 3 is assumed if the same Greek letter appears twice as a subscript and a superscript in a single term, e.g.,

$$T_{\mu\nu}\, u^\nu \equiv \sum_{\nu=0}^{3} T_{\mu\nu}\, u^\nu\,. \tag{2.7}$$

The classical action of quantum electrodynamics (QED) takes a simple form

$$\begin{aligned} S = \int d^D x \, \Bigg[& -\frac{1}{4}\, F_{\mu\nu}(x)\, F^{\mu\nu}(x) - \frac{1}{2}\left(\partial_\mu A^\mu(x)\right)^2 \\ & + \sum_{\psi=e,\,\mu,\,\tau} \overline{\psi}(x) \left\{ \gamma^\mu \left(i\partial_\mu + e\, \mu^{(4-D)/2} A_\mu(x) \right) - m_\psi \right\} \psi(x) \Bigg] \\ & + \text{counter-terms}\,. \end{aligned} \tag{2.8}$$

Several remarks are in order:

1. The field strength $F_{\mu\nu}(x)$ is given in terms of the four-dimensional vector potential $A^{\mu}(x) = (\phi(x), \mathbf{A}(x))$, $A_{\mu}(x) \equiv \eta_{\mu\nu} A^{\nu}(x)$ by

$$F_{\mu\nu} = \partial_{\mu} A_{\nu} - \partial_{\nu} A_{\mu} = -F_{\nu\mu} \,. \tag{2.9}$$

$F_{\mu\nu}(x)$ packs the electric field $\mathbf{E}(x)$ and the magnetic field $\mathbf{B}(x)$ in the following manner

$$E^{j}(x) = F^{0j}(x) \; (j = 1, 2, 3) \,, \quad B^{1}(x) = F_{23}(x) \text{ and cyclic permutation.} \tag{2.10}$$

Through this correspondence, the term bilinear with respect to $F_{\mu\nu}(x)$ in Eq. (2.8) turns out to be the familiar action for the electric and magnetic field

$$\int d^{D}x \left(-\frac{1}{4} F_{\mu\nu}(x) F^{\mu\nu}(x) \right) = \int d^{D}x \, \frac{1}{2} \left(|\mathbf{E}(x)|^{2} - |\mathbf{B}(x)|^{2} \right) . \tag{2.11}$$

2. The quantum field theory gives a "unified description of matter and force" in terms of quantum fields. Based on the particle-wave duality in quantum theory, we associate a variable called 'quantum field' to every species from the point of view of a wave. In classical electromagnetism, every component of the vector potential $A_{\mu}(x)$ takes its value in a real number at every space-time point x. After quantization, $A_{\mu}(x)$ becomes a collection of Hermitian operators labelled by (x, μ), and plays the role of creating and annihilating photons, quanta of light. To describe creation and annihilation of one lepton species ψ and its anti-particle, a four-component Dirac field $\left(\psi_{\alpha}(x), \overline{\psi}_{\alpha}(x)\right)$ $(\alpha = 1, 2, 3, 4)$ is introduced and promoted to the fields which takes its values in the operators. Gamma matrices γ^{μ} $(\mu = 0, 1, 2, 3)$ are a set of 4×4 matrices satisfying the Clifford algebra corresponding to the signature $\eta^{\mu\nu}$

$$\gamma^{\mu}\gamma^{\nu} + \gamma^{\nu}\gamma^{\mu} = 2\eta^{\mu\nu}\mathbb{I}_{4}. \tag{2.12}$$

3. In Eq. (2.8), the sum is taken over only three leptons, i.e., electron e, muon μ and tau-lepton τ, to focus on the QED contribution in a_{ψ} hereafter.
4. In quantum field theory, the amplitude corresponding to a Feynman diagram in general suffers the divergence due to short-distance singularity (ultra-violet (UV) divergence) and the divergence due to long-distance singularity (infrared (IR) divergence). To regularize these singularities, dimensional regularization is employed, which renders the amplitude well-defined by putting the space-time dimension D away from 0 and all positive integers on the complex plane [5]. A parameter μ with dimension of mass is introduced to keep e dimensionless.

5. The second term in (2.8) is the gauge-fixing term in Feynman gauge.
6. The counter-terms in Eq. (2.8) are treated as the interactions which absorb the divergences emerging from the quantum-mechanical dynamics of the terms in the first integral of Eq. (2.8), Their details will be described in Sect. 2.2.4.

2.2.2 Feynman Diagrams and Feynman Rule

Feynman diagrams provide us with a convenient tool to figure all electromagnetic processes relevant to the quantity and the perturbative order of one's interest. In perturbation theory, every term bilinear in fields in the action[2] represents the propagation of a particle (or an antiparticle), while every term trilinear or higher in fields represents the interaction between three or more particles. A Feynman diagram is a graph describing how *virtual* photons, leptons and anti-leptons are created and annihilated in the following way;

1. an undirected wavy line corresponds to the propagation of a free photon,
2. a directed line[3] corresponds to the propagation of a free lepton,
3. a vertex represents the fundamental interaction between lepton and photon.

The Feynman rule [6] derived from the action (2.8) associates the following quantities to a wave line, a directed line and a vertex, respectively where $\not{p} \equiv p_\mu \gamma^\mu$.

$$\Leftrightarrow -\mathrm{i}\frac{\eta^{\mu\nu}}{q^2+\mathrm{i}\varepsilon},$$

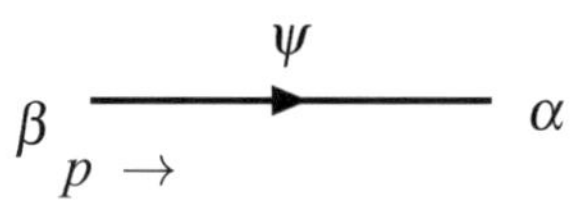

$$\Leftrightarrow \mathrm{i}\frac{(\not{p}+m_\psi)_{\alpha\beta}}{p^2-m_\psi^2+\mathrm{i}\varepsilon},$$

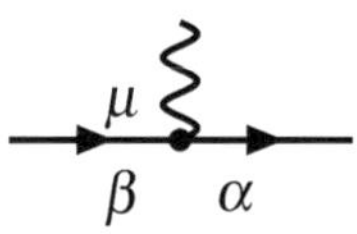

$$\Leftrightarrow \mathrm{i}e\mu^{(4-D)/2}(\gamma^\mu)_{\alpha\beta},$$

In the graph on the left hand side of the latter equation all lines just guide the eyes, and the quantity on the right hand side is associated with the vertex alone. '$\mathrm{i}\varepsilon$ with $\varepsilon > 0$' plays the role to specify the boundary condition of Green functions; ε is brought to zero after rotating the zeroth component p^0 of four-momentum to the Euclidean one p_4; $p^0 \to \mathrm{i}p_4$.

The amplitude corresponding to a Feynman diagram is obtained by

---

[2] This applies to the terms except for the counter-terms. All the counter-terms, including bilinear terms, are treated as interactions. See Sect. 2.2.4.

[3] There are actually three types of directed line depending on the species of leptons. See the corresponding Feynman rule (second of the above).

(a) assigning the quantities to the lines or vertices involved in a Feynman diagram,
(b) imposing energy-momentum conservation at every vertex,
(c) integrating n number of independent loop momenta k_r $(r = 1, \ldots, n)$ with

$$\prod_{r=1}^{n} \int \frac{d^D k_r}{(2\pi)^D} \,. \tag{2.13}$$

For instance, a Feynman diagram in Fig. 2.1 gives the amplitude $\mathscr{A}^{\mu}_{\alpha\beta}(p, q)$

$$\begin{aligned} \mathrm{i}\, e\, \mu^{(4-D)/2} \mathscr{A}^{\mu}_{\alpha\beta}(p, q) = \int \frac{d^D k}{(2\pi)^D} & \left[\mathrm{i}\, e\, \mu^{(4-D)/2}\, \gamma_\lambda \frac{\mathrm{i}}{\not{k} + \not{p} + \not{q}/2 - 1} \right. \\ & \left. \times \mathrm{i}\, e\, \mu^{(4-D)/2}\, \gamma^{\mu} \frac{\mathrm{i}}{\not{k} + \not{p} - \not{q}/2 - 1} \mathrm{i}\, e\, \mu^{(4-D)/2}\, \gamma^{\lambda} \right]_{\alpha\beta} \\ & \times \frac{-\mathrm{i}}{k^2} \,. \end{aligned} \tag{2.14}$$

Here, q is the four-momentum of the incoming photon, $p \mp q/2$ are the four-momenta of incoming and outgoing external leptons, respectively. The loop momentum is taken to be the momentum k carried by the virtual photon. Since $g - 2$ is a dimensionless quantity, it is convenient to express all dimensionful quantities in units of the mass of the external lepton, parametrizing the effect of creation and annihilation of virtual leptons by *lepton mass ratios*, which appears at the fourth and higher orders. For this reason, the masses in Eq. (2.14) are set to 1. The amplitude (2.14) contains the overall ultra-violet-divergent monopole contribution as well. The contribution to the anomalous magnetic dipole moment from Fig. 2.1 will be obtained by applying the magnetic projection to be discussed in Sect. 2.2.3, to the amplitude (2.14).

For the sake of later reference, the standard technique in quantum field theory to calculate (2.14) is explained here. We can show the identity

$$\frac{1}{ABC} = \frac{1}{2} \int_0^1 dz_1 \int_0^1 dz_2 \int_0^1 dz_a\, \delta\left(1 - (z_1 + z_2 + z_a)\right) \frac{1}{(z_a A + z_1 B + z_2 C)^3} \,. \tag{2.15}$$

Applying Eq. (2.15) to the integrand of Eq. (2.14) with $A = k^2$, $B = (k+p+q/2)^2 - 1$ and $C = (k + p - q/2)^2 - 1$, and exchanging the order of integration, the integral over the loop momentum k can be easily evaluated.[4] The resulting integral is the one with respect to Feynman parameters z_a, z_1 and z_2.

[4] One can consult with the content of Sect. 2.4.2 on this point.

2.2.3 Anomalous Magnetic Dipole Moment

In this section, we see how the contribution to the anomalous magnetic dipole moment can be extracted from the amplitude in the context of quantum field theory.

Let $\Gamma^{\mu}_{\mathbf{B},\alpha\beta}(p,q)$ be the unrenormalized vertex function, i.e., the contribution of the one-particle irreducible[5] Feynman diagrams to the Green function

$$\int d^D x_F e^{ip_F\cdot x_F} \int d^D x_I e^{-ip_I\cdot x_I} \int d^D y e^{-iq\cdot y} \langle 0|\, \psi_{\mathbf{B},\alpha}(x_I) j^{\mu}_{\rm em}(y) \overline{\psi}_{\mathbf{B},\beta}(x_F)\, |0\rangle\,, \tag{2.16}$$

where $j^{\mu}_{\rm em}(x)$ is the electromagnetic current, $p_F = p + \frac{q}{2}$ and $p_I = p - \frac{p}{2}$. The renormalized vertex function $\Gamma^{\mu}_{\alpha\beta}(p,q)$ is given in terms of the wave function renormalization constant Z_ψ of the external lepton by

$$\Gamma^{\mu}_{\alpha\beta}(p,q) = Z_\psi\, \Gamma^{\mu}_{\mathbf{B},\alpha\beta}(p,q)\,. \tag{2.17}$$

The invariance of QED under Lorentz-, charge conjugation-(C) and parity (P) transformations as well as the gauge symmetry implies that the renormalized vertex function with the external leptons put on their mass shells, but with the external photon kept off-shell, consists of two form factors $F_1(q^2)$, $F_2(q^2)$[6];

$$\Gamma^{\mu}_{\alpha\beta}(p,q)\Big|_{p_F^2=m_\psi^2=p_I^2} = F_1(q^2)\gamma^\mu + F_2(q^2)\frac{1}{2m_\psi}\, i\sigma^{\mu\nu}q_\nu\,, \tag{2.18}$$

where

$$\sigma^{\mu\nu} \equiv \frac{i}{2}\left[\gamma^\mu, \gamma^\nu\right]. \tag{2.19}$$

The on-shell renormalization condition imposes $F_1(q^2 = 0) = 1$ to the electric form factor. The magnetic form factor $F_2(q^2)$ is the quantity predictable in the renormalizable theory. The anomalous magnetic dipole moment is actually identified with $F_2(q^2 = 0)$

$$a_\psi = F_2(0)\,. \tag{2.20}$$

It is an easy exercise to see that a_ψ is obtained by applying the magnetic projection to $\Gamma^{\mu}_{\alpha\beta}(p,q)$;

---

[5] A Feynman diagram is called 'one-particle irreducible' if and only if it cannot be divided into two non-trivial connected subdiagrams when any one of the internal lines is cut off.

[6] C and P symmetries are violated in the weak and Yukawa interactions in the standard model, and thus additional form factors are actually induced. However, the classification in Table 2.1 insures that QED and QCD contributions, which respect C and P symmetry, consist of two form factors as in Eq. (2.18).

$$
\begin{aligned}
a_\psi = \lim_{q^2 \to 0} \frac{m_\psi}{4q^2 \left(p^2\right)^2} \mathrm{tr} \Bigg[\Bigg\{ m_\psi p^2 \gamma^\mu - \left(m_\psi^2 + \frac{q^2}{2} \right) p^\mu \Bigg\} \\
\times \left(\not{p} + \frac{\not{q}}{2} + m_\psi \right) \Gamma_\mu(p, q) \left(\not{p} - \frac{\not{q}}{2} + m_\psi \right) \Bigg] ,
\end{aligned} \tag{2.21}
$$

with the on-shell conditions $p \cdot q = 0$ and $p^2 + q^2/4 = m_\psi^2$ for the external leptons.

The second-order contribution to a_ψ is obtained by applying (2.21) to $\Gamma^\mu(p, q) = \mathscr{A}^\mu(p, q)$ in Eq. (2.14) as [4]

$$
a_\psi \big|_{\text{1-loop}} = \frac{\alpha}{2\pi} . \tag{2.22}
$$

2.2.4 Renormalization and Counter-Terms

In perturbation theory of quantum field theory, the notion of counter-terms provides a systematic and practical way to calculate the renormalized amplitude.

The “counter-terms” in (2.8) take the form

$$
\begin{aligned}
S_{\mathrm{c.t}} = \int d^D x \Bigg[-\frac{1}{4} \delta Z_A \, F_{\mu\nu} F^{\mu\nu} \\
+ \sum_{\psi = e, \mu, \tau} \Big\{ \delta Z_\psi \, \overline{\psi} \gamma^\mu i \partial_\mu \psi - \delta Z_{M, \psi} \, m_\psi \overline{\psi} \psi \\
+ \delta Z_{V, \psi} \, e \mu^{(4-D)/2} \overline{\psi} \gamma^\mu A_\mu \psi \Big\} \Bigg] .
\end{aligned} \tag{2.23}
$$

and are adjusted to absorb UV divergence. The constants appearing above, δZ_A, δZ_ψ, $\delta Z_{M, \psi}$ and $\delta Z_{V, \psi}$ are related to the wave function renormalization constants Z_A, Z_ψ, multiplicative mass renormalization constant $Z_{m, \psi}$[7] and the coupling renormalization constant Z_e through

$$
\begin{aligned}
&\delta Z_A = Z_A - 1 , \quad \delta Z_\psi = Z_\psi - 1 , \\
&\delta Z_{M, \psi} = Z_\psi Z_{m, \psi} - 1 , \quad \delta Z_{V, \psi} = Z_e Z_\psi Z_A^{1/2} - 1 .
\end{aligned} \tag{2.24}
$$

Perturbatively, the coefficients of counter-terms are expanded in a power series of α/π,

[7] Here we assume that the regularization preserves the non-anomalous chiral symmetry, so that no linear additive UV divergence arises in the two-point functions of leptons.

$$\delta Z_A = \sum_{n=1}^{\infty} \delta^{(n)} Z_A \left(\frac{\alpha}{\pi}\right)^n , \quad \delta Z_\psi = \sum_{n=1}^{\infty} \delta^{(n)} Z_\psi \left(\frac{\alpha}{\pi}\right)^n ,$$

$$\delta Z_{M,\psi} = \sum_{n=1}^{\infty} \delta^{(n)} Z_{M,\psi} \left(\frac{\alpha}{\pi}\right)^n , \quad \delta Z_{V,\psi} = \sum_{n=1}^{\infty} \delta^{(n)} Z_{V,\psi} \left(\frac{\alpha}{\pi}\right)^n . \quad (2.25)$$

These counter-terms generate the *interaction vertices* of various orders of perturbation theory. In particular, the renormalization of wave function or mass corresponds to a vertex from which only two lines emanate. The lower-order counter-terms together with QED interaction generate the Feynman diagrams and the subtraction terms which cancel ultra-violet divergences contained in subdiagrams. In the end, the sum of all those Feynman diagrams can have only *overall UV divergences that should be cancelled by the nth-order coefficients*, $\delta^{(n)} Z_A$, etc. Their precise values are determined from the demand that the renormalized vertex functions, $\Pi(q^2)$, $\Sigma_{\alpha\beta}(p)$ and $\Lambda^\mu(p, q)$, defined by

$$\int d^D x\, e^{iq\cdot x} \langle 0| A_\mu(x) A_\nu(0) |0\rangle = \frac{-i\eta_{\mu\nu}}{q^2 \left\{1 - \Pi(q^2)\right\}} ,$$

$$\int d^D x\, e^{ip\cdot x} \langle 0| \psi_\alpha(x) \overline{\psi}_\beta(0) |0\rangle = \left[\frac{i}{\not p - m - \Sigma(p)}\right]_{\alpha\beta} ,$$

$$\Gamma^\mu_{\alpha\beta}(p, q) = \gamma^\mu_{\alpha\beta} + \Lambda^\mu_{\alpha\beta}(p, q) , \quad (2.26)$$

satisfy certain *renormalization conditions*. The gauge symmetry guarantees that Z_e is independent of ψ. In this way, the counter-terms are determined iteratively in perturbation theory of QED.

2.2.5 Classification of Perturbative Dynamics

The quantum electrodynamics in the lepton $g - 2$, a_ψ(QED), are expected to be obtained by computing the coefficients $a_\psi^{(2n)}$ in the perturbative series

$$a_\psi(\text{QED}) = \sum_{n=1}^{\infty} a_\psi^{(2n)} \left(\frac{\alpha}{\pi}\right)^n , \quad (2.27)$$

up to the order N determined from the requirement of experimental accuracy and our theoretical interest. In what follows, the nth term in Eq. (2.27) is called 'n-loop term' or the '$2n$th-order term', as it is $O(\alpha^n) \sim O(e^{2n})$. The comparison of Eq. (2.22) with (2.27) yields the leading-order coefficient $a_\psi^{(2)}$ as

$$a_\psi^{(2)} = \frac{1}{2} \equiv A_1^{(2)} . \quad (2.28)$$

Before starting any calculations, the most dominant contribution should be identified. To do so in our context, we decompose the contributions into four types according to the dependence on lepton mass ratios (Recall the discussion below Eq. (2.14.));

$$a_e^{(2n)} = A_1^{(2n)} + A_2^{(2n)}\left(\frac{m_e}{m_\mu}\right) + A_2^{(2n)}\left(\frac{m_e}{m_\tau}\right) + A_3^{(2n)}\left(\frac{m_e}{m_\mu}, \frac{m_e}{m_\tau}\right). \tag{2.29}$$

Here, the subscript j attached to $A_j^{(2n)}$ denotes the number of leptons involved in its calculation. $A_1^{(2n)}$ is the contribution that should be computed in *QED with electron only*. It is a pure number and called 'mass-independent term', and thus universally contributes to $g-2$ of all leptons ψ. $A_1^{(2)}$ is given by Eq. (2.28). $A_2^{(2n)}\left(\frac{m_e}{m_\mu}\right)$ is the contribution to the electron $g-2$ from all Feynman diagrams with at least one muon loop but with no tau-lepton loop. $A_2^{(2n)}\left(\frac{m_e}{m_\tau}\right)$ is similarly defined. $A_3^{(2n)}\left(\frac{m_e}{m_\mu}, \frac{m_e}{m_\tau}\right)$ is the contribution of all Feynman diagrams with both muon loop(s) and tau-lepton loop(s). $A_2^{(2n)}$ appears at first at the fourth order. $A_3^{(2n)}$ appears at first at the sixth order.

Muon and tau-lepton are both much heavier than the electron. Thus, for the electron $g-2$, their virtual effects are suppressed compared to the dynamics of QED with electron only. Therefore, $A_1^{(2n)}$ is the most dominant at every order $2n$ in the electron $g-2$.

It would be instructive to compare such a perturbative feature of QED in the electron $g-2$ with that in the muon $g-2$. In the similar manner as in (2.29), each coefficient $a_\mu^{(2n)}$ in the perturbative expansion of a_μ(QED) can be decomposed into four types of terms;

$$a_\mu^{(2n)} = A_1^{(2n)} + A_2^{(2n)}\left(\frac{m_\mu}{m_e}\right) + A_2^{(2n)}\left(\frac{m_\mu}{m_\tau}\right) + A_3^{(2n)}\left(\frac{m_\mu}{m_e}, \frac{m_\mu}{m_\tau}\right). \tag{2.30}$$

The meaning of each term would now be obvious from the dependence on lepton mass ratio(s). Since the electron is lighter than the muon, $A_2^{(2n)}\left(\frac{m_\mu}{m_e}\right)$ dominates $a_\mu^{(2n)}$ for $2n \geq 6$.

2.3 Non-QED Contribution to $g-2$

The standard model contribution was decomposed into three parts as in Eq. (2.5). Table 2.2 summarizes the results for the non-QED contribution to the electron $g-2$ obtained thus far. In that table, the last one is the weak contribution, and the others are the QCD contributions relevant to the precision of our interest. Table 2.2 also

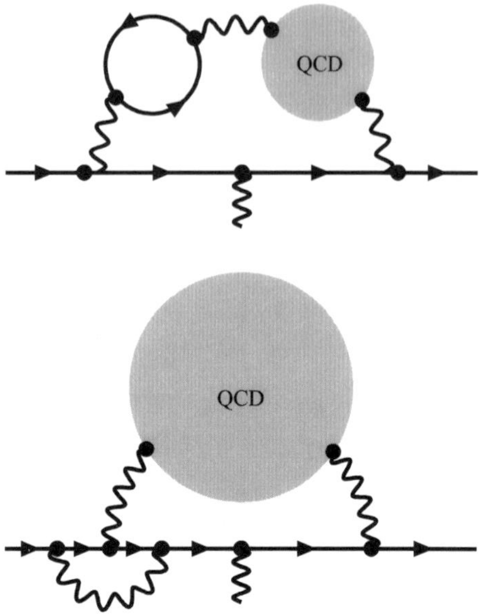

Fig. 2.3 Examples of the next-to-leading-order ($O(\alpha^3)$) hadronic vacuum polarization contribution, a_ψ(NLO had.v.p.), to the anomalous magnetic dipole moment

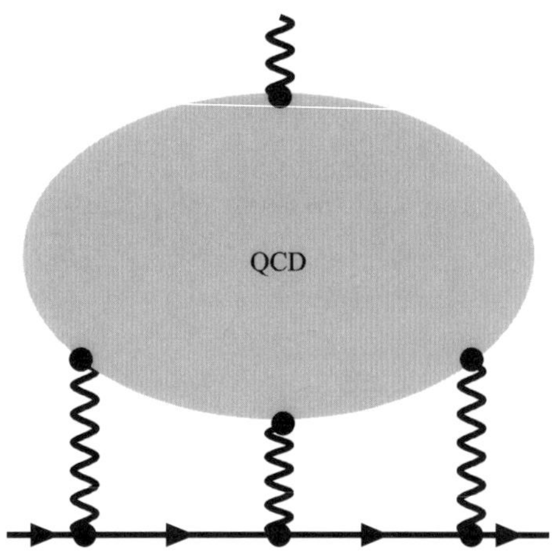

Fig. 2.4 Hadronic light-by-light scattering contribution a_ψ(had.l-l)

shows that they are only small part of a_e, but can no longer be neglected compared to the current experimental accuracy (2.3). Below we focus on the QCD contribution.

The QCD contributions relevant to the electron $g-2$ in view of the experimental accuracy are represented by the diagrams shown in Figs. 2.2, 2.3 and 2.4, and a_e(QCD) is given by their sum

$$a_e(\text{QCD}) = a_e(\text{had.v.p.}) + a_e(\text{NLO had.v.p.}) + a_e(\text{had.}l\text{-}l)\,. \tag{2.31}$$

Here, $O(\alpha^2)$-hadronic vacuum polarization contribution, a_e(had.v.p.), corresponds to the diagram in Fig. 2.2. Note that it also includes the $O(\alpha^3)$-terms caused by a single virtual photon exchange between quarks in the blob part. The QED correction to a photon line, lepton line or leptonic vertex in Fig. 2.2 gives rise to the next-to-leading-order (NLO) hadronic vacuum polarization contribution, a_e(NLO had.v.p.). Figure 2.3 illustrates a Feynman diagram contributing to a_e(NLO had.v.p.). Lastly, the hadronic light-by-light scattering contribution shown in Fig. 2.4, a_e(had.*l*-*l*), is also the order of α^3, and is induced through the elastic scattering between two photons *caused by QCD*.

Since the strongly coupled dynamics at low energy $\lesssim$1 GeV is the most important to the electron $g-2$, perturbation theory of QCD cannot account for the bulk contribution to the lepton $g-2$. In general, we must rely on some numerical means to capture the non-perturbative dynamics of QCD. For the hadronic vacuum polarization type contributions, i.e., a_ψ(had.v.p.) and a_ψ(NLO had.v.p.), we can circumvent direct computation of non-perturbative QCD dynamics as follows. The dispersive expression for the vacuum polarization function and the optical theorem, which follows from the unitary of S matrix, enables to express a_ψ(had.v.p.) as the convolution of R ratio $\widehat{R}(s)$, where $\sqrt{s} = 2E_{\rm cm}$ for the electron energy $E_{\rm cm}$ in the center of mass frame of the electron and positron beams, and a calculable function $K(s)$

$$a_\psi(\text{had.v.p.}) = \left(\frac{\alpha}{3\pi}\right)^2 \int_{E_{\rm th}^2}^{\infty} \frac{ds}{s} \frac{m_\psi^2}{s} \widehat{R}(s)\, K(s)\,, \tag{2.32}$$

with $E_{\rm th} = m_{\pi^0}$. The function $K(s)$ increases from 0.4 at $E_{\rm th}$ monotonically and approaches to 1 for $s \to \infty$. (The explicit expression of $K(s)$ is available, e.g. in [7].) To take the $O(\alpha)$ QED correction to the blob part in Fig. 2.2 into account, we focus on the the cross section $\sigma_h(s)$ in the unpolarized e^+e^- collision with the hadrons or hadrons $+\gamma$ as the final states, but with no initial state photon radiation. Note that the R ratio $\widehat{R}(s)$ required in Eq. (2.32) is not the experimentally accessible quantity $\sigma_h(s)$, but $\widehat{\sigma}_h(s)$ which is obtained from $\sigma_h(s)$ by removing $O(\alpha)$ correction to the part $e^+e^- \to \gamma^*$

$$\widehat{R}(s) = \frac{\widehat{\sigma}_h(s)}{\dfrac{4\pi\alpha^2}{3s}}\,. \tag{2.33}$$

The additional QED correction residing in $\sigma_h(s)$ compared to $\widehat{\sigma}_h(s)$ is exactly the charge renormalization. Therefore, these two quantities are related as

$$\widehat{\sigma}_h(s) = \left(\frac{\alpha}{\alpha(s)}\right)^2 \sigma_h(s)\,, \tag{2.34}$$

where, for the relevant order of α, $\sigma(e^+e^- \to \gamma^* \to \text{hadrons})$ only suffices for the QCD effect in the running gauge coupling constant $\alpha(s)$. a_ψ(NLO had.v.p.) can also be obtained with use of $\sigma(e^+e^- \to \gamma^* \to \text{hadrons})$ and the different function for $K(s)$.

Table 2.2 Summary of non-QED contribution to the electron $g-2$

Contribution X	$X \times 10^{12}$	Reference
a_e(had.v.p.)	1.866(12)	[9]
a_e(NLO had.v.p.)	−0.2234(14)	[9]
a_e(had.*l*-*l*)	0.035(10)	[10]
a_e(weak)	0.0297(5)	[11–14]

The first three are QCD contributions. a_e(weak) is estimated by scaling a_μ(weak) with the electron-muon mass ratio took into account

Practically, an inclusive cross section is impossible to directly measure; it is actually obtained by summing up all relevant exclusive cross sections. Experimental papers have reported their results for the individual cross section with or without radiative corrections, and thus careful treatment is necessary to gather and compile various data [8].

In contrast to the vacuum-polarization-type contribution, the hadronic light-by-light scattering contribution has not been successfully expressed in terms of some experimentally accessible quantities, and thus it requires explicit theoretical computation of non-perturbative QCD dynamics. The computation of this quantity is one of the remained subjects of the lepton $g-2$, in particular, the muon $g-2$. Thus far, it has been mostly evaluated according to the low-energy effective theory of QCD and/or hadronic models. a_e(had.*l*-*l*) in Table 2.2 was also obtained in this manner [10]. Even though its order in Table 2.2 is found to be much smaller than the one of our interest, it should be calculated by some other method.

2.4 Numerical Approach to Perturbative QED Calculation

The perturbative coefficients $a_\psi^{(2n)}$ have been known exactly or as an expansion in power series of mass ratios for $n = 1,\ 2,\ 3$. (See Sect. 2.5 for the literature concerning with the lower-order coefficients.) The eighth-order coefficient $a_\psi^{(8)}$ and the tenth-order coefficient $a_\psi^{(10)}$ [15–24], however, have been evaluated only by numerical means. Here, a succinct explanation is given for a particular method adopted to compute $a_\psi^{(8)}$ and $a_\psi^{(10)}$. The reader who would like to know full details on the parametric integral formalism, in particular on the ways of construction of ultraviolet and infrared subtraction terms can consult with the review article [25] or the original references [26–29].

2.4.1 Classification of Feynman Diagrams

Here we see the scheme of classification of Feynman diagrams employed in a series of works [15–24].

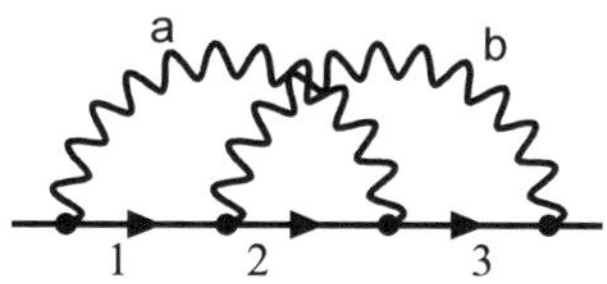

Fig. 2.5 A self-energy-like Feynman diagram (4a) corresponding to three vertex-type diagrams giving the fourth-order contribution to the anomalous magnetic moment of the lepton

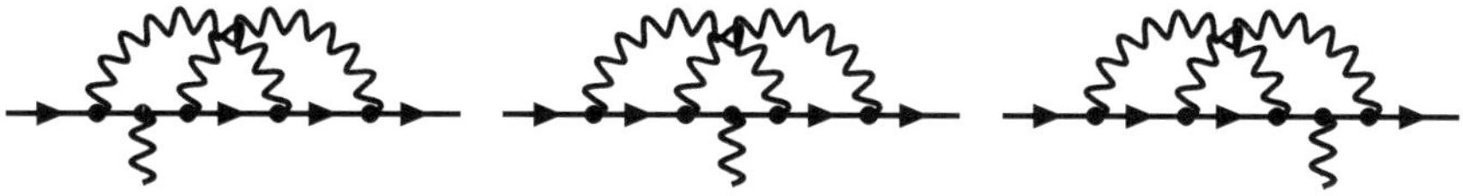

Fig. 2.6 The vertex diagrams at the fourth order related to a self-energy-like diagram in Fig. 2.5 via Ward-Takahashi identity (2.35)

We start with introducing the notion of *self-energy-like* (Feynman) diagram. A self-energy-like diagram is a diagram obtained from a vertex diagram by removing the external vertex. Note that the contribution to the self-energy function from a self-energy-like diagram may vanish identically. Such an example is the sixth-order light-by-light scattering diagram, where the external vertex lies on the virtual lepton loop. Note also that two different vertex diagrams are possibly reduced to the same self-energy-diagram if the external vertices are removed.

The reason why we focus on self-energy-like diagrams in place of vertex diagrams is based on the following important but unfamiliar fact; the Ward-Takahashi (WT) identity, which results from the gauge invariance of the system, holds between the contribution to the self-energy function $\Sigma(p)$ from a single self-energy-like Feynman diagram $\mathcal{G}$ and the contribution to the vertex function $\Lambda^{\mu}(p, q)$ from a set $\mathcal{B}_{\mathcal{G}}$ of vertex diagrams which can be obtained by inserting a single QED vertex into one of the lepton lines on the open path connected to the initial and final leptons, or the loop with odd number of internal vertices in all possible ways;

$$\sum_{\mathcal{V}\in\mathcal{B}_{\mathcal{G}}} q_{\nu}\Lambda^{\nu}_{\mathcal{V}}(p, q) = -\Sigma_{\mathcal{G}}\left(p+\frac{q}{2}\right) + \Sigma_{\mathcal{G}}\left(p-\frac{q}{2}\right). \tag{2.35}$$

For instance, to the self-energy function $\Sigma_{\mathcal{G}}$ from a single self-energy-like diagram $\mathcal{G}$ in Fig. 2.5, the sum of contributions to the vertex functions from three diagrams in Fig. 2.6 are related through Eq. (2.35).

By differentiating both sides of Eq. (2.35) with respect to the incoming photon momentum q_{μ}, and taking the limit $q \to 0$, we find

$$\sum_{\mathcal{V}\in\mathcal{B}_{\mathcal{G}}} \Lambda^{\mu}_{\mathcal{V}}(p, q\simeq 0) \simeq \sum_{\mathcal{V}\in\mathcal{B}_{\mathcal{G}}} \left\{-q_{\nu}\left[\frac{\partial\Lambda^{\nu}_{\mathcal{V}}(p, q)}{\partial q_{\mu}}\right]_{q=0}\right\} - \frac{\partial\Sigma_{\mathcal{G}}(p)}{\partial p_{\mu}}. \tag{2.36}$$

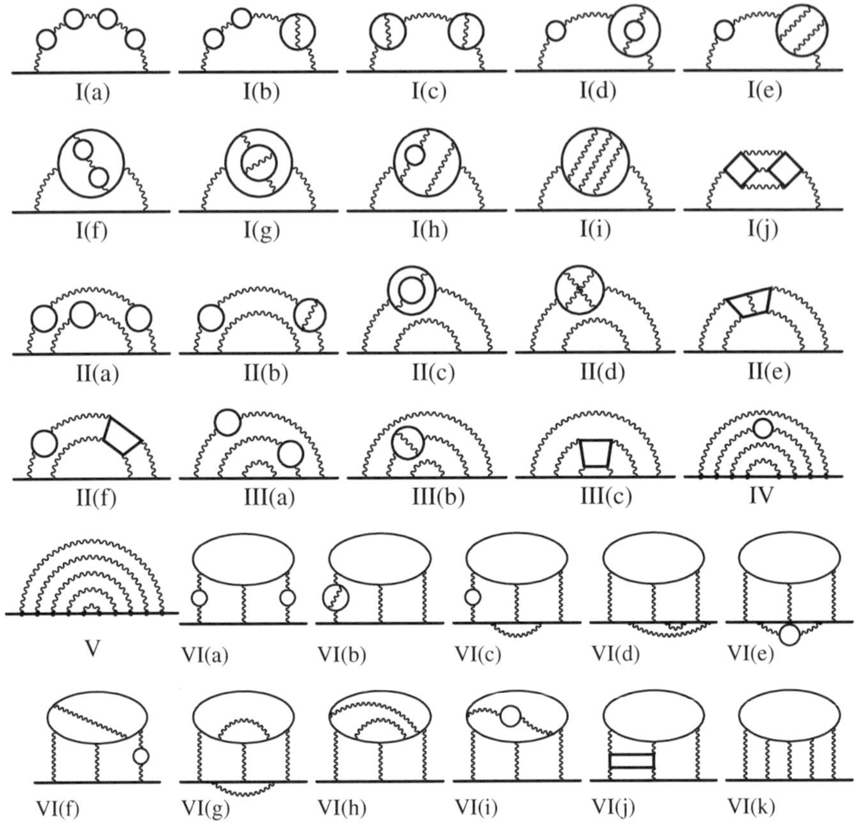

Fig. 2.7 Gauge-invariant subsets of self-energy-like diagrams at the tenth order

Via this identity, it is possible to obtain the expression of *the sum* $M_{\mathscr{G}}$ *of the bare amplitudes of* $g-2$ *induced from the vertex diagrams in* $\mathscr{B}_{\mathscr{G}}$ *which are related to a single self-energy-diagram* $\mathscr{G}$ simultaneously, once we find the expression of the integrand for the bare Feynman amplitude of $\Sigma_{\mathscr{G}}(p)$ in the momentum space.

Moreover, all the vertex diagrams belonging to $\mathscr{B}_{\mathscr{G}}$ have similar ultra-violet (UV) and infrared (IR) divergent structures, as we will see this point explicitly within the parametric integral formalism in Sect. 2.4.3. Since the number of self-energy-like diagrams is much smaller than that of the vertex diagrams by the factor $1/(2n-1)$, we first categorize the vertex diagrams at the order of our interest into the sets represented by the corresponding self-energy-like diagrams. We next classify the self-energy-like diagrams into the minimal gauge-invariant subsets. Note that a gauge-invariant set means the set of diagrams whose sum of contributions is independent of the choice of gauge fixing condition. The minimal gauge-invariant subsets can be immediately identified if one's attention is paid to the types and the number of lepton loops involved. This point can be seen in Fig. 2.7, which lists up all the gauge-invariant subsets for the self-energy-like diagrams at the tenth order [15, 28].

2.4.2 Parametric Representation of Feynman Diagrams

As introduced in Sect. 2.2.1, the amplitude can be expressed as an integral of Feynman parameters. Here, we overview a way to write down the bare amplitude in terms of building blocks which will turn out to help to construct the terms cancelling the ultra-violet and infrared divergences on the Feynman parameter space.

For that purpose, we return to Eq. (2.14), and overview the standard calculational method of the Feynman amplitude. The numerator of the integrand of Eq. (2.14) involves

$$\left\{\not{k}+\not{p}+\frac{\not{q}}{2}+1\right\}\otimes\left\{\not{k}+\not{p}-\frac{\not{q}}{2}+1\right\}, \tag{2.37}$$

which can be written in terms of $k = l - z_1\left(p+\frac{q}{2}\right) - z_2\left(p-\frac{q}{2}\right)$ with new loop momentum l as

$$\begin{aligned} f\ &\left\{\not{l}+(1-z_1)\left(\not{p}+\frac{\not{q}}{2}\right)-z_2\left(\not{p}+\frac{\not{q}}{2}\right)+1\right\} \\ \otimes&\left\{\not{l}-z_1\left(\not{p}+\frac{\not{q}}{2}\right)+(1-z_2)\left(\not{p}-\frac{\not{q}}{2}\right)+1\right\}. \end{aligned} \tag{2.38}$$

Now that the denominator is an even function of l, the terms with odd number of l in the numerator all vanish upon loop integration. Thus, the integral consists of two types

$$\int\frac{d^Dl}{(2\pi)^D}\frac{l_\mu l_\nu}{\left\{l^2-C(z_1,\,z_2)+i\varepsilon\right\}^3},\quad \int\frac{d^Dl}{(2\pi)^D}\frac{1}{\left\{l^2-C(z_1,\,z_2)+i\varepsilon\right\}^3}, \tag{2.39}$$

which can be performed easily.

Even at higher order of perturbation, the manipulation to write down the amplitude as an integral on Feynman space is essentially the same. In practice, however, it needs such devises that realize

(1) construction of the bare amplitude can be systematically done,
(2) construction of the terms to *numerically subtract* ultra-violet divergence and infrared divergence can be done systematically.

The parametric integral formalism provides a method to calculate the perturbative coefficients numerically with the properties (1) and (2). There, the numerical calculation is done on the Feynman parameter space. The loop momenta $\{k_r\}_{r=1,\cdots,n}$ (n is the total number of loops.) must thus be carried out manually. To do so, we first convert n_I number of the denominators of propagators (n_I denotes the number of internal lines in the diagram) to a single denominator through the formula

$$\prod_{j=1}^{n_I} \frac{1}{H_i} = (n_I - 1)! \prod_{j=1}^{n_I} \int_0^1 dz_j \, \delta\left(1 - \sum_{k=1}^{n_I} z_k\right) \frac{1}{\left(\sum_{i=1}^{n_I} z_i H_i\right)^{n_I}}. \tag{2.40}$$

Next, we pull out the momenta p_i appearing in the numerators of lepton propagators outside of the loop integrals by replacing p_i by the D operator

$$D_i^{\mu} \equiv \frac{1}{2} \int_{m_i^2}^{\infty} dm_i^2 \, \frac{\partial}{\partial q_{i,\,\mu}}, \tag{2.41}$$

where we decompose $p_i = k_i + q_i$ into the linear combination k_i of loop momenta k_r and the combination q_i of the external momenta. The amplitude of, say, a vertex diagram $\mathcal{G}$ with no lepton loop can be written as[8,9]

$$\begin{aligned}
\Gamma_{\mathcal{G}}^{\nu} &= \left(-\frac{1}{4}\right)^n (n-1)! \, \mathbb{F}^{\nu} \int (dz)_{\mathcal{G}} \, \frac{1}{U^2 \, V^n}, \\
(dz)_{\mathcal{G}} &\equiv \prod_{j=1}^{n_I} dz_j \, \times \delta\left(1 - \sum_{i=1}^{n_I} z_i\right), \\
V &\equiv \sum_{i=1}^{n_I} z_i \left(m_i^2 - q_i \cdot Q_i'\right), \\
Q_j'^{\mu} &\equiv -\frac{1}{U} \sum_{i=1}^{n_I} q_i^{\mu} z_i \, B_{ij}', \quad B_{ij}' \equiv B_{ij} - \delta_{ij} \, \frac{U}{z_j}. \\
\mathbb{F}^{\nu} &= \gamma^{\alpha_1} \left(\not{D}_1 + m_1\right) \gamma^{\alpha_2} \cdots \gamma^{\nu} \cdots \gamma^{\alpha_{2n-1}} \left(\not{D}_{2n} + m_{2n}\right) \gamma^{\alpha_{2n}} \prod_{k=1}^{n} \eta_{\alpha_{i_k} \alpha_{j_k}}.
\end{aligned} \tag{2.42}$$

U and B_{ij} are functions of Feynman parameters, and are determined by the *chain topology* of the diagram $\mathcal{G}$.

A *chain* is a set of consecutive internal lines with their types disregarded, which are connected by the external vertices only. A chain may consist of a single internal line, or of the internal lepton and photon lines (e.g., $\{1, a\}$ in Fig. 2.5). The chain diagram of $\mathcal{G}$ is a "vacuum diagram" consisting of the chains and the internal vertices of $\mathcal{G}$. For example, Fig. 2.8 shows the chain diagram of *all* vertex-type diagrams in Figs. 2.5 and 2.6 . One immediately understands that all vertex-type diagrams related

---

[8] We set $D = 4$ to simplify the expression.

[9] Diagrams with no lepton loop, called 'quenched-type diagrams', are completely described by a set of pairing of lepton indices $\{(i_k, j_k)\}_{k=1,\ldots,n}$ ($i_k < j_k$, $i_1 < \cdots < i_n$ and $j_{k_1} \neq j_{k_2}$ for $k_1 \neq k_2$.).

via the WT-identity to a self-energy-like diagram share a common chain diagram, in general.

For a non-self-intersecting loop $\mathscr{C}$, we introduce the incident matrix $\{\xi_{c,\mathscr{C}}\}$ associated with the chain topology of $\mathscr{G}$

$$\xi_{c,\mathscr{C}} = \begin{cases} 1 & \text{if } c \in \mathscr{C} \text{ and chain } c \text{ has the same direction as } \mathscr{C} \\ -1 & \text{if } c \in \mathscr{C} \text{ and chain } c \text{ has the direction opposite to } \mathscr{C} \\ 0 & \text{if } c \notin \mathscr{C} \end{cases}. \tag{2.43}$$

Accordingly, we assign the Feynman parameter z_c to the chain c as

$$z_c = \sum_{j \in c} z_j \,. \tag{2.44}$$

In Figs. 2.5 and 2.8,

$$z_{c_1} = z_1 + z_a\,, \quad z_{c_2} = z_2\,, \quad z_{c_3} = z_3 + z_b. \tag{2.45}$$

Starting with $U = z_c$ for a loop (the chain topology of the one-loop diagram), B_{c_1,c_2} and U are obtained recursively according to the following equations

$$\begin{aligned} B_{c_1,c_2} &= \sum_{\mathscr{C}} \xi_{c_1,\mathscr{C}} \xi_{c_2,\mathscr{C}} U_{\mathscr{G}/\mathscr{C}}\,, \\ \xi_{c,\mathscr{C}_0} U &= \sum_{c'} \xi_{c',\mathscr{C}_0} z_{c'} B_{c',c}\,. \end{aligned} \tag{2.46}$$

Here, $\mathscr{G}/\mathscr{C}$ denotes the reduced diagram which is obtained by shrinking $\mathscr{C}$ to a single vertex in $\mathscr{G}$. $\mathscr{C}_0$ and c in the second equation are arbitrarily chosen loop and a chain on it, respectively. Equation (2.46) shows that B_{c_1,c_2} and U are functions of Feynman parameters of homogeneous degrees $n-1$ and n, respectively. $B_{ij} = B_{ji}$ for a pair of the internal lines i, j is equal to $+B_{c_1,c_2}$ for $i \in c_1$, $j \in c_2$ if both lines have the same or the opposite directions as the chains, and to $-B_{c_1,c_2}$ otherwise.

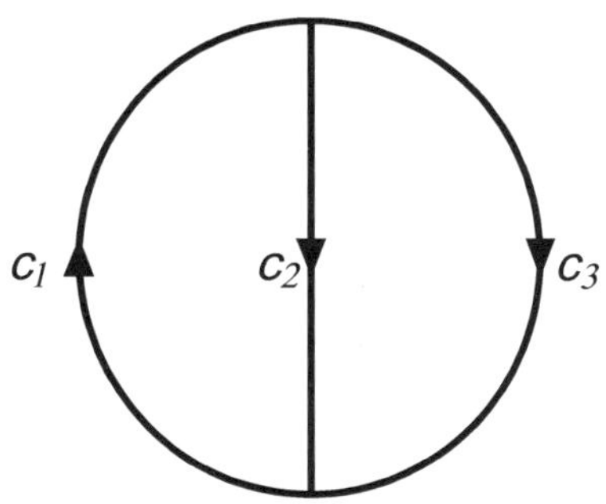

Fig. 2.8 Chain diagram of the fourth-order Feynman diagram (4a)

For the chain diagram in Fig. 2.8,

$$B_{c_1,c_2} = z_{c_2} + z_{c_3}, \quad B_{c_2,c_1} = z_{c_3}, \tag{2.47}$$

and we have for the choice $c = c_1$ and $\mathscr{C}_0 = \{c_1, c_2\}$ in the second equation of Eq. (2.46)

$$U = z_{c_1} B_{c_1,c_1} + z_{c_2} B_{c_2,c_1} = z_{c_1} z_{c_2} + z_{c_2} z_{c_3} + z_{c_3} z_{c_1} \,. \tag{2.48}$$

The application of D operators contained in $\mathbb{F}^\nu$ of Eq. (2.42) will yield the terms as in Eq. (2.39). It is straightforward to see that the manipulation of D operators is equivalent to performing all possible ways of following *contraction operations*:

(a) In case that D_i^μ and D_j^ν are "contracted", they are turned into $\left(-\frac{1}{2} B'_{ij}\, \eta^{\mu\nu}\right)$.
(b) The uncontracted D_i^μ is replaced by $Q_i'^\mu$.
(c) The terms obtained by k-time contraction is multiplied by

$$\frac{1}{(n-1)\cdots(n-k)} \frac{1}{U^{k+2}\, V^{n-k}}, \tag{2.49}$$

in place of $\dfrac{1}{U^2 V^n}$.

The denominator of the resulting integrand is now written in terms of B_{ij} and $Q_i'^\mu$. A vectorial quantity $Q_i'^\mu$ is expressed as a linear combination of the external momenta with the *scalar currents* A_i^p, A_i^q as the coefficients

$$Q_i'^\mu = A_i^p p^\mu + A_i^q q^\mu \,. \tag{2.50}$$

(See the explanation just below Eq. (2.14) for the meaning of p, q.) No A_i^q appears in the expression of the integrand of $g-2$ amplitude. A_i^p is given in terms of B'_{ij} and U as [27]

$$A_i^p = -\frac{1}{U} \sum_{j=1}^{n_I} \eta_{j,\mathscr{P}}\, z_j\, B'_{ji}, \tag{2.51}$$

for the path $\mathscr{P}$ consisting of all consecutive lepton lines connected to the external leptonic states, and $\eta_{j,\mathscr{P}}$ is 1 (-1) if the line j lies on $\mathscr{P}$ with the same (opposite) direction, and 0 otherwise.

Obviously, use of U, B_{ij} and A_i^p as "building blocks" of the integrand[10] reduces the size of the expression of the integrand, compared to the one written solely in

[10] The WT-summed amplitude consists of four types of terms. One is written by

$$\widetilde{C}_{ij} = \frac{1}{U} \sum_{k<l} z_k z_l \left(B'_{ik} B'_{jl} - B'_{il} B'_{jk}\right), \tag{2.52}$$

terms of $\{z_i\}$, and thus reduces the computational cost significantly. Furthermore, it will turn out that they also play the essential role in the systematic construction of the integrands of the terms to subtract divergence. In order to see a little bit about this point, we end this section with observing when ultra-violet (UV) divergence and infrared (IR) divergence can arise. From Eq. (2.51), it turns out that the scalar currents are rational functions of Feynman parameters with homogeneous degree 0. Thus, V in Eq. (2.42) is a rational function with homogeneous degree 1. Hence, the terms obtained by the same number of contractions, whose integrands are all proportional to the quantity (2.49), have the same homogeneous degree. Note that the terms obtained by more and more number of contraction operations correspond to the ones with more and more loop momenta in the numerators of the integrands, and thus become dominated by larger and larger momenta. As seen in Eq. (2.49), contraction increases the degree of U while it decreases that of V in the denominator. Hence, UV divergences can arise on some boundary of Feynman parameter space where $U \to 0$. Likewise, IR divergences can arise where $V \to 0$.

2.4.3 Subtraction of UV and IR Divergences

One of the distinguishing features of the numerical method [26–29] adopted to compute the high-order QED contribution to a_ψ is the *numerical subtraction* of ultra-violet (UV) divergences in the Feynman amplitudes. We suppose that those amplitudes are written in the form of the parametric integrals, as was explained in Sect. 2.4.2.

The numerical subtraction of divergence can be attained only if the bare amplitude and all the required subtraction terms are prepared on the common Feynman parameter space, and singularities are organized to be cancelled in a pointwise way. This way of singularity cancellation, however, requires that a single vertex-type Feynman diagram, or some set of vertex-type Feynman diagrams sharing a common UV structure must be dealt with separately to prepare the subtraction terms. Here we start with observing that a collection of vertex subdiagrams associated with one self-energy-like diagram is a reasonable example of such sets.

As already remarked, the definition of a self-energy-like diagram in Sect. 2.4.1 and the definition of the chain in Sect. 2.4.2 implies that a collection of vertex-type Feynman diagrams related to a given self-energy-like diagram shares the same chain topology. Therefore, those vertex-type diagrams share common U and B_{ij}. Since UV subdivergence appears on the boundary of Feynman parameter space where U vanishes, these diagrams have the same UV divergent structure.

As already mentioned above, the use of the Ward-Takahashi (WT) identity (2.36) to collectively deal with a series of vertex diagrams reduces the total task needed to prepare the integrals significantly. For instance, at the tenth order, there are

(Footnote 10 continued)
where the sum is taken over the lepton lines where an external photon vertex can be inserted in a self-energy-like diagram. $\widetilde{C}_{ij}$ is actually a homogeneous polynomial of degree n. We must deal with these $\widetilde{C}_{ij}$ in addition to U, B_{ij} and A_i.

6,354 quenched-type diagrams. Since nine vertex diagrams are related to each single quenched-type self-energy-like diagram, the WT-identity (2.36) thus reduces them to 706 self-energy-like diagrams. Time reflection invariance in QED further reduces the number of independent integrals to 389 [28].

We call the sum of the $g-2$ amplitudes over the set of vertex diagrams related via the WT-identity to a single self-energy-like diagram $\mathcal{G}$ as the *WT-summed amplitude*. The bare WT-summed amplitude, denoted by $M_{\mathcal{G}}$, in general contains infrared (IR) divergence(s). Therefore, the additional subtraction terms must be added to remove those IR divergences and to get the integral which can be carried out with a supercomputer system. Symbolically, such a finite integral takes the form

$$\Delta \mathcal{M}_{\mathcal{G}} = \int (dz)_{\mathcal{G}} \left(f_{\mathcal{G}}^{\text{bare}}(z) + \sum_{\mathfrak{F}} f_{\mathfrak{F}}^{\text{UV}}(z) + \sum_{\mathfrak{G}} f_{\mathfrak{G}}^{\text{IR}}(z) \right), \tag{2.53}$$

Each of UV subtraction terms necessary for the muon $g-2$ is in one-to-one correspondence with a normal forest $\mathfrak{F}$. A forest of a diagram $\mathcal{G}$ is a set of UV-divergent one-particle irreducible subdiagrams, any two of which are not overlapped with each other.[11] A forest is normal if it does not contain $\mathcal{G}$ as its element.

Before proceeding further, a comment on the necessity of IR subtraction is in order here. The sum of the on-mass-shell subtracted amplitudes over each gauge-invariant subset $\{\mathcal{G}\}'$, i.e.,

$$a_{\psi}\left(\{\mathcal{G}\}'\right) = \sum_{\{\mathcal{G}\}'} \left\{ M_{\mathcal{G}} + \sum_{\mathfrak{F}_{\mathcal{G}}} F_{\mathfrak{F}_{\mathcal{G}}}^{\text{UV, on-shell}} \right\}, \tag{2.54}$$

is finite. The expression on the right-hand side of Eq. (2.54) can be cast into the form required for the numerical subtraction

$$a_{\psi}\left(\{\mathcal{G}\}'\right) = \sum_{\{\mathcal{G}\}'} \int (dz)_{\mathcal{G}} \left\{ f_{\mathcal{G}}^{\text{bare}} + \sum_{\mathfrak{F}_{\mathcal{G}}} f_{\mathfrak{F}_{\mathcal{G}}}^{\text{UV, on-shell}} \right\}. \tag{2.55}$$

The essence of derivation of this form can be seen in the case where $\mathcal{G}$ have a vertex subdiagram $\mathcal{S}$, and focus on the forest $\mathfrak{F} = \{\mathcal{S}\}$, for simplicity. The on-mass-shell subtraction term corresponding to $\{\mathcal{S}\}$ is given by the product

$$F_{\{\mathcal{S}\}}^{\text{UV, on-shell}} = M_{\mathcal{G}/\mathcal{S}} \, L_{\mathcal{S}} , \tag{2.56}$$

of the vertex renormalization constant $L_{\mathcal{S}}$ on mass shells and the WT-summed magnetic moment $M_{\mathcal{G}/\mathcal{S}}$ of the reduced (self-energy-like) diagram $\mathcal{G}/\mathcal{S}$ which

[11] Two subdiagrams $\mathcal{S}_1$ and $\mathcal{S}_2$ are said to be *overlapped* if $\mathcal{S}_1 \cap \mathcal{S}_2 \neq \emptyset$, but neither $\mathcal{S}_1 \subset \mathcal{S}_2$ nor $\mathcal{S}_1 \supset \mathcal{S}_2$.

is obtained by shrinking $\mathcal{S}$ to a point-like vertex in $\mathcal{G}$. The parametric integral formalism allows to express $L_{\mathcal{S}}$ and $M_{\mathcal{G}/\mathcal{S}}$ as the integrals of *respective* sets of Feynman parameters. Thus, their product takes the form

$$M_{\mathcal{G}/\mathcal{S}}\, L_{\mathcal{S}} = \sum_{\alpha,\,\beta} \Gamma(\alpha) \int (dz)_{\mathcal{G}/\mathcal{S}} \frac{g_\alpha\,[\mathcal{G}/\mathcal{S}]}{\left(V_{\mathcal{G}/\mathcal{S}}\right)^\alpha} \times \Gamma(\beta) \int (dz)_{\mathcal{S}} \frac{g_\beta\,[\mathcal{S}]}{(V_{\mathcal{S}})^\beta}\,, \tag{2.57}$$

where, for instance, $g_\alpha[\mathcal{S}]$ is a function with equal numbers of $U_{\mathcal{S}}$, $B'^{\mathcal{S}}_{ij}$, $A^{\mathcal{S}}_i$. The application of the formula

$$\frac{\Gamma(\alpha)}{H^\alpha}\,\frac{\Gamma(\beta)}{K^\beta} = \Gamma(\alpha+\beta) \int_0^1 ds \int_0^1 dt\, \delta(1-s-t)\, \frac{s^{\alpha-1}\, t^{\beta-1}}{(sH+tK)^{\alpha+\beta}}\,, \tag{2.58}$$

and rescaling of the Feynman parameters will reduce Eq. (2.57) to the desired form [29]

$$M_{\mathcal{G}/\mathcal{S}}\, L_{\mathcal{S}} = \int (dz)_{\mathcal{G}} \sum_{\alpha,\,\beta} \Gamma(\alpha+\beta) \frac{g_\alpha\,[\mathcal{G}/\mathcal{S}]\; g_\beta\,[\mathcal{S}]}{\left(V_{\mathcal{G}/\mathcal{S}} + V_{\mathcal{S}}\right)^{\alpha+\beta}}\,. \tag{2.59}$$

The integrand on the right-hand side is exactly $f^{\text{UV, on-shell}}_{\{\mathcal{S}\}}$ in Eq. (2.55). By construction, the integral in Eq. (2.55) for each $\mathcal{G}$ is free from UV divergence. However, it contains IR subdivergence, in general. The form (2.55) demands that cancellation of IR subdivergence be realized when the summation is taken over $\{\mathcal{G}\}'$, not at the stage of integration. This implies that the form (2.55) may allow cancellation of UV divergence, but never does so for IR divergence *at the numerical level*. This is the reason why explicit construction of IR subtraction is necessary in the numerical approach.

In what follows, we will see the construction of UV subtraction terms first, and that of IR subtraction terms next.

We first concentrate on the construction of UV subtraction terms. As was stressed above, the numerical subtraction can be attained only if the subtraction terms are prepared on the same Feynman parameter space as the bare amplitude so that the UV singularities of the bare amplitude are cancelled in a pointwise way. As we have seen already above, the on-mass-shell subtraction term can be cast into such a required form, and the maximally contracted term involved in $L_{\mathcal{S}}$ in fact cancels the UV divergence. However, the on-shell subtraction term in general brings the additional IR divergence not present in the bare amplitude through non-maximally contracted terms. As discussed in the last paragraph in Sect. 2.4.2, lesser and lesser contracted terms tend to be dominated by the dynamics of degrees of freedom with longer and longer wavelength.

For this reason, renormalization is processed in two steps. That is, the UV subtraction terms $f^{\text{UV}}_{\mathfrak{F}}(z)$ in $\Delta\mathcal{M}_{\mathcal{G}}$ in Eq. (2.53) are constructed according to the *K operation* [27], which provides the UV subtraction term free from IR subdivergence. The second step, called the *residual renormalization*, calculates the difference between the renormalization condition corresponding to K operation and the on-shell condition.[12] This step adds the finite term $R\left(\{\mathcal{G}\}'\right)$ to the sum of $\Delta\mathcal{M}_{\mathcal{G}}$ over a gauge-invariant subset $\{\mathcal{G}\}'$

$$a_{\psi}^{(2n)}\left(\{\mathcal{G}\}'\right) = \sum_{\{\mathcal{G}\}'} \Delta\mathcal{M}_{\mathcal{G}} + R\left(\{\mathcal{G}\}'\right) . \tag{2.60}$$

To state the content of the K operation, we define the *UV limit*. For simplicity, we deal with the forest $\{\mathcal{S}\}$ consisting of a single vertex or self-energy subdiagram $\mathcal{S}$. We suppose that the Feynman parameters are organized to parametrize the corresponding UV singularity on the boundary $z_i \sim 0\,(i \in \mathcal{S})$. The UV limit $[A]_{\mathcal{S}}^{\mathrm{UV}}$ of the quantity A is defined to gather all the leading terms of ϵ $(\epsilon \ll 1)$ for

$$z_j = \begin{cases} O(\epsilon) & \text{if } j \in \mathcal{S} \\ O(1) & \text{otherwise} \end{cases} . \tag{2.61}$$

Now, K operation is defined as the following manipulation *on the integrand*:

(1) It extracts only the maximally contracted terms.
(2) B_{ij}, A_j appearing in the numerator of the integrand and U are replaced by the corresponding UV limits.
(3) V is replaced by $V_{\mathcal{S}} + V_{\mathcal{R}}$, where $\mathcal{R}$ is the reduced diagram, and $V_{\mathcal{S}}$ and $V_{\mathcal{R}}$ are written in terms of Feynman parameters and scalar currents in $\mathcal{S}$ and $\mathcal{R}$, respectively.

(1) and (3), where V is not replaced by its UV limit, are necessary in order not to introduce any additional IR divergences.

The most important feature of the integral obtained via K operation is the factorization property; the subtraction terms obtained via K operation are given by the sum of the products of the quantities associated with the subdiagrams $\mathcal{S}_i$ of the forest $\{\mathcal{S}_i\}$ and the magnetic moment of the reduced self-energy-like diagram. Again, let's consider a forest $\{\mathcal{S}\}$ consisting of a single UV-divergent subdiagram. If $\mathcal{S}$ is a vertex subdiagram, $\mathcal{R} = \mathcal{G}/\mathcal{S}$ and

$$\mathsf{K}_{\mathcal{S}} M_{\mathcal{G}} = L_{\mathcal{S}}^{\mathrm{UV}}\, M_{\mathcal{G}/\mathcal{S}} , \tag{2.62}$$

where $L_{\mathcal{S}}^{\mathrm{UV}}$ denotes the maximally contracted terms of $L_{\mathcal{S}}$. If $\mathcal{S}$ is a self-energy subdiagram, the result of K operation consists of two terms

$$\mathsf{K}_{\mathcal{S}} M_{\mathcal{G}} = \delta m_{\mathcal{S}}^{\mathrm{UV}}\, M_{\mathcal{G}/\mathcal{S}(i^*)} + B_{\mathcal{S}}^{\mathrm{UV}}\, M_{\mathcal{G}/\{\mathcal{S},i\}} . \tag{2.63}$$

In $\mathcal{G}/\mathcal{S}(i^*)$, the remnant of $\mathcal{S}$ is left as the mass insertion vertex $(*)$ on a lepton line i connecting $\mathcal{S}$ to the rest of $\mathcal{G}$. Note that $\mathcal{G}/\mathcal{S}(i^*)$ and $\mathcal{G}/\{\mathcal{S}, i\}$ have the same

[12] The masses of leptons and the electric charge can be used only for the amplitude renormalized in the on-shell renormalization condition.

chain topology so that $V_{\mathcal{G}/\{\mathcal{S},i\}}$ can be used in Sect. 2.4.3. The UV-finite amplitude $M^{\mathrm{R}}_{\mathcal{G}}$, which may contain IR divergence, is obtained from the forest formula with use of K operation

$$M^{\mathrm{R}}_{\mathcal{G}} = M_{\mathcal{G}} + \sum_{\mathfrak{F}} \prod_{\mathcal{S}\in\mathfrak{F}} (-\mathsf{K}_{\mathcal{S}})\, M_{\mathcal{G}}\,, \tag{2.64}$$

whose integral is organized in the same form as in Eq. (2.54).

Because the on-shell subtraction terms by definition exhibit the same factorization as above, the factorization property of K operation (and IR-subtraction scheme explained below) guarantees that *the residual renormalization term* $R\left(\{\mathcal{G}\}'\right)$ *in Eq.* (2.60) *can be written as the sum of the products of finite quantities at lower orders.* The factorization property of the amplitude obtained by K operation follows from the factorization of U, B_{ij} and A_i under the UV limit [27, 28].

Next, we turn our attention to the construction of IR-subtraction terms. The method we seek should be systematic enough to be implemented as a code to produce those terms as numerical programs.

IR subtraction is not as simple as UV subtraction in various respects. First, no general formula like Zimmermann's forest formula has been found. This may be due to the fact that IR subtraction is a tentatively required operation. The method to construct IR-subtraction terms must therefore be invented for the individual problems.

Another important difference from UV subtraction is that the IR singularities involved in the bare amplitude $M_{\mathcal{G}}$ can become harder than logarithmic. Indeed, there are eighth- and tenth-order diagrams that contain linear IR divergence. To recognize that the subtraction of such hard singularities is a complicated problem, we recall the fact that UV divergence is at most logarithmic at any order of perturbation. Thus, only leading singularity needs to be subtracted for UV. Indeed, this is attained by K operation, whose main part is composed of the power-counting operation, UV limit. However, if the quantity diverges linearly, the next-to-leading-order singularity must also be extracted from it. This is actually hard to do along the lines of any power-counting scheme. We thus have invented an IR-subtraction scheme for the high-order QED contribution to $g-2$, which is systematic enough to be implemented as the code generating numerical programs [29].

IR divergence arises basically in the following way. Let's consider the vertex diagram in Fig. 2.1 with the external leptons on their mass shells, $p_F^2 = m^2 = p_I^2$. The amplitude contains the IR-divergent piece which is proportional to

$$\begin{aligned}\int \frac{d^4k}{k^2}\frac{1}{(k+p_F)^2-m^2}\,\frac{1}{(k+p_I)^2-m^2} \\ = \int \frac{d^4k}{k^2}\frac{1}{k^2+2p_F\cdot k}\,\frac{1}{k^2+2p_I\cdot k}\,.\end{aligned} \tag{2.65}$$

If a mass insertion vertex is inserted into one of the internal lepton lines in Fig. 2.1, IR divergence becomes harder. This indicates the mechanism of how worse IR divergence can arise at higher orders. That is, a diagram with more self-energy

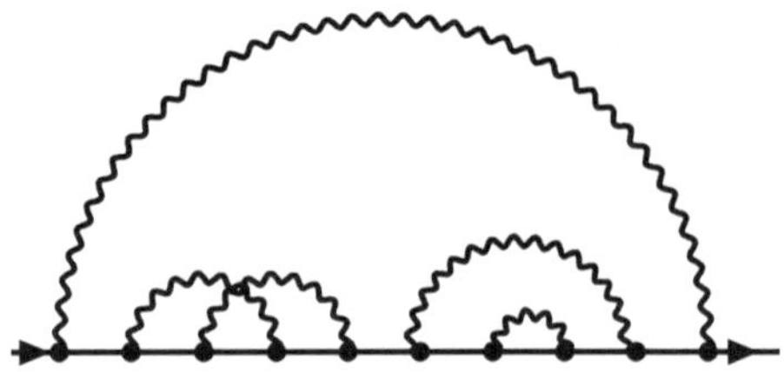

Fig. 2.9 A self-energy-like diagram having IR singularities which are logarithmic and harder at the tenth order

subdiagrams which are disconnected to one another can have harder IR divergence. Figure 2.9 illustrates such a self-energy-like diagram. The hardest IR divergence arises when only the outmost photon carries longer wavelength compared to the others and all the self-energy subdiagrams involved become relatively point-like.

One may imagine that the hard IR divergences arising through such a mechanism will be removed by the UV-subtraction terms corresponding to the self-energy diagrams. It is correct. But, as remarked at the stage of UV subtraction, full on-mass-shell renormalization can bring additional IR subdivergence through the wave function renormalization constant. For this reason, K operation is used for the construction of UV-subtraction terms. Suppose that the self-energy diagrams are all second-order. Then, the on-shell mass renormalization constant δm is equal to its UV limit δm^{UV}; $\delta m = \delta m^{\mathrm{UV}}$ and the amplitude which is made UV-finite by K operation becomes free from hard IR subdivergence. In general, in order to completely eliminate hard IR subdivergence, say for a forest consisting of a single self-energy subdiagram $\{\mathcal{S}\}$, it suffices to subtract the term in proportion to the difference $\delta m^{\mathrm{R}}_{\mathcal{S}} \equiv \delta m_{\mathcal{S}} - \delta m^{\mathrm{UV}}_{\mathcal{S}} -$ (all UV subdivergences) from the UV-finite amplitude $M^{\mathrm{R}}_{\mathcal{G}}$. This is exactly the idea of the R operation defined in [29]

$$\mathsf{R}_{\mathcal{S}} M^{\mathrm{R}}_{\mathcal{G}} \equiv M^{\mathrm{R}}_{\mathcal{G}/\mathcal{S}(i^*)}\, \delta m^{\mathrm{R}}_{\mathcal{S}} . \tag{2.66}$$

Equation (2.59) and its generalization to the product of more than two terms will convert the result of $\mathsf{R}_{\mathcal{S}} M^{\mathrm{R}}_{\mathcal{G}}$ into such a form of the integral that allows to cancel IR subdivergence in a pointwise way on the same Feynman parameter space as the bare amplitude.

The amplitude obtained by applying required R operations may be logarithmically divergent. They can be removed by I operation, say, for a single self-energy diagram $\mathcal{S}$ of a self-energy-like diagram $\mathcal{G}$ [29]

$$\mathsf{I}_{\mathcal{S}} M^{\mathrm{R}}_{\mathcal{G}} \equiv L^{\mathrm{R}}_{\mathcal{G}/\mathcal{S}(k)}\, M^{\mathrm{R}}_{\mathcal{S}} . \tag{2.67}$$

Here, $\mathcal{G}/\mathcal{S}(k)$ is obtained by inserting an external vertex k into one of the internal lepton lines in $\mathcal{S}$ which lies on the open path connected to the external leptons, and shrinking it to a point. $L^{\mathrm{R}}_{\mathcal{G}/\mathcal{S}(k)}$ is UV-finite but contains IR divergence; it is obtained by applying required K-operations to the on-mass-shell vertex renormalization constant $L_{\mathcal{G}/\mathcal{S}(k)}$.

The required IR-subtraction terms are characterized by the *annotated forests*. An annotated forest α is a set of pairs $(\mathcal{S}, \mathsf{O})$ of a self-energy subdiagram $\mathcal{S}$ of a self-energy-like diagram $\mathcal{G}$ and the operation $\mathsf{O} = \mathsf{R}$ or I with one restriction; if $\mathcal{S}_1 \subset \mathcal{S}_2$, $(\mathcal{S}_1, \mathsf{I})$ and $(\mathcal{S}_2, \mathsf{R})$ are prohibited as the elements of an annotated forest. If $\mathfrak{A}$ denotes the set of all normal annotated forests, the finite amplitude $\Delta M_{\mathcal{G}}$ is obtained as

$$\Delta M_{\mathcal{G}} = M_{\mathcal{G}}^{\mathrm{R}} + \sum_{\alpha \in \mathfrak{A}} \prod_{(\mathcal{S}, \mathsf{O}) \in \alpha} (-\mathsf{O}_{\mathcal{S}})\, M_{\mathcal{G}}^{\mathrm{R}}, \tag{2.68}$$

where, if $\mathcal{S}_1 \subset \mathcal{S}_2$, the operators act according to the following order

(i) $\mathsf{R}_{\mathcal{S}_2} \mathsf{R}_{\mathcal{S}_1}$.
(ii) $\mathsf{I}_{\mathcal{S}_1} \mathsf{I}_{\mathcal{S}_2}$.
(iii) $\mathsf{R}_{\mathcal{S}_1} \mathsf{I}_{\mathcal{S}_2}$.

2.5 Result for QED Contribution

By combining Eqs. (2.5) with (2.27), a_e is found to be represented as a power series with respect to the fine structure constant α

$$a_e(\alpha) = \text{non-QED contribution} + \sum_{n=1}^{\infty} a_e^{(2n)} \left(\frac{\alpha}{\pi}\right)^n . \tag{2.69}$$

Table 2.2 implies that the theoretical ambiguity in the "non-QED contribution" is less than the experimental uncertainty of a_e, but cannot be negligible. Therefore, if the relevant coefficients $a_e^{(2n)}$ in the perturbative expansion of the QED contribution can be computed with sufficient precision, equating such obtained $a_e(\alpha)$ as a function of α to the experimental value $a_e(\exp)$ will yield the value of α, referred to as $\alpha(a_e)$.

Table 2.3 summarizes the terms $A_j^{(2n)}$ in each coefficients $a_e^{(2n)}$ of $a_e(\text{QED})$. See Eq. (2.29) for the definition of $A_j^{(2n)}$. $A_1^{(2)}$ [4], $A_1^{(4)}$ [30, 31], and $A_1^{(6)}$ [32] are known exactly. We also have the analytic expression for $A_2^{(2)}$ [33], $A_2^{(4)}$ [34, 35] and the asymptotic expansion in mass ratios for $A_3^{(6)}$ [36, 37]. The uncertainties of these terms are attributed to the ones in lepton mass ratios. The values in Table 2.3 were recalculated using the newest values [38].

$$\begin{aligned} \frac{m_e}{m_\mu} &= 4.83633166\,(12) \times 10^{-3}, \\ \frac{m_e}{m_\tau} &= 2.87592\,(26) \times 10^{-4}, \\ \frac{m_\mu}{m_\tau} &= 5.94649\,(54) \times 10^{-2}. \end{aligned} \tag{2.70}$$

All of $A_j^{(8)}$ and $A_j^{(10)}$ have been obtained only by the numerical means explained in Sect. 2.4.

From Table 2.3, we can explicitly see that the mass-independent term $A_1^{(2n)}$ dominates the coefficient $a_e^{(2n)}$ of the perturbative expansion of the QED contribution to the electron $g-2$, and the orders of magnitude of them are all unity.

Now, by using the result in Table 2.3 for $a_e^{(2n)}$ in Eq. (2.69), we get [39]

$$\alpha^{-1}(a_e) = 137.035\,999\,1727\,(68)_{8\mathrm{th}}(46)_{10\mathrm{th}}(19)_{\mathrm{QCD}}(331)_{\mathrm{exp}}\,, \tag{2.71}$$

where the uncertainties come from the eighth-order QED coefficient $a_e^{(8)}$, the tenth-order coefficient $a_e^{(10)}$, the QCD contribution, and the experiment (2.3) of the electron $g-2$, respectively.

Recall that the value of α has also been determined by various other methods such as quantum Hall effect, etc [38, 40]. In particular, besides the electron $g-2$, the precise α has been obtained by determination of the ratio h/m_A due to the atom recoil-velocity measurement for $A =$ cesium [41] and $A =$ rubidium [42] combined with the Rydberg constant and m_{Rb}/m_e in [38]

$$\alpha^{-1}(\mathrm{Rb}) = 137.035\,999\,049\,(90)\,. \tag{2.72}$$

Within the present precision, Eq. (2.71) is compatible with Eq. (2.72)

$$\alpha^{-1}(a_e) - \alpha^{-1}(\mathrm{Rb}) = 124(96) \times 10^{-9}\,. \tag{2.73}$$

In this way, check of consistency of the values of α thus obtained provides us a cross-sectional understanding on wide range of physical phenomena.

Instead, if we use the value (2.72) for the fine structure constant, we obtain the following prediction for $a_e(\mathrm{SM})$

Table 2.3 The terms $A_j^{(2n)}$ in the coefficients $a_e^{(2n)}$

$2n$	$A_1^{(2n)}$	$A_2^{(2n)}\left(\frac{m_e}{m_\mu}\right)$	$A_2^{(2n)}\left(\frac{m_e}{m_\tau}\right)$	$A_3^{(2n)}\left(\frac{m_e}{m_\mu}, \frac{m_e}{m_\tau}\right)$
2	0.5	–	–	–
4	−0.328478965579...	$5.19738667\ (26) \times 10^{-7}$	$1.83798\ (34) \times 10^{-9}$	–
6	1.181241456...	$-7.37394155\ (27) \times 10^{-6}$	$-6.5830\ (11) \times 10^{-8}$	$0.1909\ (1) \times 10^{-12}$
8	−1.9106 (20)	$9.222\ (66) \times 10^{-4}$	$8.24\ (12) \times 10^{-6}$	$7.465\ (18) \times 10^{-7}$
10	9.168 (571)	−0.00382 (39)	◇	◇

There are no Feynman diagrams for $A_2^{(2)}\left(\frac{m_e}{m_\mu}\right)$, $A_2^{(2)}\left(\frac{m_e}{m_\tau}\right)$ and $A_3^{(2n)}\left(\frac{m_e}{m_\mu}, \frac{m_e}{m_\tau}\right)$ $(2n = 2, 4)$. $A_2^{(10)}\left(\frac{m_e}{m_\tau}\right)$ and $A_3^{(10)}\left(\frac{m_e}{m_\mu}, \frac{m_e}{m_\tau}\right)$ have not been computed as they are much smaller than $A_2^{(10)}\left(\frac{m_e}{m_\mu}\right)$, and is thus negligible for the precision of our interest

$$a_e(\text{SM, Rb}) = 1\;159\;652\;180.07\;(6)_{8\text{th}}(8)_{10\text{th}}(3)_{\text{QCD}}(77)_{\alpha(\text{Rb})} \times 10^{-12}. \quad (2.74)$$

The comparison of this with Eq. (2.3) is less impressive due to the uncertainty of α^{-1}(Rb).

Acknowledgments The author thanks T. Aoyama, T. Kinoshita and M. Nio for the collaboration over a long period of time, and K. Asano and N. Watanabe for their intensive works in the short term. The numerical integration of the eighth and tenth order contribution has been carried out using supercomputer systems, RSCC and RICC. This work is supported in part by the JSPS Grant-in-Aid for Scientific Research (C)20540261.

References

1. D. Hanneke, S. Fogwell, G. Gabrielse, New measurement of the electron magnetic moment and the fine structure constant. Phys. Rev. Lett. **100**, 120801 (2008). doi:DOIurl10.1103/PhysRevLett.100.120801
2. D. Hanneke, S. Fogwell Hoogerheide, G. Gabrielse, Cavity control of a single-electron quantum cyclotron: measuring the electron magnetic moment. Phys. Rev. A **83**, 052122 (2011)
3. R. Van Dyck, P. Schwinberg, H. Dehmelt, New high-precision comparison of electron and positron g-factors. Phys. Rev. Lett. **59**, 26 (1987). doi:10.1103/PhysRevLett.59.26
4. J.S. Schwinger, On quantum-electrodynamics and the magnetic moment of the electron. Phys. Rev. **73**, 416 (1948). doi:10.1103/PhysRev.73.416
5. G. 't Hooft, M. Veltman, Regularization and renormalization of gauge fields. Nucl. Phys. **B44**, 189 (1972). doi:10.1016/0550-3213(72)90279-9
6. R. Feynman, Space-time approach to quantum electrodynamics. Phys. Rev. **76**, 769 (1949). doi:10.1103/PhysRev.76.769
7. F. Jegerlehner, A. Nyffeler, The muon $g-2$. Phys. Rep. **477**, 1 (2009). doi:10.1016/j.physrep.2009.04.003
8. K. Hagiwara, A. Martin, D. Nomura, T. Teubner, Predictions for $g-2$ of the muon and $\alpha_{QED}(M_Z^2)$. Phys. Rev. **D69**, 093003 (2004). doi:10.1103/PhysRevD.69.093003
9. D. Nomura, T. Teubner, Hadronic contributions to the anomalous magnetic moment of the electron and the hyperfine splitting of muonium. Nucl. Phys. **B867**, 236 (2013). doi:10.1016/j.nuclphysb.2012.10.001
10. J. Prades, E. de Rafael, A. Vainshtein, in *Hadronic Light-by-Light Scattering Contribution to the Muon Anomalous Magnetic Moment*. ed. by B. Lee Roberts, William J. Marciano. Lepton Dipole Moments (Advanced Series on Directions in High Energy Physics), vol 20 (World Scientific, Singapore, 2009)
11. K. Fujikawa, B. Lee, A. Sanda, Generalized renormalizable Gauge formulation of spontaneously broken Gauge theories. Phys. Rev. **D6**, 1972 (2923). doi:10.1103/PhysRevD.6.2923
12. A. Czarnecki, B. Krause, W.J. Marciano. Electroweak corrections to the muon anomalous magnetic moment. Phys. Rev. Lett. **76**, 3267 (1996). doi:10.1103/PhysRevLett.76.3267
13. M. Knecht, S. Peris, M. Perrottet, E. De Rafael, Electroweak hadronic contributions to the muon $(g-2)$. JHEP **0211**, 003 (2002)
14. A. Czarnecki, W.J. Marciano, A. Vainshtein, Refinements in electroweak contributions to the muon anomalous magnetic moment. Phys. Rev. **D67**, 073006 (2003). doi:10.1103/PhysRevD.67.073006, doi:10.1103/PhysRevD.73.119901
15. T. Kinoshita, M. Nio, Tenth-order QED contribution to the lepton $g-2$: evaluation of dominant α^5 terms of muon $g-2$. Phys. Rev. **D73**, 053007 (2006). doi:10.1103/PhysRevD.73.053007
16. T. Aoyama, M. Hayakawa, T. Kinoshita, M. Nio, N. Watanabe, Eighth-order vacuum-polarization function formed by two light-by-light-scattering diagrams and its contribution to

the tenth-order electron $g-2$. Phys. Rev. **D78**, 053005 (2008). doi:10.1103/PhysRevD.78.053005
17. T. Aoyama, M. Hayakawa, T. Kinoshita, M. Nio, Tenth-order lepton anomalous magnetic moment: Second-order vertex containing two vacuum polarization subdiagrams, one within the other. Phys. Rev. **D78**, 113006 (2008). doi:10.1103/PhysRevD.78.113006
18. T. Aoyama, K. Asano, M. Hayakawa, T. Kinoshita, M. Nio, et al., Tenth-order lepton $g-2$: contribution from diagrams containing sixth-order light-by-light-scattering subdiagram internally. Phys. Rev. **D81**, 053009 (2010). doi:10.1103/PhysRevD.81.053009
19. T. Aoyama, M. Hayakawa, T. Kinoshita, M. Nio, Tenth-order lepton $g-2$: contribution of some fourth-order radiative corrections to the sixth-order $g-2$ containing light-by-light-scattering subdiagrams. Phys. Rev. **D82**, 113004 (2010). doi:10.1103/PhysRevD.82.113004
20. T. Aoyama, M. Hayakawa, T. Kinoshita, M. Nio, Proper eighth-order vacuum-polarization function and its contribution to the tenth-order lepton $g-2$. Phys. Rev. **D83**, 053003 (2011). doi:10.1103/PhysRevD.83.053003
21. T. Aoyama, M. Hayakawa, T. Kinoshita, M. Nio, Tenth-order QED contribution to lepton anomalous magnetic moment: fourth-order vertices containing sixth-order vacuum-polarization subdiagrams. Phys. Rev. **D83**, 053002 (2011). doi:10.1103/PhysRevD.83.053002
22. T. Aoyama, M. Hayakawa, T. Kinoshita, M. Nio, Tenth-order lepton anomalous magnetic moment sixth-order vertices containing vacuum-polarization subdiagrams. Phys. Rev. **D84**, 053003 (2011). doi:10.1103/PhysRevD.84.053003
23. T. Aoyama, M. Hayakawa, T. Kinoshita, M. Nio, Tenth-order QED lepton anomalous magnetic moment: eighth-order vertices containing a second-order vacuum polarization. Phys. Rev. **D85**, 033007 (2012). doi:10.1103/PhysRevD.85.033007
24. T. Aoyama, M. Hayakawa, T. Kinoshita, M. Nio, Tenth-order QED contribution to the lepton anomalous magnetic moment: sixth-order vertices containing an internal light-by-light-scattering subdiagram. Phys. Rev. **D85**, 093013 (2012). doi:10.1103/PhysRevD.85.093013
25. T. Aoyama, M. Hayakawa, T. Kinoshita, M. Nio, Quantum electrodynamics calculation of lepton anomalous magnetic moments: Numerical approach to the perturbation theory of QED. Prog. Theor. Exp. Phys. **2012**, 01A107 (2012). doi:10.1093/ptep/pts030
26. P. Cvitanovic, T. Kinoshita, Feynman-Dyson rules in parametric space. Phys. Rev. **D10**, 3978 (1974). doi:10.1103/PhysRevD.10.3978
27. P. Cvitanovic, T. Kinoshita, New approach to the separation of ultraviolet and infrared divergences of Feynman-parametric integrals. Phys. Rev. **D10**, 3991 (1974). doi:10.1103/PhysRevD.10.3991
28. T. Aoyama, M. Hayakawa, T. Kinoshita, M. Nio, Automated calculation scheme for α^n contributions of QED to lepton $g-2$: generating renormalized amplitudes for diagrams without lepton loops. Nucl. Phys. **B740**, 138 (2006). doi:10.1016/j.nuclphysb.2006.01.040
29. T. Aoyama, M. Hayakawa, T. Kinoshita, M. Nio, Automated calculation scheme for α^n contributions of QED to lepton g-2: new treatment of infrared divergence for diagrams without lepton loops. Nucl. Phys. **B796**, 184 (2008). doi:10.1016/j.nuclphysb.2007.12.013
30. A. Petermann, Fourth-order magnetic moment of the electron. Helv. Phys. Acta **30**, 407 (1957)
31. C.M. Sommerfield, Magnetic dipole moment of the electron. Phys. Rev. **107**, 328 (1957). doi:10.1103/PhysRev.107.328
32. S. Laporta, E. Remiddi, The analytical value of the electron $(g-2)$ at order α^3 in QED. Phys. Lett. B379, **283** (1996). doi:10.1016/0370-2693(96)00439-X
33. H. Elend, On the anomalous magnetic moment of the muon. Phys. Lett. 20, **682** (1966). doi:10.1016/0031-9163(66)91171-1. Erratum: ibid., **21**, 720 (1966)
34. S. Laporta, E. Remiddi, The analytical value of the electron light-light graphs contribution to the muon $(g-2)$ in QED. Phys. Lett. **B301**, 440 (1993). doi:10.1016/0370-2693(93)91176-N
35. S. Laporta, The analytical contribution of the sixth-order graphs with vacuum polarization insertions to the muon $(g-2)$ in QED. Nuovo Cim. **A106**, 675 (1993). doi:10.1007/BF02787236

36. A. Czarnecki, M. Skrzypek, The muon anomalous magnetic moment in QED: three-loop electron and tau contributions. Phys. Lett. **B449**, 354 (1999). doi:10.1016/S0370-2693(99)00076-3
37. S. Friot, D. Greynat, E. De Rafael, Asymptotics of Feynman diagrams and the Mellin Barnes representation. Phys. Lett. **B628**, 73 (2005). doi:10.1016/j.physletb.2005.08.126
38. Rev. Mod. Phys. CODATA recommended values of the fundamental physical constants: 2010. **84**(1527), 2012 (2010). doi:10.1103/RevModPhys.84.1527
39. T. Aoyama, M. Hayakawa, T. Kinoshita, M. Nio, Tenth-order QED contribution to the electron $g-2$ and an improved value of the fine structure constant. Phys. Rev. Lett. **109**, 111807 (2012). doi:10.1103/PhysRevLett.109.111807.
40. T. Kinoshita, The fine structure constant. Rept. Prog. Phys. **59**, 1459 (1996). doi:10.1088/0034-4885/59/11/003.
41. A. Wicht, J.M. Hensley, E. Sarajlic, S. Chu, A preliminary measurement of the fine structure constant based on atom interferometry. Physica Scripta **T106**, 82 (2002). doi:10.1238/Physica.Topical.102a00082
42. R. Bouchendira, P. Clade, S. Guellati-Khelifa, F. Nez, F. Biraben, New determination of the fine structure constant and test of the quantum electrodynamics. Phys. Rev. Lett. **106**, 080801 (2011). doi:10.1103/PhysRevLett.106.080801

Chapter 3
Magnetic Moment of the Bound Electron

Manuel Vogel and Wolfgang Quint

Abstract The magnetic moment of the *bound* electron is significantly different from that of the *free* electron. This is true for any binding situation, but particularly pronounced when the electron is bound in a highly charged ion, where it is subject to extreme electromagnetic fields in the close vicinity of the nucleus. In such a situation, it represents a sensitive probe of nuclear properties and serves as a link between nuclear and atomic physics. Few-electron ions, especially hydrogen-like ones, brilliantly serve as 'microscopic high-field laboratories' which can be investigated with a broad range of techniques for ion confinement, manipulation and in-trap spectroscopy. Ion cooling methods and long observation times open the way to high precision, especially in single-ion experiments. As calculations of the bound electron magnetic moment in the framework of quantum electrodynamics advance, a comparison between experiment and theory allows stringent benchmarks of models or in turn gives access to fundamental quantities such as the electron mass and the fine structure constant. In this chapter, we present corresponding experimental concepts and discuss results obtained so far.

3.1 The Case of the Bound Electron

Generally, the magnetic moment $\boldsymbol{\mu}$ of a particle is referred to in terms of the dimensionless proportionality factor g ('g-factor') as given in the relation

$$\boldsymbol{\mu} = -g\mu_B \frac{\mathbf{J}}{\hbar}, \tag{3.1}$$

M. Vogel (✉)
Technische Universität Darmstadt, Darmstadt, Germany
e-mail: m.vogel@gsi.de

W. Quint
Universität Heidelberg, Heidelberg, Germany
e-mail: w.quint@gsi.de

W. Quint and M. Vogel (eds.), *Fundamental Physics in Particle Traps*,
Springer Tracts in Modern Physics 256, DOI: 10.1007/978-3-642-45201-7_3,

where the minus sign expresses that the magnetic moment $\boldsymbol{\mu}$ and the angular momentum $\mathbf{J}$ have opposite orientations and μ_B is the Bohr magneton $\mu_B = e\hbar/(2m_e) = 9.27400912(23) \times 10^{-24}$ J/T. While for the free electron, the spin $\mathbf{s}$ is the only angular quantity, the bound electron is characterized by both spin and orbital angular momentum $\mathbf{L}$ which couple to $\mathbf{J} = \mathbf{L} + \mathbf{s}$ such that the spin alone is no longer an observable. Hence, the g-factor is now labelled g_J. Other than in the case of the free electron, the deviation of the bound electron g-factor from 2 is not dominantly due to QED effects, but due to binding and also nuclear and relativistic effects.[1] Yet, the QED effects (which roughly scale with Z^2) are far from being negligible and their predictability together with the possibility of testing those predictions form a good part of the motivation for precision measurements of magnetic moments in highly charged ions [6–10]. The theoretical concepts for calculation of bound-electron magnetic moments will be given in the following book chapter.

Experimentally, the magnetic moment (or g-factor) can be determined by a measurement of the precession frequency of the electron spin in an external magnetic field B_0. This so-called 'Larmor frequency' ω_L is given by

$$\hbar\omega_L = \mu B_0 = g_J \mu_B B_0. \tag{3.2}$$

Hence, a determination of the magnetic moment (or g-factor) requires knowledge of the magnetic field strength B_0 at the position and time of the measurement. This is possible via a determination of the free ion cyclotron frequency in the external magnetic field given by $\omega_c = qB_0/m_i$ where q is the ion's electric charge and m_i is its atomic mass. Figure 3.1 schematically shows the involved quantities for the example of a hydrogen-like ion in an external magnetic field B_0. Note, that $\omega_L \gg \omega_c$.

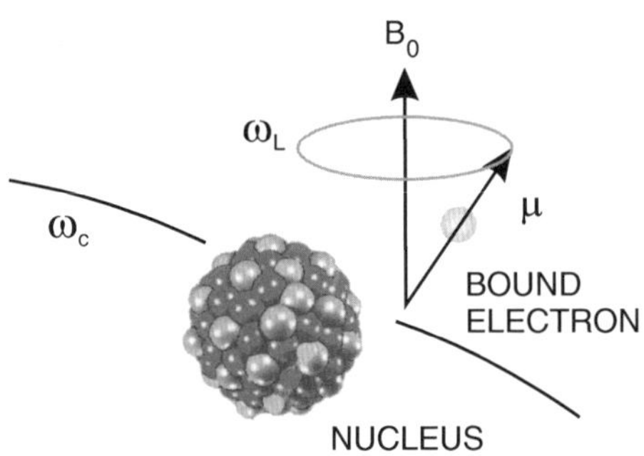

Fig. 3.1 Schematic of the involved quantities. The electron spin performs a precession about the external magnetic field B_0 as used for ion confinement with the Larmor frequency ω_L. In the same field, the ion performs a cyclotron oscillation at a frequency ω_c which is also proportional to B_0. For the sake of simplicity, we have discarded the electric field used for ion confinement, however this does not change the validity of this picture

[1] We should note that the g-factors of the proton (first measured in 1933 [1–3]) and neutron (first measured in 1940 [4, 5]) are already in the free state much different from 2 due to their compositeness ($g_P = 5.585694713(46)$ and $g_N = -3.82608545(90)$).

3.2 Why the Bound Electron is Interesting

Already for atoms and singly charged ions, significant deviations from the free electron value of the electronic magnetic moment have been observed, both for light hydrogen-like systems and for many-electron systems such as Be^+, Mg^+, Ca^+, Ba^+ and Hg^+ [11–13]. The magnetic moment of the bound electron, particularly in a highly charged ion, is drastically different from that of the free electron. The magnitude of the deviation with respect to the free electron value is up to 10 orders of magnitude larger than the current experimental uncertainty. This is based on the fact that in such highly charged ions, the electron under investigation is spatially close to the ionic nucleus and experiences extreme electric and magnetic fields of up to about 10^{16} V/cm and 10^5 T [14]. In hydrogen-like ions in the ground state, the expectation value of the electron-nucleus distance $< r >$ is inversely proportional to the nuclear charge number Z, and the probability $|\Psi(r = 0)|^2$ to find the electron at the nuclear centre is proportional to Z^3, which yields a significant difference when going from hydrogen ${}^1_1$H ($Z = 1$) to hydrogen-like uranium ${}^{238}_{92}U^{91+}$ ($Z = 92$). Figure 3.2 shows the probability density $r^2|\Psi(r)|^2$ according to [14] of the 1s-electron in a hydrogen-like ion ${}^A_Z X^{(Z-1)+}$ of nuclear charge eZ as a function of the distance to the nucleus (measured in units of the Bohr radius $a_0 = 0.529\,177\,210\,92\,(17) \times 10^{-10}$ m) together with the expectation value $< r >$ which is relevant for the bound electron's magnetic moment. The right hand scale shows the corresponding values of the nucleus' electric field strength per nuclear charge number Z, which is about 2.5×10^{13} V/cm/Z at the relevant distance.

Figure 3.3 shows the deviation of the bound electron g-factor g_J from the free electron g-factor g_e as a function of the nuclear charge number Z for hydrogen-like

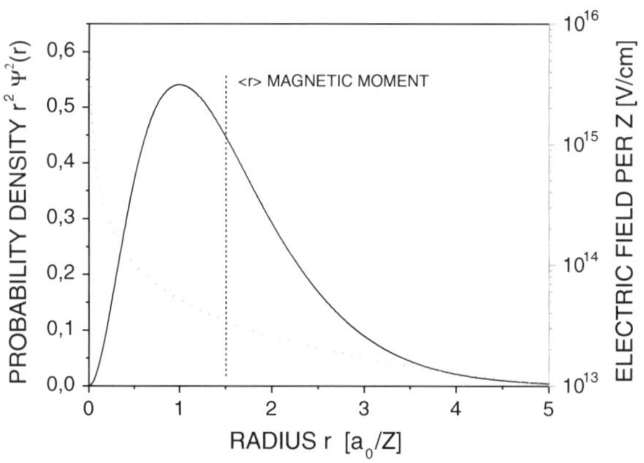

Fig. 3.2 Probability density of the 1s electron in a hydrogen-like ion as a function of the distance in units of the Bohr radius (*left scale*). Electric field strength of the nucleus per nuclear charge Z as a function of radius (*right scale*). The expectation value of the bound electron distance $< r >$ is indicated as this is the relevant quantity for the bound electron magnetic moment

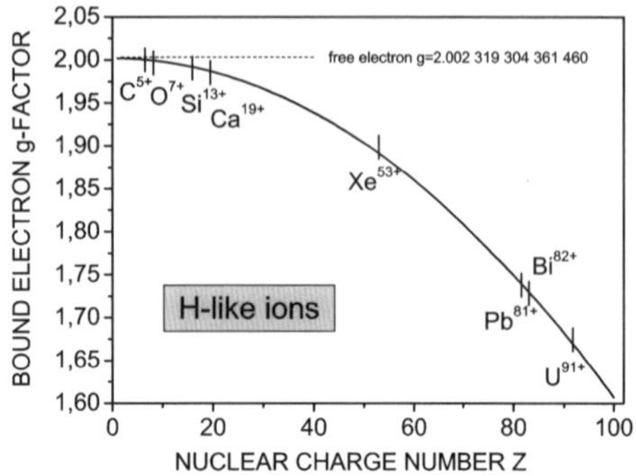

Fig. 3.3 Dependence of the electron magnetic moment on the nuclear charge number Z of the H-like ion it is bound to. The line is a fit to the theory values listed in [14]. The main contribution to the deviation from the free electron value comes from the interaction with the nuclear Coulomb potential (Breit term) stating $g_J = \frac{2}{3} + \frac{4}{3}\sqrt{1 - Z^2\alpha^2}$ [18]

ions. It uses a fit to the theory values given in [14] and highlights ions of particular relevance for recent, current and future measurements. The main part of this deviation is due to the so-called 'Breit term', which is an analytical solution within Dirac theory for the g-factor of the electron bound in the ground state of hydrogen-like systems with nuclear charge Z (point-like motionless nucleus) and reads

$$g_J = \frac{2}{3}\left(1 + 2\sqrt{1 - (Z\alpha)^2}\right) = 2\left[1 - \frac{1}{3}(Z\alpha)^2 - \frac{1}{12}(Z\alpha)^4 - \frac{1}{24}(Z\alpha)^6 \dots\right]. \tag{3.3}$$

Using this, the ratio of g-factors of the bound and the free electron can be written as

$$\frac{g_J}{g_e} = 1 - \frac{(Z\alpha)^2}{3} + \frac{\alpha(Z\alpha)^2}{4\pi} + \dots \tag{3.4}$$

and leads to significant deviations from 1 particularly for increasing values of Z. Of second-largest influence are contributions from bound-state QED as will be discussed in the following book chapter. Nuclear size and recoil effects also play a role within experimental resolution, see for example [14–16]. Particularly, the nuclear recoil contribution Δg_{rec} to the bound electron g-factor is given by

$$\Delta g_{rec} = (Z\alpha)^2\left[\left(\frac{m_e}{m_i} - (1+Z)\frac{m_e^2}{m_i^2}\right) + \frac{\alpha}{\pi}\left(-\frac{1}{3}\frac{m_e}{m_i} + \frac{3-2Z}{6}\frac{m_e^2}{m_i^2}\right)\right] \tag{3.5}$$

and is the leading nuclear-mass dependent term. The currently obtained experimental accuracy in g of the order of ppb is sufficient to distinguish isotopes of the same ion species based on this contribution [17].

Selected theory values of the bound electron magnetic moment as expressed by the g-factor are given in Table 3.1 for ions of immediate experimental relevance. Modern calculations in the framework of bound state quantum electrodynamics have partially reached a relative accuracy level of ppb and better [21–26]. Currently, the

Table 3.1 Selected theory values of bound electron magnetic moments in hydrogen-like ions of nuclear charge number Z

Z	Ion	Magnetic Moment (g_J)	Reference
1	Hydrogen	2.002 283 853	[14]
6	Carbon	2.001 041 590 18(3)	[19]
8	Oxygen	2.000 047 020 32(11)	[19]
14	Silicon	1.995 348 958 0(17)	[20]
20	Calcium	1.988 056 946 6(100)	[19]
54	Xenon	1.894 640 57(19)	[16]
82	Lead	1.738 282 7(11)	[16]
83	Bismuth	1.731 013 38	[14]
92	Uranium	1.659 208 9(27)	[16]

A complete set can be found in [14]

experimental accuracy has reached the permille level of the magnitude of bound-state QED contributions to the bound electromagnetic moment. Hence, measurements of the bound electron magnetic moment particularly in its binding to highly charged ions serve as a stringent benchmark for the vast set of existing theory predictions. The connection of magnetic moments to fundamental symmetries and constants can be used for a determination of these and their possible variation with time and space. Additionally, the closely bound electron serves as an excellent probe of nuclear properties without the need for diamagnetic shielding corrections. The connection between the electron magnetic moment and other quantities will be discussed in Sect. 3.10.

3.3 A Brief Look Back

The scientific history of the electron magnetic moment goes back to 1916 when Sommerfeld and Debye postulated quantized angular momenta in atoms [27, 28]. In 1921, Compton assumed an inner magnetic moment of the electron based on the behaviour of ferromagnetic materials [29]. In the same year, Alfred Landé introduced the concept of vector addition of quantum-mechanical angular momenta and the phenomenological quantity called gyromagnetic factor ('g-factor', 'Landé g-factor') to express the ratio of actual magnetic moment and angular momentum of particles like the electron, all in relation to his work on the anomalous Zeeman effect [30, 31]. One year later, Stern and Gerlach performed their famous experiment with an atomic beam flying perpendicular to a magnetic field gradient resulting in a spatial splitting of the silver atom beam [32, 33]. This spatial quantization was first attributed to the silver atom as a whole. In 1925, Ralph Kronig (assistant to Alfred Landé) and later Uhlenbeck and Goudsmit proposed the electron spin to explain a number of spectroscopic observations made at that time in atomic physics [34, 35], and following Pauli's comment on the doublet structure in alkaline spectra [36] which gave a further hint towards half-integer angular momenta. When in 1927 Fraser found that the ground-state orbital angular momentum and associated magnetic moments of silver, hydrogen, and sodium are zero, this was the missing proof that the observed momentum quantization had indeed been due to the electron spin [31, 37] and new light was shed on the Stern-Gerlach experiment. In the same year, Phipps and Taylor reproduced the Stern-Gerlach experiment with hydrogen atoms in their ground state, which was known to have zero magnetic moment. Thus, any influence of the silver atom was ruled out and spatial quantization due to the electron spin was corroborated [38]. Theoretical justification for such a quantized electron spin was then given by Dirac in 1928. The Dirac equation predicted a value of $g = 2$ for the free electron, as phenomenologically introduced by Landé in 1923.

In 1947, Kusch and Foley measured the ratio of bound electron magnetic moments in the $^2P_{3/2}$ and $^2P_{1/2}$ states in atomic gallium by spectroscopy of the respective Zeeman substates at a magnetic field strength of 0.038 T. This measurement resulted in $g_{J=3/2}/g_{J=1/2}(\mathrm{Ga}) = 2.003\,44\,(12)$ [39, 40]. Based on the assumption that the electronic coupling in these states is correctly described by Russel-Saunders coupling, they derived a value of the free electron magnetic moment significantly different from 2, namely $g_s(e) = 2.00229(8)$, which was soon found to agree with theoretical calculations in the framework of QED by Schwinger [41]. This agreement was a demonstration of the relatively new field of quantum electrodynamics and gave it an enormous impetus[2], together with Bethe's calculation [45] of the newly discovered Lamb shift [46] during the course of the same year. Referring to the publication by Kusch and Foley [39], Schwinger said 'It is indeed gratifying that recently acquired data confirm this prediction' [41]. 'For his precision determination of the magnetic moment of the electron' Polykarp Kusch received the 1955 Nobel Prize in Physics together with Eugene Lamb. Note, that both the 1947 measurements of the electron's anomalous magnetic moment and of the Lamb shift were made possible by the newly developed microwave spectroscopy, whereas the earlier findings as discussed above mainly go back to optical spectroscopy.

In 1972, Walther et al. measured contributions to the magnetic moment of the electron due to the finite nuclear mass by a comparison of the g_J-factors of the neutral hydrogen and deuterium atom using a hydrogen maser. They measured the ratio $g_J(\mathrm{H})/g_J(\mathrm{D})$ to be $1 + (7.22 \pm 0.03) \times 10^{-9}$ and thus found a mass dependence on the ppb level with a relative accuracy of 3×10^{-11} [47].

In 1977, Tiedeman and Robinson compared the g-factor of the free electron to the g_J-factor of the neutral hydrogen atom using a method based on optical pumping and spin exchange collisions [48]. These measurements were the first observations of a mass-independent contribution to the bound electron magnetic moment. The result is $g_J(\mathrm{H})/g_s(e) = 1 - 17.709(13) \times 10^{-6}$ and thus represents a test of relativistic and radiative corrections to the bound state of the electron with a relative accuracy of about 3×10^{-8}.

[2] Similar findings in a comparison of the free muon magnetic moment [42] with the bound muon magnetic moment were made in the early 1960s by Hutchinson et al. [43], and were shortly after described theoretically by Ford et al. [44].

Johnson et al. in 1980 used similar optical pumping methods in measurements of the g_J-factor of hydrogen-like helium $^4He^+$ with a relative accuracy of 6×10^{-7} [49]. The result was $g_J(^4\mathrm{He}^+)/g_s(e) = 1 - 70.87(30) \times 10^{-6}$ and represents the first precision measurement of the g_J-factor of a simple atomic ion, which means that it comprises both charge and mass effects to the bound state of the electron.

An indirect determination of the g_J-factor of the electron bound in hydrogen-like lead $^{207}Pb^{81+}$ and bismuth $^{209}Bi^{82+}$ has been undertaken in 1998 by use of a measurement of the excited hyperfine state lifetime obtained from laser spectroscopy [50]. However, the relative accuracy was limited to about 10^{-3}, see also the discussion in [14] and Table 3.3.

In all these spectroscopy experiments, the measurement is conceptually restricted to systems with a suitable electronic level scheme. Such a restriction is not present when use is made of the so-called 'continuous Stern-Gerlach effect' which has first been applied to single free electrons in a Penning trap by Dehmelt et al. [51–56]. Later, this concept has successfully been transferred to magnetic moment measurements of electrons bound in highly charged ions [57] such as C^{5+} [58], O^{7+} [59, 60], Si^{13+} [20] and Si^{11+} [61], as will be discussed below.

3.4 The Continuous Stern-Gerlach Effect

The 'classical' Stern-Gerlach effect [32] has been used in numerous atomic beam experiments to determine the magnetic moments of electrons bound in atomic systems. Typically, an atomic beam enters an inhomogeneous magnetic field which separates the spin directions, then enters a region with a homogeneous magnetic field where a radio-frequency (rf) field interacts with the spin, and a subsequent inhomogeneous magnetic field analyses the spin direction. The spin flip probability is measured as a function of the radio frequency and yields a resonance curve. From this and from knowledge of the field strength in the homogeneous magnetic field part, the magnetic moment of the bound electron is derived. Such measurements of magnetic moments represent critical tests of atomic physics calculations for complex systems [62–64].

Over a long period of time it has been discussed whether the Stern-Gerlach effect can be applied to neutral atoms only as for charged particles the Lorentz force acting on a moving charged particle in a magnetic field is considered to mask the force due to the magnetic moment. Proposals to separate the spin directions by the acceleration of electrons moving along the field lines of an inhomogeneous magnetic field [65] were rejected both by Bohr [66] and Pauli [67] arguing with the Heisenberg uncertainty principle $\Delta x \Delta p \geq h$. Attempts to separate the spin states in an electron beam by using the different sign of the force on the spin evoked by a longitudinal inhomogeneous magnetic field have been unsuccessful [68]. New proposals have been brought forward to perform Stern-Gerlach experiments on electron beams which would, in contrast to the analysis by Bohr and Pauli, result in a high degree of spin separation under carefully chosen initial conditions [69–71]. A detailed discussion of a moving electron's spin and an account of the historic debate is given in [72].

In 1963, Byrne studied the requirements for measuring the electron magnetic moment in a constant homogeneous magnetic field by inducing transitions between particular states of the electron's motion with a perturbing electric field. He showed the necessity of 'some form of trapping device' that allows for a sufficiently long time period for the transitions to be detected [73].

In 1973, Dehmelt and Eckstrom proposed to use the force of an inhomogeneous magnetic field on the spin of an electron confined in a Penning trap [51]. They found that confining a charged particle by electromagnetic fields provides a way to circumvent Bohr's and Pauli's reasoning since for a particle which oscillates in a parabolic potential well, the presence of a magnetic inhomogeneity leads to a measurable difference of its oscillation frequency for different orientations of the spin [74]. Therefore, a precise measurement of the oscillation frequency yields information on the spin direction. Dehmelt et al. used this effect for the detection of induced changes of the spin direction of an electron by observing the corresponding changes in the electron's oscillation frequency in a Penning trap. Since the confined particle's spin direction is monitored continuously, Dehmelt called this the 'continuous Stern-Gerlach effect'. It has since been successfully applied to a number of magnetic moment studies on both free and bound electrons as will be discussed below. It was mainly this development which won Dehmelt the 1989 Nobel Prize in physics together with Wolfgang Paul 'for the development of the ion trap technique'.

Figure 3.4 illustrates the relation of the continuous Stern-Gerlach effect to its classical counterpart. While in the Stern-Gerlach experiment [32], two different orientations of the magnetic moment (spin) are separated in position space by a

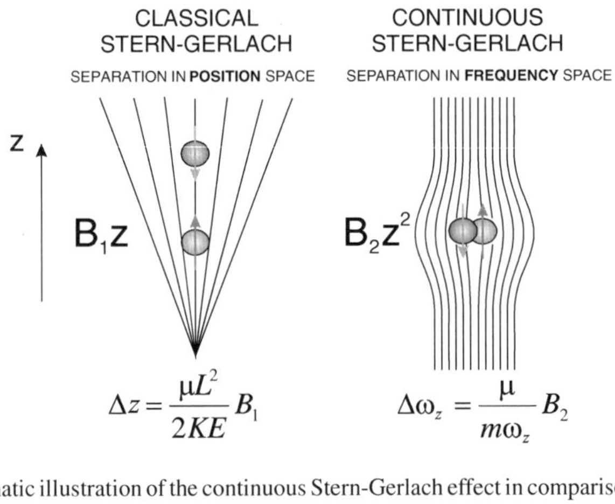

Fig. 3.4 Schematic illustration of the continuous Stern-Gerlach effect in comparison to the classical Stern-Gerlach effect. While the linear gradient B_1 separates spin states in position space along a traversing beam, the magnetic field curvature B_2 (magnetic bottle) separates spin states in frequency space when the ion oscillation in a Penning trap is considered

linear magnetic field gradient B_1, the continuous Stern-Gerlach effect separates the different orientations in frequency space by a magnetic field curvature B_2 (magnetic bottle).

3.5 Measurement Principle and Ion Confinement

The comprehension of the experimental application of the continuous Stern-Gerlach effect and related techniques requires a brief summary of some Penning trap principles. Here, we restrict ourselves to the necessary topics and their relevance for the presented experiments, a broad discussion of ion trap physics can be found for example in monographs by Ghosh [75] and Werth [11, 76], and to some extent also in the textbook by Foot [77].

In a Penning trap, charged particles are confined by a superposition of a homogeneous magnetic field B_0 with a harmonic electrostatic potential. Presently, we are concerned with the trapping of a single ion of mass m_i and charge q in a cylindrical Penning trap with open endcaps as described in detail in [78, 79]. Such a trap has cylindrical symmetry along the field lines (z-axis) of the homogeneous magnetic field which confines the ion in the radial dimensions. Axially, the ion is confined by a constant voltage U_0 applied between the central ring electrode and the endcap electrodes. Additionally, compensation electrodes with voltage U_C are placed on either side of the ring between the ring and the endcaps to create a harmonic potential around the trap centre. A schematic of such a trap is given in Fig. 3.5.

3.5.1 Measurement Principle and Ideal Confinement

Experimentally, the ion is confined by the combination of magnetic and electric fields such that the free ion cyclotron frequency $\omega_c = qB_0/m_i$ is perturbed by the presence of the electric field and cannot be measured directly. In a Penning trap configuration as shown in Fig. 3.5, an ion performs three independent oscillatory motions with well-defined frequencies ω_z ('axial frequency'), ω_+ ('modified cyclotron frequency') and ω_- ('magnetron frequency'). For any configuration, the oscillation frequencies obey the hierarchy $\omega_c > \omega_+ \gg \omega_z \gg \omega_-$, for highly charged ions under the present conditions we can additionally write $\omega_L \gg \omega_c$. These frequencies are given by

$$\omega_z^2 = \frac{qU_0C_2}{m_id^2} \quad \text{and} \quad \omega_\pm = \frac{\omega_c}{2} \pm \left(\frac{\omega_c^2}{4} - \frac{\omega_z^2}{2}\right)^{1/2} \tag{3.6}$$

where q is the particle charge, U_0 is the trapping voltage, m_i is the ion mass, d is the effective trap size $d^2 = (z_0^2 + \rho_0^2/2)/2$ with z_0 half the endcap separation and ρ_0 the inner trap radius, C_2 is a dimensionless coefficient of order 1 as will be discussed below.

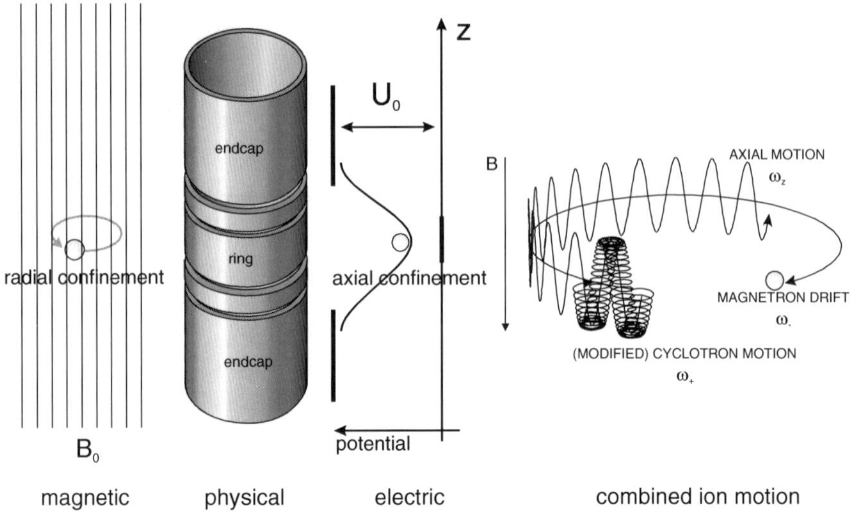

Fig. 3.5 Schematic of a five-pole open-endcap cylindrical Penning trap as used in the present experiments. The confining electric and magnetic fields are indicated as well as an enlarged illustration of the resulting motion of a confined ion

The required free ion cyclotron frequency ω_c can be determined from the above observable oscillation frequencies. In principle, this can done by use of the relation $\omega_c = \omega_+ + \omega_-$, which follows directly from Eq. (3.6). This, however, is only true in the ideal case. In reality, it is favourable to use the so-called 'invariance theorem' $\omega_c^2 = \omega_-^2 + \omega_+^2 + \omega_z^2$ [78], in which shifts of the individual oscillation frequencies caused by a misalignment of the experimental axis with respect to the magnetic field axis cancel, see the discussion in Sect. 3.5.2. The g_J-factor of the bound electron is thus calculated from the measured Larmor and cyclotron frequencies by

$$g_J = 2\frac{\omega_L}{\omega_c}\frac{q}{e}\frac{m_e}{m_i} \quad \text{where we use} \quad \omega_c^2 = \omega_-^2 + \omega_+^2 + \omega_z^2 \tag{3.7}$$

and the charge ratio q/e is assumed integer. The electron-to-ion mass ratio m_e/m_i needs to be measured independently, as will be discussed in Sect. 3.7.2.

3.5.2 Imperfections

In precision measurements of bound electron magnetic moments with ions in Penning trap, a number of imperfections are of relevance as they influence the ion oscillation frequencies and hence the applicability of the continuous Stern-Gerlach effect and the magnetic field measurement by use of the invariance theorem. The most prominent imperfections in experimental setups are non-ideal electric and magnetic fields, misalignments of the trap axis with respect to the axis of the magnetic field and mechanical imperfections of trap electrodes. Such imperfections can *in principle* be avoided by appropriate choice of experimental parameters, in reality however, they are present and need to be taken into account in the analysis of the data. The effects of such deviations are described in detail for example in [76, 78–81].

3.5.2.1 Field Distortions

Since imperfections of the electric and magnetic fields are not only present, but can be artificially created and used for various experimental techniques, they shall briefly be discussed here. We can write the electrostatic potential near the trap centre by the expansion

$$U = \frac{1}{2} U_0 \sum_{k=0}^{\infty} C_k \left(\frac{\rho}{d}\right)^k P_k(\cos\theta) \tag{3.8}$$

where ρ is the radial distance to the trap centre and $P_k(cos\theta)$ are Legendre polynomials of the k-th degree with the argument $\cos(\theta) = z/(\rho^2+z^2)^{1/2}$, z being the axial distance to the trap centre. If the trap has electric symmetry with respect to the trap centre, only even terms have non-zero values. The coefficient C_0 is an overall potential offset and hence irrelevant for the ion motion. C_2 (as in Eq. 3.6) is the quadrupole term relevant for electrostatic confinement and represents the 'efficiency' of the trap in creating a potential well from the applied voltages. A hyperbolic Penning trap has $C_2 = 1$ by design. For cylindrical Penning traps C_2 is typically about 0.5. In cylindrical Penning traps like the present ones, the dominant electric imperfection is characterized by the term C_4. The next term C_6 is suppressed with respect to the term in C_4 by a factor of $(r/d)^2$, which typically is of order 10^{-4} or smaller. Usually, terms above C_6 may be neglected as for cold ions their contribution is far below the experimental resolution. The coefficients including C_2 and C_4 depend on applied voltages and can be written as [79]

$$C_2 = C_2^{(0)} + D_2 \frac{U_C}{U_0} \quad \text{and} \quad C_4 = C_4^{(0)} + D_4 \frac{U_C}{U_0} \tag{3.9}$$

where the $C_k^{(0)}$ and D_k are given by the trap geometry [79] and U_C is the voltage applied to the correction electrodes of the trap. It is hence possible to minimize the effect of certain imperfections by appropriate choice of U_C/U_0, the so-called 'tuning ratio'. We speak of a 'compensated trap' if at least the terms C_4 and C_6 can be tuned to zero simultaneously. The trap is additionally 'orthogonal' if the oscillation frequencies are independent of the applied tuning ratio. The requirements and possibilities for this are discussed in detail in [79]. While C_2 only represents a linear scaling of the trapping potential, a non-vanishing term $C_4 \neq 0$ leads to frequency dependences described by

$$\begin{pmatrix} \Delta\omega_+/\omega_+ \\ \Delta\omega_z/\omega_z \\ \Delta\omega_-/\omega_- \\ \Delta\omega_L/\omega_L \end{pmatrix} = M_E \begin{pmatrix} \Delta E_+ \\ \Delta E_z \\ \Delta E_- \end{pmatrix} \quad \text{with} \quad M_E = \frac{6C_4}{qU_0} \begin{pmatrix} \eta^4/4 & -\eta^2/2 & -\eta^2 \\ -\eta^2/2 & 1/4 & 1 \\ -\eta^2 & 1 & 1 \\ 0 & 0 & 0 \end{pmatrix}, \tag{3.10}$$

where M_E is a 3×4 matrix containing the dependences of all oscillation frequencies on all oscillation energies and $\eta = \omega_z/\omega_+$ (which is typically of order 10^{-2}). The bottom line of the matrix M_E contains only zeros since the Larmor frequency is a purely magnetic property and thus not affected by electric anharmonicities.

The corresponding effect of non-ideal magnetic fields will be discussed in connection with magnetic bottles in Sect. 3.5.3.1.

3.5.2.2 Field Alignment

The most prominent field imperfections apart from non-vanishing higher-order field components such as C_4 and so forth are a tilt of the trap axis $\mathbf{e_z}$ with respect to the magnetic field axis by an angle θ, and a non-vanishing ellipticity ε of the electric field (which breaks its rotational symmetry). The angle θ represents a misalignment of the electric field configuration with respect to the magnetic field and results in a shift of the oscillation frequencies. The same is true for a non-vanishing value of ε which adds a term $-m_i\omega_z^2\varepsilon(x^2 - y^2)/4$ to the confining electrostatic potential $m_i\omega_z^2(z^2 - (x^2 + y^2)/2)/2$. This situation has been carefully discussed by Brown and Gabrielse [82], in particular, one finds the modified oscillation frequencies to be

$$\omega_z' \approx \omega_z \left(1 - \frac{1}{4}(3+\varepsilon)\sin^2\theta\right) \tag{3.11}$$

$$\omega_\pm' \approx \omega_\pm + \frac{1}{2}\omega_-(3+\varepsilon)\sin^2\theta \tag{3.12}$$

where both the tilt θ and the ellipticity ε are assumed small. But most importantly one finds

$$\omega_+'^2 + \omega_z'^2 + \omega_-'^2 = \omega_c^2 \tag{3.13}$$

which means that the invariance theorem is valid also in presence of small tilts and ellipticities [83]. As such imperfections are practically unavoidable and non-negligible, this is of particular importance since the unperturbed cyclotron frequency ω_c directly enters the value of the magnetic moment (see Eq. 3.7) which is to be determined with ppb accuracy. Special properties of an artificially elliptical trap, however, may be used for special applications, see for example [84].

Clearly, Eq. (3.11) can be used to measure trap misalignments and possibly correct for them. Such measurements have been performed for various setups and typically reveal tilt angles of the order of 0.1°.

There are, however, also imperfections which are introduced by the mere presence or motions of ions in the trap and hence are *inherent* to the method. Such effects cannot be avoided irrespective of the efforts and hence may need to be corrected for. Presently, for ions, the most important effects are image charge and space charge shifts of the ion oscillation frequencies.

3.5.2.3 Image Charge and Image Current Effects

The presence of image charges in the conducting electrodes of the trap as induced by the ion charge effectively changes the confining potential at the position of the ion and thus alters its oscillation frequencies [85, 86]. Hence, the true free ion cyclotron frequency in the magnetic field is shifted by the presence of the trap electrodes and needs to be recovered as it directly enters the determination of the magnetic moment when the continuous Stern-Gerlach effect is employed, see Eq. (3.7).

Measurements of the image charge shift of confined particles have been performed by Van Dyck et al. for $^1H^+$, $^2H^+$, $^3He^+$, $^3He^{2+}$, and $^{12}C^{4+}$ ions [80]. The result for that trap was an observed absolute shift of the reduced cyclotron and magnetron frequencies of about 20 mHz per confined charge, and a relative shift of these frequencies by about 0.25 ppb per atomic mass unit of the confined ion species, in good agreement with theoretical expectations [80]. Detailed calculations of the influence of image charges on the oscillation frequencies of confined particles have also been performed by Fischbach [87], Boulware [88] and Porto [86], the latter including more general trap geometries.

To describe the effect of image charges in a cylindrical Penning trap of inner radius ρ_0, it is convenient to separate the transversal from the longitudinal component of the electric field of the image charge at the position of the particle. For ions, the longitudinal image field effect is significant and causes frequency shifts given by [89]

$$\Delta\omega_{\pm} = \mp \frac{q}{4\pi\varepsilon_0\rho_0^3 B_0} \quad \text{and} \quad \Delta\omega_z = 0, \tag{3.14}$$

which via the invariance theorem lead to [89]

$$\frac{\Delta\omega_c}{\omega_c} = \left(\frac{\omega_-}{\omega_c} - \frac{\omega_+}{\omega_c}\right) \frac{m_i}{4\pi\varepsilon_0\rho_0^3 B_0^2} \approx -\frac{m_i}{4\pi\varepsilon_0\rho_0^3 B_0^2}. \tag{3.15}$$

This shift is usually relevant for precision ion frequency measurements due to the scaling with the particle mass. From this, the corresponding effects on the bound electron magnetic moment in $^{12}C^{5+}$, $^{16}O^{7+}$, $^{28}Si^{13+}$ and $^{28}Si^{11+}$ have been estimated to −583(47) ppt [89], −777(63) ppt, −686(34) ppt [20] and −658(33) ppt [61], respectively. In all cases but the latter, this shift is within the experimental accuracy and hence had to be accounted for. The effect may also need to be considered in the design phase of precision traps, see for example [90].

Apart from the image *charge* effect which is due to the presence of conducting electrodes, there is also an image *current* effect which is due to the presence of a resonant circuit connecting these electrodes. In the present experiments, both the radial and axial motions interact with respective circuits. A resonant RLC-circuit as used for resistive cooling or ion detection has an impedance which consists of a real and an imaginary part. In resonance, the imaginary part vanishes and the real part gives rise to an exponential damping of the ion motion as will be discussed in Sect. 3.6.1. While the real part is not reason of any considerable shift of the respective oscillation frequency, the imaginary part is. It gives rise to an effective electrostatic potential which originates from the image current and is phase-shifted with respect to the ion motion. This shift in phase effectively reduces the respective frequency. The relative shift is given by

$$\frac{\Delta\omega_+}{\omega_+} \approx -\frac{1}{\tau_+}\frac{Q_+\delta\omega_+}{\omega_+^2 + 4Q_+^2\delta\omega_+^2} \quad \text{and} \quad \frac{\Delta\omega_z}{\omega_z} \approx -\frac{1}{\tau_z}\frac{Q_z\delta\omega_z}{\omega_z^2 + 4Q_z^2\delta\omega_z^2} \tag{3.16}$$

where $\tau_{+,z}$ is the cooling time constant of the circuit, $Q_{+,z}$ its quality factor (see Eq. 3.46) and $\delta\omega_{+,z}$ is the detuning of the actual oscillation frequency with respect to the resonance frequency of the circuit. The effect on the free cyclotron frequency ω_c may be calculated as the square sum of these shifts. Alternatively, since the reduced cyclotron frequency ω_+ is by far the dominant contribution to the free cyclotron frequency ω_c as calculated by the invariance theorem, Eq. (3.16) may be considered a good approximation of the cyclotron frequency shift, i.e. $\Delta\omega_c/\omega_c \approx \Delta\omega_+/\omega_+$. Depending on the actual detuning $\delta\omega$, for typical parameters of these experiments, this effect may amount up to about 10^{-10} of thc frequency and hence lies within achievable experimental resolutions.

3.5.2.4 Space Charge Effect

For single-ion measurements like in the present Stern-Gerlach experiments [20, 58–61] space charge effects are not relevant. However, in experiments with many ions as will be discussed in Sect. 3.8, they need to be taken into account. When many ions are confined, the corresponding space charge lifts the trapping potential and changes the oscillation frequencies. This also sets a limit to the amount of charge that can be confined in a given trap. Assuming cold ions in a nearly spherical cloud, the space charge potential has the same quadratic behaviour as the external confining potential. This results in a shift of the oscillation frequencies. Yu et al. have found equations for the space charge shifted frequencies [91]

$$\omega_z' = \omega_z\sqrt{1 - \frac{\omega_p^2}{3\omega_z^2}} \quad \text{and} \quad \omega_\pm' = \frac{\omega_c}{2}\left(1 \pm \sqrt{1 - (1 + \frac{2\omega_p^2}{3\omega_z^2})\frac{2\omega_z^2}{\omega_c^2}}\right) \tag{3.17}$$

where ω_p is the plasma frequency given by $\omega_p^2 = q^2n/\varepsilon_0 m_i$, n being the ion number density and ε_0 the permittivity of free space. The square roots in Eq. (3.17) translate into two conditions, which read

$$\frac{\omega_p^2}{\omega_z^2} \leq 3 \quad \text{and} \quad \frac{\omega_p^2}{\omega_z^2} \leq \frac{3}{2}\left(\frac{\omega_c^2}{2\omega_z^2} - 1\right). \tag{3.18}$$

The first condition refers to electrostatic quantities only, but the second one links electrostatic and magnetostatic properties. When these conditions are written in terms of experimental parameters and solved for the ion number density n they yield

$$n \leq \frac{3\varepsilon_0 U_0}{qd^2} \quad \text{and} \quad n \leq \frac{3}{4}\frac{\varepsilon_0 B_0^2}{m_i} - \frac{3}{2}\frac{\varepsilon_0 U_0}{qd^2}. \tag{3.19}$$

For a trapping voltage of

$$U_0 = \frac{1}{6}\frac{qB_0^2 d^2}{m_i} \tag{3.20}$$

the right hand sides of both Eq. (3.19) become equal to the Brillouin limit

$$n^{(\mathrm{max})} = \frac{\varepsilon_0 B_0^2}{2m_i}, \tag{3.21}$$

which is the maximum achievable ion density, typically of order 10^9 charges per cm^3. Dropping the condition of a spherical cloud, the Brillouin limit can be reached for any value of the trapping potential obeying the confinement condition $2\omega_z^2 \leq \omega_c^2$. For higher trapping potentials, the terms on the right hand sides of Eq. (3.19) become identical and confinement is lost.

3.5.2.5 Other Inherent Effects

There are other inherent effects with influence on the present kind of measurements, which are not due to the presence of charge, but due to the necessary or desired motion of the confined particle. Since they are often points of discussion we give them brief consideration. These effects are of fundamental importance, however they are (yet) insignificant for ions in experiments like presently discussed and at the present experimental resolutions.

Radiation Damping

Radiation damping, i.e. energy loss due to radiation of the accelerated charge according to $E(t) = E_0 \cdot \exp(-\gamma^{(R)} t)$, is important for trapped electrons but becomes negligible for ions since the damping constant $\gamma^{(R)}$ of the perturbed cyclotron motion (for which the effect is most pronounced) and the axial motion is given by [78]

$$\gamma_+^{(R)} = \frac{4q^2\omega_+^2}{3m_i c^3} \approx \frac{4q^4 B_0^2}{3m_i^3 c^3} \propto \frac{q^4}{m_i^3} \quad \text{and} \quad \gamma_z^{(R)} = \frac{2q^2\omega_z^2}{3m_i c^3} = \frac{2q^3 U_0^2}{3m_i^3 c^3} \propto \frac{q^3}{m_i^3}, \tag{3.22}$$

respectively, and is hence even for highly charged ions suppressed by a factor of about 10^9 with respect to electrons where typical damping times are of the order of milliseconds [78].

Power Shift of the Larmor Frequency

As the Larmor frequency $\hbar\omega_L = g_J \mu_B B_0$ is the precession frequency of the electron spin in the external magnetic field, any contribution which adds to the confining field B_0 will shift it. In principle, this is true also for the magnetic field component of the microwave radiation used to find the true Larmor frequency as the maximum of the spin transition probability as a function of the microwave frequency, as will be discussed in detail in Sect. 3.6.2. The Larmor resonance is shifted by the magnetic field component B_{MW} of the microwave signal ('power shift') according to [92]

$$\frac{\Delta\omega_L}{\omega_L} = \left(\frac{B_{MW}}{2B_0}\right)^2 \quad \text{with} \quad B_{MW}^2 = \frac{1}{c^2}\sqrt{\frac{\mu_0}{\varepsilon_0}\frac{P_{MW}}{A_{MW}}}. \tag{3.23}$$

Typical irradiated areas A_{MW} are of order mm^2 in the present experiments. The microwave power in these experiments [20, 58–60] is limited to values of the order of μW and below to avoid undesired broadening and distortion of the lineshape of the Larmor resonance, as will be discussed in detail in Sect. 3.6.2. For these values, the resulting relative shift of the Larmor frequency is of order 10^{-15} and hence outside of the present experimental accuracy.

Sokolov-Ternov Effect

The Sokolov-Ternov effect describes the self-polarization of charged particles moving in a magnetic field. The self-polarization occurs through the emission of spin-flip synchrotron radiation. It was predicted in 1963 by Igor Ternov and then justified by Arsenij Sokolov [93] using exact solutions of the Dirac equation. Potentially, it obstructs measurements of magnetic moments which use the continuous Stern-Gerlach effect, because the spin orientation of the confined particle in a Penning trap is affected during the measurement time by the magnetic field used for radial confinement. For an ensemble of trapped ions one would observe an increasing orientation ξ of the spins antiparallel to the magnetic field with an exponential time behaviour given by

$$\xi(t) = A\left(1 - e^{-t/\tau_{ST}}\right) \tag{3.24}$$

where the saturation $A = 8\sqrt{3}/15 \approx 0.924$ is slightly below 1 and the orientation time constant τ_{ST} is given by

$$\tau_{ST} = A\frac{\hbar^2}{m_i c q^2}\left(\frac{m_i c^2}{E_k}\right)^2\left(\frac{B_S}{B_0}\right)^3. \tag{3.25}$$

Here, $B_S = m_e^2 c^2/e\hbar \approx 4.41 \times 10^9$ T is the Schwinger field [94], B_0 is the magnetic field strength in the trap and E_k is the kinetic particle energy. While this effect of radiative polarization provides a unique capability for creating polarized beams of high-energy particles in storage ring experiments, in typical trap experiments it can be ignored as for any particle and any realistic set of trapping parameters the time constant τ_O exceeds the lifetime of the experiment(er). A thorough discussion of this effect and its close relation to the Unruh effect is given in [95].

Relativistic Frequency Shifts

The relativistic shifts of the frequencies in the external degrees of freedom (ion oscillation frequencies) can be understood in a straight-forward way in terms of the relativistic mass effect due to the motional energy of the ion. These shifts can be written in analogy to the magnetic and electric field effects in Eqs. (3.10) and (3.30) as

$$\begin{pmatrix} \Delta\omega_+/\omega_+ \\ \Delta\omega_z/\omega_z \\ \Delta\omega_-/\omega_- \\ \Delta\omega_L/\omega_L \end{pmatrix} = M_R \begin{pmatrix} \Delta E_+ \\ \Delta E_z \\ \Delta E_- \end{pmatrix} \quad \text{with} \quad M_R = -\frac{1}{m_i c^2}\begin{pmatrix} 1 & 1/2 & -\eta^2 \\ 1/2 & 3/8 & -\eta^2/4 \\ -\eta^2 & -\eta^2/4 & -\eta^4/4 \\ 2/9 & 1/2 & -\eta^2 \end{pmatrix}. \tag{3.26}$$

At typical kinetic ion energies in the present experiments of below 1 meV, all of the relative oscillation frequency shifts are smaller than 10^{-13}.

For the frequency in the internal degree of freedom (i.e. the spin precession frequency), this is not so easy. Following the discussions in [96–98], for the relativistic effect on the Larmor precession frequency, two cases need to be distinguished, depending on the spin orientation relative to the magnetic field. Writing $\beta = v/c$ with the ion velocity v and $\gamma = (1 - \beta^2)^{-1/2}$, the free ion cyclotron frequency is given by $\omega_c = qB/(\gamma m_i)$. Denoting the bound electron's magnetic moment anomaly by $a = g/2 - 1$ one finds the Larmor frequency to be $\omega_L = (1 + a)\omega_c^e$ for an orientation parallel to the magnetic field, and $\omega_L = (1 + \gamma a)\omega_c^e$ for perpendicular orientation, where ω_c^e is the true free electron cyclotron frequency. For ions at motional temperatures of about 4 K and for an anomaly of about 10^{-3} the factor γ is of order 10^{-15}, and hence still far outside of experimental resolutions. Even if ions need to be excited to motional energies of several eV in order to be properly detected, the magnitude of this effect is below 10^{-12} [89].

Note, however, that the above said is true only for the relativistic effect of typical *kinetic* energies of confined ions. When ions change their internal (electronic excitation) energy, e.g. by absorption or emission of photons, the relativistic mass effect may be significant as will be discussed in Sect. 3.9.

Spontaneous Decay of the Spin State

As the measurement of the magnetic moment as described here relies on manipulation and detection of the bound electron's spin state in the external magnetic field, a spontaneous radiative decay of a higher-energy to a lower-energy spin state during the measurement would obstruct the applicability of the continuous Stern-Gerlach effect and of Zeeman spectroscopy. A decay from a state $|M_J + 1>$ to $|M_J>$ is energetically possible via a magnetic dipole (M1) transition at a rate given by

$$\gamma_s = \frac{4}{3}\frac{\omega^3 \mu_B^2}{\mu_0 \hbar c^3} X; \quad X = \frac{1}{2}(J - M_J)(J + M_J + 1)\left[\frac{3J(J+1) - L(L+1) + \frac{3}{4}}{2J(J+1)}\right]^2 \tag{3.27}$$

where ω is the transition frequency, in the present experiments about $2\pi \times 105$ GHz, and for the s-state $X = 2$. Note, that Eq. (3.27) is only a non-relativistic approximation excluding QED effects. According to this equation, the natural lifetime of the electron's upper spin state in the measurements with $^{12}C^{5+}$, $^{16}O^{7+}$ and $^{28}Si^{13+}$ [20, 58, 59] is of the order of years and can safely be ignored.

The same formula applies to the bound electron in excited Zeeman substates of the fine structure as employed in measurements by use of laser-microwave double-resonance spectroscopy of ions such as Ar^{13+} as will be detailed out in Sect. 3.8, see Fig. 3.23. Here, $X = 2/9$ for $p_{1/2}$-states, $X = 32/9$ for $p_{3/2}$ states ($M_J = +1/2$ to $M_J = -1/2$) and $X = 24/9$ for $p_{3/2}$ states ($M_J = +3/2$ to $M_J = +1/2$). For these transitions at 65 and 130 GHz, the corresponding lifetimes of the upper Zeeman substates are also of the order of years and thus also outside of experimental time windows.

3.5.3 Magnetic Bottle

3.5.3.1 General Effects

The determination of the bound electron magnetic moment follows the idea of the 'continuous Stern-Gerlach effect' as used for the free electron. A single ion is stored in a Penning trap with an artificial inhomogeneity of the magnetic field ('magnetic bottle'). Let the symmetry axis $\hat{e}_z$ of the trap be parallel to the magnetic field B_0 and the radial coordinate be ρ. The value of the magnetic field strength near the trap centre can then be written as

$$B(z,\rho) = B_0 - 2B_1 z + B_2\left(z^2 - \frac{1}{2}\rho^2\right) + \dots \qquad (3.28)$$

where B_0 is the homogeneous part of the field, B_1 describes a linear gradient along $\hat{e}_z$ and the curvature B_2 characterizes a so-called 'magnetic bottle' and its dependence on the axial and radial coordinates. Higher-order terms are not of relevance for the present discussion. Hence, a magnetic bottle is magnetic field inhomogeneity of the kind

$$\mathbf{B}'(z,\rho) = B_2\left((z^2 - \frac{\rho^2}{2})\mathbf{e_z} - \rho z \mathbf{e}_\rho\right) \qquad (3.29)$$

superimposed on the magnetic trapping field $\mathbf{B} = B_0\mathbf{e_z}$ with radial symmetry around the central trap axis.

The presence of $B_2 \neq 0$ results in a dependence of the oscillation frequencies on the motional energies. Hence, the frequencies shift with the energies (amplitudes) of the motions. Making use of the hierarchy $\omega_- \ll \omega_z \ll \omega_+$ of the oscillation frequencies [79], this dependence can be expressed in a classical formulation by the matrix equation [78]

$$\begin{pmatrix} \Delta\omega_+/\omega_+ \\ \Delta\omega_z/\omega_z \\ \Delta\omega_-/\omega_- \\ \Delta\omega_L/\omega_L \end{pmatrix} = M_B \begin{pmatrix} \Delta E_+ \\ \Delta E_z \\ \Delta E_- \end{pmatrix} \quad \text{with} \quad M_B = \frac{1}{m_i\omega_z^2}\frac{B_2}{B_0}\begin{pmatrix} -\eta^2 & 1 & 2 \\ 1 & 0 & -1 \\ 2 & -1 & -2 \\ -\eta^2 & 1 & 2 \end{pmatrix}, \qquad (3.30)$$

where M_B again is a 3×4 matrix containing the dependences of all oscillation frequencies on all oscillation energies and $\eta = \omega_z/\omega_+$. It expresses that all frequencies depend linearly on all energies except for the axial frequency which does not depend on the axial energy since the corresponding matrix element is zero (due to the purely electrostatic nature of the motion). Here, ω_L is the Larmor frequency of the particle which is not a classical oscillation but can be described within the same formalism.

3.5.3.2 Magnetic Bottle and the Invariance Theorem

The invariance theorem $\omega_c^2 = \omega_+^2 + \omega_z^2 + \omega_-^2$ is not exactly valid in the presence of a magnetic bottle. The individual oscillation frequencies are shifted by finite motional energies E_+, E_z, E_- or by M_J-transitions in a way which does not to all orders fulfil the invariance theorem. Expressing the effect of non-zero motional energies by the perturbed cyclotron frequency shift $\Delta\omega_+$ as obtained from Eq. (3.30), we have $\Delta\omega_- = -\Delta\omega_+$ and $\Delta\omega_z = -\Delta\omega_+(\omega_+ - \omega_-)/\omega_z$. Inserting this into the invariance theorem yields a second-order cyclotron frequency shift given by

$$\frac{\Delta\omega_c}{\omega_c}(\Delta\omega_+) \approx \left(1+\frac{1}{2}\left(\frac{\omega_+-\omega_-}{\omega_z}\right)^2\right)\left(\frac{\Delta\omega_+}{\omega_+}\right)^2, \tag{3.31}$$

of which the leading contribution expressed in terms of experimental quantities is

$$\frac{\Delta\omega_c}{\omega_c} \approx \frac{\omega_+^2}{2\omega_z^2}\frac{1}{m_i^2\omega_z^4}\frac{B_2^2}{B_0^2}\left(E_z^2+2E_zE_-+4E_-^2\right). \tag{3.32}$$

In the experiments discussed here, this shift is commonly of order 10^{-13} and hence still outside the spectroscopic resolution. With the advent of detection techniques which do not require significant excitation of the ion motion (see Sect. 3.6.3), this effect will be rendered irrelevant as the energies on the right hand side will be negligible.

3.5.3.3 Spin State Transitions in a Magnetic Bottle

Of particular interest for the application of the continuous Stern-Gerlach effect is, that in the presence of a magnetic bottle the intrinsic spin orientation is linked to external degrees of freedom of the ion, namely to the ion oscillation frequencies. Using the orientation energy $E_J = g_J\mu_B B_0 M_J$, we find the dependence of the axial and modified cyclotron frequencies on the M_J orientation by [81]

$$\Delta\omega_z(\Delta M_J) \approx \frac{\hbar}{2m_e}\frac{\omega_+}{\omega_z}\frac{B_2}{B_0}\frac{e}{q}g_J\Delta M_J \tag{3.33}$$

$$\Delta\omega_+(\Delta M_J) \approx -\frac{\hbar}{2m_e}\frac{\omega_+}{\omega_+-\omega_-}\frac{B_2}{B_0}\frac{e}{q}g_J\Delta M_J \approx \frac{\omega_z}{\omega_+}\Delta\omega_z(\Delta M_J) \tag{3.34}$$

$$\Delta\omega_-(\Delta M_J) \approx \frac{\hbar}{2m_e}\frac{\omega_+}{\omega_+-\omega_-}\frac{B_2}{B_0}\frac{e}{q}g_J\Delta M_J \approx -\Delta\omega_+(\Delta M_J) \tag{3.35}$$

where m_e is the mass of the electron, g_J is the g-factor corresponding to the magnetic moment of interest and M_J is the angular momentum quantum number. Figure 3.6 schematically shows the energy levels of the single ion in the total confining potential (dashed lines). The full lines indicate the situation when a magnetic bottle evokes an energy difference between the two M_J states ('spin' states of the bound electron) according to Eqs. (3.33)–(3.35). Typical scales for ω_+, ω_z, ω_-, $\Delta\omega_z(\Delta M_J = 1)$ and $\Delta\omega_\pm(\Delta M_J = 1)$ are 10 MHz, 1 MHz, 10 kHz, 100 mHz and 10 mHz, respectively.

In all experiments discussed here, the axial motion has been used as the effect is highest (radial frequency shifts are smaller by a factor of about $\omega_z/\omega_+ \approx 1/25$). These equations can similarly be used for free particles like electrons, protons (or antiprotons) by inserting the corresponding masses, charges and magnetic moments. For simplicity, a change of $\Delta M_J = \pm 1$ is often called a 'spin flip' of the bound electron.

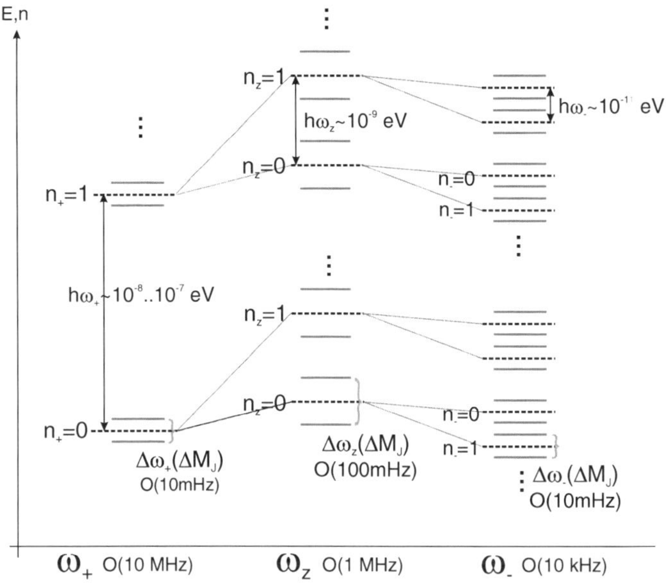

Fig. 3.6 Scheme of the energy levels of a single ion in the total confining potential of the trap (*dashed lines*). Full lines indicate the situation when a magnetic bottle evokes an energy difference between the two M_J states ('spin' states of the bound electron) according to Eqs. (3.33)–(3.35). Not true to scale

In typical experiments, also the relative shifts $\Delta\omega_+/\omega_+$ and $\Delta\omega_z/\omega_z$ of the oscillation frequencies ω_+ and ω_z due to corresponding finite motional energies E_+ and E_z are relevant, either as an unwanted residual effect or as a working principle for measurement techniques. According to Eq. (3.30), they are given by

$$\begin{pmatrix} \Delta\omega_+/\omega_+ \\ \Delta\omega_z/\omega_z \end{pmatrix} = \frac{1}{m_i\omega_z^2}\frac{B_2}{B_0}\begin{pmatrix} -(\omega_z/\omega_+)^2 & 1 \\ 1 & 0 \end{pmatrix}\begin{pmatrix} \Delta E_+ \\ \Delta E_z \end{pmatrix}. \tag{3.36}$$

Throughout this discussion we use classical equations with continuous ion energies, all of which have counterparts using quantized energies, see for example [76, 78, 81]. Those are appropriate for ions of very low motional energies close to the quantum mechanical ground state. In the experiments discussed here, however, the quantum numbers for the reduced cyclotron motion are typically of order 10^3, for the other motions again two to three orders of magnitude larger. Table 3.2 gives typical parameters of the ion motion for the example of $^{16}O^{7+}$ in a magnetic field of $B_0 = 3.796830$ T. From the measured frequencies, the invariance theorem yields an unperturbed cyclotron frequency of $\omega_c/2\pi = 25{,}306{,}484$ Hz, and at a measured frequency ratio ω_L/ω_c of 4164.3761834 (upon corrections [59]), the true Larmor frequency of the bound electron in the external magnetic field is $\omega_L/2\pi = 105{,}385{,}720{,}421$ Hz.

Table 3.2 Ion motion parameters for the measurements with $^{16}O^{7+}$ in a magnetic field of $B_0 = 3.796830\,T$ as performed in [59]

Frequency	Energy	Amplitude	Quantum number
$\omega_+/2\pi = 25\,289\,562$ Hz	$E_+ = 0.36$ meV	$A_+ = 0.41$ µm	$n_+ = 3460$
$\omega_z/2\pi = 925\,156$ Hz	$E_z = 5.29$ meV	$A_z = 43.3$ µm	$n_z = 1385100$
$\omega_-/2\pi = 16\,942$ Hz	$E_- = -0.097$ meV	$A_- = 8.29$ µm	$n_- = 1385100$ (coupled)
$\omega_c/2\pi = 25\,306\,484$ Hz			
$\omega_L/2\pi = 105\,385\,720\,421$ Hz			

The quantum numbers result from $E_k = (n_k + 1/2)\hbar\omega_k$

3.5.3.4 Experimental Implementation of a Magnetic Bottle

The usual way to implement such a magnetic bottle in a Penning trap experiment like presently discussed is to use a ferromagnetic central trap electrode in the shape of an annular disc with inner radius r_1, outer radius r_2 and a thickness of $2a$ in an external homogeneous magnetic field B_0, as depicted in Fig. 3.7. Such an arrangement distorts the homogeneity of the external magnetic field used for confinement to form a magnetic bottle of the kind given in Eq. (3.29). The coefficient B_2 has a unit of Tesla per square metre. It can be calculated by considering the scalar magnetic potential Φ with $\Delta\Phi = -4\pi\rho_M = 4\pi\nabla\cdot\mathbf{M}$, where $\mathbf{M}$ is the magnetisation which is assumed to be homogeneous over the ring volume and has the value M_0. Inserting the ring geometry and comparing coefficients with the general multipole expansion of the magnetic field [78], the strength B_2 of the magnetic bottle at its centre position $(z = 0, \rho = 0)$ is given by

$$B_2 = 3\mu_0 M_0 \left(\frac{a r_1^2}{2(a^2 + r_1^2)^{5/2}} - \frac{a r_2^2}{2(a^2 + r_2^2)^{5/2}} \right). \tag{3.37}$$

The linear gradient B_1 is zero at the centre position. The homogeneous part B_0 of the field at the centre position is modified by an amount

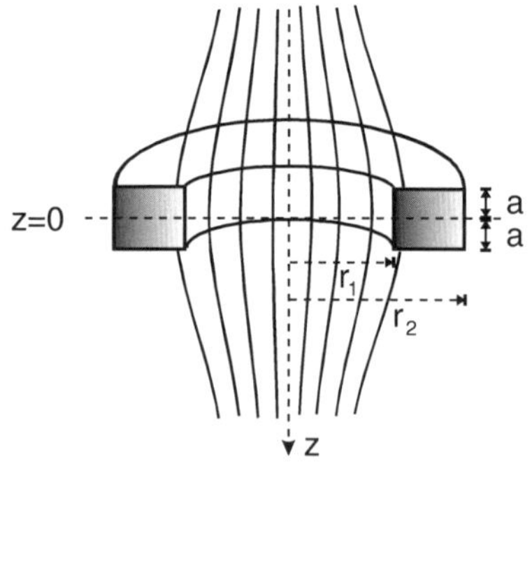

Fig. 3.7 Schematic view of a magnetic bottle formed by a ferromagnetic ring electrode of a Penning trap with inner radius r_1, outer radius r_2 and a thickness of $2a$ inside an otherwise homogeneous magnetic field along the z-direction. Only the field distortion inside the ring is indicated

$$\Delta B_0 = -\frac{\mu_0 M_0}{2}\left(\frac{2a}{a^2+r_1^2} - \frac{2a}{a^2+r_2^2}\right). \tag{3.38}$$

Typical materials for ring electrodes to form a magnetic bottle come from the Nickel, Cobalt, Iron, CoFe, SmCo, NdFeB and AlNiCo families. The saturation field strength $\mu_0 M_0$ of specific high-permeability iron alloys such as Hisat50 reaches up to 2.44 T for a specific CoFe alloy with vanadium. With a CoFe alloy, values of B_2 up to around 400 mT/mm^2 have been achieved by use of optimized ring geometries [99–101].

For the ferromagnetic annular disc ring electrode, and at a given magnetization M_0, the maximum of the magnetic bottle strength B_2 is reached when the ratio a/r_1 is chosen such that the Legendre polynomial $P_4(\cos\beta) = 35/8\cos^4\beta - 30/8\cos^2\beta + 3/8$ for $\beta = \tan^{-1}(a/r_1)$ vanishes, see also the discussion in [78]. This leads to the condition $a/r_1 \approx 0.577$ (and $r_2 \gg r_1$). Higher values of B_2 can be reached for geometries which more closely follow the conditions imposed by the behaviour of P_4, for details see the discussion in [78].

It should not go unnoted that also non-ferromagnets such as the often-used copper and MACOR® have finite magnetizations in the external magnetic field. At the typical fields strengths of 6 T and liquid helium temperature, copper shows an M_0 of $-50\,\mu$T and MACOR® of $780\,\mu$T [78]. In principle, when calculating the total magnetic bottle strength, these need to be taken into account as well, however, at the present machining precisions and experimental resolutions of magnetic bottle strength measurements, these contributions may well be neglected. The same is true for the temperature-dependence of the ring's saturation magnetization, which e.g. for nickel is $\Delta M_0/M_0 \approx 2.4\times10^{-5}$/K, and for typical temperature fluctuations of much below one Kelvin leads to relative changes of the magnetic field at the position of the oscillation frequency measurement of much below 10^{-10}.

3.5.3.5 Measurement of the Magnetic Bottle Strength

An experimental determination of the magnetic bottle strength B_2 can be performed by a position-dependent measurement of the magnetic field strength, for example a measurement of the cyclotron frequency as a function of the axial ion position. To this end, the trap can be made electrically asymmetric by applying an additional small voltage U_a across the endcaps. This adds a small uniform axial electric field $E_z = c_1 U_a/2z_0$ close to the trap centre and shifts the centre position of the axial oscillation by an amount Δz given by [102]

$$\Delta z = \frac{1}{2}\frac{d^2}{z_0}\frac{c_1}{C_2}\frac{U_a}{U_0}, \tag{3.39}$$

where c_1 is a dimensionless geometry coefficient of order unity [78, 79, 102]. In the present configuration, it is identical to the coefficient κ as used in Eq. (3.46). Note, that a non-zero value of c_1 shifts the axial frequency ω_z by an amount

$$\frac{\Delta\omega_z}{\omega_z} = -\frac{3}{4}\frac{d^4}{z_0^4}c_1c_3\left(\frac{U_a}{U_0}\right)^2, \tag{3.40}$$

where again c_3 is a dimensionless geometry coefficient which for realistic trap geometries obeys $c_1 + c_3 \approx 1$ [78, 102]. This shift needs to be considered when the magnetic field strength is deduced from the measured oscillation frequencies. Also, the range of possible centre shifts Δz is limited by the potential asymmetry U_a giving rise to additional anharmonicities. The leading contribution is given as a change of the C_4-coefficient (see Sect. 3.5.2) where we have [79]

$$C_4 \rightarrow C_4 - \frac{5}{4}c_3^2\left(\frac{U_a}{U_0}\right)^2. \tag{3.41}$$

The general recipe for the calculation of such potential influences is given in [78]. A measurement of the magnetic bottle strength B_2 for the trap in [58, 59, 89] has been performed [89] and yielded results for the actual B_2 in fair agreement with the prediction by Eq. (3.37).

3.6 Experimental Setups and Techniques

The measurements of bound electron magnetic moments by use of the continuous Stern-Gerlach effect performed so far have a common experimental outline. The respective details have been given in a number of publications [58–60, 89, 103, 104]. Here a brief overview of the general idea and experimental realization is given.

The ions of interest are produced in a miniature electron beam ion source ('mini-EBIS' [105, 106]) which is part of the trap setup as shown in Fig. 3.8. To that end, a cold electron source (field emission point) produces an electron beam that sputters material from a target which is subsequently ionized by electron impact ionization in the same beam [105]. The beam energy is chosen for optimized production of the charge state of interest. The electron impact ionization cross section typically has a maximum at about two to three times the ionization potential of the desired state [107, 108]. Generally, the ionization potential for the hydrogen-like state can be approximated by the equation

$$I \approx m_e c^2 \left[1 - \left(1 - \hat{q}^2\alpha^2\right)^{1/2}\right], \tag{3.42}$$

where $\hat{q} = q/e = Z - 1$ is the charge state of the desired ion and $\alpha \approx 1/137$ is the fine structure constant. To reach $^{12}C^{5+}$, $^{16}O^{7+}$ and $^{28}Si^{13+}$, respective minimum impact energies of 392, 739 and 2438 eV are necessary.

The highly charged ions are then transferred into the precision trap, selected and singlified by resonant ejection of unwanted ions [89]. The remaining single ion of

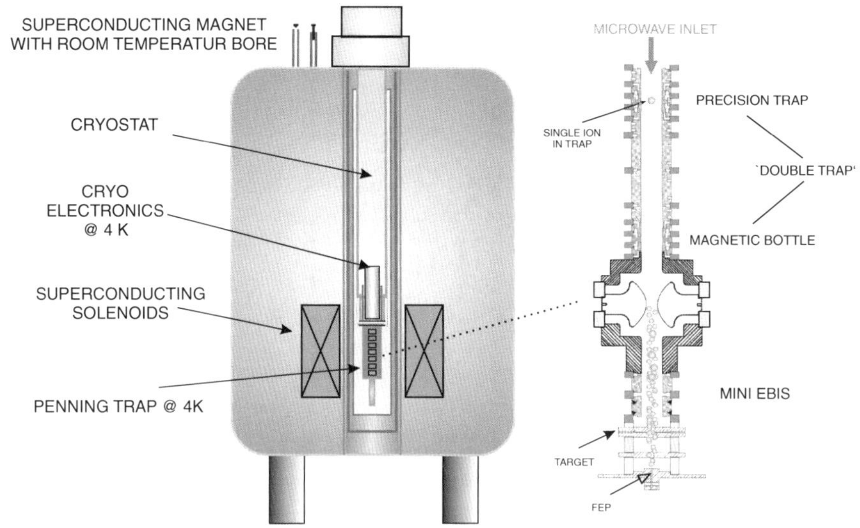

Fig. 3.8 Schematic view of a setup for bound electron magnetic moment measurements by use of the continuous Stern-Gerlach effect. It features a miniature cold ion source and a double Penning trap arrangement attached to a cryostat inside the bore of a superconducting magnet

interest is cooled resistively. Microwaves of a tunable frequency ω_{MW} around the expected Larmor frequency of the electron spin are irradiated. At the same time, the ion oscillation frequencies in the trap are measured by electronic detection of image currents induced in resonant RLC circuits attached to respective trap electrodes. Figure 3.9 schematically shows the double-trap arrangement which is the central part of the complete trap arrangement shown in Fig. 3.8 together with the magnetic and electric fields applied. The free cyclotron frequency ω_c at the position and time of the microwave measurement is calculated from the measured ion oscillation frequencies ω_+, ω_z and ω_- by use of the invariance theorem (Eq. 3.7). This way, the result is not subject to fluctuations and drifts of the confining fields on time scales longer than one individual cycle. The ion is then adiabatically transported to the magnetic bottle for an analysis of the spin direction. The ion is transported back and another microwave frequency is irradiated. This cycle is repeated to produce a resonance curve of the spin flip probability as a function of the frequency ratio ω_{MW}/ω_c (see Fig. 3.16). Upon systematic corrections mainly regarding the line shape (see for example [89]), the centre of this resonance curve is the desired frequency ratio ω_L/ω_c as needed in Eq. (3.7) to calculate the bound electron g-factor g_J.

To obtain good statistics of the resonance curve, a large number of repetitions of the individual cycle of several hundreds of times is necessary. This requires the ability to store the ion for a sufficiently long time, at least on the time scale of hours. Hence, the residual gas pressure in the trap needs to be low, which for closed systems like the present ones is assured by the cryogenic surrounding which acts as a cryo-pump. Particularly for highly charged ions, the storage time of a certain ion is limited by

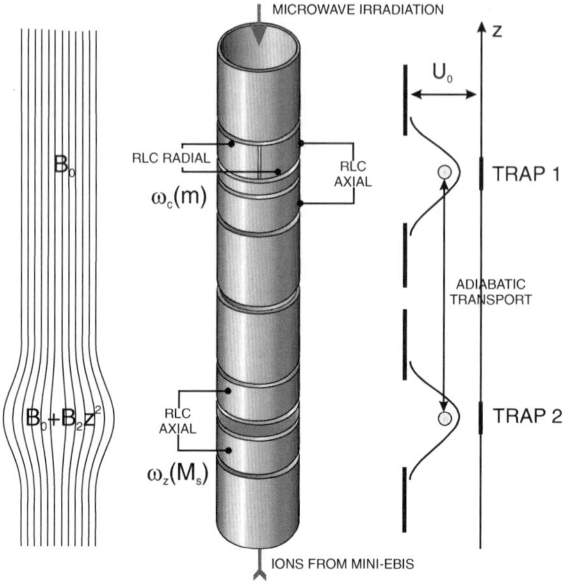

Fig. 3.9 Schematic of the double trap arrangement and the applied magnetic and electric fields as well as the pick-up for ion detection and oscillation frequency measurement by resonant RLC-circuits. It consists of two nearly identical 5-pole Penning traps, the lower of which features a ferromagnetic ring electrode (magnetic bottle) for spin state determination while the upper is used to induce spin flips by microwave irradiation and is also used for a measurement of the free cyclotron frequency via the invariance theorem

charge exchange with residual gas, which at ambient temperatures of 4 K is mainly helium and hydrogen. The exponential time constant t_c of charge state loss due to a charge-exchange collision can be calculated by

$$t_c = \frac{1}{\sigma p}\sqrt{\frac{k_B \mu_m}{3}} \quad \text{with} \quad \mu_m = \frac{m_R T_R m_i T_i}{m_R T_R + m_i T_i} \tag{3.43}$$

where m_R is the mass of the residual gas atom or molecule, m_i is the ion mass, p is the residual gas pressure, T_R and T_i are the corresponding temperatures and k_B is the Boltzmann constant. In the experiments discussed here, the temperatures are typically assumed equal.

The simplest model describing electron capture processes in low-energy ion-neutral collisions is the 'classical over-the-barrier model' which assumes the cross section σ to be given by [109]

$$\sigma = \frac{1}{2}\pi R_C^2 \quad \text{with} \quad R_C = \frac{27.2 a_0 (2\sqrt{\bar{q}} + 1)}{I[\text{eV}]}, \tag{3.44}$$

where $a_0 = 5.2917720859(36) \times 10^{-11}$ m is the Bohr radius (atomic length unit), I is the ionisation potential of the neutral given in eV and $\hat{q} = q/e$ is the ion charge state. For helium, $I = 24.587387$ eV, for atomic hydrogen $I = 13.598433$ eV, for molecular hydrogen H_2, $I = 15.4257$ eV.

The electron capture cross section σ may for low and medium charge states be taken from the classical over-the-barrier model (see for example Mann [109]), for highly charged ions it can be calculated according to a semi-empirical formula by Müller and Salzborn [110]

$$\sigma[\mathrm{cm}^2] \approx 1.43 \times 10^{-12}\, \hat{q}^{1.17}\, I[\mathrm{eV}]^{-2.76} + 1.08 \times 10^{-12}\, \hat{q}^{0.71}\, I[\mathrm{eV}]^{-2.80}, \quad (3.45)$$

where the first term describes single electron capture, and the second one double-electron capture. Similar terms exist for higher-order electron capture processes, but contribute to a lesser extent [110]. Equation (3.45) is a fit to data taken in the keV regime of collision energies, however, for high charge states such as Th^{80+} it agrees well with measurements at collision energies of few eV [111]. Detailed studies of low-energy cross sections can be found in [109, 111, 112].

From the observed storage time of a single C^{5+} ion of longer than 273 days, and from an assumed charge exchange cross section of $\sigma = 1.35 \cdot 10^{-14}\,\mathrm{cm}^2$ with residual helium, an upper limit for the residual gas pressure of about $9 \cdot 10^{-17}$ hPa has been derived for the experiments [58–60] performed at the University of Mainz [89].

3.6.1 Ion Cooling and Oscillation Frequency Measurement

In an ideal Penning trap, i.e. for vanishing inhomogeneities and anharmonicities, there is no energy transfer between the motional degrees of freedom. Hence, their oscillation energies (amplitudes) are independent and different temperatures can be assigned to the different motional degrees of a single particle [113]. This also means that motional degrees can be cooled or excited individually. In the absence of imperfections, the motions are harmonic and thus the oscillation frequencies are independent of the motional energies.

In the present experiments, the ion is cooled by resistive cooling [114] with an external resonance circuit in the thermal bath of liquid helium. The ion oscillation causes an image current proportional to q^2/m_i across the trap electrodes connected by an RLC circuit which extenuates the current and hence the ion oscillation. Such cooling leads to an exponential energy loss of the single ion oscillation according to $E(t) = E_0 \cdot \exp(-t/\tau)$ where E_0 is the initial ion energy upon creation or capture and τ is the so-called 'cooling time constant'. Regarding the axial and radial motions, it is given by

$$\tau_z = m_i \frac{(2z_0)^2}{R_z q^2 \kappa_z^2} \quad \text{and} \quad \tau_\pm = m_i \frac{(2\rho_0)^2}{R_\pm q^2 \kappa_\pm^2} \frac{\omega_+ - \omega_-}{\omega_\pm}, \tag{3.46}$$

respectively, where R is the resonance resistance of the RLC-circuit used for cooling. κ is a dimensionless coefficient reflecting the geometry of the cooling electrode arrangement and its calculation has been discussed in detail in [78]. Regarding the axial motion, typical values are 0.75 to 0.8. It approaches unity for $\rho_0^2/2z_0^2 > 1$ [102].

While τ_z and τ_+ characterize an exponential decrease of the respective energy and hence of the motional amplitude, for the magnetron motion (which is an unstable drift in the $\mathbf{E} \times \mathbf{B}$ field) such an energy dissipation with τ_- means an *increase* in amplitude and is thus not of use when ions are to be centred in the trap and their storage time hence prolonged. The magnetron motion can instead be 'cooled' (meaning a decrease of amplitude) by a coupling to another motional degree of freedom. This has first been demonstrated by Wineland and Dehmelt, when a quadrupolar excitation with either the frequency $\omega_z + \omega_-$ or $\omega_+ + \omega_-$ was used to centre the magnetron motion of electrons in a Penning trap [115, 116].

In resonance, the impedance of an RLC-circuit acts as an ohmic resistor with resistance $R = QL\omega$, where Q is the quality factor [117] of the circuit and L is its inductance. The bandwidth is given by $\Delta\omega = \omega/Q$ and for typical values of the axial motion ($\omega_z \approx$ 1MHz, $Q \approx 1000$) the bandwidth is $\approx$ 1 kHz and cooling time constants are of order 100 ms. Resistive cooling of ion clouds is much more complicated due to the ion-ion interaction and the few existing measurements [89, 118] are not yet fully explained.

The actual final ion temperature is governed by the noise in the corresponding electronics. At a given temperature T, the so-called 'Johnson noise' induces a root mean square (RMS) voltage across the resistor R given by $U_T^2 = 4k_B TR\Delta\omega$. Additional noise sources may increase this value. Measurements by Djekic et al. have shown that while the radial motion of a single $^{12}C^{5+}$ ion was cooled to a temperature close to liquid helium temperature, the axial oscillation had a temperature of about 60 K which was identical to the measured noise temperature in that circuit [113]. The single ion in electromagnetic interaction with an external heat bath such as the cooling circuit is an ergodic system and hence consecutive measurements of the single ion energy lead to a Boltzmann distribution of energies to which a temperature can be assigned in the same way as to an ion ensemble. Since in the absence of significant field imperfections, energy transfer between oscillations is negligible, each motion keeps its individual temperature on time scales much larger than the measurement times.

In recent experiments, also negative feedback cooling of the ion has been employed similarly to feedback cooling of a single confined electron as demonstrated by D'Urso et al. [119]. Within this technique, the ion can be electronically cooled to temperatures well below the ambient temperature, values below 1 K have been achieved.

The measurement of ion oscillation frequencies relies on the same principles as the resistive cooling, which is the use of ion image charges induced in trap electrodes.

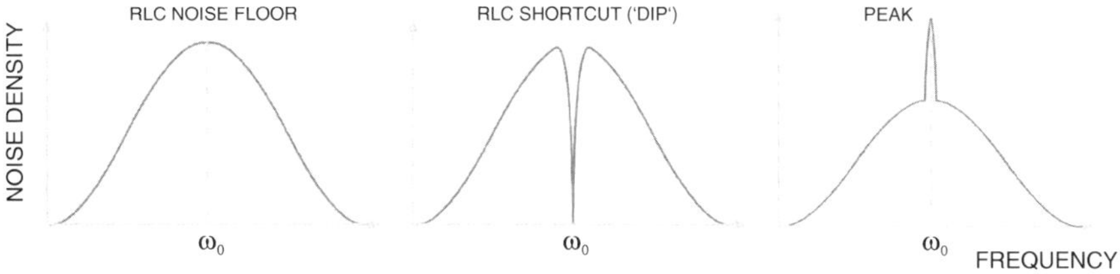

Fig. 3.10 Electronic noise density as a function of frequency of a free RLC-circuit with resonance frequency ω_0 (*left*), the 'dip' (shortcut) created by a resonant ion which is motionally colder than the electronic noise temperature (*middle*) and the 'peak' created by a resonant ion of higher motional temperature (*right*)

When appropriate trap electrodes are connected by a resonance RLC-circuit as above, the ion oscillation leads to an image current through the circuit which leads to a time-dependent voltage signal. A Fourier-transform of that signal yields the motional frequency spectrum. Figure 3.10 schematically illustrates the resulting electronic noise density as a function of frequency of a free RLC-circuit with resonance frequency ω_0 (left), the 'dip' (shortcut) created by a resonant ion which is motionally colder than the electronic noise temperature (middle) and the 'peak' created by a resonant ion of higher motional energy (right). The frequency resolution of these spectra are given by the Fourier limit, i.e. the inverse data acquisition time. The dip signal width is given by the coupling constant γ of the respective circuit, i.e. the inverse cooling time constant $\gamma = \tau^{-1}$. With a single ion, peak signal widths of order 10 mHz have been achieved. The signal voltages V across axial and radial RLC-circuits for a single confined ion are given by

$$V_z = q\frac{A_z Q_z \kappa_z}{z_0 C_z} \quad \text{and} \quad V_\pm = q\frac{A_\pm Q_\pm \kappa_\pm}{\rho_0 C_\pm}, \tag{3.47}$$

respectively, where A are the motional amplitudes and κ are the geometry factors[3] as discussed for Eq. (3.46). Typical voltages are in the nV regime and hence need to be amplified in several stages. In case of an ion cloud, the right hand sides of these equations have to be multiplied by a factor N for coherent ion motion and $\sqrt{N}$ for incoherent ion motion. Anyway, the signal increases linearly with the charge state of the ion. The same is true for the achievable signal-to-noise ratio of such a resonant pick-up given by

$$\chi_z = q\frac{A_z\kappa_z}{z_0}\sqrt{\frac{\omega_z\pi}{4\Delta\omega_z}}\sqrt{\frac{Q_z}{k_B T_z C_z}} \quad \text{and} \quad \chi_\pm = q\frac{A_\pm\kappa_\pm}{\rho_0}\sqrt{\frac{\omega_\pm\pi}{4\Delta\omega_\pm}}\sqrt{\frac{Q_\pm}{k_B T_\pm C_\pm}}, \tag{3.48}$$

[3] Please note that sometimes in literature $2z_0/\kappa$ is called D which is sometimes confused with the trap size d as defined in Sect. 3.5.1.

where $\Delta\omega$ is the frequency bandwidth of the respective resonant circuit at the oscillation frequency ω, Q is the quality factor of the circuit and C is its capacitance. The equation shows that the signal-to-noise ratio increases linearly with the charge state of the ion such that resonant detection is easiest for highly charged ions. A thorough discussion of resistive cooling and other ion cooling methods in Penning traps can be found in [76, 114, 120].

3.6.2 Larmor Frequency Measurement

The Larmor frequency $\hbar\omega_L = g_J \mu_B B_0$ is the precession frequency of the electron spin in the external magnetic field and hence does not produce a detectable electronic signal by which it could be measured directly (like in the case of the oscillation frequencies $\omega_\pm$ and ω_z). It is therefore determined indirectly from the probability resonance of spin state transitions ('spin flips') when microwave radiation is used to find the Larmor frequency as the microwave frequency for which the probability of induced spin flips is highest. To this end, microwaves of a known and fixed frequency ω_{MW} are irradiated for a defined time span t of order seconds. The magnetic field component of the microwave radiation drives the precession of the electron spin and Rabi oscillations of the spin state occur at a frequency

$$\Omega = \omega_L \frac{B_{MW}}{B_0} \tag{3.49}$$

where B_{MW} is the magnetic field component of the microwave radiation as given in Eq. (3.23). The probability for a spin flip as a function of the irradiated frequency ω_{MW} is given by [78]

$$P(\omega_{MW}) = \frac{\pi}{2}\Omega^2 t \chi(\omega_{MW}) \tag{3.50}$$

where χ is the lineshape of the resonance which is asymmetric and broadened in a non-trivial fashion by finite motional temperatures of the ion, by residual field inhomogeneities and by the power of the microwave radiation as discussed in detail for example in [78, 89]. Experimentally, one can only distinguish between 'spin up' and 'spin down' after a given irradiation time $t > \Omega^{-1}$ by a method to be described in the following section. This means that it is only possible to distinguish between an odd and an even number of spin flips which have occurred during the irradiation time t. An even number will be seen as 'no spin flip' while an odd number will be counted as 'spin flip'. Hence, it is necessary to take this saturation effect into account by use of the relation between the true spin flip probability P and the observed probability $\hat{P}$ as given by [78]

$$\hat{P} = \frac{1}{2}\left(1 - e^{-2Pt}\right). \tag{3.51}$$

Combining this saturation correction with Eq. (3.50), we find the experimentally observed probability for a spin flip to be given by

$$\hat{P}(\omega_{MW}) = \frac{1}{2}\left(1 - \exp\left(-\pi \Omega^2 t \chi(\omega_{MW})\right)\right). \tag{3.52}$$

This limits the maximum observable spin flip probability to 0.5, under optimum conditions resulting from a trade-off between microwave power, irradiation time and lineshape broadening, the typical maximum is around 0.4, as will be seen in the measured resonances, see Fig. 3.16.

3.6.3 Spin State Determination

The measurement of the Larmor resonance requires a determination of the spin state of the bound electron after each microwave irradiation to yield information about the success in driving a spin flip. This is achieved in a separate trap featuring a magnetic bottle. The magnetic bottle term B_2 evokes a difference of the oscillation frequencies between the ion with 'spin up' and 'spin down' according to Eqs. (3.33) and (3.34). Hence, it allows a determination of the spin quantum number without the need to manipulate it and thus represents a quantum non-demolition measurement. This is true both for the axial and radial motions. In the experiments [20, 58, 59] the axial motion has been used. The effect can be illustrated by writing the axial component of the total magnetic field in the magnetic bottle $B_z = B_0 + B_2(z^2 - r^2/2)$ such that the magnetic potential energy is given by $\mu_z B_0 + \mu_z B_2(z^2 - r^2/2)$. The latter term adds the magnetic contribution to the electric trap potential to form the total potential $U = U_0 + U_m = (qU_0 + \mu_z B_2)(z^2 - r^2/2)$. Inserting this total potential into the equation for the axial oscillation frequency yields

$$\Delta\omega_z = \omega_z(U) - \omega_z(U_0) \approx \frac{\mu_z B_2}{m_i \omega_z(U_0)}, \tag{3.53}$$

which for hydrogen-like ions can be written as

$$\Delta\omega_z \approx \frac{\hbar}{2m_e} g \frac{1}{Z-1} \frac{\omega_c}{\omega_z} \frac{B_2}{B_0}. \tag{3.54}$$

Hence, the frequency difference is largest for the electron bound in a light ion confined in a shallow trap with a strong magnetic bottle. This becomes particularly obvious when the electron is compared to other particles. The free particle magnetic moment in units of the Bohr magneton is 0.00484197049(12) for the muon, 0.00104187563(25) for the neutron, 0.001521032209(12) for the proton, 0.0004669754556(39) for the deuteron, and 0.001622393657(21) for the triton [121]. Hence, muonic ions as well as free protons and antiprotons under similar

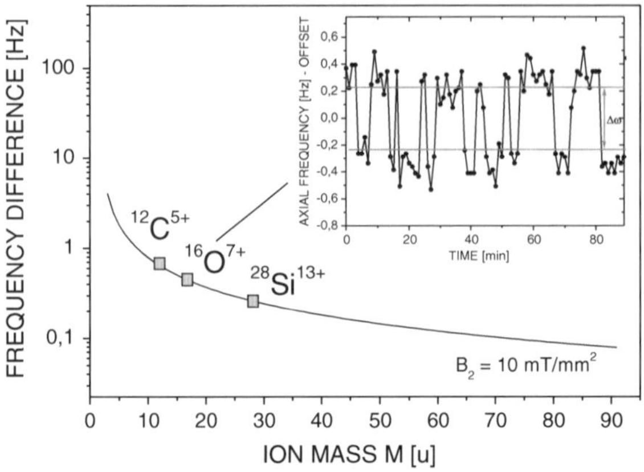

Fig. 3.11 Difference of the axial ion oscillation frequencies between electron spin up and down as a function of the ion mass. Indicated are the values for $^{12}C^{5+}$, $^{16}O^{7+}$ and $^{28}Si^{13+}$. *Inset*: measured axial frequency of $^{16}O^{7+}$ as a function of time, indicating the spin flips as sudden changes of the oscillation frequency between the two levels indicated by the *horizontal lines* separated by 460 mHz

conditions cause a frequency difference due to the change of the spin direction which is about three orders of magnitude smaller and hence pose serious experimental challenges to the application of the continuous Stern-Gerlach effect [122]. For the magnetic bottle with $B_2 = 10\,\mathrm{mT/mm}^2$ as used in [20, 58, 59] with $^{12}C^{5+}$, $^{16}O^{7+}$ and $^{28}Si^{13+}$ ions, the resulting frequency difference $\Delta\omega_z/2\pi$ was about 700, 460 and 240 mHz, respectively, which is a relative difference of order 10^{-7} to 10^{-6}. Its unambiguous detection is an experimental challenge. This is even more true for heavier ions, as Eq. (3.53) shows that $\Delta\omega_z \propto \sqrt{1/(qm_i)}$ (Fig. 3.11).

As both the magnetic bottle strength and the minimum trapping potential have close boundaries in realistic experimental setups, sensitive detection schemes have been conceived. They circumvent the classical Fourier limit of frequency analysis by relating only to the frequency difference. These are phase-sensitive detection schemes in which the frequency difference between an ion with electron spin up and one with spin down is seen as a phase difference of the two cases after a given time of free phase evolution, as depicted in Fig. 3.12. The details of these methods ('PnP':Pulse and Phase, and 'PnA':Pulse and Amplify) have been given in [123, 124] and [104], respectively. They are the non-destructive counterparts of the phase-imaging technique as described in [125]. The current relative frequency resolution is of order 10^{-10} with measurement times of the order of seconds. The result of such a repeated PnP measurement is depicted in Fig. 3.13, where the phase information evaluated after a given free evolution time shows a clear difference, indicating that the two frequencies are different and thus allowing to distinguish between the two spin states.

The unambiguous distinction between spin states by use of the continuous Stern-Gerlach effect is potentially obstructed by fluctuations of the electric potential U_0,

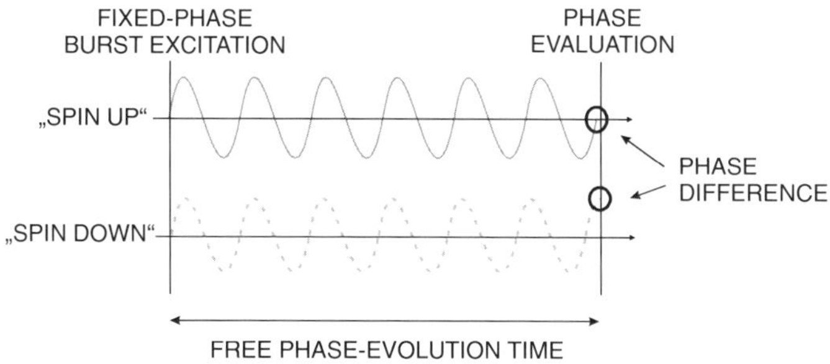

Fig. 3.12 Schematic principle of the phase-sensitive detection of frequency differences by the different phases after a defined time of free phase evolution following a fixed-phase excitation

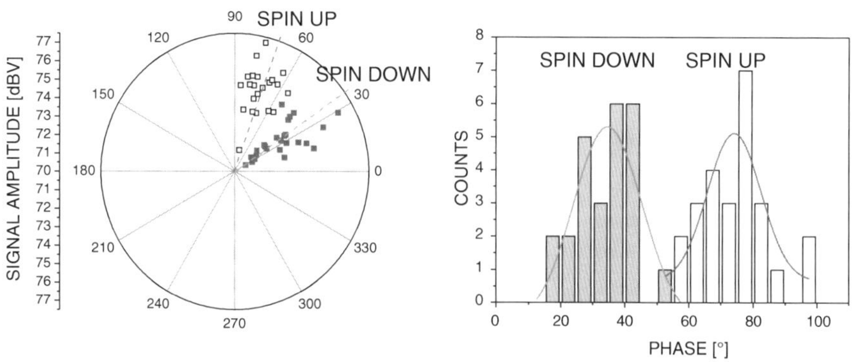

Fig. 3.13 Repeated measurement and Fourier analysis of the phase information only leads to two distinguishable sets of data which represent the two possible spin states and hence allows their distinction. Data from [123]

since they may cause shifts of the axial oscillation frequency according to Eq. (3.6) which are of the same magnitude or larger than the axial frequency difference due to the spin state as given by Eq. (3.54). The stability of available voltage sources hence limits the detectable frequency difference due to a spin flip. Using Eq. (3.54), the voltage change ΔU_0 which affects the axial frequency exactly like a spin flip can be written as

$$\Delta U_0 = \frac{\hbar}{m_e} g \frac{1}{Z-1} d^2 B_2. \tag{3.55}$$

In realistic experiments, ΔU_0 must be significantly smaller than this limit. For the present values of $\Delta\omega$ of few hundreds of mHz, the necessary voltage stability is of order μV and below. Hence, such experiments feature special voltage sources [126].

3.6.4 Double-Trap Technique

For means of an easy analysis of the electron spin direction, it is advantageous to choose the strength B_2 of the magnetic bottle as high as possible. This however limits the overall accuracy which can be reached by use of the continuous Stern-Gerlach effect [127], since from Eq. (3.30) it follows that the cyclotron frequency depends on the motional energies (and hence motional amplitudes A_z,$A_\pm$) in the presence of a magnetic bottle with strength B_2 according to

$$\frac{\Delta\omega_c}{\omega_c} = B_2\left(A_z^2 - \frac{A_+^2}{4}\left(1 - \frac{\omega_+ + \omega_-}{\omega_+ - \omega_-}\right) - \frac{A_-^2}{4}\left(1 + \frac{\omega_+ + \omega_-}{\omega_+ - \omega_-}\right)\right), \quad (3.56)$$

in which the relations between motional energy and motional amplitude are given by

$$A_z^2 = \frac{E_z}{qC_2U_0}, \quad A_+^2 = \frac{2E_+}{m_i(\omega_+^2 - \omega_z^2/2)} \text{ and } A_-^2 = \frac{2E_-}{m_i(\omega_-^2 - \omega_z^2/2)}. \quad (3.57)$$

Therefore, thermal and any other fluctuations of these energies (amplitudes) broaden the Larmor-to-cyclotron resonance and hence increase the uncertainty of the resulting g-factor. Proposals to solve this problem by use of an adjustable magnetic bottle have been brought forward and initial tests were successful [128], however this idea was not used in such experiments. A measurement of the magnetic moment of the electron bound in hydrogen-like carbon $^{12}C^{5+}$ by Hermanspahn et al. [129] was limited due to the continuous presence of the magnetic bottle with $B_2 = 10\,\text{mT/mm}^2$ and resulted in $g_J = 2.001042(2)$ which corresponds to a relative accuracy of 1×10^{-6}.

In all successor experiments, the location of the precision oscillation frequency measurement has been separated from the location of the spin analysis ('double trap technique', see Fig. 3.9) which helped to shift the relative accuracy to the region of 10^{-9} and better [20, 58–60, 130]. The spatial separation of electron spin analysis and precision oscillation frequency measurement has first been used by Häffner et al. in 2000 and resulted in an improvement of the bound electron magnetic moment in $^{12}C^{5+}$ by about three orders of magnitude [58], see also Fig. 3.14. For the application of the double-trap technique, the Penning trap basically features two trapping regions separated by transport electrodes (see Fig. 3.9). One trapping region is the centre of a magnetic bottle, while the other is designed for high precision of the oscillation frequency measurement. Ideally, in this trap the magnetic inhomogeneity is zero. However, the remaining field distortion of the magnetic bottle at the position of the precision trap does not vanish. When the axial separation of the two trap centres is given by Δz, the residual axial magnetic field component of the bottle at the position of the precision trap reduces the effective axial magnetic field there. The magnitude of this contribution is given by

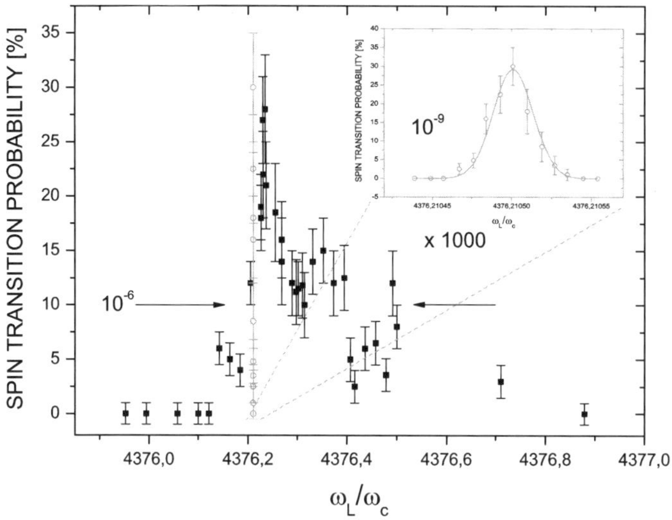

Fig. 3.14 Comparison of the Larmor-to-cyclotron resonance for $^{12}C^{5+}$ measured without (*solid squares*) and with the double trap technique (*open circles*), yielding a resonance sharper by about three orders of magnitude. The inset shows the latter in its normal scaling. Data taken from [129] and [58]

$$B_z(\Delta z) = \frac{\mu_0 M_0}{2}\left(\frac{z_+}{\sqrt{r_2^2+z_+^2}} - \frac{z_-}{\sqrt{r_2^2+z_-^2}} - \frac{z_+}{\sqrt{r_1^2+z_+^2}} + \frac{z_-}{\sqrt{r_1^2+z_-^2}}\right), \quad (3.58)$$

where $z_\pm = \Delta z \pm a$. The higher-order components of this residual distortion are given by

$$B_{(n)}(\Delta z) = \frac{\partial^n B_z(\Delta z)}{\partial z^n} \quad (3.59)$$

and cause a linear gradient $B_1 \neq 0$ and a residual magnetic bottle $B_2 \neq 0$ at the position of the precision trap. The first derivative reads

$$B_1(\Delta z) = \frac{\partial B}{\partial z}(\Delta z) = -\frac{\mu_0 M_0}{2} G_1(\Delta z), \quad (3.60)$$

where the geometry factor $G_1(\Delta z)$ is given by

$$G_1(\Delta z) = \frac{r_1^2}{\left(r_1^2+z_+^2\right)^{3/2}} - \frac{r_1^2}{\left(r_1^2+z_-^2\right)^{3/2}} + \frac{r_2^2}{\left(r_2^2+z_-^2\right)^{3/2}} - \frac{r_2^2}{\left(r_2^2+z_+^2\right)^{3/2}}, \quad (3.61)$$

where we use $z_\pm = a \pm \Delta z$. As discussed above, for $z = 0$ the linear gradient B_1 vanishes as the geometry factor $G_1(z = 0)$ vanishes. The residual magnetic bottle strength B_2 at the position Δz along the central axis is given by the second derivative of B with respect to z

$$B_2(\Delta z) = \frac{\partial^2 B}{\partial z^2}(\Delta z) = -\frac{\mu_0 M_0}{2} G_2(\Delta z) \tag{3.62}$$

where $G_2(\Delta z)$ is given by

$$G_2(\Delta z) = -\frac{3r_1^2 z_+}{\left(r_1^2+z_+^2\right)^{5/2}} - \frac{3r_1^2 z_-}{\left(r_1^2+z_-^2\right)^{5/2}} + \frac{3r_2^2 z_-}{\left(r_2^2+z_-^2\right)^{5/2}} + \frac{3r_2^2 z_+}{\left(r_2^2+z_+^2\right)^{5/2}}. \tag{3.63}$$

For typical trap separations Δz of few cm, the residual magnetic bottle strength there is suppressed by three to four orders of magnitude. As an example, in [58] and [59] the magnetic bottle with $B_2 = 10\,\text{mT/mm}^2$ created a residual magnetic bottle at the position of the precision trap with $B_2 = 4\,\mu\text{T/mm}^2$, a linear gradient B_1 of $66\,\mu\text{T/mm}$ and a fourth-order term of $B_4 = 11\,\text{nT/mm}^4$.

The effect of a non-zero term B_2 has been discussed in Sect. 3.5.3.1 and residual magnetic bottles at the position of the cyclotron frequency measurement typically are the dominating source of systematic uncertainties over residual contributions B_1 and B_4. The effect of a linear magnetic field gradient B_1 on the cyclotron frequency is given by

$$\frac{\Delta\omega_c}{\omega_c} \approx -\frac{1}{m_i \omega_z^2}\left(\frac{B_1}{B_0}\right)^2 E_c. \tag{3.64}$$

For the present parameters, this effect is much smaller than the residual magnetic bottle effect due to the B_1^2-dependence. The corresponding effect of a non-zero B_4 term on an ion with motional radius ρ_c is given by

$$\frac{\Delta\omega_c}{\omega_c} \approx \frac{3}{8}\left(\frac{B_4}{B_0}\right)\rho_c^4 = \frac{3}{2m_i^2\omega_c^4}\left(\frac{B_4}{B_0}\right)E_c^2. \tag{3.65}$$

and can be neglected in the present experiments due to the small value of B_4.

During the measurement cycle, the ion is transported adiabatically between the two regions, such that the spin state is preserved. A detailed account of the experimental details is given in [89]. This method has also been used in the measurements by Verdú et al. [59], Sturm et al. [20] and Wagner et al. [61].

3.6.5 Mode Coupling Techniques

When the bound electron magnetic moment is determined by application of the continuous Stern-Gerlach effect, the frequency ratio ω_L/ω_c of Larmor to cyclotron frequency is the key experimental quantity. While the Larmor frequency is determined by microwave spectroscopy, the free cyclotron frequency is reconstructed from measurements of all three oscillation frequencies of the ion in the trap. As obvious from Eq. (3.48), the detection signal depends linearly on the amplitude of the corresponding motion and can hence be increased to a detectable limit by excitation of the ion motion. When it is necessary to significantly excite the motions in order to be able to detect the ion oscillation by Fourier transform of the induced voltage in the trap electrodes, residual inhomogeneities of the magnetic field at the position of the ion where ω_L and ω_c are measured cause a dependence of the observed frequencies on the energy of the stored ion. This is typically relevant for the perturbed cyclotron motion. Therefore, measurements need to be performed at several cyclotron excitation energies of several eV to several tens of eV and the frequency ratio ω_L/ω_c is then extrapolated to zero cyclotron energy. This has a number of disadvantages:

- the line shape of the extrapolation is *a priori* unknown, giving rise to systematic uncertainties
- the measurement has to be performed at several cyclotron energies and hence needs a comparatively long total measuring time
- for large amplitudes in the magnetic field inhomogeneity, the resonance of the spin flip probability as a function of the frequency ratio is asymmetrically broadened and shifted

When two of the independent oscillations are coupled by irradiation of a suited electric rf field, the corresponding motions are amplitude-modulated by the interaction with the external field [131]. The equations of motion of those coupled oscillations can formally be described in analogy to a driven quantum-mechanical two-level system [132]. If the external rf frequency is identical to a sideband frequency (e.g. the difference or sum frequency) of the two oscillations, the coupling is resonant. Typically, the coupling frequency is detuned by a frequency δ relative to the sideband frequency, and the frequency components $\omega_{r,l} = \omega + \varepsilon_{r,l}$ of an oscillation with uncoupled frequency ω appear in the spectrum. The split frequencies are given by

$$\omega_{r,l} = \omega + \varepsilon_{r,l} = \omega - \frac{\delta}{2} \pm \sqrt{\delta^2 + A^2} \tag{3.66}$$

where A measures the strength of the coupling field amplitude in units of a frequency. A specific measurement of ω_l and ω_r leads to a determination of δ and thus to ω [60]. Figure 3.15 (right) shows the frequency difference $\omega_r - \omega_l$ in the axial motion as a function of the coupling frequency detuning δ. For zero detuning, i.e. when the coupling frequency is identical to the sideband frequency, the frequency splitting is minimal. In case of symmetric splitting the widths of the two resonances at ω_l and ω_r are identical. By use of a time-dependent or modulated coupling rf frequency, the resulting spectrum may contain not only the two split components of a certain frequency, but also the frequency itself, see Fig. 3.15 (left). The split components can again be coupled to the third motion such that five signals appear in the spectrum. From these, all three frequencies ω_+, ω_z and ω_- can be determined simultaneously, as has been demonstrated for a single confined proton by Ulmer et al. [133]. Sideband coupling of motions circumvents the above disadvantages and has been used

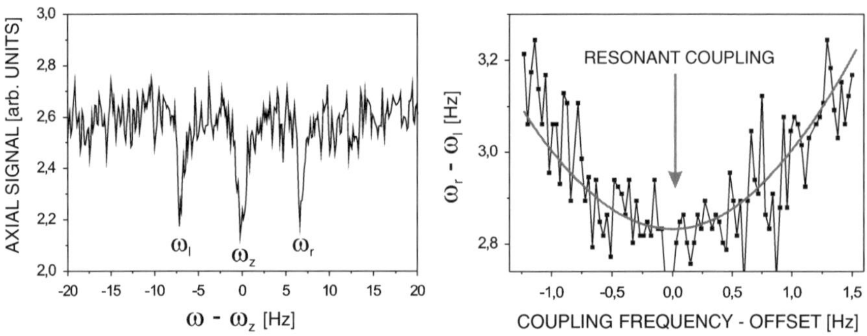

Fig. 3.15 *Left*: Axial signal as a function of frequency for resonant coupling of the reduced cyclotron and axial motions. The original frequency ω_z is visible as well since the coupling was modulated. *Right*: frequency difference $\omega_r - \omega_l$ in the axial motion as a function of the coupling frequency detuning δ. Data taken from [60]

to determine the reduced cyclotron frequency ω_+ by a coupling of ω_+ to ω_z in a measurement of the electron magnetic moment in $^{16}O^{7+}$ by Verdú et al. [60]. The result was $g_J = 2.0000470208(24)(44)$ where the first uncertainty is statistical and the second uncertainty is due to the uncertainty of the electron mass at that time [60].

Another feature of sideband coupling can be used for cooling of a lower-frequency degree of freedom below the ambient temperature when it is coupled to a cooled higher-frequency degree of freedom. Typically, in the case of a coupling between the perturbed cyclotron motion and the axial motion, this is achieved by irradiation of the sum frequency $\omega_+ + \omega_z$. Such coupling leads to an equalization of the quantum numbers of the two oscillations [114], see also Table 3.2 where the magnetron motion was coupled to the axial motion. Let the higher-frequency motion at ω_+ be cooled e.g. by a liquid helium heat bath at $T_+ \approx 4\,\text{K}$. Then the sideband coupling cooling limit for the axial temperature T_z is given by

$$T_z = \frac{\omega_z}{\omega_+} T_+ \approx \frac{1}{B_0 d}\sqrt{\frac{m_i U_0}{q}} T_+. \tag{3.67}$$

For heavy few-electron ions in typical traps, the frequency ratio ω_z/ω_+ is of order 10^{-2} to 10^{-1}, such that for the axial temperature values of the order of 100 mK are achieved. Due to the high charge state of such ions, not only is the final temperature low, but also the cooling time constant is small even for weak coupling. Sideband coupling leads to an exponential energy loss in full analogy to the resistive cooling discussed in Sect. 3.6.1, however with a time constant τ_c which in case of weak resonant coupling is given by

$$\tau_c^{-1} = 4\pi\varepsilon_0 \frac{3m_i a^2 c^3 \omega_z^3}{16q^2\omega_c^2(\omega_+ - \omega_-)} \approx 4\pi\varepsilon_0 \frac{3a^2c^3}{16B_0^3}\frac{U_0^{3/2}}{d^3}\sqrt{\frac{m_i^5}{q^7}}, \tag{3.68}$$

where $a = \eta U_d/U_0$ is a dimensionless constant of order 10^{-4} given by a trap geometry factor η and the amplitude U_d of the coupling drive [114]. For heavy few-electron ions confined at typical trap parameters, the cooling time constant τ_c is below seconds. Thus, if the coupling is sufficiently effective, the cooling of the axial motion at ω_z is limited by the resistive cooling time constant τ_+ of the radial motion (as given by Eq. 3.46) rather than by the time constant τ_c itself.

3.7 Results

3.7.1 Larmor Resonances

Figure 3.16 shows the obtained resonances of the Larmor-to-cyclotron frequency ratio ω_L/ω_c in the bound electron magnetic moment measurements with $^{12}C^{5+}$, $^{16}O^{7+}$ (with and without mode coupling as described in Sect. 3.6.5) and $^{28}Si^{13+}$.

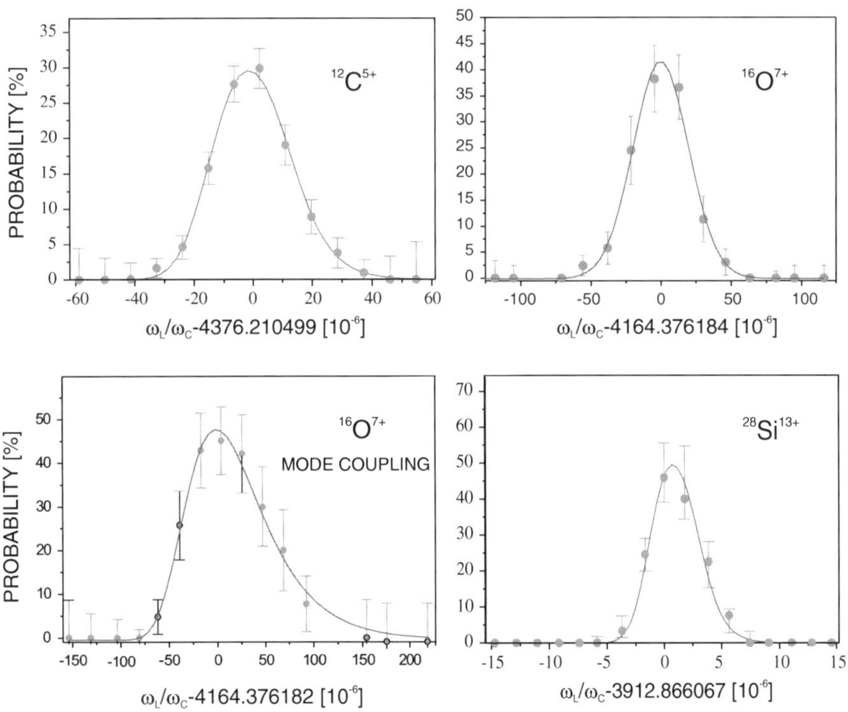

Fig. 3.16 Measured resonances of the Larmor-to-cyclotron frequency ratio ω_L/ω_c for the electron bound in $^{12}C^{5+}$, $^{16}O^{7+}$ (with and without mode coupling) and $^{28}Si^{13+}$. The data for these graphs were taken from [20, 58–60]

The respective data were taken from [20, 58–60]. In the lower right hand side graph, the error bars shown are 1-σ uncertainties corresponding to the 68 % prediction band of the original maximum-likelihood fit for better comparison. Note, that the scaling is different throughout the four plots with $^{28}Si^{13+}$ being the sharpest resonance. Application of a mode coupling technique to the latter measurement is expected to significantly increase the relative precision [104]. The lineshapes of the resonances are non-Gaussian mainly due to the Boltzmann distribution of motional ion energies corresponding to the finite heat bath temperature the ions are coupled to. Further line shape aspects have been discussed in [78].

3.7.2 Resulting Magnetic Moments and Uncertainties

From the measured resonances of the ω_L/ω_c frequency ratios, and upon systematic lineshape corrections, the bound electron magnetic moments are calculated according to Eq. 3.7. Table 3.3 gives an overview of the corrected frequency ratios $\Gamma = \omega_L/\omega_c$ and the resulting values for g_J. In cases where two individual uncertainties are given, the first one refers to the statistics and systematics of the Larmor-to-cyclotron resonance measurement, while the second one refers to the uncertainty of the electron mass as known at the respective time of publication. For silicon, the first uncertainty refers to the statistics, the second to the systematics and the third to the electron mass.

It should be noted, that these values in the middle section of Table 3.3 are not completely independent, but due to their common experimental outline, there is a finite correlation. For the measurements with $^{12}C^{5+}$ [58] and $^{16}O^{7+}$ [59], a correlation coefficient of 0.035 has been given [135], for the remaining measurements this has not been studied so far, but it may be assumed of similar magnitude. The correlation

Table 3.3 Measured values of the bound electron magnetic moment (g_J) in hydrogen-like ions

Z	Ion	$\Gamma = \omega_L/\omega_c$	Magnetic Moment (g_J)	Reference
0	free electron		2.002 319 304 361 460(56)	[134]
1	Hydrogen ^1H		2.002 283 853(26)	[48]
4	Helium $^4He^+$		2.002 259 33(60)	[49]
6	Carbon $^{12}C^{5+}$	4376.2095(37)	2.001 042(2)	[129]
6	Carbon $^{12}C^{5+}$	4376.2104989(23)	2.001 041 596 3 (10)(44)	[58, 89]
8	Oxygen $^{16}O^{7+}$	4164.3761834(7)	2.000 047 026 0 (15)(44)	[59]
8	Oxygen $^{16}O^{7+}$	4164.3761844(49)	2.000 047 020 8 (24)(44)	[60]
14	Silicon $^{28}Si^{13+}$	3912.866064(1)	1.995 348 958 77(50)(30)(80)	[103]
14	Silicon $^{28}Si^{11+}$	4637.318949(4)	2.000 889 889 9 (19)(5)(8)	[61]
82	Lead $^{207}Pb^{81+}$		1.78(12)	[14, 50]
83	Bismuth $^{209}Bi^{82+}$		1.734 1(35)	[14, 50]

The middle section lists all values obtained by application of the continuous Stern-Gerlach effect and the corresponding value of the Larmor resonance $\Gamma = \omega_L/\omega_c$ (upon corrections) is indicated is of relevance particularly when the results are combined as in the derivation of the value of the electron mass, as will be discussed in Sect. 3.10.

While the influence of the finite statistics (finite number of trials to induce spin flips) can be quantified in a straight-forward way, the systematic uncertainties hold more challenges. The most prominent systematic influences are

- spatial and temporal imperfections of the confining fields
- finite motional energies of the confined ion
- cavity and charge effects (image charge)
- systematics of the underlying time base (frequency standard)
- line shape distortions due to microwave intensity, inhomogeneity, impurity
- saturation of the Larmor transition
- uncertainty of the ion-to-electron mass ratio

For given experimental imperfections, the desired magnetic moment is typically obtained from an extrapolation of the measured resonance data to zero motional energies ($E_+, E_z, E_- \rightarrow 0$) and zero microwave power ($P_{MW} \rightarrow 0$). The remaining systematics are corrected for based on theoretical models.

Equation (3.7) requires the atomic mass of the ion to be known. For carbon, this is trivial since the carbon atom defines the atomic mass. In this case, only the masses and the relativistic equivalent of the electron binding energies of the five missing electrons needs to be taken into account according to the mass equation for an ion with charge state k

$$m_{ion^{k+}} = m_{atom} - k \cdot m_e + \frac{1}{c^2} \sum_{p=1}^{k} E_b^{(p)} \tag{3.69}$$

where $E_b^{(p)}$ is the binding energy of the pth electron. In [58], the ratio of free cyclotron frequencies of the ion [58] and the free electron [136] has been used, $\omega_c(^{12}C^{5+})/\omega_c(e^-) = 0.000228627 21033(50)$, which is identical to the desired mass ratio m_e/m_i.

For oxygen $^{16}O^{7+}$, the accepted values for the mass of the neutral oxygen atom [137] and the sum of the binding energies of the seven missing electrons of $m_b = 1.25866(9) \times 10^{-6}$ u [138] were used to find the mass of the hydrogen-like oxygen as $m(^{16}O^{7+}) = 15.991075819(2)$ u [59].

The mass of $^{28}Si^{13+}$ was calculated from the atomic mass of 27.9769265350(6) u, as measured by Redshaw et al., [139] corrected by the masses and binding energies of the missing 13 electrons as given by Martin et al. [140]. The result was $m(^{28}Si^{13+}) = 27.9698005949(7)$ u [20].

For $^{28}Si^{11+}$ an electron mass of $m_e = 5.4857990946(22) \times 10^{-4}$ u [141] was used, and the silicon atomic mass of 27.9769265350(6) u [139] was corrected by the masses and binding energies [140] of the eleven missing electrons to yield $m(^{28}Si^{11+}) = 27.97089457581(66)$ u [61].

Figure 3.17 shows a timeline with important experiments (black) and methods (grey) concerning the magnetic moment of the free electron (left hand side) and bound

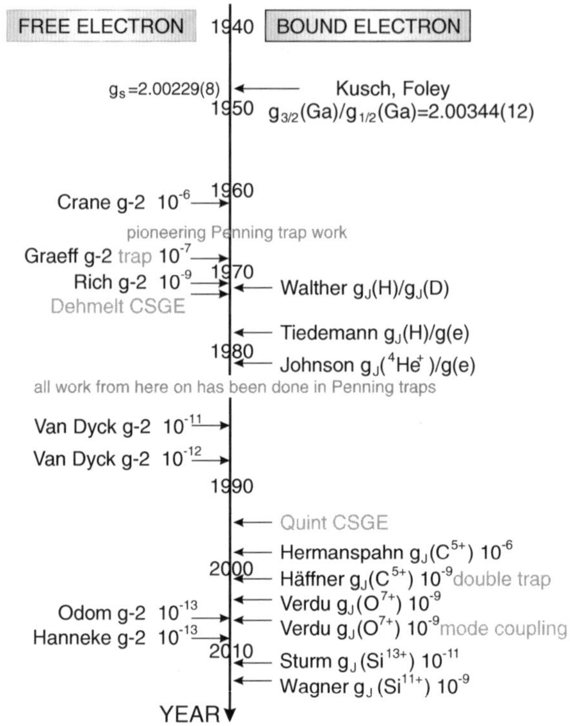

Fig. 3.17 Timeline summarizing important experiments (*black*) and methods (*grey*) for the free electron magnetic moment (*left hand side*) and for the bound electron magnetic moment (*right hand side*). The numbers indicated give the order of magnitude of the achieved relative accuracies. CGSE is for 'continuous Stern-Gerlach effect'. For references see text

electron (right hand side), indicating typical relative accuracies of the measurements. 'CSGE' is short for 'continuous Stern-Gerlach effect' and its mention in the timeline indicates when its respective application has first been proposed. The mentioned work is (chronologically) Kusch and Foley (1947) [39, 40], Crane et al. (1961) [142], Graeff et al. (1969) [143], Rich et al. (1971) [144], Walther et al. (1972) [47], Dehmelt CSGE (1973) [51], Tiedeman et al. (1977) [48], Johnson et al. (1980) [49], van Dyck et al. (1987) [136], Quint CSGE (1995) [57], Hermanspahn et al. (2000) [129], Häffner et al. double trap (2000) [58], Verdú et al. (2004) [59], Verdú et al. mode coupling (2004) [60], Sturm et al. (2011) [20] and Wagner et al. (2012) [61].

3.8 Double-Resonance Spectroscopy

While measurements of the bound electron magnetic moment by the continuous Stern-Gerlach effect combine microwave probing of the Larmor frequency with a radio-frequency detection of the success by an ion oscillation frequency measurement in the trap, it is also possible to combine the microwave probing of the Larmor frequency with optical detection of the success by laser spectroscopy of the fine or hyperfine structure of the ion. As the laser and microwave excitations are performed simultaneously, this technique is called laser-microwave double resonance spectroscopy. It circumvents the need for a magnetic bottle and trades this for the availability of an appropriate laser and optical detection.

Double resonance spectroscopy has previously been used to perform precision measurements of hyperfine and Zeeman splittings in atoms and singly charged ions such as Hg^+, Ba^+, Pb^+, Yb^+ and Be^+, partially with sub-Hertz resolution [145–150]. For a thorough discussion of this application and an overview of the measurements, see [11].

3.8.1 Application to Highly Charged Ions

In highly charged ions, the energy of fine structure and hyperfine structure transitions are shifted to much higher values as compared to singly charged ions by the extreme electromagnetic fields in the vicinity of the nucleus.

In hydrogen-like ions, the energy of the ground state hyperfine structure (HFS) splitting depends as Z^3 on the nuclear charge number, for other few-electron ions the dependence is similar. It can be seen from the r^{-3}-dependence of the magnetic field of a dipole keeping in mind the length scale for r is given by a_0/Z, so $E_{HFS} \propto < r^{-3} > \propto Z^3$. Ignoring relativistic, nuclear and QED effects, the 'classical' ground-state HFS transition energy in a hydrogen-like ion can be used as a fair approximation[4] of the real value and reads [14]

$$E_{\text{HFS}} \approx \frac{4}{3}\alpha(Z\alpha)^3 g_I \frac{m_e}{m_p}\frac{2I+1}{2} m_e c^2 \quad \propto Z^3, \tag{3.70}$$

where g_I is the nuclear g-factor which measures the nuclear magnetic moment in units of the nuclear magneton $\mu_N = e\hbar/2m_p = 5.05078324(13) \times 10^{-27}$ J/T. The same equation applies also to other charge states, the factor $4/3$ then changes to $1/6$ for lithium-like ions, to $1/18$ for boron-like ions and so forth [14]. The ground-state hyperfine splitting in hydrogen-like ions has so far been measured in $^{165}Ho^{66+}$ [151], $^{185,187}Re^{74+}$ [152], $^{203,205}Tl^{80+}$ [153], $^{207}Pb^{81+}$ [154, 155] and $^{209}Bi^{82+}$ [155, 156] as well as in $^{209}Bi^{80+}$ [157]. Detailed calculations of the hyperfine energies in highly charged ions have been performed in all rigour, see for example [25, 158–166].

For the fine structure (FS) splitting, the Z-dependence is even stronger and scales with Z^4. This is obvious from the fine structure energy $E_{FS} \propto \mu B \propto Z^2/r^2$ where $< r^{-2} > \propto Z^2$ such that we obtain $E_{FS} \propto Z^4$, see for example [167]. The energy of

[4] This approximation is usually not good within a few percent, but a valuable tool for judging whether the transition is at all in the laser-accessible region.

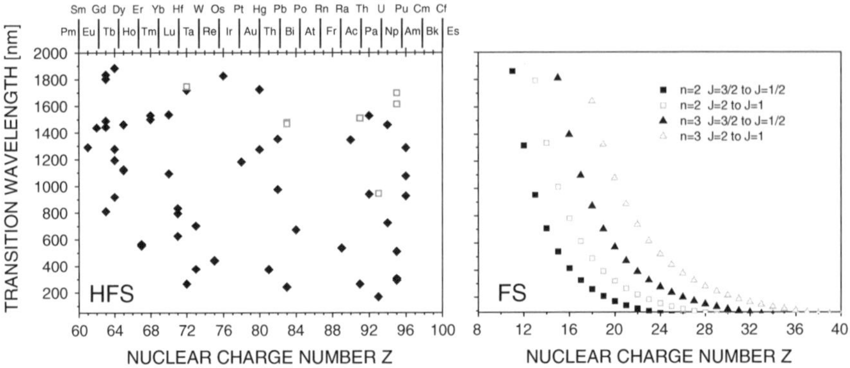

Fig. 3.18 Overview of hyperfine- and fine-structure transitions in the laser-accessible wavelength region: HFS in hydrogen- and lithium-like ions (full and open symbols, *left graph*), and FS in hydrogen-like ions (*right graph*)

a fine-structure transition can be approximated by use of the Dirac equation for the energy of the single electron in hydrogenic systems with quantum numbers n and j. For example, the energy of a fine structure transition in a hydrogen-like ion from a state (n, j) to $(n, j-1)$ is given by $E_{\text{FS}} = E(n, j) - E(n, j-1)$ with the energy of a state (n, j) given by the equation

$$E(n,j) \approx \left[1 + Z^2\alpha^2\left(n - j - \frac{1}{2} + \sqrt{(j + \frac{1}{2})^2 - Z^2\alpha^2}\right)^{-2}\right]^{-1/2} m_e c^2. \tag{3.71}$$

This leads to the splittings being shifted to the laser-accessible regime for certain values of Z, see Fig. 3.18. For fine structure transitions, the optical frequency regime is reached for medium charge states, while the weaker dependence of the hyperfine transition energy requires higher charge states. For a number of such ions, laser spectroscopic measurements of energy splittings have been performed, an overview is given for example in [168].

Fine and hyperfine structure transitions are both magnetic dipole (M1) transitions with correspondingly long lifetimes and hence much narrower linewidths than the principal transitions, which are electric dipole (E1) transitions and for highly charged ions in the XUV to x-ray regime of energies, see also Fig. 3.20.

Magnetic dipole transitions are caused by the oscillating magnetic field $\mathbf{B} = \mathbf{B_0}\cos\omega t$ of the laser radiation. The transition matrix element between two levels 1 and 2 is given by $\mu_{21} \propto < 2|\boldsymbol{\mu} \cdot \mathbf{B}|1 >$ where $\boldsymbol{\mu}$ is the magnetic dipole operator. The transition rate of spontaneous decay is given by $A_{21} \propto \omega^3 |\mu_{21}|^2$. It is smaller than the rate of an electric dipole (E1) transition by a factor of about

$$\frac{| < 2|\boldsymbol{\mu} \cdot \mathbf{B}|1 > |^2}{| < 2|e\mathbf{r} \cdot \mathbf{E}|1 > |^2} \approx \left(\frac{\mu_B/c}{ea_0/Z}\right)^2 \approx (Z\alpha)^2. \tag{3.72}$$

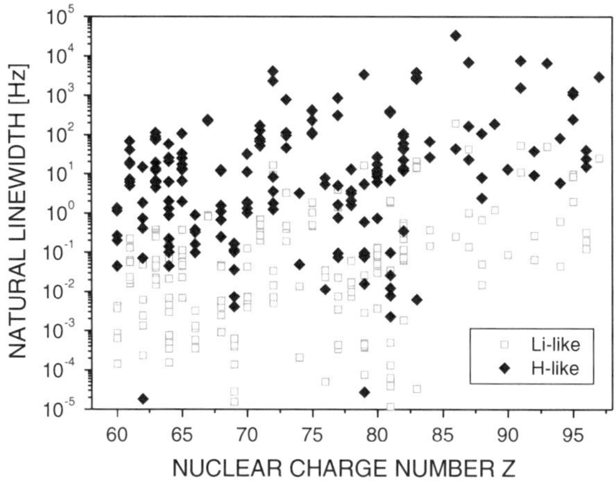

Fig. 3.19 Calculated upper-state lifetimes (and hence transition linewidths) according to Eq. (3.73) of ground-state hyperfine transitions in hydrogen- and lithium-like ions

The transition rate (linewidth) for a ground-state ($n = 1$) magnetic dipole transition from the excited to the lower hyperfine state is given by [14]

$$\gamma_{\rm HFS} = \frac{4\alpha\omega^3\hbar^2 I\,(2k+1)^2}{27m_e^2c^4\,(2I+1)}, \tag{3.73}$$

where α is the fine structure constant, I is the nuclear spin and $k = \sqrt{1 - Z^2\alpha^2}$. For hydrogen-like ions, this linewidth is of order kHz and below [168], see also Fig. 3.19. The ions with highest optical transition rates $\gamma_{\rm HFS}$ are $^{233}\mathrm{Pa}^{90+}$ ($7480\,\mathrm{s}^{-1}$), $^{212}\mathrm{Fr}^{86+}$ ($6890\,\mathrm{s}^{-1}$), $^{237}\mathrm{Np}^{92+}$ ($6600\,\mathrm{s}^{-1}$), $^{231}\mathrm{Pa}^{85+}$ ($3750\,\mathrm{s}^{-1}$), $^{196}\mathrm{Au}^{78+}$ ($3550\,\mathrm{s}^{-1}$) and all odd isotopes of bismuth $^{odd}\mathrm{Bi}^{82+}$ (2500–$3700\,\mathrm{s}^{-1}$).

For the transition rate (linewidth) of fine structure transitions in hydrogen-like ions, a semi-classical calculation [169] yields the expression

$$\gamma_{\rm FS} \approx \frac{2Z^4\alpha^5 m_e c^2}{3\hbar n^5}\,\frac{n^2}{l(l+1)} \tag{3.74}$$

for principal quantum numbers $n \geq 2$ and non-vanishing orbital quantum number l. This expression is good within a few percent of accuracy and agrees with experimental findings on that level [168]. It is especially well-suited for high quantum numbers n and l. More detailed calculations employing full quantum mechanics have been performed [170] and lead to values with accuracies of some hundreds of ppm. They also give a better approximation to the one shown in Eq. (3.74) by replacing $l(l+1)$ by $l^2 + l + (l+1)/(8l)$. A more detailed discussion can be found in [170]. The linewidth of Zeeman substate transitions is typically of order mHz and below, as has been presented in Sect. 3.5.2.5.

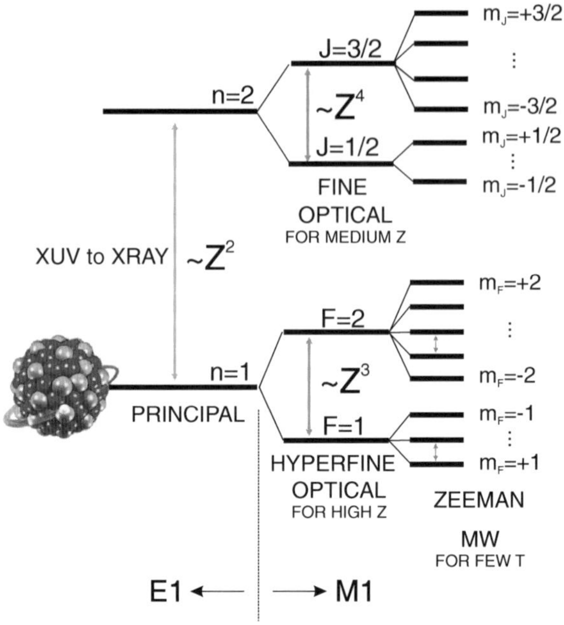

Fig. 3.20 Level scheme of a hydrogen-like ion indicating a hyperfine splitting in the electronic ground state (for nuclear spin $I = 3/2$, resulting in $F = 1$ and $F' = 2$) and a fine splitting for the 2P state (for nuclear spin $I = 0$, resulting in $J = 1/2$ and $J' = 3/2$), both with the corresponding Zeeman splitting of states due to a magnetic field as present in a Penning trap. Not to scale. All transitions of present relevance are magnetic dipole (M1) transitions. Indicated is the scaling of the transition energy with the nuclear charge number of the hydrogen-like ion

Figure 3.20 schematically shows the level splittings of a hydrogen-like ion with non-zero nuclear spin in the ground state including the Zeeman splitting due to a magnetic field as in the Penning trap, indicating the above mentioned dependences on the nuclear charge number and the typical transition frequency domain. It is indicated that in the presence of a magnetic field as used for confinement of ions in a Penning trap, the fine structure or hyperfine structure levels are subject to the Zeeman effect which lifts the degeneracy with respect to the magnetic quantum numbers of the states. The result is a splitting into $2J + 1$ fine structure substates, or, if present, $2F + 1$ hyperfine sublevels where $F = I \pm J$. This Zeeman splitting is nearly uniform within a given state and well within the microwave regime for typical magnetic fields of a few Tesla. To first order, the splitting is given by $\hbar\omega_{MW} = g\mu_B B_0$, where $g\mu_B$ is the ion magnetic moment. Thus, a measurement of the microwave frequency is a measurement of the ionic magnetic moment for a given value of the magnetic field strength. Experimentally, this can again be determined from a measurement of the free ion cyclotron frequency as discussed above.

The present concept of double resonance spectroscopy is to use a change in the optical signal (fluorescence from the FS or HFS transition) as an indication for the

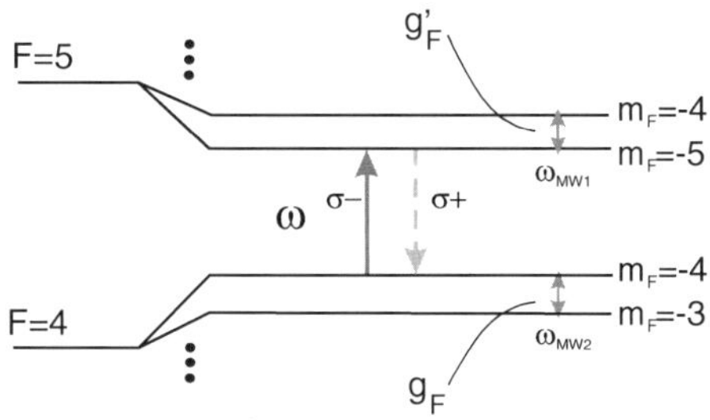

Fig. 3.21 Relevant hyperfine levels and their Zeeman splitting in $^{209}Bi^{82+}$ for a measurement of the ionic g-factors g'_F and g_F. Laser light at frequency ω is observed while microwaves around ω_{MW1} or ω_{MW2} are irradiated. In resonance, the optical signal is reduced and the irradiated MW frequencies yield the ionic g-factors

irradiated MW frequency to be in resonance with the Zeeman transition. Figure 3.21 shows the relevant part of the level scheme of hydrogen-like bismuth where an optical transition takes place between extremal Zeeman sublevels of the hyperfine components with $F = 4$ and $F = 5$. The ionic g-factors g_F and $g_{F'}$ are determined from a scan of the irradiated MW frequencies ω_{MW1} and ω_{MW2} while the closed optical cycle at frequency ω is observed. When one of the MW transitions is in resonance, the optical signal is reduced due to population being extracted from the otherwise closed cycle. Hence, a minimum in the optical signal indicates the Zeeman resonance and allows determinations of the ionic magnetic moments. Details and further applications are given in [171]. It can be shown, that for ions with non-vanishing nuclear spin, the ionic magnetic moment is a non-trivial composition of the magnetic moments of both the bound electron and the nucleus. Generally,

$$g_F = g_J \frac{F(F+1) + J(J+1) - I(I+1)}{2F(F+1)} - \frac{m_e}{m_p} g_I \frac{F(F+1) + I(I+1) - J(J+1)}{2F(F+1)} \tag{3.75}$$

describes the general relation between the g_F-factor, the g_J-factor of the bound electron and the nuclear g-factor $g_I = \mu/(\mu_N I)$ where μ_N is the nuclear magneton and m_e and m_p are the electron and proton mass, respectively. The g-factors of the bound electron g_J and of the nucleus g_I can be disentangled to a high degree by use of theory [171]. To this end, the ionic magnetic moments need to be measured in two different hyperfine states F and F'. If two g-factors g_F and $g_{F'}$ are measured for states with different F in one ion ($F = I - 1/2$ and $F' = F + 1$), the bound electron and nuclear g-factors g_J and g_I can be expressed independently in terms of the experimentally obtained values by means of [171]

$$g_J = (I+1) g_{F'} - I g_F - \delta_Q Q \left(\frac{m_e c}{\hbar}\right)^2 \frac{2(I+1)}{2I-1} \tag{3.76}$$

and

$$g_I = -\frac{m_p}{m_e} \frac{g_{F'} + g_F + \delta_Q Q \left(\frac{m_e c}{\hbar}\right)^2 \frac{3}{I(2I-1)}}{2(1-\delta_\mu)}, \tag{3.77}$$

where Q is the electric quadrupole moment of the nucleus and δ_μ and δ_Q are small corrections which can be obtained from theory [171], see also Sect. 3.10.3. This concept, however, can only be applied to ions with $I > 1/2$ since for an ion with $I = 1/2$ the lower HFS level has $F = 0$ and does not split in the magnetic field.

Due to the factor m_e/m_p in Eq. (3.75) the obtainable accuracy in g_I is about three orders of magnitude smaller than the ones in g_J and g_F. However, given an assumed relative accuracy of the microwave frequency measurement of ppb, a determination of the nuclear magnetic moment is possible at the ppm level of accuracy and thus competitive to most other measurements. Note, that in this kind of measurement, diamagnetic shielding of the nucleus is absent due to only one or few electrons being present. This is an advantage with respect to nuclear magnetic resonance (NMR) measurements and allows a benchmarking of shielding models.

A corresponding experiment (ARTEMIS, Fig. 3.22) is currently being set up at GSI, Germany, where either internally or externally produced ions of medium to high charge states can be confined in a Penning trap suited for laser-microwave double-resonance spectroscopy, see [172, 173].

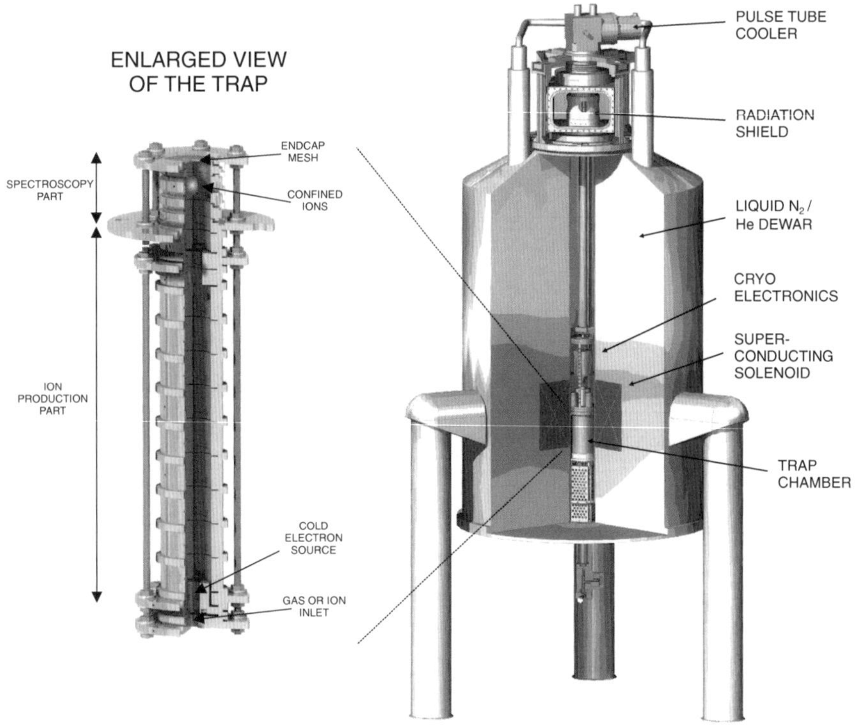

Fig. 3.22 Schematic of the ARTEMIS experimental setup (*right*) with an enlarged view of the ARTEMIS Penning trap (*left*). Details can be found in [172, 173]

3.8.2 Double-Resonance Spectroscopy and the Zeeman Effect

While the concept of double-resonance spectroscopy is conceived to allow measurements of magnetic moments of bound electrons and ionic nuclei, its application to certain level schemes yields access to higher-order Zeeman effects. Since the discovery of the quadratic Zeeman effect by Segré and Jenkins in the 1930s [174, 175], there have been numerous studies, both experimental and theoretical, of higher-order Zeeman contributions in atoms, molecules, and singly charged ions in laboratory magnetic fields, see for example [176–178]. However, higher-order contributions to the Zeeman effect have not yet been observed in highly charged ions. Among others, boron-like argon is a good candidate for such studies, since its P doublet structure offers experimental access and fair theoretical prediction. Figure 3.23 indicates the Zeeman sublevels of the fine structure in Ar^{13+} and their behaviour when higher-order Zeeman effects are considered. When writing the energy of the Zeeman splitting in the form

$$\Delta E_A = \Delta E_A^{(1)} + \Delta E_A^{(2)} + \Delta E_A^{(3)} + \dots, \tag{3.78}$$

the first term

$$\Delta E_A^{(1)}(B_0) = g_J \mu_B B_0 \tag{3.79}$$

is the usual linear Zeeman splitting with its g-factor g_J. The second- and third-order terms in the Zeeman splitting can be represented by

$$\Delta E_A^{(2)}(B_0) = g_J^{(2)}(M_J)(\mu_B B_0)^2/E_0 \quad \text{and} \quad \Delta E_A^{(3)}(B_0) = g_J^{(3)}(M_J)(\mu_B B_0)^3/E_0^2, \tag{3.80}$$

where $E_0 = mc^2$ is the electron rest energy and $g_J^{(2)}$ and $g_J^{(3)}$ are again dimensionless coefficients. A fully relativistic calculation in the framework of quantum electrodynamics yields

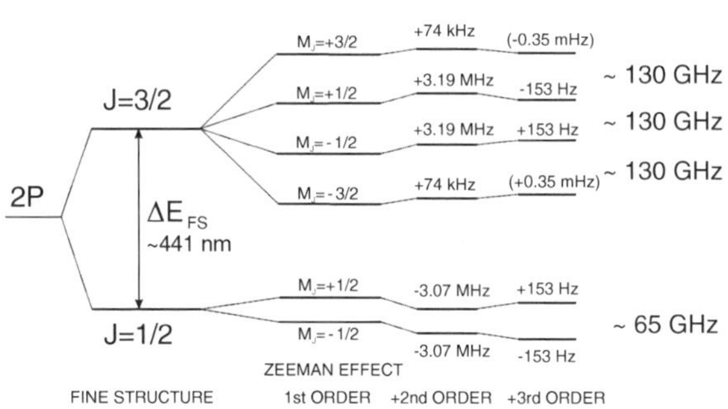

Fig. 3.23 Level scheme of the 2p state in Ar^{13+} indicating the Zeeman sublevels of the fine structure and their behaviour when higher-order Zeeman effects are taken into consideration. Not true to scale

$$\frac{\Delta E_A^{(2)}(B_0)}{\Delta E_A^{(1)}(B_0)} \approx \frac{\Delta E_A^{(1)}(B_0)}{\Delta E_{FS}} \approx 10^{-4} \quad \text{and} \quad \frac{\Delta E_A^{(3)}(B_0)}{\Delta E_A^{(1)}(B_0)} \approx \left(\frac{\Delta E_A^{(1)}(B_0)}{\Delta E_{FS}}\right)^2 \approx 10^{-8}, \tag{3.81}$$

where $\Delta E_{FS} = E_{3/2}^{(0)} - E_{1/2}^{(0)}$ is the optical fine-structure interval at about 441 nm. Hence, the second-order Zeeman effect has a 10^{-4} contribution to the microwave Zeeman splitting and thus needs to be taken into account in such measurements. Also the 10^{-8} contribution of the third order may be within the experimental resolution. Currently, an experiment is being prepared for a detailed study of these effects [173].

3.9 Comment on Trap-Specific Spectroscopy

It is a common aspect both of the continuous Stern-Gerlach effect as described in Sect. 3.5.3.1 and the laser-microwave double-resonance spectroscopy as discussed in Sect. 3.8 to make use of the predictable influence of one experimental quantity which is inaccessible on another experimental quantity which can be observed. In some sense, the same is true also for resonant motional coupling as described in Sect. 3.6.5.

In case of the classical Stern-Gerlach effect, the spin orientation M_S is translated via a magnetic field gradient B_1 into a change Δz in position, we could write $\Delta M_S \overset{B_1}{\rightarrow} \Delta z$. For the continuous Stern-Gerlach effect it is the intrinsic electron spin orientation which by the magnetic bottle is projected onto a macroscopic ion oscillation in the trap which can be observed, so we may write $\Delta M_J \overset{B_2}{\rightarrow} \Delta \omega_z$. In case of the laser-microwave double-resonance spectroscopy it is the resonant Zeeman transition which is found as a minimum of observed fluorescence light I when the irradiated microwave frequency is varied in the presence of the constant homogeneous magnetic field B_0, so we may denote $\Delta \omega_{MW} \overset{B_0}{\rightarrow} \Delta I$.

It has previously been demonstrated in the decay of metastable nuclear states of ions that the emission (or absorption) of such photons with energy E can be measured by the relativistic mass equivalent $m = E/c^2$, by which the ion mass is changed during the process. In a trap, this is for example seen as a change of the cyclotron frequency, we can write $\Delta m \overset{B_0}{\rightarrow} \Delta \omega_c$. Corresponding measurements of long-lived nuclear isomers have been performed, e.g., on ^{65m}Fe [179] and ^{68m}Cu [180]. Also, the Q_{EC}-value of the super-allowed β-emitter ^{26}Si has been determined by such a measurement with a relative accuracy of about 10^{-5} [181].

Taking the well-known relations of the ion oscillation frequencies for specific non-ideal confining fields as given in Sects. 3.5.2 and 3.5.3.1 for granted, also defined magnetic inhomogeneities or electrostatic anharmonicities of the trapping fields can be used for translations of optical or x-ray photons being absorbed or emitted into changes of ion oscillation frequencies. For example, when a confined ion is laser-cooled, the change in motional energy may be translated via a magnetic bottle B_2 or an electrostatic anharmonicity C_4 into a change of an oscillation frequency, for example axial cooling or heating changing the radial oscillation frequency, which may be written as $\Delta E_z \overset{B_2}{\rightarrow} \Delta\omega_+$, or $\Delta E_z \overset{C_4}{\rightarrow} \Delta\omega_+$, as discussed in [182]. In such experiments, classical spectroscopy can be performed without the need for corresponding fluorescence detectors, since the ion oscillation frequency serves as the measurable quantity. Hence, such specific field distortions may been seen as a kind of 'transducer' from the optical and microwave domains into the radio-frequency domain where non-destructive detection is possible. A detailed account on well-controlled field distortions and the related experimental possibilities is given in [81, 182].

3.10 Relation of the Bound Electron Magnetic Moment to Other Quantities

The bound electron magnetic moment is closely related to other ionic quantities. This is particularly obvious for heavy, highly charged ions where the electron is very close to the ionic nucleus and therefore serves as an excellent probe of its properties.

While the free electron magnetic moment is most closely linked to the fine structure constant, the bound electron magnetic moment is also linked with the hyperfine structure splitting in the ion, the electron mass and several nuclear properties of the ion such as mass, extension, shape, charge and magnetization and their spatial distributions.

3.10.1 Fine Structure Constant

As far as the fine structure constant is concerned, the most notable precisions come from atom recoil measurements, atom interferometry and the electron magnetic moment. Figure 3.24 gives a compilation of fine structure constant determinations as a function of the relative accuracy in $1/\alpha$, details can be found in [183, 184] and references therein.

Atom recoil measurements for a determination of α make use of the fact that the Rydberg constant $R_\infty = 1.0973731568525(73) \times 10^7\,\mathrm{m}^{-1}$ [141] is known to seven parts in a trillion, and that the mass of the ^{87}Rb atom is known to a few parts in 10^{10}. Thus, when the mass of the ^{87}Rb atom is measured by the recoil speed of the atom after it emits a photon of known wavelength in an atomic transition, the electron mass can be inferred precisely. This is done with ultra-cold ^{87}Rb atoms in an optical lattice. Inserting the electron mass into

$$R_\infty = \frac{\alpha^2 m_e c}{4\pi\hbar} \tag{3.82}$$

yields a value of $\alpha^{-1} = 137.03599878(91)$ [185]. A comparable measurement using the recoil of a ^{133}Cs atom yields a value of $\alpha^{-1} = 137.0360000(11)$ [169]. The hence

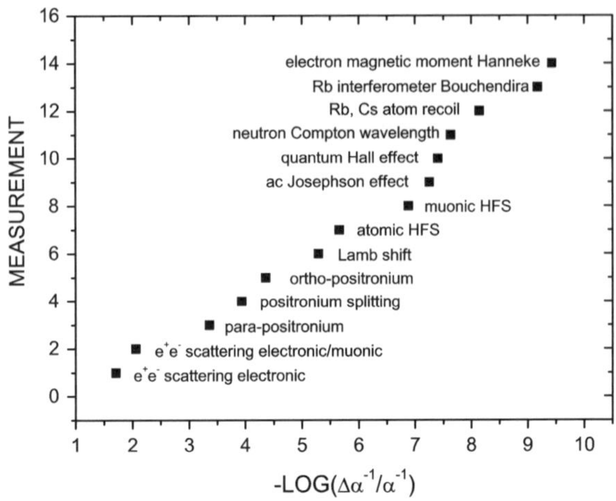

Fig. 3.24 Experimental determinations of the fine structure constant as a function of the relative accuracy in $1/\alpha$. Please note that the graph contains no chronological order

determined value of α agrees within one standard deviation to that found from the electron's anomalous magnetic moment, an agreement to within ten parts in a billion.

Recently, the fine structure constant has been measured independently with competitive accuracy by Bouchendira et al. [183] in an atom interferometry experiment. These measurements give the value $\alpha^{-1} = 137.035999037(91)$ and based on this, QED theory predicts $g_e = 2.00231930436226(168)$.

QED theory provides the magnetic moment of the free electron by its g-factor $g_s(e)$ as a function of the fine structure constant α. A comparison of such theoretical predictions with experimental values of the free electron magnetic moment have yielded $\alpha^{-1} = 137.035\,998\,83(50)$ from the result by van Dyck et al. [136] and $\alpha^{-1} = 137.035\,999\,084(51)$ from the result by Hanneke et al. [134, 186]. The latter has a relative uncertainty of 3.7×10^{-10} and is the most precise value of the fine structure constant to date.

Corresponding determinations of the fine structure constant from measurements of bound electron magnetic moments are favoured by the accuracy relation

$$\frac{\Delta\alpha}{\alpha} \approx \frac{\Delta g_J}{g_J} \frac{1}{(Z\alpha)^2} \tag{3.83}$$

which for high Z shows that the relative accuracy in α is similar to that of g_J. However, this requires a theoretical understanding of the relation between bound electron magnetic moment and fine structure constant on at least the same level. Currently, however, such information is not available. The sensitivity of the bound electron magnetic moment to changes of the value of the fine structure constant as given by Eq. (3.83) is much smaller than the sensitivity of certain optical transitions in highly charged ions, see the discussion in Chap. 6. Laboratory experiments for the detection of a spatial and/or temporal variation of the value of the fine structure constant [187] therefore concentrate on optical clocks [188–190] and it is desired to extend this to specific highly charged ion optical clocks like Pm^{14+} and others [191, 192] or nuclear optical clocks such as ^{229}Th [193, 194].

3.10.2 Electron Mass

By the time the first bound electron magnetic moment measurements in $^{12}C^{5+}$ and $^{16}O^{7+}$ [58, 59] were performed, the uncertainty of the electron mass relative to the ion mass was the largest contribution to the error budget. At that time, the CODATA recommended value of the electron mass was mainly given by a 1995 measurement by Farnham et al. [195]. Making use of the good agreement between theory and the measured g_J-factors in $^{12}C^{5+}$ and $^{16}O^{7+}$, a new value of the electron's atomic mass was derived. It was $m_e = 0.00054857990945(24)$ u [196] and hence about five times more accurate than the previous value. From this determination of the electron mass, a new ratio of proton to electron mass $m_p/m_e = 1836.15267247(80)$ has been obtained [196]. Table 3.4 lists the values of the electron mass obtained from the different measurements and the respective appearances in the CODATA tables of recommended values. Taking into account the experimental advances made since the measurements with $^{12}C^{5+}$ [58], it appears highly attractive to revisit this ion as no independent ion mass measurement limits the accuracy when the electron mass is derived from the measured magnetic moment of the bound electron. It is currently deemed possible to decrease the uncertainty of the electron mass by about one order of magnitude.

3.10.3 Relations to Nuclear Properties

Recent theoretical work shows that from precision measurements of bound electron magnetic moments, it is possible to determine the ionic nuclear dipole moment [197] as well as the ionic nuclear shape [198]. As already discussed in Sect. 3.8, few-electron ions, particularly hydrogen-like ions, are highly attractive for such studies since the bound electron is located close to the nucleus and hence probes its properties with high sensitivity. In the Bohr picture, the orbital radius a of the electron bound to a nucleus with charge number Z is given by $a \approx a_0/Z$ with the Bohr radius $a_0 = 0.52917721092(17) \times 10^{-10}$ m, such that in hydrogen-like uranium ($Z=92$), the electron is closer to the nucleus by about two orders magnitude. Also, neither the nucleus nor the bound electron are effectively shielded by other electrons and can thus be interrogated directly. This avoids the need for diamagnetic corrections which are based on models [197, 199–201]. In turn, such models can be benchmarked with high stringency.

Table 3.4 Values of the electron mass resulting from different measurements of the bound electron magnetic moment and the respective entries in the CODATA tables of recommended values

Measurement	Electron mass [u]	Reference
Farnham et al. 1995	0.000 548 579 911 10(120)	[195]
CODATA 1998 (2000)	0.000 548 579 911 00(120)	[202]
$^{12}C^{5+}$ Häffner et al. 2000	0.000 548 579 909 31(29)	[203]
$^{16}O^{7+}$ Verdú et al. 2004	0.000 548 579 909 57(43)	[203]
CODATA 2002 (2005)	0.000 548 579 909 45(24)	[203]
CODATA 2010 (2012)	0.000 548 579 909 46(22)	[141]

As far as the ionic g-factor g_F in the ground state of ions with non-zero nuclear spin is concerned, we have seen that it is related to the nuclear magnetic dipole moment μ as well as the nuclear electric quadrupole moment Q. The total one-electron ground-state g-factor value of the ion can be represented by [171]

$$g_F = g_J Y_{\text{el}}(F) - \frac{m_e}{m_p} g_I Y_{\text{nuc}}^{(\mu)}(F) + \delta g_{\text{HFS}}^{(\mu)}(F) + \delta g_{\text{HFS}}^{(\text{Q})}(F), \tag{3.84}$$

where

$$\delta g_{\text{HFS}}^{(\mu)}(F) = \delta_\mu \frac{m_e}{m_p} g_I Y_{\text{nuc}}^{(\mu)}(F) \tag{3.85}$$

and

$$\delta g_{\text{HFS}}^{(Q)}(F) = \delta_Q Q \left(\frac{m_e c}{\hbar}\right)^2 Y_{\text{nuc}}^{(Q)}(F) \tag{3.86}$$

are the magnetic dipole and electric quadrupole contributions to the ionic g-factor, respectively. The quantities

$$Y_{\text{el}}(F) = \frac{F(F+1) + 3/4 - I(I+1)}{2F(F+1)}, \tag{3.87}$$

$$Y_{\text{nuc}}^{(\mu)}(F) = \frac{F(F+1) + I(I+1) - 3/4}{2F(F+1)}, \tag{3.88}$$

and

$$Y_{\text{nuc}}^{(Q)}(F) = -\frac{(I+1)(2I+3)}{I(2I-1)(2I+1)} \tag{3.89}$$

are angular factors and hence exact. Further, Q is the electric quadrupole moment of the nucleus, and δ_μ and δ_Q are corrections given by

$$\delta_\mu = \alpha^2 Z \frac{1}{3} S^{(t)}(\alpha Z) \approx \frac{1}{3}\alpha^2 Z \left(1 + \frac{97}{36}\alpha^2 Z^2\right) \tag{3.90}$$

$$\delta_Q = \alpha^4 Z^3 \frac{11}{135} T^{(t)}(\alpha Z) \approx \frac{11}{135}\alpha^4 Z^3 \left(1 + \frac{43}{33}\alpha^2 Z^2\right) \tag{3.91}$$

where the approximations hold true for light ions. The full functions $S^{(t)}(\alpha Z)$ and $T^{(t)}(\alpha Z)$ for an *ns* state have been considered in detail in [204]. For lithium-like ions these functions must also include interelectronic-interaction corrections [205]. The shifts of the ionic *g*-factor $\delta g^{(\mu)}_{\text{HFS}}(F)$ and $\delta g^{(\text{Q})}_{\text{HFS}}(F)$ due to finite magnetic dipole moment and electric quadrupole moment take values of up to ppm and several tens of ppb (with uncertainties on the ppb level), respectively [204].Hence, they are within experimental resolution. The effect of the nuclear quadrupole moment is too small for a precise determination by either Stern-Gerlach or double-resonance spectroscopy based on a measurement of electronic or ionic *g*-factors, respectively. The effect of the nuclear dipole moment, however, is deemed to be accessible to high accuracy [197]. It is furthermore possible to determine the nuclear charge radius by a comparison of the experimental and the theoretical values of the electronic g-factor, as has been discussed for hydrogen-like silicon [20]. The extracted value is not competitive with the established one, but may be seen as a proof of principle. Effects of the finite nuclear electric polarisability and magnetic susceptibility are currently still outside of the experimental resolution.

Acknowledgments We wish to express our gratitude to everyone who has contributed to the work we have discussed in this chapter. We would also like to thank everybody who has taken upon himself to contribute to this book, our co-authors and their working groups.

References

1. R. Frisch, O. Stern, Über die magnetische Ablenkung von Wasserstoffmolekülen und das magnetische moment des protons. Z. Phys. **85**, 4 (1933)
2. R. Frisch, O. Stern, Über die magnetische Ablenkung von Wasserstoffmolekülen und das magnetische moment des protons. Helv. Phys. Acta **6**, 426 (1933)
3. A. Landé, The magnetic moment of the proton. Phys. Rev. **44**, 1028 (1933)
4. L.W. Alvarez, F. Bloch, A quantitative determination of the neutron magnetic moment in absolute nuclear magnetons. Phys. Rev. **57**, 111 (1940)
5. K.M. Case, The magnetic moments of the neutron and proton. Phys. Rev. **74**, 1884 (1948)
6. G. Werth et al., Precision studies in traps: measurement of fundamental constants and tests of fundamental theories. Nucl. Inst. Meth. B **205**, 1 (2003)
7. M. Vogel, J. Alonso, S. Djekic, W. Quint, S. Stahl, M. Vogel, G. Werth, Towards electronic *g*-factor measurements in medium-heavy hydrogen-like and lithium-like ions. Nucl. Inst. Meth. B **235**, 7 (2005)
8. M. Vogel et al., The anomalous magnetic moment of the electron in hydrogenlike ions, a test of bound-state QED: recent, present and future precision experiments. Eur. Phys. J. Special Topics **163**, 113 (2008)
9. S. Sturm et al., *g*-factor measurement of hydrogen-like $^{28}\text{Si}^{13+}$ as a challenge to QED calculations. Phys. Rev. A **87**, 030501(R) (2013)

10. S. Sturm, G. Werth, K. Blaum, Electron g-factor determinations in Penning traps, Ann. Phys. (Berlin) **525**, 620–635 (2013)
11. G. Werth, V.N. Gheorghe, F.G. Major, *Charged Particle Traps II* (Springer, Heidelberg, 2009)
12. K. Böklen, W. Dankwort, E. Pitz, S. Penselin, High precision measurements of the g_J-factors of the alkalis using the atomic beam magnetic resonance method. Phys. Lett. **21**, 294 (1966)
13. D.J. Wineland, J.J. Bollinger, W.M. Itano, Laser-fluorescence mass spectroscopy. Phys. Rev. Lett. **50**, 628 (1983)
14. T. Beier, The g_j factor of a bound electron and the hyperfine structure splitting in hydrogen-like ions. Phys. Rep. **339**, 79 (2000)
15. V.M. Shabaev, QED theory of the nuclear recoil effect on the atomic g-factor. Phys. Rev. A **64**, 052104 (2001)
16. V.M. Shabaev, V.A. Yerokhin, Recoil correction to the bound-electron g-factor in H-like atoms to all orders in αZ. Phys. Rev. Lett. **88**, 091801 (2002)
17. K. Blaum, High-accuracy mass spectrometry with stored ions. Phys. Rep. **425**, 1 (2006)
18. G. Breit, The magnetic moment of the electron. Nature **122**, 649 (1928)
19. K. Pachucki, A. Czarnecki, U.D. Jentschura, V.A. Yerokhin, Complete two-loop correction to the bound-electron g factor. Phys. Rev. A **72**, 022108 (2005)
20. S. Sturm et al., g-factor of hydrogen-like $^{28}Si^{13+}$. Phys. Rev. Lett. **107**, 023002 (2011)
21. V.A. Yerokhin, P. Indelicato, V.M. Shabaev, Self-energy correction to the bound-electron g factor in H-like ions. Phys. Rev. Lett. **89**, 143001 (2002)
22. V.A. Yerokhin, P. Indelicato, V.M. Shabaev, Evaluation of the self-energy correction to the g factor of S states in H-like ions. Phys. Rev. A **69**, 052503 (2004)
23. D.A. Glazov, V.M. Shabaev, I.I. Tupitsyn, A.V. Volotka, V.A. Yerokhin, G. Plunien, G. Soff, Relativistic and QED corrections to the g factor of Li-like ions. Phys. Rev. A **70**, 062104 (2004)
24. A.V. Volotka, D.A. Glazov, V.M. Shabaev, I.I. Tupitsyn, G. Plunien, Screened QED corrections in lithiumlike heavy ions in the presence of magnetic fields. Phys. Rev. Lett. **103**, 033005 (2009)
25. D.A. Glazov, A.V. Volotka, V.M. Shabaev, I.I. Tupitsyn, G. Plunien, Evaluation of the screened QED corrections to the g factor and the hyperfine splitting of lithiumlike ions. Phys. Rev. A **81**, 062112 (2010)
26. V.M. Shabaev, O.V. Andreev, A.I. Bondarev, D.A. Glazov, Y.S. Kozhedub, A.V. Maiorova, G. Plunien, I.I. Tupitsyn, A.V. Volotka, Quantum electrodynamics effects in heavy ions and atoms. AIP Conf. Proc. **1344**, 60 (2011)
27. A. Sommerfeld, Zur theorie des zeemaneffektes der wasserstofflinien, mit einem anhang über den starkeffekt. Physikalische Zeitschrift **17**, 491 (1916)
28. P. Debye, Quantenhypothese und zeeman-effekte. Physikalische Zeitschrift **17**, 507 (1916)
29. A.H. Compton, The magnetic electron. J. Franklin Inst. **192**, 144 (1921)
30. P. Forman, Alfred landé and the anomalous Zeeman effect 1919–1921. Hist. Stud. Phy. Sci. **2**, 153 (1970)
31. J. Mehra, H. Rechenberg, *The Historical Development of Quantum Theory*, vol. 1, part 2, (Springer, New York, 1982)
32. O. Stern, W. Gerlach, Das magnetische moment des silberatoms [The magnetic moment of the silver atom]. Z. Phys. **9**, 349 (1922) and Z. Phys. **9**, 353 (1922)
33. W. Gerlach, O. Stern, Über die richtungsquantelung im magnetfeld. Ann. Phys. **379**, 673 (1924)
34. G.E. Uhlenbeck, S. Goudsmit, Ersetzung der Hypothese vom unmechanischen Zwang durch eine Forderung bezüglich des inneren Verhaltens jedes einzelnen Elektrons. Naturwiss **13**, 953 (1925)
35. G.E. Uhlenbeck, S. Goudsmit, Spinning electrons and the structure of spectra. Nature **127**, 264 (1926)
36. W. Pauli, Über den zusammenhang des abschlusses der elektronengruppen im atom mit der komplexstruktur der spektren. Z. Phys. **31**, 373 (1925)

37. F. Hund, *The History of Quantum Theory* (Bibliographisches Institut, Mannheim, Germany, 1975)
38. T.E. Phipps, J.B. Taylor, The magnetic moment of the hydrogen atom. Phys. Rev. **29**, 309 (1927)
39. P. Kusch, H.M. Foley, Precision measurement of the ratio of the atomic g values in the $^2P_{3/2} and ^2P_{1/2}$ states of gallium. Phys. Rev. **72**, 1256 (1947)
40. P. Kusch, H.M. Foley, On the intrinsic moment of the electron. Phys. Rev. **74**, 250 (1948)
41. J. Schwinger, On quantum-electrodynamics and the magnetic moment of the electron. Phys. Rev. **73**, 416 (1948)
42. G. Charpak, F.J.M. Farley, R.L. Garwin, T. Muller, J.C. Sens, V.L. Telegdi, A. Zichichi, Measurement of the anomalous magnetic moment of the muon. Phys. Rev. Lett. **6**, 128 (1961)
43. D.P. Hutchinson, J. Menes, G. Shapiro, A.M. Patlach, S. Penman, Magnetic moment of positive and negative muons. Phys. Rev. Lett. **7**, 129 (1961)
44. K.W. Ford et al., Theoretical values for magnetic moments of mu-mesonic atoms. Phys. Rev. **129**, 194 (1963)
45. H. Bethe, The electromagnetic shift of energy levels. Phys. Rev. **72**, 339 (1947)
46. W.E. Lamb, R.C. Retherford, Fine structure of the hydrogen atom by a microwave method. Phys. Rev. **72**, 241 (1947)
47. F.G. Walther, W.D. Phillips, D. Kleppner, Effect of nuclear mass on the bound-electron g factor. Phys. Rev. Lett. **28**, 1159 (1972)
48. J.S. Tiedeman, H.G. Robinson, Determination of $g_J(^1\mathrm{H}, 1^2S_{1/2)}/g_s(e)$: test of mass-independent corrections. Phys. Rev. Lett. **39**, 602 (1977)
49. C.E. Johnson, H.G. Robinson, g_J factor of an ion: determination of $g_J(^4\mathrm{He}^+, 1^2S_{1/2)}/g_J(^4\mathrm{He}, 2^3S_1)$. Phys. Rev. Lett. **45**, 250 (1980)
50. V.M. Shabaev, Transition probability between the hyperfine structure components of hydrogenlike ions and bound-electron g-factor. Can. J. Phys. **76**, 907 (1998)
51. H. Dehmelt, P. Ekstrom, Proposed $g-2$ experiment on stored single electron or positron. Bull. Am. Phys. Soc. **18**, 727 (1973)
52. P.B. Schwinberg, R.S. Van Dyck, Jr., H.G. Dehmelt, Preliminary comparison of the positron and electron spin anomalies. Phys. Rev. Lett. **47**, 1679 (1981)
53. H. Dehmelt, g-factor of electron centered in symmetric cavity. Proc. Natl. Acad. Sci. USA **81**, 8037 (1984)
54. H. Dehmelt, g-factor of electron centered in symmetric cavity. Proc. Natl. Acad. Sci. USA **82**, 6366 (1985)
55. H. Dehmelt, Continuous Stern Gerlach effect: principle and idealized apparatus, Proc. Natl. Acad. Sci. USA **83**, 2291 (1986)
56. H. Dehmelt, Continuous Stern Gerlach effect: principle and idealized apparatus, Proc. Natl. Acad. Sci. USA **83**, 3074 (1986)
57. W. Quint, The g_j factor of hydrogenic ions. Phys. Scr. **T59**, 203 (1995)
58. H. Häffner, T. Beier, N. Hermanspahn, H.J. Kluge, W. Quint, S. Stahl, J. Verdu, G. Werth, High-accuracy measurement of the magnetic moment anomaly of the electron bound in hydrogen-like carbon. Phys. Rev. Lett. **85**, 5308 (2000)
59. J. Verdú, S. Djekic, S. Stahl, T. Valenzuela, M. Vogel, G. Werth, T. Beier, H.J. Kluge, W. Quint, Electronic g factor of hydrogen-like oxygen $^{16}\mathrm{O}^{7+}$. Phys. Rev. Lett. **92**, 093002 (2004)
60. J. Verdú et al., Determination of the g-factor of single hydrogen-like ions by mode coupling in a penning trap. Phys. Scr. **T112**, 68 (2004)
61. A. Wagner et al., g-factor of lithium-like silicon $^{28}\mathrm{Si}^{11+}$. Phys. Rev. Lett. **110**, 033003 (2013)
62. L. Veseth, Magnetic moment of atomic lithium. Phys. Rev. A **11**, 421 (1980)
63. L. Veseth, Many-body calculations of atomic properties. I. g_j factors. J. Phys. B **16**, 2891 (1983)
64. E. Lindroth, A. Ynnerman, Ab initio calculations of g_j factors for Li, Be^+, and Ba^+. Phys. Rev. A **47**, 961 (1993)
65. L. Brillouin, Is it possible to test by a direct experiment the hypothesis of the spinning electron? Proc. Natl. Acad. Sci. USA **14**, 755 (1928)

66. N. Bohr, The Magnetic Electron, in *Collected Works of Niels Bohr*, ed. by J. Kalckar, vol. 6, North-Holland, Amsterdam, p. 333 (1985)
67. W. Pauli, Die allgemeinen Prinzipien der Wellenmechanik, in *Handbuch der Physik*, ed. by S. Flügge, vol. 5, Springer, Berlin, p. 167 (1958)
68. F. Bloch, Experiments on the g-factor of the electron. Physica **19**, 821 (1953)
69. H. Batelaan, T.J. Gay, J.J. Schwendiman, Stern-Gerlach effect for electron beams. Phys. Rev. Lett. **79**, 4517 (1997)
70. B.M. Garraway, S. Stenholm, Observing the spin of a free electron. Phys. Rev. A **60**, 63 (1999)
71. G.A. Gallup, H. Batelaan, T.J. Gay, Quantum-mechanical analysis of a longitudinal Stern-Gerlach effect. Phys. Rev. Lett. **86**, 4508 (2001)
72. B.M. Garraway, S. Stenholm, Does a flying electron spin? Contemp. Phys. **43**, 147 (2002)
73. J. Byrne, Study of a proposal for determining the g-factor anomaly for electrons by resonance excitation in a magnetic field. Can. J. Phys. **41**, 1571 (1963)
74. H.G. Dehmelt, New continuous Stern-Gerlach effect and a hint of "the" elementary particle. Z. Phys. D **10**, 127 (1988)
75. P. Ghosh, *Ion Traps* (Oxford University Press, Oxford, 1995)
76. G. Werth, V.N. Gheorghe, F.G. Major, *Charged Particle Traps* (Springer, Heidelberg, 2005)
77. C.J. Foot, *Atomic Physics, Oxford Master Series in Atomic, Optical and Laser Physics* (Oxford University Press, New York, 2005) reprint 2009
78. L.S. Brown, G. Gabrielse, Geonium theory: physics of a single electron or ion in a Penning trap. Rev. Mod. Phys. **58**, 233 (1986)
79. G. Gabrielse, L. Haarsma, S.L. Rolston, Open-endcap Penning traps for high precision experiments. Int. J. Mass Spectr. Ion Proc. **88**, 319 (1989)
80. R.S. Van Dyck Jr., F.L. Moore, D.L. Farnham, P.B. Schwinberg, Number dependency in the compensated Penning trap. Phys. Rev. A **40**, 6308 (1989)
81. M. Vogel, W. Quint, W. Nörtershäuser, Trapped ion oscillation frequencies as sensors for spectroscopy. Sensors **10**, 2169 (2010)
82. L.S. Brown, G. Gabrielse, Precision spectroscopy of a charged particle in an imperfect Penning trap. Phys. Rev. A **25**, 2423 (1982)
83. G. Gabrielse, Why is sideband mass spectrometry possible with ions in a Penning trap? Phys. Rev. Lett. **102**, 172501 (2009)
84. M. Breitenfeldt, S. Baruah, K. Blaum, A. Herlert, M. Kretzschmar, F. Martinez, G. Marx, L. Schweikhard, N. Walsh, The elliptical Penning trap: experimental investigations and simulations. Int. J. Mass Spectrom. **275**, 34 (2008)
85. D.F.A. Winters, M. Vogel, D. Segal, R.C. Thompson, Electronic detection of charged particle effects in a Penning trap. J. Phys. B **39**, 3131 (2006)
86. J.V. Porto, Series solution for the image charge fields in arbitrary cylindrically symmetric Penning traps. Phys. Rev. A **64**, 023403 (2001)
87. E. Fischbach, N. Nakagawa, Apparatus-dependent contributions to g-2 and other phenomena. Phys. Rev. D **30**, 2356 (1984)
88. D.G. Boulware, L.S. Brown, T. Lee, Apparatus-dependent contributions to $g-2$? Phys. Rev. D **32**, 729 (1985)
89. H. Häffner et al., Double Penning trap technique for precise g factor determinations in highly charged ions. Eur. Phys. J. D **22**, 163 (2003)
90. C. Roux, Ch. Böhm, A. Dörr, S. Eliseev, S. George, Yu. Novikov, J. Repp, S. Sturm, S. Ulmer, K. Blaum, The trap design of PENTATRAP. Appl. Phys. B **107**, 997 (2012)
91. J. Yu, M. Desaintfuscien, F. Plumelle, Ion density limitation in a Penning trap due to the combined effect of asymmetry and space charge. Appl. Phys. B **48**, 51 (1989)
92. A. Abragam, *The Principles of Nuclear Magnetism* (Clarendon Press, Oxford, 1962)
93. A.A. Sokolov, I.M. Ternov, Dokl. Akad. Nauk SSSR **153**, 1052 (1963)
94. J. Schwinger, On gauge invariance and vacuum polarization. Phys. Rev. **82**, 664 (1951)
95. E.T. Akhmedov, D. Singleton, On the relation between Unruh and Sokolov-Ternov effects. Int. J. Mod. Phys. A **22**, 4797 (2007)

96. H. Mendlowitz, Double scattering of electrons with magnetic interaction. Phys. Rev. **97**, 33 (1955)
97. V. Bargmann, L. Michel, V.L. Telegdi, Precession of the polarization of particles moving in a homogeneous electromagnetic field. Phys. Rev. Lett. **2**, 435 (1958)
98. F. Combley, F.J.M. Ferley, J.H. Field, E. Picasso, $g - 2$ experiments as a test of special relativity. Phys. Rev. Lett. **42**, 1383 (1979)
99. J. DiSciacca, G. Gabrielse, Direct measurement of the proton magnetic moment. Phys. Rev. Lett. **108**, 153001 (2012)
100. A. Mooser et al., Resolution of single spin flips of a single proton. Phys. Rev. Lett. **110**, 140405 (2013)
101. J. Verdú et al., Calculation of electrostatic fields using quasi-Green's functions: application to the hybrid Penning trap. New J. Phys. **10**, 103009 (2008)
102. G. Gabrielse, F.C. Macintosh, Cylindrical Penning traps with orthogonalized anharmonicity compensation. Int. J. Mass. Spec. Ion Proc. **57**, 1 (1984)
103. B. Schabinger et al., Experimental g factor of hydrogenlike silicon-28. Eur. Phys. J. D **66**, 71 (2012)
104. S. Sturm, A. Wagner, B. Schabinger, K. Blaum, Phase-sensitive cyclotron frequency measurements at ultralow energies. Phys. Rev. Lett. **107**, 143003 (2011)
105. J. Alonso, K. Blaum, S. Djekic, H.J. Kluge, W. Quint, B. Schabinger, S. Stahl, J. Verdu, M. Vogel, G. Werth, A miniature electron beam ion source for in-trap creation of highly charged ions. Rev. Sci. Inst. **77**, 03A901 (2006)
106. B. Schabinger, S. Sturm, K. Blaum, W. Quint, A. Wagner, G. Werth, Creation of highly-charged calcium ions for the g-factor determination of the bound electron. J. Phys. Conf. Ser. **163**, 012108 (2009)
107. W. Lotz, An empirical formula for the electron-impact ionization cross-section. Z. Phys. **206**, 205 (1967)
108. V.A. Bernshtam, YuV Ralchenko, Y. Maron, Empirical formula for cross section of direct electron-impact ionization of ions. J. Phys. B: At. Mol. Opt. Phys. **33**, 5025 (2000)
109. R. Mann, Total one-electron capture cross sections for Ar^{q+} and I^{q+} ions in slow collisions on H_2 and He. Z. Phys. D **3**, 85 (1986)
110. A. Müller, E. Salzborn, Scaling of cross sections for multiple electron transfer to highly charged ions colliding with atoms and molecules. Phys. Lett. A **62**, 391 (1977)
111. G. Weinberg et al., Electron capture from H_2 to highly charged Th and Xe ions trapped at center-of-mass energies near 6 eV. Phys. Rev. A **57**, 4452 (1998)
112. H. Knudsen, H.K. Haugen, P. Hvelplund, Single-electron-capture cross section for medium- and high-velocity, highly charged ions colliding with atoms. Phys. Rev. A **23**, 597 (1981)
113. S. Djekic et al., Temperature measurement of a single ion in a Penning trap. Eur. Phys. J. D **31**, 451 (2004)
114. D.J. Wineland, H.G. Dehmelt, Principles of the stored ion calorimeter. J. Appl. Phys. **46**, 919 (1975)
115. D.J. Wineland, H.G. Dehmelt, Line shifts and widths of axial, cyclotron and g-2 resonances in tailored, stored electron (ion) cloud. Int. J. Mass Spectrom. Ion Proc. **16**, 338 (1975)
116. D.J. Wineland, H.G. Dehmelt, Line shifts and widths of axial, cyclotron and g-2 resonances in tailored, stored electron (ion) cloud. Int. J. Mass Spectrom. Ion Proc. **19**, 251 (1976)
117. S. Ulmer, H. Kracke, K. Blaum, S. Kreim, A. Mooser, W. Quint, C.C. Rodegheri, J. Walz, The quality factor of a superconducting rf resonator in a magnetic field. Rev. Sci. Instrum. **80**, 123302 (2009). doi.org/10.1063/1.3271537
118. L. Gruber, J.P. Holder, D. Schneider, Formation of strongly coupled plasmas from multi-component ions in a Penning trap. Phys. Scr. **71**, 60 (2005)
119. B. D'Urso, B. Odom, G. Gabrielse, Feedback cooling of a one-electron oscillator. Phys. Rev. Lett. **90**, 043001 (2003)
120. W.M. Itano, J.C. Bergquist, J.J. Bollinger, D.J. Wineland, Cooling methods in ion traps. Phys. Scr. **T59**, 106 (1995)

121. P.J. Mohr, B.N. Taylor, D.B. Newell, CODATA recommended values of the fundamental physical constants 2006. Rev. Mod. Phys. **80**, 633 (2008)
122. A. Mooser, H. Kracke, K. Blaum, S.A. Bräuninger, K. Franke, C. Leiteritz, W. Quint, C.C. Rodegheri, S. Ulmer, J. Walz, Resolution of single spin flips of a single proton. Phys. Rev. Lett. **110**, 140405 (2013)
123. S. Stahl et al., Phase-sensitive measurement of trapped particle motions. J. Phys. B **38**, 297 (2005)
124. E.A. Cornell et al., Single-ion cyclotron resonance measurement of M(CO^+)/M(N_2^+). Phys. Rev. Lett. **63**, 1674 (1989)
125. S. Eliseev, K. Blaum, M. Block, C. Droese, M. Goncharov, E. Minaya Ramirez, D.A. Nesterenko, Yu.N. Novikov, L. Schweikhard, Phase-imaging ion-cyclotron-resonance measurements for short-lived nuclides. Phys. Rev. Lett. **110**, 082501 (2013)
126. D.B. Pinegar, K. Blaum, T.P. Biesiadzinski, S.L. Zafonte, R.S. Van Dyck, Jr., Stable voltage source for Penning trap experiments. Rev. Sci. Instrum. **80**, 064701 (2009)
127. R.S. Van Dyck Jr., P.B. Schwinberg, H.G. Dehmelt, Electron magnetic moment from geonium spectra: early experiments and background concepts. Phys. Rev. D **34**, 722 (1986)
128. R.S. Van Dyck Jr., F.L. Moore, D.L. Farnham, P.B. Schwinberg, Variable magnetic bottle for precision geonium experiments. Rev. Sci. Inst. **57**, 593 (1986)
129. N. Hermanspahn et al., Observation of the continuous Stern-Gerlach effect on an electron bound in an atomic ion. Phys. Rev. Lett. **84**, 427 (2000)
130. A. Mooser, S. Bräuninger, K. Franke, H. Kracke, C. Leiteritz, C.C. Rodegheri, H. Nagahama, G. Schneider, C. Smorra, K. Blaum, Y. Matsuda, W. Quint, J. Walz, Y. Yamazaki, S. Ulmer, Demonstration of the double Penning Trap technique with a single proton. Phys. Lett. B **723**, 78 (2013)
131. E.A. Cornell, R.M. Weisskoff, K.R. Boyce, D.E. Pritchard, Mode coupling in a Penning trap: π pulses and a classical avoided crossing. Phys. Rev. A **41**, 312 (1990)
132. M. Kretzschmar, in *Trapped Charged Particles and Fundamental Physics* (1999), p. 242
133. S. Ulmer et al., Direct measurement of the free cyclotron frequency of a single particle in a Penning trap. Phys. Rev. Lett. **107**, 103002 (2011)
134. D. Hanneke, S. Fogwell, G. Gabrielse, New measurement of the electron magnetic moment and the fine structure constant. Phys. Rev. Lett. **100**, 120801 (2008)
135. G.W.F. Drake (ed.), *Handbook of Atomic, Molecular and Optical Physics* (Spinger, Heidelberg, 2006)
136. R.S. van Dyck, P.B. Schwinberg, H.G. Dehmelt, New high-precision comparison of electron and positron g factors. Phys. Rev. Lett. **59**, 26 (1987)
137. G. Audi, A.H. Wapstra, C. Thibault, The AME2003 atomic mass evaluation (II). Tables, graphs, and references. Nucl. Phys. **A729**, 337 (2003)
138. R.L. Kelly, NIST atomic spectra database: unpublished level list prepared for line classifications published in atomic and ionic spectrum lines below 2000 angstroms: hydrogen through krypton. J. Phys. Chem. Ref. Data **16**(1), 1 (1987)
139. M. Redshaw, J. McDaniel, E.G. Myers, Dipole moment of PH+ and the atomic masses of ^{28}Si, ^{31}P by comparing cyclotron frequencies of two ions simultaneously trapped in a Penning trap. Phys. Rev. Lett. **100**, 093002 (2008)
140. W. Martin, R. Zalubas, NIST atomic spectra database: energy levels of silicon, Si I through Si XIV. J. Phys. Chem. Ref. Data **12**, 323 (1983)
141. P.J. Mohr, B.N. Taylor, D.B. Newell, CODATA recommended values of the fundamental physical constants 2010, URL http://physics.nist.gov/constants (2012)
142. A.A. Schupp, R.W. Pidd, H.R. Crane, Measurement of the g factor of free, high-energy electrons. Phys. Rev. **121**, 1 (1961)
143. G. Graeff, E. Klempt, G. Werth, Method for measuring the anomalous magnetic moment of free electrons. Z. Phys. A **222**, 201 (1969)
144. J.C. Wesley, A. Rich, High-field electron $g-2$ measurement. Phys. Rev. A **4**, 1341 (1971)
145. F.G. Major, G. Werth, High-resolution magnetic hyperfine resonance in harmonically bound ground-state ^{199}Hg ions. Phys. Rev. Lett. **30**, 1155 (1973)

146. M. McGuire, R. Petsch, G. Werth, Precision determination of the ground-state hyperfine separation in $^{199}Hg^{+}$ using the ion-storage technique. Phys. Rev. A **17**, 1999 (1978)
147. R. Blatt, G. Werth, Precision ground state Hfs-separation of ^{137}Ba. Z. Phys. A **299**, 93 (1981)
148. R. Blatt, H. Schnatz, G. Werth, Ultrahigh-resolution microwave spectroscopy on trapped $^{171}Yb^{+}$ ions. Phys. Rev. Lett. **48**, 1601 (1982)
149. X. Feng, G.Z. Li, G. Werth, High-precision hyperfine spectroscopy in M1–M1 double-resonance transitions on trapped $^{207}Pb^{+}$. Phys. Rev. A **46**, 2959 (1992)
150. T. Nakamura et al., Precision spectroscopy of the Zeeman splittings of the $^{9}Be^{+}$ $2^2S_{1/2}$ hyperfine structure for nuclear structure studies. Opt. Commun. **205**, 329 (2002)
151. J.R. Crespo López-Urrutia, P. Beiersdorfer, D.W. Savin, K. Widmann, Direct observation of the spontaneous emission of the hyperfine transition $F = 4$ to $F = 3$ in ground state hydrogenlike $^{165}Ho^{66+}$ in an electron beam ion trap. Phys. Rev. Lett. **77**, 826 (1996)
152. J.R. Crespo López-Urrutia, P. Beiersdorfer, K. Widmann, B.B. Birkett, A.-M. Martensson-Pendrill, M.G.H. Gustavsson, Nuclear magnetization distribution radii determined by hyperfine transitions in the 1s level of H-like ions $^{185}Re^{74+}$ and $^{187}Re^{74+}$. Phys. Rev. A **57**, 879 (1998)
153. P. Beiersdorfer et al., Hyperfine structure of hydrogenlike thallium isotopes. Phys. Rev. A **64**, 032506 (2001)
154. P. Seelig et al., Ground state hyperfine splitting of hydrogenlike $^{207}Pb^{81+}$ by laser excitation of a bunched ion beam in the GSI experimental storage ring. Phys. Rev. Lett. **81**, 4824 (1998)
155. S. Borneis et al., Ground state hyperfine structure of heavy hydrogen like ions. Hyp. Int. **127**, 305 (2000)
156. I. Klaft et al., Precision laser spectroscopy of the ground state hyperfine splitting of hydrogenlike $^{209}Bi^{82+}$. Phys. Rev. Lett. **73**, 2425 (1994)
157. P. Beiersdorfer, A.L. Osterheld, J.H. Scofield, J.R. Crespo López-Urrutia, K. Widmann, Measurement of QED and hyperfine splitting in the $2s_{1/2}$ - $2p_{3/2}$ X-ray transition in Li-like $^{209}Bi^{80+}$. Phys. Rev. Lett. **80**, 3022 (1998)
158. A.N. Artemyev, V.M. Shabaev, G. Plunien, G. Soff, V.A. Yerokhin, Vacuum-polarization corrections to the hyperfine splitting in heavy ions and to the nuclear magnetic moments. Phys. Rev. A **63**, 062504 (2001)
159. V.M. Shabaev, A.N. Artemyev, V.A. Yerokhin, O.M. Zherebtsov, G. Soff, Towards a test of QED in investigations of the hyperfine splitting in heavy ions. Phys. Rev. Lett. **86**, 3959 (2001)
160. V.A. Yerokhin, A.N. Artemyev, V.M. Shabaev, G. Plunien, All-orders results for the one-electron QED correction to the hyperfine structure in light H-like ions. Phys. Rev. A **72**, 052510 (2005)
161. A.A. Elizarov, V.M. Shabaev, N.S. Oreshkina NS, I.I. Tupitsyn, T. Stoehlker, The hyperfine structure of heavy hydrogen-like ions: calculation based on experimental data on muonic atoms. Opt. Spectrosc. **100**, 361 (2006)
162. D.L. Moskovkin, V.M. Shabaev, Zeeman effect of the hyperfine-structure levels in hydrogen-like ions. Phys. Rev. A **73**, 052506 (2006)
163. D.L. Moskovkin, V.M. Shabaev, W. Quint, Zeeman effect of the hyperfine structure levels in lithiumlike ions. Phys. Rev. A **77**, 063421 (2008)
164. A.V. Volotka, D.A. Glazov, I.I. Tupitsyn, N.S. Oreshkina, G. Plunien, V.M. Shabaev, Ground-state hyperfine structure of H-, Li-, and B-like ions in the intermediate-Z region. Phys. Rev. A **78**, 062507 (2008)
165. N.S. Oreshkina, D.A. Glazov, A.V. Volotka, V.M. Shabaev, I.I. Tupitsyn, G. Plunien, Radiative and interelectronic-interaction corrections to the hyperfine splitting in highly charged B-like ions. Phys. Lett. A **372**, 675 (2008)
166. A.V. Volotka, D.A. Glazov, O.V. Andreev, V.M. Shabaev, I.I. Tupitsyn, G. Plunien, Test of many-electron QED effects in the hyperfine splitting of heavy high-Z ions. Phys. Rev. Lett. **108**, 073001 (2012)
167. D. Budker, D.F. Kimball, D.P. DeMille, *Atomic Physics* (Oxford University Press, Oxford, 2004)

168. M. Vogel, W. Quint, Trap-assisted precision spectroscopy of forbidden transitions in highly charged ions. Phys. Rep. **490**, 1 (2010)
169. V. Gerginov, K. Calkins, C.E. Tanner, J.J. McFerran, S. Diddams, A. Bartels, L. Hollberg, Optical frequency measurements of $6s^2S_{1/2}$ to ÂŰ$6p^2P_{1/2}(D1)$ transitions in ^{133}Cs and their impact on the fine-structure constant. Phys. Rev. A **73**, 032504 (2006)
170. M.W. Horbatsch, M. Horbatsch, E.A. Hessels, A universal formula for the accurate calculation of hydrogenic lifetimes. J. Phys. B **38**, 1765 (2005)
171. W. Quint, D. Moskovkin, V.M. Shabaev, M. Vogel, Laser-microwave double-resonance technique for g-factor measurements in highly charged ions. Phys. Rev. A **78**, 032517 (2008)
172. D. Von Lindenfels et al., Bound electron g-factor measurement by double-resonance spectroscopy on a fine-structure transition. Can. J. Phys. **89**, 79 (2011)
173. D. von Lindenfels et al., Experimental access to higher-order Zeeman effects by precision spectroscopy of highly charged ions in a Penning trap. Phys. Rev. A **87**, 023412 (2013)
174. F.A. Jenkins, E. Segré, The quadratic Zeeman effect. Phys. Rev. **55**, 52 (1939)
175. L.I. Schiff, H. Snyder, Theory of the quadratic Zeeman effect. Phys. Rev. **55**, 59 (1939)
176. W.R.S. Garton, F.S. Tomkins, Diamagnetic Zeeman effect and magnetic configuration mixing in long spectral series of BA I. Astrophys. J. **158**, 839 (1969)
177. G. Feinberg, A. Rich, J. Sucher, Quadratic Zeeman effect in positronium. Phys. Rev. A **41**, 3478 (1990)
178. M. Raoult, S. Guizard, D. Gauyacq, A. Matzkin, Quadratic Zeeman effect in Rydberg states of NO. J. Phys. B: At. Mol. Opt. Phys. **38**, S171 (2005)
179. M. Block et al., Discovery of a nuclear isomer in ^{65}Fe with Penning trap mass spectrometry. Phys. Rev. Lett. **100**, 132501 (2008)
180. K. Blaum et al., Population inversion of nuclear states by a Penning trap mass spectrometer. Europhys. Lett. **67**, 586 (2004)
181. T. Eronen et al., Mass and QEC value of ^{26}Si. Phys. Rev. C **79**, 032802 (2009)
182. M. Vogel, W. Quint, Laser spectroscopy by a radiofrequency measurement on a single ion in a Penning trap. New J. Phys. **11**, 013024 (2009)
183. R. Bouchendira et al., New determination of the fine structure constant and test of the quantum electrodynamics. Phys. Rev. Lett. **106**, 080801 (2011)
184. M. Vogel, The anomalous magnetic moment of the electron. Contemp. Phys. **50**, 437 (2009)
185. P. Clade, E. deMirandes, M. Cadoret, S. Guellati-Khelifa, C. Schwob, F. Nez, L. Julien, F. Biraben, Determination of the fine structure constant based on bloch oscillations of ultracold atoms in a vertical optical lattice. Phys. Rev. Lett. **96**, 033001 (2006)
186. D. Hanneke, S. Fogwell Hoogerheide, G. Gabrielse, Cavity control of a single-electron quantum cyclotron: measuring the electron magnetic moment. Phys. Rev. A **83**, 052122 (2011)
187. J.K. Webb, J.A. King, M.T. Murphy, V.V. Flambaum, R.F. Carswell, M.B. Bainbridge, Indications of a spatial variation of the fine structure constant. Phys. Rev. Lett. **107**, 191101 (2011)
188. R. Li, K. Gibble, K. Szymaniec, Improved accuracy of the NPL-CsF_2 primary frequency standard: evaluation of distributed cavity phase and microwave lensing frequency shifts. Metrologia **48**, 283 (2011)
189. T.L. Nicholson, M.J. Martin, J.R. Williams, B.J. Bloom, M. Bishof, M.D. Swallows, S.L. Campbell, J. Ye, Comparison of two independent Sr optical clocks with $1\text{x}10^{-17}$ stability at 10^3s. Phys. Rev. Lett. **109**, 230801 (2012)
190. C.W. Chou, D.B. Hume, J.C.J. Koelemeij, D.J. Wineland, T. Rosenband, Frequency comparison of two high-accuracy Al^+ optical clocks. Phys. Rev. Lett. **104**, 070802 (2010)
191. J.C. Berengut, V.A. Dzuba, V.V. Flambaum, Enhanced laboratory sensitivity to variation of the fine-structure constant using highly charged ions. Phys. Rev. Lett. **105**, 120801 (2010)
192. J.C. Berengut, V.A. Dzuba, V.V. Flambaum, A. Ong, Highly charged ions with E1, M1, and E2 transitions within laser range. Phys. Rev. A **86**, 022517 (2012)
193. E. Peik, C. Tamm, Nuclear laser spectroscopy of the 3.5 eV transition in Th-229. Europhys. Lett. **61**, 181 (2003)

194. M.P. Hehlen, R.R. Greco, W.G. Rellergert, S.T. Sullivan, D. DeMille, R.A. Jackson, E.R. Hudson, J.R. Torgerson, Optical spectroscopy of an atomic nucleus: progress toward direct observation of the 229Th isomer transition. J. Lumin. **133**, 91 (2013)
195. D.L. Farnham, R.S. van Dyck, P.B. Schwinberg, Determination of the electron's atomic mass and the proton/electron mass ratio via Penning trap mass spectroscopy. Phys. Rev. Lett. **75**, 3598 (1995)
196. T. Beier et al., New determination of the electron's mass. Phys. Rev. Lett. **88**, 011603 (2002)
197. V.A. Yerokhin, K. Pachucki, Z. Harman, C.H. Keitel, QED theory of the nuclear magnetic shielding in hydrogenlike ions. Phys. Rev. Lett. **107**, 043004 (2011)
198. J. Zatorski, N.S. Oreshkina, C.H. Keitel, Z. Harman, Nuclear shape effect on the g factor of hydrogenlike ions. Phys. Rev. Lett. **108**, 063005 (2012)
199. W.E. Lamb, Internal diamagnetic fields. Phys. Rev. **60**, 817 (1941)
200. F.D. Feiock, W.R. Johnson, Relativistic evaluation of internal diamagnetic fields for atoms and ions. Phys. Rev. Lett. **21**, 785 (1968)
201. W.C. Dickinson, Hartree computation of the internal diamagnetic field for atoms. Phys. Rev. **80**, 563 (1950)
202. P.J. Mohr, B.N. Taylor, CODATA recommended values of the fundamental physical constants 1998. Rev. Mod. Phys. **72**, 351 (2000)
203. P.J. Mohr, B.N. Taylor, CODATA recommended values of the fundamental physical constants 2002. Rev. Mod. Phys. **77**, 1 (2005)
204. D.L. Moskovkin, N.S. Oreshkina, V.M. Shabaev, T. Beier, G. Plunien, W. Quint, G. Soff, g factor of hydrogen-like ions with nonzero nuclear spin. Phys. Rev. A **70**, 032105 (2004)
205. D.L. Moskovkin, V.M. Shabaev, W. Quint, g factor of Li-like ions with a nonzero nuclear spin. Opt. Spectrosc. **104**, 637 (2008)

Chapter 4
QED Theory of the Bound-Electron Magnetic Moment

D. A. Glazov, A. V. Volotka, V. M. Shabaev and G. Plunien

Abstract The bound electron in the few-electron ion experiences strong electromagnetic interaction with the nucleus. The adequate description of its properties is only possible within the quantum electrodynamics for bound states. Both high-precision measurements and accurate theoretical prediction are available today for the magnetic moments of few-electron systems and the ongoing progress is observed. These investigations result in stringent tests of the theoretical methods, in determination of the fundamental constants and nuclear properties. Application of the quantum electrodynamical theory to the bound-electron magnetic moment is presented in this chapter.

4.1 Introduction

The bound-electron magnetic moment manifests itself in Zeeman effect—a splitting of energy levels of an atom placed in magnetic field **B** according to the projection M_J of the angular momentum **J** onto the direction of the field,

$$\Delta E = g\mu_{\mathrm{B}} B M_J, \qquad (4.1)$$

D. A. Glazov (✉) · A. V. Volotka · V. M. Shabev
Department of Physics, St. Petersburg State University, St Petersburg, Russia
e-mail: glazov.d.a@gmail.com

A. V. Volotka
e-mail: volotka@pcqnt1.phys.spbu.ru

V. M. Shabaev
e-mail: shabaev@pcqnt1.phys.spbu.ru

D. A. Glazov · A. V. Volotka · G. Plunien
Institut für Theoretische Physik, Technische Universität Dresden, Dresden, Germany
e-mail: Guenter.Plunien@tu-dresden.de

W. Quint and M. Vogel (eds.), *Fundamental Physics in Particle Traps*,
Springer Tracts in Modern Physics 256, DOI: 10.1007/978-3-642-45201-7_4,

where $\mu_{\rm B} = |e|\hbar/(2m_e)$ is the Bohr magneton, and the g-factor is the dimensionless parameter, which is the subject to calculation for the particular system under consideration.

Below we restrict ourselves with the case of one electron over closed shells, if any,—e.g. hydrogen-, lithium- or boron-like systems. In this case, the g-factor is primarily determined by the angular quantum numbers of the valence electron, and the interaction with the nucleus and with the closed electron shells results in relatively small corrections. An important property of this corrections is that almost all of them have the relativistic origin and therefore should be considered within the appropriate relativistic theory, which is the bound-state QED. Below we consider how the most accurate values of the g-factor in few-electron ions can be found within the BS-QED.

The relativistic units ($\hbar = c = m_e = 1$) and the Heaviside charge unit ($\alpha = e^2/(4\pi),\ e < 0$) are used throughout the chapter.

4.2 Furry Picture of QED

Theoretical description of atomic system may start from the approximate picture, where the nucleus is an infinitely heavy source of a static electric field. Let $V_{\rm nucl}(r)$ be the corresponding spherically symmetric potential for the electron. Within the model of point-like nucleus it is the Coulomb potential,

$$V_{\rm nucl}(r) = -\frac{\alpha Z}{r}. \tag{4.2}$$

In the Furry picture of quantum electrodynamics assumes the nuclear potential is taken into account non-perturbatively. In zeroth approximation the electron wave function obeys the Dirac equation,

$$h\,\psi(\mathbf{r}) = \varepsilon\,\psi(\mathbf{r})\,, \tag{4.3}$$

$$h = -i\,\boldsymbol{\alpha}\cdot\nabla + \beta + V(r). \tag{4.4}$$

The spherically symmetric binding potential $V(r)$ can be the nuclear potential $V_{\rm nucl}(r)$ or some effective screening potential, as considered in Sect. 4.2.1. We designate as $|a\rangle = |n\kappa m_j\rangle$ the solutions to the Dirac Eq. (4.3) with principle quantum number n, the relativistic angular quantum number $\kappa = (-1)^{j+l+1/2}\,(j+1/2)$, the total angular momentum j, its projection m_j, and the parity $l = j \pm 1/2$. The corresponding bispinor wave function can be presented as

$$\psi(\mathbf{r}) = \begin{pmatrix} g_{n\kappa}(r)\Omega_{\kappa m}(\mathbf{n}) \\ if_{n\kappa}(r)\Omega_{-\kappa m}(\mathbf{n}) \end{pmatrix} = \frac{1}{r}\begin{pmatrix} G_{n\kappa}(r)\Omega_{\kappa m}(\mathbf{n}) \\ iF_{n\kappa}(r)\Omega_{-\kappa m}(\mathbf{n}) \end{pmatrix}. \tag{4.5}$$

For the Coulomb potential (4.2) the solutions to the Dirac equation can be found analytically. In particular, the energies of the bound states ($0 < \varepsilon < 1$) are given by

$$\varepsilon_{n\kappa} = \sqrt{1 - \frac{(\alpha Z)^2}{(n_r + \sqrt{\kappa^2 - (\alpha Z)^2})^2 + (\alpha Z)^2}}, \tag{4.6}$$

where $n_r = n - j - 1/2$ is the radial quantum number.

The homogeneous magnetic field can be described by the vector potential

$$\mathbf{A}(\mathbf{r}) = \frac{1}{2}\,[\mathbf{B} \times \mathbf{r}]\,. \tag{4.7}$$

It is introduced in the Dirac equation by means of the minimal coupling, which yields the following interaction operator

$$V_{\rm magn} = -e\alpha \cdot \mathbf{A}(\mathbf{r}) = -\frac{e}{2}\,\mathbf{B} \cdot [\mathbf{r} \times \alpha]. \tag{4.8}$$

The leading-order contribution to the Zeeman effect is represented by the diagram at Fig. 4.1. According to the Feynman rules, the corresponding energy shift is

$$\Delta E = \langle a|V_{\rm magn}|a\rangle. \tag{4.9}$$

The g-factor can be found through (4.1), where the projection M_J is the one-electron projection m_j in the case of one electron over closed shells,

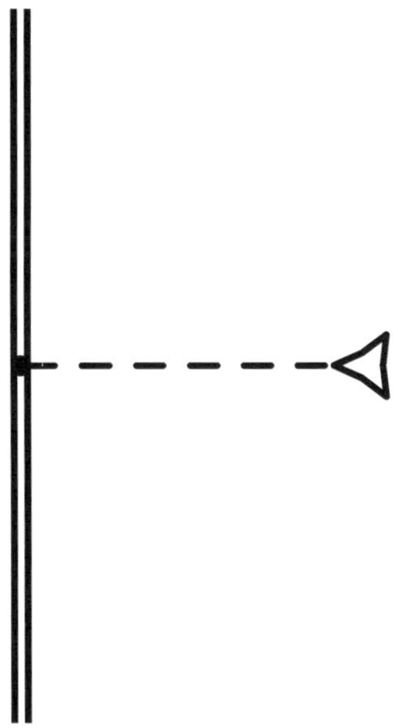

Fig. 4.1 Leading-order diagram for the g factor, representing the interaction of the bound electron with the external magnetic field

$$g = \frac{1}{m_j} \langle a|[\mathbf{r} \times \alpha]_z|a\rangle. \tag{4.10}$$

The angular integration yields

$$g = \frac{2\kappa}{j(j+1)} \int_0^\infty dr\, r\, G_{n\kappa}(r)\, F_{n\kappa}(r). \tag{4.11}$$

For the Coulomb potential it can be evaluated analytically to yield the so-called Dirac value

$$g_{\mathrm{D}} = \frac{\kappa}{2j(j+1)} (2\kappa\varepsilon_{n\kappa} - 1), \tag{4.12}$$

where $\varepsilon_{n\kappa}$ is the Dirac energy (4.6). This expression can be expanded in powers of αZ with the zeroth-order term given by the well known Landé formula,

$$g_{\mathrm{L}} = \frac{2\kappa}{2\kappa+1} = 1 + \frac{j(j+1) - l(l+1) + \frac{3}{4}}{2j(j+1)}, \tag{4.13}$$

In particular, for $1s$, $2s$, and $2p$ states one finds

$$\begin{aligned} g_{\mathrm{D}}[1s_{1/2}] &= \frac{2}{3}\left[2\sqrt{1-(\alpha Z)^2}+1\right] \\ &= 2 - \frac{2}{3}(\alpha Z)^2 - \frac{1}{6}(\alpha Z)^4 - \cdots, \end{aligned} \tag{4.14}$$

$$\begin{aligned} g_{\mathrm{D}}[2s_{1/2}] &= \frac{2}{3}\left[\sqrt{2\left(1+\sqrt{1-(\alpha Z)^2}\right)}+1\right] \\ &= 2 - \frac{1}{6}(\alpha Z)^2 - \frac{5}{96}(\alpha Z)^4 - \cdots, \end{aligned} \tag{4.15}$$

$$\begin{aligned} g_{\mathrm{D}}[2p_{1/2}] &= \frac{2}{3}\left[\sqrt{2\left(1+\sqrt{1-(\alpha Z)^2}\right)}-1\right] \\ &= \frac{2}{3} - \frac{1}{6}(\alpha Z)^2 - \frac{5}{96}(\alpha Z)^4 - \cdots, \end{aligned} \tag{4.16}$$

$$\begin{aligned} g_{\mathrm{D}}[2p_{3/2}] &= \frac{4}{15}\left[2\sqrt{4-(\alpha Z)^2}+1\right] \\ &= \frac{4}{3} - \frac{2}{15}(\alpha Z)^2 - \frac{1}{120}(\alpha Z)^4 - \cdots. \end{aligned} \tag{4.17}$$

The quantum electrodynamics predicts deviations of the g factors from the Dirac values (4.12, 4.14). In particular, the free electron g factor is given by

$$g_{\mathrm{free}} = 2\left[1 + \sum_{i=1}^{\infty} A^{(2i)} \left(\frac{\alpha}{\pi}\right)^i\right], \tag{4.18}$$

where the coefficient $A^{(2)}$, $A^{(4)}$, and $A^{(6)}$ are known analytically, while $A^{(8)}$ is evaluated numerically,

$$\begin{aligned} A^{(2)} &= \frac{1}{2}, \\ A^{(4)} &= -0.328\,478\,965\,579\ldots, \\ A^{(6)} &= 1.181\,241\,456\ldots, \\ A^{(8)} &= -1.9144(35). \end{aligned} \tag{4.19}$$

For further detail on these outstanding research we refer the reader to the Chaps. 1 and 2 of the present book.

For low-Z systems the g factor, as well as any other quantity, like e.g. binding energy, can be presented as an expansion in two parameters α and αZ

$$g = \sum_{i,k=0}^{\infty} \alpha^i (\alpha Z)^k B_{i,k}. \tag{4.20}$$

Since the formulation of relativistic quantum mechanics till present a lot of theoretical efforts were devoted to analytical and numerical evaluation of the coefficients $B_{i,k}$. However, the expansion in αZ is poorly convergent in general. For middle-Z and high-Z systems the complete αZ-dependent results are preferable. On the other hand, in this case, $1/Z$ is suitable expansion parameter for many-electron contributions alongside with α. It represents the ratio of electron–electron to electron–nucleus interaction strengths. Among other quantities, related to the bound electron, like e.g. binding energy or hyperfine splitting, the g factor has a specific property, that all the many-electron contributions possess the $\sim(\alpha Z)^2$ behaviour, i.e. vanish in the non-relativistic limit.

The complexity of QED treatment grows very rapidly with each order of the perturbation theory. It is manifold: the derivation of general formulas, the identification of the divergences and their proper regularization, time consumption of the numerical implementation.

In the following paragraphs of this section we consider the evaluation of the QED corrections to the g-factor within the Furry picture. A summary of the results obtained within the αZ expansion is given as well.

4.2.1 Screening Potential

The potential $V(r)$ in (4.4) can be not just the nuclear potential $V_{\rm nucl}(r)$, but some effective spherically symmetric potential $V_{\rm eff}(r)$ that approximately takes into account the interelectronic interaction in many-electron systems. Evaluations of some contribution to the g-factor with this potential provides an approximate account for the screening effect on this contribution. An ultimate simplification of this method is the replacement of the nuclear charge Z by the 'effective charge' $Z_{\rm eff}$ A well-known choice of $V_{\rm eff}(r)$ is the core-Hartree (CH) potential

$$V_{\rm eff}(r) = V_{\rm nucl}(r) + \alpha \int_0^\infty dr' \frac{1}{r_>} \rho_c(r'). \tag{4.21}$$

Here ρ_c is the density of the core (closed shells) electrons,

$$\rho_c(r) = \sum_{\kappa_c, n_c} (2j_c + 1) \left(G_c^2(r) + F_c^2(r)\right), \tag{4.22}$$

where κ_c and n_c are the quantum numbers of the closed shells, G_c and F_c are the corresponding radial components of the wave function (4.5), and $2j_c + 1$ reflects the number of electrons in the closed shell. The potential derived from the density-functional theory reads [16, 84]

$$V_{\rm eff}(r) = V_{\rm nucl}(r) + \alpha \int_0^\infty dr' \frac{1}{r_>} \rho_t(r') - x_\alpha \frac{\alpha}{r} \left(\frac{81}{32\pi^2} r \rho_t(r) \right)^{1/3}. \tag{4.23}$$

Here ρ_t is the total electron density, including the closed shells and the valence electron,

$$\rho_t(r) = \left(G_a^2(r) + F_a^2(r)\right) + \sum_{\kappa_c, n_c} (2j_c + 1) \left(G_c^2(r) + F_c^2(r)\right). \tag{4.24}$$

The parameter x_α can be varied from 0 to 1. The cases of values $x_\alpha = 0,\ 2/3$ and 1 are referred to as the Dirac-Hartree, the Kohn-Sham and the Dirac-Slater potentials, respectively. The potentials of these form (4.23) lack proper asymptotic behaviour $-\alpha(Z - N + 1)/r$ at large r, where N is the total number of electrons. It can be restored with the Latter correction [61]. The effective potential $V_{\rm eff}(r)$ (4.21, 4.23) implies self-consistency, i.e. the wave functions used to construct the potential should be solutions to the Dirac equation with the same potential. In practice it can be achieved by iterations until the convergence is reached.

4.2.2 *Effective Hamiltonian*

The value of the QED correction for free electron (4.18) can be employed for approximate evaluation of various QED contributions to the bound-electron g factor. Below, the corresponding effective Hamiltonian is derived.

The operator of the electromagnetic interaction

$$j^\mu A_\mu = ie\overline{\psi}\gamma^\mu \psi A_\mu \tag{4.25}$$

can be transformed to the following form

$$j^\mu A_\mu = j_1^\mu A_\mu + j_2^\mu A_\mu \tag{4.26}$$

$$j_1^\mu A_\mu = \frac{ie}{2}\left(\partial_\mu \overline{\psi}\psi - \overline{\psi}\partial_\mu \psi\right) A^\mu - e^2 \overline{\psi}\psi A^\mu A_\mu \tag{4.27}$$

$$j_2^\mu A_\mu = -\frac{1}{4}\, e\,\overline{\psi}\sigma^{\mu\nu}\psi F_{\mu\nu} = -\frac{1}{2}\, e\,\overline{\psi}\left[\beta(\Sigma\cdot\mathbf{B}) - i\beta(\boldsymbol{\alpha}\cdot\mathbf{E})\right]\psi, \tag{4.28}$$

where $\sigma^{\mu\nu} = \frac{1}{2}(\gamma^\mu\gamma^\nu - \gamma^\nu\gamma^\mu)$ and $F_{\mu\nu} = \partial_\mu A_\nu - \partial_\nu A_\mu$ is the tensor of electromagnetic field. This transformation is known as Gordon decomposition. The first part corresponds to the electromagnetic interaction of a spinless particle. The second part in the non-relativistic limit simplifies to

$$j_2^\mu A_\mu = -g\,\mu_\mathrm{B}\,\phi^\dagger \frac{\boldsymbol{\sigma}}{2}\phi\cdot\mathbf{B}, \tag{4.29}$$

with $g = 2$. Here, ϕ is the large component of ψ, which coincides with the Schrödinger wave function in the non-relativistic limit. This corresponds to the classical formula $-\mu\cdot\mathbf{B}$ for the particle with arbitrary g factor. Therefore, one can substitute $g = 2$ with $2 + \Delta g$ to obtain the additional term to the Dirac Hamiltonian, that approximately accounts for the anomalous magnetic moment,

$$h_\mathrm{rad} = -\Delta g\,\frac{e}{4}\,\left[\beta(\Sigma\cdot\mathbf{B}) - i\beta(\boldsymbol{\alpha}\cdot\mathbf{E})\right]. \tag{4.30}$$

This expression was proposed by Wolfgang Pauli for the nucleons, which obey the Dirac equation at low energies, but exhibit large deviations of the g factors from 2: $g_p \approx 2.793$, $g_n \approx -1.913$. For the free electron $\Delta g = g_\mathrm{free} - 2$ has the quantum electrodynamical origin, and it can be written as an expansion in α/π (4.18).The expression (4.30) can be employed for approximate treatment of QED effects, which is especially useful for evaluation of the higher-order contributions. Since (4.30) is the non-relativistic limit of the full Hamiltonian, the approximation works better for lower Z. Brodsky and Primack [13] showed that the Breit equation with h_rad (4.30) leads to the proper radiative corrections to the bound-electron g factor in the lowest order in αZ. Hegstrom and Grotch used it to evaluate various contributions to the g factor [34, 35, 37–40]. Relativistic many-electron calculations with the Hegstrom Hamiltonian were performed in [3, 27, 30, 105, 114–116].

4.2.3 *One-Electron QED Effects*

In this section we consider quantum electrodynamical corrections to the g-factor, which do not involve the interaction between electrons. These corrections are relevant both for hydrogen-like and many-electron systems. The leading-order one-electron

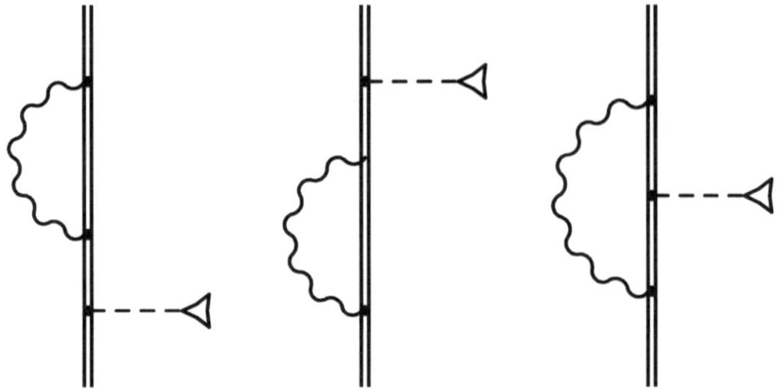

Fig. 4.2 One-loop self-energy diagrams for the g factor

corrections to the g factor arise from the diagrams of self-energy (Fig. 4.2) and vacuum polarization (Fig. 4.5).

4.2.3.1 Self Energy

The contribution of the self-energy diagrams is given by the sum of so-called 'irreducible', 'reducible', and 'vertex' parts:

$$\Delta E_{\mathrm{SE}} = \Delta E_{\mathrm{SE}}^{\mathrm{irr}} + \Delta E_{\mathrm{SE}}^{\mathrm{red}} + \Delta E_{\mathrm{SE}}^{\mathrm{ver}}. \tag{4.31}$$

The irreducible part originates from diagrams (a) and (b) of Fig. 4.2 where the energy of the intermediate state between the magnetic vertex and the self-energy loop is not equal to the energy of the reference state. The corresponding expressions reads

$$\Delta E_{\mathrm{SE}}^{\mathrm{irr}} = 2 \sum_{n_1}^{\varepsilon_{n_1} \neq \varepsilon_a} \left[\frac{\langle a | \Sigma_R(\varepsilon_a) | n_1 \rangle \langle n_1 | V_{\mathrm{magn}} | a \rangle}{\varepsilon_a - \varepsilon_{n_1}} \right]. \tag{4.32}$$

Here $\Sigma_R(E) = \Sigma(E) - \gamma^0 \delta m$, while δm is the mass counterterm and $\Sigma(E)$ is the self-energy operator defined by

$$\langle p | \Sigma(\varepsilon) | q \rangle = \frac{i}{2\pi} \int_{-\infty}^{\infty} d\omega \sum_n \frac{\langle pn | I(\omega) | nq \rangle}{\varepsilon - \omega - u\varepsilon_n}, \tag{4.33}$$

where $u = 1 - i0$. The operator $I(\omega, \mathbf{r}_1, \mathbf{r}_2)$ is derived from the photon propagator, see Sect. 4.2.4. The self-energy operator and the mass counterterm both possess ultraviolet (UV) divergences, which are partly cancelled in the difference $\Sigma_R(E)$. The remainder can be written as

$$\langle p|\Sigma_R(\varepsilon)|q\rangle = B^{(1)}(\varepsilon - \varepsilon_p)\delta_{pq} + \text{finite part}, \tag{4.34}$$

in assumption that $|p\rangle$ and $|q\rangle$ are solutions to the Dirac Eq. (4.3). $B^{(1)}$ is the UV-divergent constant, which can be evaluated explicitly within the particular regularization approach, e.g. dimensional regularization. It is easy to see, that the UV-divergent part of $\Delta E_a^{(\text{SE, irr})}$ appears to be zero.

The 'reducible' part originates partly from diagrams (a) and (b) of Fig. 4.2 and partly from the additional non-diagram terms of the perturbation theory,

$$\begin{aligned}\Delta E_{\text{SE}}^{\text{red}} &= \langle a|\Sigma'(\varepsilon_a)|a\rangle\langle a|V_{\text{magn}}|a\rangle \\ &= -\frac{i}{2\pi}\int_{-\infty}^{\infty} d\omega \sum_{n_1} \frac{\langle an_1|I(\omega)|n_1 a\rangle}{(\varepsilon_a - \omega - u\varepsilon_{n_1})^2}\langle a|V_{\text{magn}}|a\rangle.\end{aligned} \tag{4.35}$$

With the help of Eq. (4.34) one finds the UV-divergent part of $\Delta E_a^{(\text{SE,red})}$ to be $B^{(1)}\langle a|V_{\text{magn}}|a\rangle$. The 'vertex' part corresponds to the diagram (c) of Fig. 4.2,

$$\Delta E_{\text{SE}}^{\text{ver}} = \frac{i}{2\pi}\int_{-\infty}^{\infty} d\omega \sum_{n_1,n_2} \frac{\langle an_2|I(\omega)|n_1 a\rangle\langle n_1|V_{\text{magn}}|n_2\rangle}{(\varepsilon_a - \omega - u\varepsilon_{n_1})(\varepsilon_a - \omega - u\varepsilon_{n_2})}. \tag{4.36}$$

The UV-divergent part of $\Delta E_a^{(\text{SE,ver})}$ is $-B^{(1)}\langle a|V_{\text{magn}}|a\rangle$, so the sum of the reducible and vertex parts is UV-finite. In practical calculations, the isolation of the UV-divergences is performed after the separation of the zero- and one-potential terms, which are treated in the momentum space. It was first proposed by Snyderman [96] for the self-energy correction to the energy levels and implemented numerically by Blundell and Snyderman [8]. This separation is based on the so-called 'potential expansion' schematically depicted in Fig. 4.3 for the self-energy diagram ($\Delta E_{\text{SE}}^{\text{irr}}$, $\Delta E_{\text{SE}}^{\text{red}}$). Similar expansion for the vertex diagram ($\Delta E_{\text{SE}}^{\text{ver}}$) is shown in Fig. 4.4. Strictly speaking, only the zero-potential term possesses UV-divergencies and requires separate treatment for the reducible and vertex parts. However, separation of the one-potential term significantly facilitates the convergence of the partial-wave expansion of the remaining part, evaluated in coordinate space. It was

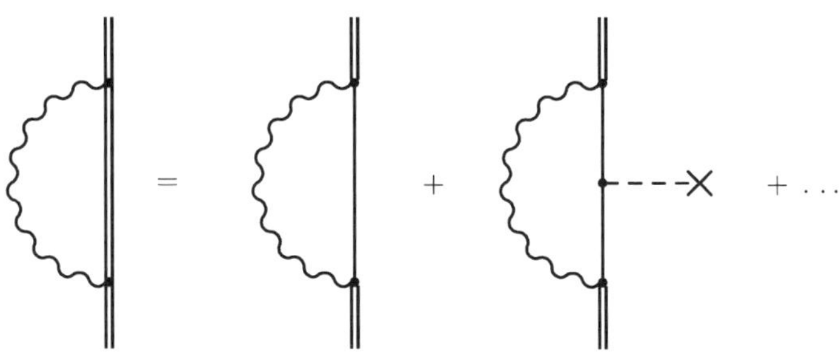

Fig. 4.3 Potential expansion for the self-energy diagram

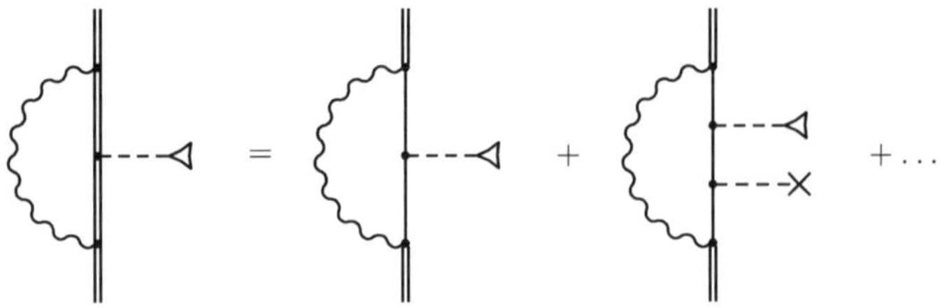

Fig. 4.4 Potential expansion for the vertex diagram

first employed in [119], together with a special treatment of the momentum-space integration, which utilizes the particular form of the operator $V_{\rm magn}$ in the momentum space.

Vertex and reducible parts suffer from infrared (IR) divergences, which arise in terms with $\varepsilon_{n_1} = \varepsilon_{n_2} = \varepsilon_a$. These terms can be isolated and evaluated introducing the non-zero photon mass to yield the finite result in total. Another way to treat IR-divergences is to evaluate $\Delta E_{\rm SE}^{\rm red}$ and $\Delta E_{\rm SE}^{\rm ver}$ together to ensure that the divergent terms are summed up before the ω-integration is performed.

The first evaluation of the self-energy correction to the g-factor to all orders in αZ was done by Blundell, Cheng and Sapirstein [9] and by Persson, Salomonson and Lindgren [80] for the $1s$ state. Yerokhin, Indelicato and Shabaev [119, 120] extended the calculation to the $2s$ state and strongly improved the numerical accuracy. This was achieved with the analytical representation of the Coulomb-Green function and with the specific evaluation procedure for the zero- and one-potential terms, as has been mentioned above. Later Yerokhin and Jentschura [121, 122] extended the calculation to the $3s$, $2p_{1/2}$ an $2p_{3/2}$ states and further improved the accuracy for low Z.

4.2.3.2 Vacuum Polarization

There are two types of the vacuum-polarization diagrams: the 'electric-loop' (Fig. 4.5a, b) and the 'magnetic-loop' (Fig. 4.5c),

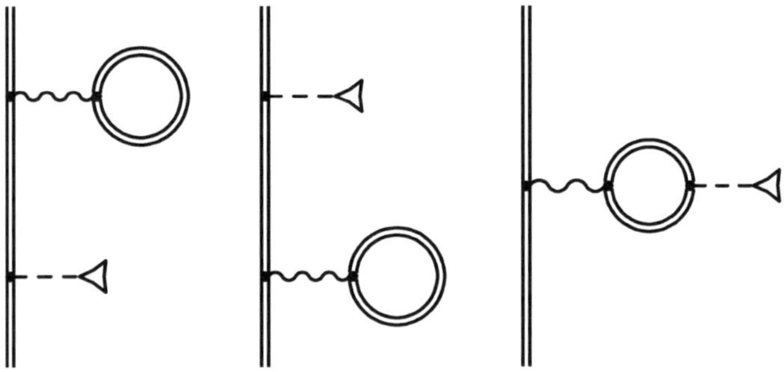

Fig. 4.5 One-loop vacuum polarization diagrams for the g factor

$$\Delta E_{\rm VP} = \Delta E^{\rm el}_{\rm VP} + \Delta E^{\rm ml}_{\rm VP}. \tag{4.37}$$

The contribution of the electric-loop diagrams is given by

$$\Delta E^{\rm el}_{\rm VP} = 2 \sum_{n_1}^{\varepsilon_{n_1} \neq \varepsilon_a} \left[\frac{\langle a | V^{\rm el}_{\rm VP} | n_1 \rangle \langle n_1 | V_{\rm magn} | a \rangle}{\varepsilon_a - \varepsilon_{n_1}} \right], \tag{4.38}$$

where the electric-loop vacuum-polarization potential $V_{\rm VP,el}$ is defined as

$$\langle p | V^{\rm el}_{\rm VP} | q \rangle = -\frac{i}{2\pi} \int_{-\infty}^{\infty} d\omega \sum_n \frac{\langle pn | I(0) | qn \rangle}{\omega - u\varepsilon_n}. \tag{4.39}$$

This divergent expression becomes finite after the charge renormalization procedure [71]. The potential expansion, similar to that for the self-energy diagram, is employed to isolate the divergency (Fig. 4.6). The renormalized expression for the leading term of the expansion, the Uehling potential, reads

$$\begin{aligned} V^{\rm el-Ue}_{\rm VP}(r) = &-\alpha Z \frac{\alpha}{3\pi} \int_1^{\infty} dt \sqrt{t^2 - 1} \left[\frac{2}{t^2} + \frac{1}{t^4} \right] \\ &\times \int d^3\mathbf{r}' \exp(-2|\mathbf{r} - \mathbf{r}'|) \frac{\rho(r')}{|\mathbf{r} - \mathbf{r}'|}, \end{aligned} \tag{4.40}$$

where $\rho(r)$ is the density of the nuclear charge distribution. The remainder, called Wichmann-Kroll potential, is finite. Based on the detailed investigation of the electric-loop vacuum-polarization potential [67, 87, 97, 102, 113], the corresponding correction to the g factor was evaluated for $1s$ [5, 6] and for $2s$ [27] states.

The contribution of the magnetic-loop diagram is given by

$$\Delta E^{\rm ml}_{\rm VP} = \langle a | V^{\rm ml}_{\rm VP} | a \rangle, \tag{4.41}$$

where the magnetic-loop potential $V^{\rm ml}_{\rm VP}$ is defined as

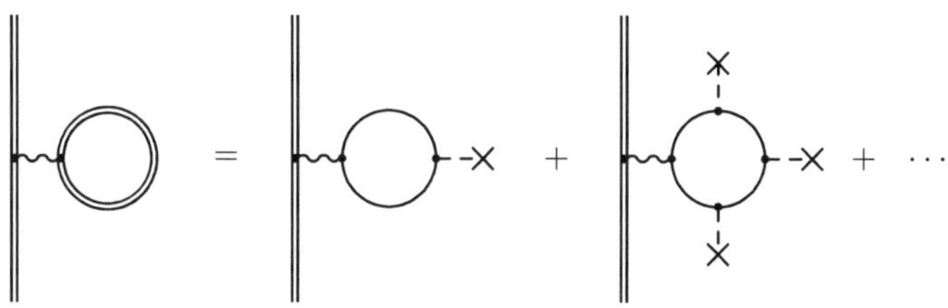

Fig. 4.6 Potential expansion for the electric-loop vacuum-polarization diagram

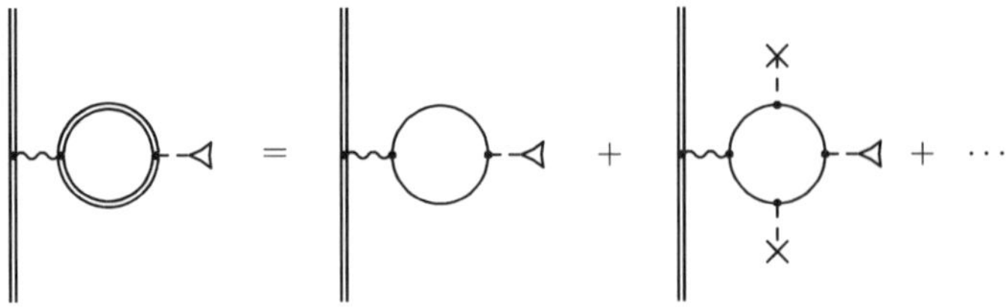

Fig. 4.7 Potential expansion for the magnetic-loop vacuum-polarization diagram

$$\langle p|V_{\mathrm{VP}}^{\mathrm{ml}}|q\rangle = -\frac{i}{2\pi}\int_{-\infty}^{\infty} d\omega \sum_{n_1,n_2} \frac{\langle pn_2|I(0)|qn_1\rangle\langle n_1|V_{\mathrm{magn}}|n_2\rangle}{(\omega - u\varepsilon_{n_1})(\omega - u\varepsilon_{n_2})}. \tag{4.42}$$

The corresponding potential expansion is depicted in Fig. 4.7. Due to the particular properties of the operator V_{magn} (4.8) the Uehling term of $V_{\mathrm{VP,ml}}$ vanishes. The remaining Wichmann-Kroll part has not been calculated rigorously up-to-date. In [54] the leading-order term in αZ was derived for ns states. In [62] the next-order term and more accurate, although not exact, αZ-dependent results were obtained.

4.2.3.3 Expansion in αZ

The diagrams of the second and higher orders in α give rise to the corresponding corrections to the g factor. The all-order in αZ results are not available up-to-date, even for two-loop diagrams. To zeroth order in αZ these terms are related to the free-electron values (4.18, 4.19) [12],

$$\Delta g_{\mathrm{QED}} = -\frac{1}{2\kappa_a + 1}(g_{\mathrm{free}} - 2). \tag{4.43}$$

Moreover, for ns-states the $(\alpha Z)^2$-term is given by

$$\Delta g_{\mathrm{QED}} = \left[1 + \frac{(\alpha Z)^2}{6n^2}\right](g_{\mathrm{free}} - 2). \tag{4.44}$$

This relation was derived for the first order in α for $1s$ state by Grotch [33]. It was generalized to ns state by Grotch and Kashuba [31, 32], while the expression derived in that chapter for arbitrary states was later found to be incomplete for $l \neq 0$. In [17] it was argued that the Eq. (4.44) for $n = 1$ is valid to all orders α, and in [90] this statement was generalized to ns state. This result is based on the approximate Hamiltonian, considered in Sect. 4.2.2.

In [14] the complete expression for $\alpha(\alpha Z)^2$ correction for arbitrary hydrogenic state was derived. This expression involves summation over the intermediate states and is hardly evaluable analytically. An attempt to calculate it numerically for $2p$ state yielded wrong result due to incomplete basis set employed. Recently, Jentschura

[46] recalculated this term within an independent approach, confirmed the expression obtained in [14], and found the correct coefficients of the $\alpha(\alpha Z)^2$ terms for np_j states with $n = 2, 3, 4$, $j = 1/2, 3/2$. It was confirmed by numerical evaluation for $n = 2$ in [121, 122].

One-loop contributions of the higher orders in αZ, which complement the all-order in αZ results, were addressed in [76, 77] (self-energy through the order $\alpha(\alpha Z)^4$ for ns states) and in [49–52, 54, 62] (vacuum polarization through the order $\alpha(\alpha Z)^6$).

Two-loop contributions of the order $\alpha^2(\alpha Z)^4$ were derived by Pachucki and co-authors [76–78] for ns states, while a certain part of the $\alpha^2(\alpha Z)^5$ term (vacuum-polarization diagrams with two closed electric loops) was evaluated by Jentschura [45] for $1s$ state.

We mention here some general properties of the αZ-expansion. There are no terms of the orders αZ, $(\alpha Z)^3$, and the $(\alpha Z)^2$-term is of purely kinematic origin. Terms starting from $(\alpha Z)^4$ are also accompanied by the powers of $\ln(\alpha Z)$, and there are odd powers as $(\alpha Z)^5$ and higher. As noticed in [78], the αZ-expansion appears to be poorly convergent, e.g. for $Z = 20$ for $1s$ state the $(\alpha Z)^2$ and $(\alpha Z)^4$ terms are of the same magnitude due to the large value of the coefficient in the latter.

4.2.4 *Many-Electron QED Effects*

The diagrams with two or more electron lines, connected by photon lines, lead to the contributions of the order $1/Z$ and higher. As mentioned above, all of these terms are of pure relativistic origin and, accordingly, have additional smallness of $(\alpha Z)^2$.

4.2.4.1 Interelectronic Interaction

The contribution of the diagrams without self-energy or vacuum polarization loops is referred to as the interelectronic-interaction correction $\Delta g_{\rm int}$. The diagrams with n photon lines (n-photon exchange) lead to the correction of the order $(\alpha Z)^2/Z^n$,

$$\Delta g_{\rm int} = g_{\rm L}(\alpha Z)^2 \left[\frac{1}{Z} B(\alpha Z) + \frac{1}{Z^2} C(\alpha Z) + \frac{1}{Z^3} D(\alpha Z) + \cdots \right], \tag{4.45}$$

where $g_{\rm L}$ is the non-relativistic value (4.13). One should mention, that the complete evaluation of these terms is only possible within the QED framework. The leading $(\alpha Z)^2$ contribution can be obtained within the Breit approximation employing the quantum mechanical approaches (MBPT, DF, DFS), however the contribution of the negative-energy states requires a special treatment.

The first-order term is represented by the one-photon-exchange diagram (Fig. 4.8). The corresponding contribution to the g factor reads

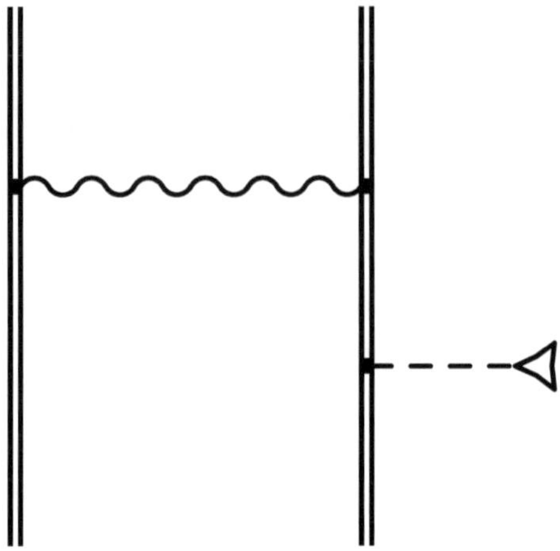

Fig. 4.8 One-photon exchange contribution to the g factor

$$\Delta E_{\rm int}^{(1)} = 2\sum_{m_b}\Bigg\{\sum_{n}^{\varepsilon_n\neq\varepsilon_a} \frac{\langle a|V_{\rm magn}|n\rangle\,[\langle nb|I(0)|ab\rangle - \langle nb|I(\Delta_{ab})|ca\rangle]}{\varepsilon_a-\varepsilon_n}$$
$$+\sum_{n}^{\varepsilon_n\neq\varepsilon_b} \frac{\langle b|V_{\rm magn}|n\rangle\,[\langle na|I(0)|ba\rangle - \langle na|I(\Delta_{ab})|ab\rangle]}{\varepsilon_b-\varepsilon_n}$$
$$-\big[\langle a|V_{\rm magn}|a\rangle - \langle b|V_{\rm magn}|b\rangle\big]\,\langle ab|I'(\Delta_{ab})|ba\rangle\Bigg\}. \qquad (4.46)$$

Here $\Delta_{ab} = \varepsilon_a - \varepsilon_b$, and $I(\omega)$ is the QED interelectronic-interaction operator, directly related to the photon propagator $D_{\mu\nu}(\omega)$,

$$I(\omega,\mathbf{r}_1,\mathbf{r}_2) = e^2\alpha^\mu\alpha^\nu D_{\mu\nu}(\omega,\mathbf{r}_1-\mathbf{r}_2), \qquad I'(\omega) = \frac{\mathrm{d}I(\omega)}{\mathrm{d}\omega}. \qquad (4.47)$$

In the Feynman gauge it is given by

$$I(\omega,\mathbf{r}_1,\mathbf{r}_2) = \alpha(1-\boldsymbol{\alpha}_1\cdot\boldsymbol{\alpha}_2)\frac{\exp(i\omega r_{12})}{r_{12}}. \qquad (4.48)$$

In the Coulomb gauge it is given by

$$I(\omega,\mathbf{r}_1,\mathbf{r}_2) = \alpha\Bigg(\frac{1}{r_{12}} - \boldsymbol{\alpha}_1\cdot\boldsymbol{\alpha}_2\frac{\exp(i\omega r_{12})}{r_{12}}$$
$$+\Big[\boldsymbol{\alpha}_1\cdot\boldsymbol{\nabla}_1,\Big[\boldsymbol{\alpha}_1\cdot\boldsymbol{\nabla}_1\frac{\exp(i\omega r_{12})-1}{r_{12}}\Big]\Big]\Bigg). \qquad (4.49)$$

For details see e.g. [92]. The one-photon-exchange correction to the g factor was calculated in [90] for lithium-like ions in a wide range of Z and in [25, 95] for boron-like argon and lead. The finite-basis-set method with the basis functions constructed from B-splines was employed to perform the infinite summation over the Dirac spectrum.

A distinct property of the magnetic interaction operator V_{magn} (4.8) is the mixing of the 'large' and 'small' components of the wave function (see Eq. (4.11)). While for positive-energy Dirac states the mean magnitude of $F(r)$ is roughly the mean magnitude of $G(r)$ times αZ, for negative-energy states this relation is opposite. As a consequence, the matrix elements of V_{magn} between the positive-energy states are generally smaller, with the factor αZ, than those between the positive- and negative-energy states. Due to this, and due to the fact that the non-relativistic limit of the interelectronic-interaction correction is zero, the contribution of the negative-energy states to $\Delta E^{(1)}_{\text{int}}$ is not small as compared to the positive-energy states. In particular, for the pure Coulomb potential in the non-relativistic limit these parts amount to -0.5 and 1.5 of the total result, respectively.

Rigorous evaluation of the two-photon exchange contribution to the g factor has been accomplished only recently for the case of Li-like silicon [111]. The corresponding diagrams can be classified into two-electron and three-electron ones, as presented in Figs. 4.9 and 4.10, respectively.

As has been mentioned, the rigorous QED treatment is much more complicated for each next order of the perturbation theory. Still, the contribution of higher orders can

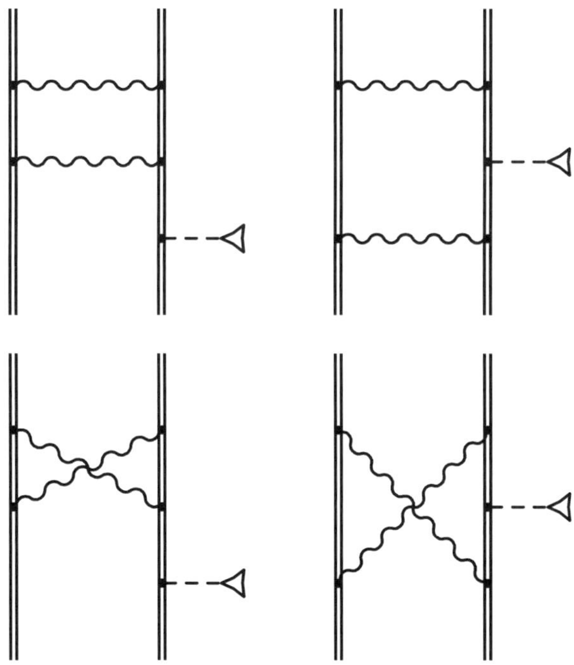

Fig. 4.9 Two-photon exchange contribution to the g factor: two-electron diagrams

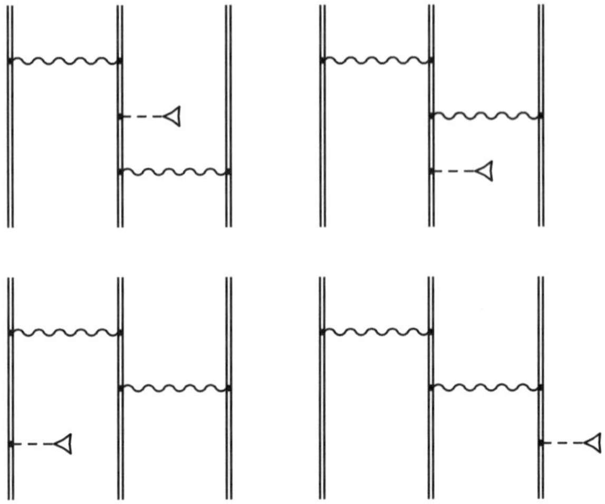

Fig. 4.10 Two-photon exchange contribution to the g factor: three-electron diagrams

be large enough, as compared to the uncertainty achieved in recent measurements. Therefore, an approximate treatment is highly desirable.

4.2.4.2 QED Screening Effects

There are many-electron QED diagrams which contain self-energy and/or vacuum-polarization loops. In order to distinguish them from the 'interelectronic- interaction' diagrams considered in Sect. 4.2.4.1, these ones are referred to as 'screened QED' diagrams, since they correspond to the screening of the one-electron QED effects.

The leading order is represented by the two-electron self-energy (or screened self-energy) diagrams (Fig. 4.11) and by the two-electron vacuum-polarization (or screened vacuum-polarization) diagrams (Fig. 4.12). The first set was evaluated rigorously for $Z = 82, 92$ [29, 110]. For lower Z the convergence of the partial-wave expansion worsen, and the accuracy becomes unacceptable. In addition, the accuracy suffers from the large cancellation of individual terms. The dominant part of the two-electron vacuum-polarization contribution was evaluated in [29, 110] as well.

As an alternative to the rigorous evaluation of the two-electron QED diagrams, one can compute the one-electron QED corrections with the screening potential, described in Sect. 4.2.1. This approach was employed in [28] for lithium-like ions and in [25] for boron-like argon.

Another method to evaluate the QED screening effects in lithium-like ions is the perturbation theory based on the approximate non-relativistic Hamiltonian, introduced in Sect. 4.2.2. This method is much more simple in numerical implementation than the rigorous evaluation of the two-electron QED diagrams, and is free from the problems of large cancellations and partial-wave expansion convergence. At the same time, it allows to account for the leading order in αZ exactly, so it fits best the low-Z ions.

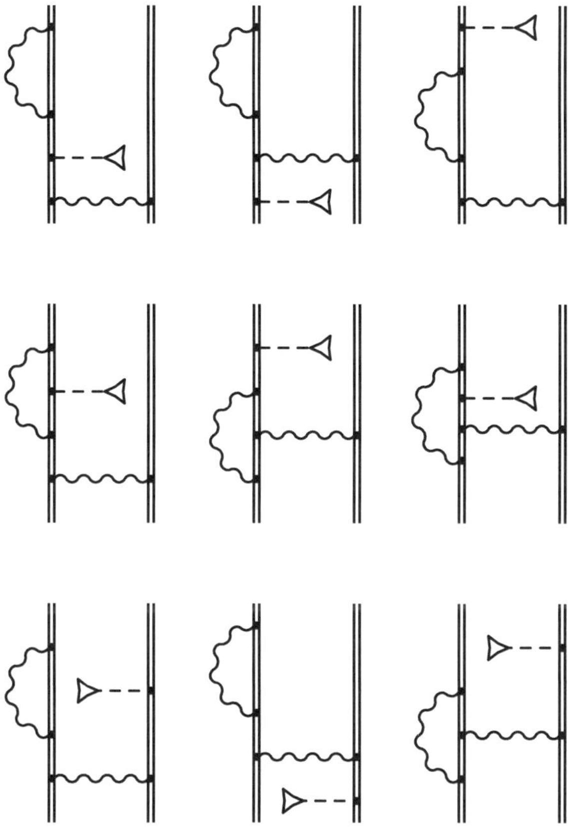

Fig. 4.11 Two-electron self-energy contribution to the g factor

4.3 Nuclear Recoil Effect

Consideration of the finite nuclear mass goes beyond the Furry picture of QED. Due to the smallness of the electron-to-nucleus mass ratio m/M the nucleus can be considered as a non-relativistic particle. On the other hand, for highly charged ions the results without expansion in αZ are desirable, which means the summation of the infinite number of diagrams of the electron-nucleus interaction. To the first order in m/M the complete αZ-dependent formulae were derived in [89]. It can be presented in the following form:

$$\Delta g_{\text{rec}} = \Delta g_{\text{rec}}^{\text{L}} + \Delta g_{\text{rec}}^{\text{H}}. \tag{4.50}$$

$\Delta g_{\text{rec}}^{\text{L}}$ was evaluated analytically for arbitrary state in case of the pure Coulomb potential with the help of generalized virial relations [88]. The result for $1s$ state reads

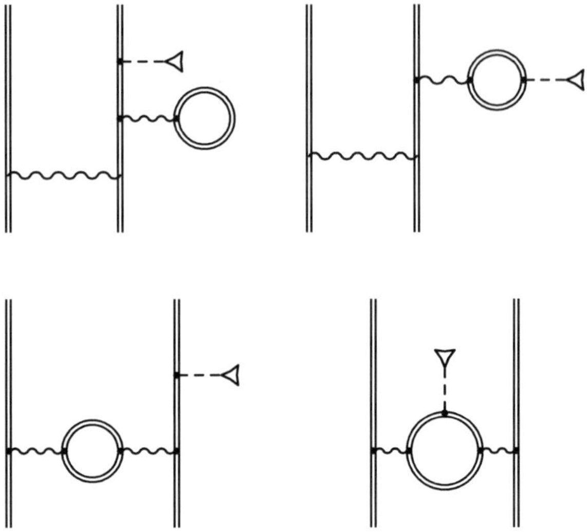

Fig. 4.12 Two-electron vacuum-polarization contribution to the g factor

$$\Delta g^{\mathrm{L}}_{\mathrm{rec}}[1s] = \frac{m}{M}(\alpha Z)^2 \left[1 - \frac{(\alpha Z)^2}{3(1+\sqrt{1-(\alpha Z)^2})^2}\right]. \tag{4.51}$$

The higher-order part involves the integration over the photon energy, similar to the self-energy correction. The corresponding numerical evaluation for $1s$ state was performed in [91]. The lowest-order terms in αZ were derived previously in [21, 22, 34, 35].

4.4 Nuclear Size and Polarization Effects

The effect of finite nuclear size can be introduced as a modification of the nuclear potential according to some model of the nuclear charge distribution. The parameters of this distribution, firstly the root-mean-square radius $\langle r^2\rangle^{1/2}$, are to be determined from experiment. In [49] the analytical formula for the leading-order term for the $1s$ state was derived,

$$\Delta g_{\mathrm{ns}}[1s] = \frac{8}{3}(\alpha Z)^4 \langle r^2\rangle. \tag{4.52}$$

In [26] a relativistic formula for arbitrary state was derived. For medium and high nuclear charge the numerical evaluation according to (4.11) with the finite nucleus potential delivers more accurate results. The Fermi model and the model of homogeneously charged sphere are mostly used. Since more detailed knowledge about the nuclear charge distribution is hardly available, the uncertainty of this contribution sets the ultimate limit for the *ab initio* theoretical prediction of the g factor. This uncertainty grows with Z faster than Z^4 and for H-like uranium amounts to 2.5×10^{-6}, which is comparable with the two-loop QED correction and with the recoil correction. In order to tackle this problem it was proposed in [90] to consider the specific difference of the g-factor values of H-like and Li-like ions of the same isotope,

$$g' = g[(1s)^2 2s] - \xi g[1s]\,, \tag{4.53}$$
$$\xi = \Delta g_{\rm ns}[(1s)^2 2s]/\Delta g_{\rm ns}[1s]. \tag{4.54}$$

The parameter ξ is chosen so as to cancel the nuclear size effect in g'. As was shown numerically [90] ξ is much less sensitive to the model of the nuclear charge distribution. Therefore the differences g' is advantageous, as it can be theoretically evaluated to much higher accuracy than the g factors $g[1s]$ and $g[(1s)^2 2s]$ themselves. This idea motivated further investigations of the g factor of Li-like ions. In [95] it was proposed to consider the specific difference of the g-factor values of H-like and B-like ions of the same isotope,

$$g' = g[(1s)^2(2s)^2 2p_{1/2}] - \xi g[1s]\,, \tag{4.55}$$
$$\xi = \Delta g_{\rm ns}[(1s)^2(2s)^2 2p_{1/2}]/\Delta g_{\rm ns}[1s]. \tag{4.56}$$

The nuclear polarization effect was investigated by Nefiodov, Plunien and Soff [75]. This contribution is small, but it limits the theoretical accuracy of the specific differences (4.53, 4.55), as well as for the separate g-factor values.

4.5 Zeeman Splitting in Few-Electron Ions

The magnetic moment of the atom manifests itself in the Zeeman level splitting whenever it is exposed to an external magnetic field. In case of the zero nuclear spin, the Zeeman effect is determined by the bound electrons. The Zeeman shift of each level can be evaluated within the perturbation theory,

$$E_A(B) = E_A^{(0)} + \Delta E_A^{(1)}(B) + \Delta E_A^{(2)}(B) + \Delta E_A^{(3)}(B) + \cdots\,, \tag{4.57}$$

where $|A\rangle = |\gamma, J, M_J\rangle$ is the state with the total angular momentum J and its projection M_J, while γ stands for all other quantum numbers. Each term of the perturbation expansion is proportional to the magnetic field strength to the corresponding power, $\Delta E_A^{(n)}(B) \sim B^n$. $E_A^{(0)}$ is just the energy in the absence of magnetic field. The first-order term is directly related to the g factor (4.1) Theoretical value of the g factor is conveniently presented in the form,

$$g = g_{\rm D} + \Delta g_{\rm QED} + \Delta g_{\rm int} + \Delta g_{\rm SQED} + \Delta g_{\rm nuc}. \tag{4.58}$$

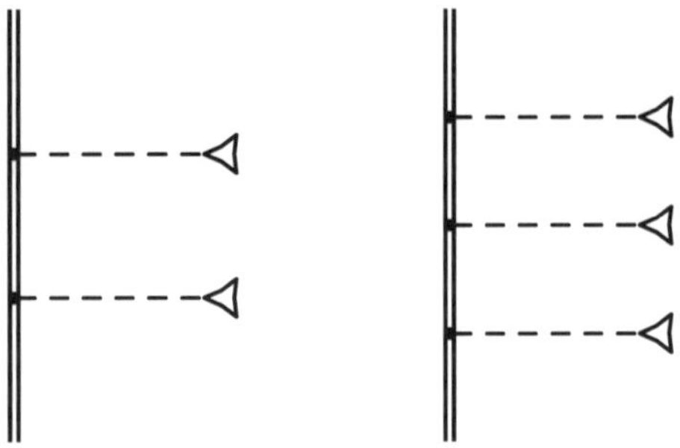

Fig. 4.13 Leading-order diagrams for the second- and third-order zeeman effect

Here g_{D} is the Dirac value (4.12), Δg_{QED} is the one-electron QED correction (see Sect. 4.2.3), Δg_{int} is the interelectronic-integration correction, (see Sect. 4.2.4.1), Δg_{QED} is the screened QED correction (see Sect. 4.2.4.2), Δg_{nuc} is the sum of the nuclear recoil, finite nuclear size and nuclear polarization corrections (see Sect. 4.4). The QED theory of the g factor has been considered in detail in previous sections of this chapter.

In the next subsection we consider the non-linear in magnetic field contributions $\Delta E_A^{(2,3)}(B)$.

4.5.1 Non-linear in Magnetic Field Effects

The second-order term in the Zeeman splitting can be presented in the following form:

$$\Delta E_A^{(2)}(B) = g^{(2)}(M_J)(\mu_{\mathrm{B}} B)^2/(m_e c^2), \tag{4.59}$$

where the dimensionless coefficient $g^{(2)}$ is introduced, by analogy with the g factor. The dependence of this term on M_J is not as simple as for the first-order effect. However, $g^{(2)}$ obeys the symmetry relation $g^{(2)}(-M_J) = g^{(2)}(M_J)$. Due to this relation the quadratic effect does not affect the Zeeman splitting for $J = 1/2$. So, it is not observable in the ground-state g-factor measurements in hydrogen-like and lithium-like ions. It can be observed e.g. in boron-like ions, if the first excited state $P_{3/2}$ is available for measurement, as it was proposed in [64] for boron-like argon. The leading-order contribution to $\Delta E_A^{(2)}$ can be calculated according to the formula:

$$\Delta E_A^{(2)}(B) = \sum_n{}' \frac{\langle a|V_{\mathrm{magn}}|n\rangle\langle n|V_{\mathrm{magn}}|a\rangle}{\varepsilon_a - \varepsilon_n}. \tag{4.60}$$

The corresponding diagram is depicted in Fig. 4.13.

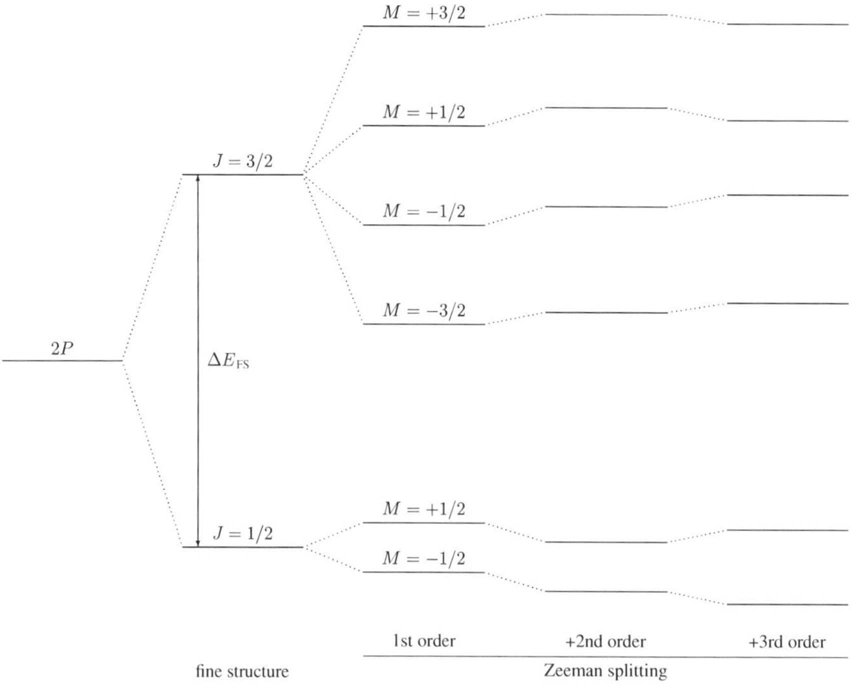

Fig. 4.14 Level scheme of the $2\,^2P_J$ states of boron-like ion in external magnetic field with higher-order contributions to the zeeman effect (not true to scale)

Typically, for the magnetic fields of several Tesla the quadratic effect is rather small. However, it can be strongly enhanced due to the small energy difference between the levels mixed by the magnetic interaction. Such situation takes place in boron-like ions, especially for low nuclear charge. In this case a simple order-of-magnitude estimation is valid,

$$\frac{\Delta E_A^{(2)}(B)}{\Delta E_A^{(1)}(B)} \sim \frac{\Delta E_A^{(1)}(B)}{\Delta E_{\mathrm{FS}}}, \tag{4.61}$$

where $\Delta E_{\mathrm{FS}} = E_{3/2}^{(0)} - E_{1/2}^{(0)}$ is the fine-structure interval Fig. 4.14.

The third-order term in the Zeeman splitting can be presented in the following form:

$$\Delta E_A^{(3)}(B) = g^{(3)}(M_J)(\mu_{\mathrm{B}} B)^3/(m_e c^2)^2, \tag{4.62}$$

where the dimensionless coefficient $g^{(3)}$ obeys the relation $g_J(-M_J) = -g_J(M_J)$. The leading-order contributions to $\Delta E_A^{(3)}$ can be calculated according to the formula:

$$\Delta E_A^{(3)}(B) = \sum_{n_1,n_2}{}' \frac{\langle a|V_{\rm magn}|n_1\rangle\langle n_1|V_{\rm magn}|n_2\rangle\langle n_2|V_{\rm magn}|a\rangle}{(\varepsilon_a-\varepsilon_{n_1})(\varepsilon_a-\varepsilon_{n_2})} - \sum_n{}' \frac{\langle a|V_{\rm magn}|n\rangle\langle n|V_{\rm magn}|a\rangle}{(\varepsilon_a-\varepsilon_n)^2}\langle a|V_{\rm magn}|a\rangle, \qquad (4.63)$$

The third-order effect is strongly enhanced in boron-like ions as well. Order-of-magnitude estimation in this case reads

$$\frac{\Delta E_A^{(3)}(B)}{\Delta E_A^{(1)}(B)} \sim \left(\frac{\Delta E_A^{(1)}(B)}{\Delta E_{\rm FS}}\right)^2. \qquad (4.64)$$

References

1. O.V. Andreev, D.A. Glazov, A.V. Volotka, V.M. Shabaev, G. Plunien, Evaluation of the screened vacuum-polarization corrections to the hyperfine splitting of Li-like bismuth. Phys. Rev. A **85**, 022510 (2012)
2. I. Angeli, A consistent set of nuclear rms charge radii: properties of the radius surface R(N;Z). At. Data Nucl. Data Tables **87**, 185 (2004)
3. J.M. Anthony, K.J. Sebastian, Relativistic corrections to the zeeman effect in heliumlike atoms. Phys. Rev. A **48**, 3792 (1993)
4. A.N. Artemyev, V.M. Shabaev, I.I. Tupitsyn, G. Plunien, V.A. Yerokhin, QED Calculation of the $2p_{3/2}$-$2p_{1/2}$ transition energy in boronlike argon. Phys. Rev. Lett. **98**, 173004 (2007)
5. T. Beier, The g_j-factor of a bound electron and the hyperfine structure splitting in hydrogenlike ions. Phys. Rep. **339**, 79 (2000)
6. T. Beier, I. Lindgren, H. Persson, S. Salomonson, P. Sunnergren, H. Häffner, N. Hermanspahn, g-Factor of an electron bound in a hydrogenlike ion. Phys. Rev. A **62**, 032510 (2000)
7. T. Beier, S. Djekic, H. Häffner, P. Indelicato, H.-J. Kluge, W. Quint, V.M. Shabaev, J. Verdú, T. Valenzuela, G. Plunien, V.A. Yerokhin, Determination of the electron's mass from g-factor experiments on $^{12}C^{5+}$ and $^{16}O^{7+}$. Nucl. Instrum. Methods Phys. Res. B **205**, 15 (2003)
8. S.A. Blundell, N.J. Snyderman, Basis-set approach to calculating the radiative self-energy in highly ionized atoms. Phys. Rev. A **44**, R1427 (1991)
9. S.A. Blundell, K.T. Cheng, J. Sapirstein, Radiative corrections in atomic physics in the presence of perturbing potentials. Phys. Rev. A **55**, 1857 (1997)
10. G. Breit, The magnetic moment of the electron. Nature **122**, 649 (1928)
11. S.J. Brodsky, J.R. Primack, The electromagnetic interactions of composite systems. Ann. Phys. (N.Y.) **52**, 315 (1969)
12. S.J. Brodsky, R.G. Parsons, Precise theory of the zeeman spectrum for atomic hydrogen and deuterium and the lamb shift. Phys. Rev. **163**, 134 (1967)
13. S.J. Brodsky, J.R. Primack, Electromagnetic interactions of loosely-bound composite systems. Phys. Rev. **174**, 2071 (1968)
14. J. Calmet, H. Grotch, D.A. Owen, Comment on radiative magnetic energy shifts in hydrogen. Phys. Rev. A **17**, 1218 (1978)
15. F.E. Close, H. Osborn, Relativistic extension of the electromagnetic current for composite systems. Phys. Lett. B **34**, 400 (1971)
16. R. Cowan, *The Theory of Atomic Spectra* (University of California Press, Berkeley, 1981)
17. A. Czarnecki, K. Melnikov, A. Yelkhovsky, Anomalous magnetic moment of a bound electron. Phys. Rev. A **63**, 012509 (2001)
18. I. Draganić, J.R. Crespo, High precision wavelength measurements of QED-sensitive forbidden transitions in highly charged argon ions. Phys. Rev. Lett. **91**, 183001 (2003)

19. M.I. Eides, H. Grotch, Gyromagnetic ratios of bound particles. Ann. Phys. (N.Y.) **260**, 191 (1997)
20. A.G. Fainshtein, N.L. Manakov, A.A. Nekipelov, Vacuum polarization by a Coulomb field. Analytical approximation of the polarization potential. J. Phys. B **24**, 559 (1991)
21. R.N. Faustov, Magnetic moment of the hydrogen atom. Nuovo Cimento A **69**, 37 (1970)
22. R.N. Faustov, Magnetic moment of the hydrogen atom. Phys. Lett. B **33**, 422 (1970)
23. W.H. Furry, On bound states and scattering in positron theory. Phys. Rev. **81**, 115 (1951)
24. M. Gell-Mann, F. Low, Bound states in quantum field theory. Phys. Rev. **84**, 350 (1951)
25. D.A. Glazov, A.V. Volotka, A.A. Schepetnov, M.M. Sokolov, V.M. Shabaev, I.I. Tupitsyn, G. Plunien, Phys. Scr., in press
26. D.A. Glazov, V.M. Shabaev, Finite nuclear size correction to the bound-electron g-factor in a hydrogenlike atom. Phys. Lett. A **297**, 408 (2002)
27. D.A. Glazov, V.M. Shabaev, I.I. Tupitsyn, A.V. Volotka, V.A. Yerokhin, G. Plunien, G. Soff, Relativistic and QED corrections to the g-factor of Li-like ions. Phys. Rev. A **70**, 062104 (2004)
28. D.A. Glazov, A.V. Volotka, V.M. Shabaev, I.I. Tupitsyn, G. Plunien, Screened QED corrections to the g factor of Li-like ions. Phys. Lett. A **357**, 330 (2006)
29. D.A. Glazov, A.V. Volotka, V.M. Shabaev, I.I. Tupitsyn, G. Plunien, Evaluation of the screened QED corrections to the g-factor and the hyperfine splitting of lithiumlike ions. Phys. Rev. A **81**, 062112 (2010)
30. I. Gonzalo, E. Santos, Radiative corrections in the zeeman effect of 2^3P states of helium. Phys. Rev. A **56**, 3576 (1997)
31. H. Grotch, R. Kashuba, Magnetic interactions of one-electron atoms and of positronium. Phys. Rev. A **7**, 78 (1973)
32. H. Grotch, R. Kashuba, Magnetic interactions of one-electron atoms and of positronium. Phys. Rev. A **8**, 575(E) (1973)
33. H. Grotch, Electron g-factor in hydrogenic atoms. Phys. Rev. Lett. **24**, 39 (1970)
34. H. Grotch, Nuclear mass corrections to the electron g-factor. Phys. Rev. A **2**, 1605 (1970)
35. H. Grotch, R.A. Hegstrom, Hydrogenic atoms in a magnetic field. Phys. Rev. A **4**, 59 (1971)
36. H. Häffner, T. Beier, N. Hermanspahn, H.-J. Kluge, W. Quint, S. Stahl, J. Verdú, G. Werth, High-accuracy measurement of the magnetic moment anomaly of the electron bound in hydrogenlike carbon. Phys. Rev. Lett. **85**, 5308 (2000)
37. R.A. Hegstrom, Relativistic treatment of the shielding of the electron and proton magnetic dipole moments in atomic hydrogen. Phys. Rev. A **184**, 17 (1969)
38. R.A. Hegstrom, Relativistic treatment of the shielding of the electron and proton magnetic dipole moments in atomic hydrogen. Errata Phys. Rev. A **1**, 536 (1970)
39. R.A. Hegstrom, Nuclear-mass and anomalous-moment corrections to the Hamiltonian for an atom in a constant external magnetic field. Phys. Rev. A **7**, 451 (1973)
40. R.A. Hegstrom, Magnetic moment of atomic lithium. Phys. Rev. A **11**, 421 (1975)
41. N. Hermanspahn, H. Häffner, H.-J. Kluge, W. Quint, S. Stahl, J. Verdú, G. Werth, Observation of the continuous Stern-Gerlach effect on an electron bound in an atomic ion. Phys. Rev. Lett. **84**, 427 (2000)
42. V.W. Hughes, T. Kinoshita, Anomalous g values of the electron and muon. Rev. Mod. Phys. **71**, S133 (1999)
43. P. Indelicato, A.-M. Mårtensson-Pendrill, W. Quint, J.-P. Desclaux, Correlation and relativistic effects on lande g_J factors of atomic ions. Hyperfine Interact. **146/ 147**, 127 (2003)
44. P. Indelicato, J.P. Desclaux, Multiconfiguration Dirac-Fock calculations of transition energies with QED corrections in three-electron ions. Phys. Rev. A **42**, 5139 (1990)
45. U.D. Jentschura, Binding two-loop vacuum-polarization corrections to the bound-electron g-factor. Phys. Rev. A **79**, 044501 (2009)
46. U.D. Jentschura, Self-energy correction to the bound-electron g-factor of P states. Phys. Rev. A **81**, 012512 (2010)
47. W.R. Johnson, J. Sapirstein, Computation of second-order many-body corrections in relativistic atomic systems. Phys. Rev. Lett. **57**, 1126 (1986)

48. W.R. Johnson, S.A. Blundell, J. Sapirstein, Finite basis sets for the Dirac equation constructed from B splines. Phys. Rev. A **37**, 307 (1988)
49. S.G. Karshenboim, Non-relativistic calculations of the g-factor of a bound electron. Phys. Lett. A **266**, 380 (2000)
50. S.G. Karshenboim, V.G. Ivanov, V.M. Shabaev, Some analytic results on the Uehling correction to the g-factor of a bound electron (muon). Can. J. Phys. **79**, 81 (2001)
51. S.G. Karshenboim, V.G. Ivanov, V.M. Shabaev, Some analytic results on the Uehling correction to the g-factor of a bound electron (muon). Zh. Eksp. Teor. Fiz. **120**, 546 (2001)
52. S.G. Karshenboim, V.G. Ivanov, V.M. Shabaev, Some analytic results on the uehling correction to the g-factor of a bound electron (muon). Sov. Phys. JETP **93**, 477 (2001)
53. S.G. Karshenboim, in *The Hydrogen Atom*, ed. by S.G. Karshenboim, et al. (Springer, Berlin, 2001), p. 651
54. S.G. Karshenboim, A.I. Milstein, Delbruck scattering and the g-factor of a bound electron. Phys. Lett. B **549**, 321 (2002)
55. I.B. Khriplovich, Fundamental symmetries and atomic physics. Phys. Scr. **T112**, 52 (2004)
56. T. Kinoshita, M. Nio, Revised α^4 Term of Lepton $g-2$ from the Feynman diagrams containing an internal light-by-light scattering subdiagram. Phys. Rev. Lett. **90**, 021803 (2003)
57. H.-J. Kluge, T. Beier, K. Blaum, L. Dahl, S. Eliseev, F. Herfurth, B. Hofmann, O. Kester, S. Koszudowski, C. Kozhuharov, G. Maero, W. Nörtershäuser, J. Pfister, W. Quint, U. Ratzinger, A. Schempp, R. Schuch, T. Stöhlker, R.C. Thompson, M. Vogel, G. Vorobjev, D.F.A. Winters, G. Werth, HITRAP: a facility at GSI for highly charged ions. Adv. Quantum Chem. **53**, 83 (2008)
58. P. Kusch, H.M. Foley, Precision measurement of the ratio of the atomic 'g values' in the $^2P_{3/2}$ and $^2P_{1/2}$ states of gallium. Phys. Rev. **72**, 1256 (1947)
59. W.E. Lamb Jr, Fine structure of the hydrogen atom III. Phys. Rev. **85**, 259 (1952)
60. A. Lapierre, U.D. Jentschura, J.R. Crespo, Relativistic electron correlation, quantum electrodynamics, and the lifetime of the $1s^2 2s^2 2p\ ^2P^0_{3/2}$ level in boronlike argon. Phys. Rev. Lett. **95**, 183001 (2005)
61. R. Latter, Atomic energy levels for the Thomas-Fermi and Thomas-Fermi-Dirac potential. Phys. Rev. **99**, 510 (1955)
62. R.N. Lee, A.I. Milstein, I.S. Terekhov, S.G. Karshenboim, Virtual light-by-light scattering and the g factor of a bound electron. Phys. Rev. A **71**, 052501 (2005)
63. D. von Lindenfels, N. Brantjes, G. Birkl, W. Quint, V. Shabaev, M. Vogel, Bound electron g-factor measurement by doubleresonance spectroscopy on a fine-structure transition. Can. J. Phys. **89**, 79 (2011)
64. D. von Lindenfels, W. Quint, D.A. Glazov, A.V. Volotka, M.M. Sokolov, V.M. Shabaev, G. Plunien, M. Wiesel, G. Birkl, A. Martin, M. Vogel, Experimental access to higher-order zeeman effects by precision spectroscopy of highly charged ions in a Penning trap. Phys. Rev. A. **87**, 023412 (2013)
65. E. Lindroth, A. Ynnerman, Ab initio calculations of g_j factors for Li, Be^+, and Ba^+. Phys. Rev. A **47**, 961 (1993)
66. V. Mäckel, R. Klawitter, G. Brenner, J.R. Crespo, Laser spectroscopy on forbidden transitions in trapped highly charged Ar^{13+} ions. Phys. Rev. Lett. **107**, 143002 (2011)
67. N.L. Manakov, A.A. Nekipelov, A.G. Fainstein, Vacuum polarization by a strong coulomb field and its contribution to the spectra of multiply-charged ions. Sov. Phys. JETP **68**, 673 (1989)
68. A.P. Martynenko, R.N. Faustov, Magnetic moment of a two-particle bound state in quantum electrodynamics. Zh. Eksp. Teor. Fiz. **120**, 539 (2001)
69. A.P. Martynenko, R.N. Faustov, Magnetic moment of a two-particle bound state in quantum electrodynamics. Sov. Phys. JETP **93**, 471 (2001)
70. A.P. Martynenko, R.N. Faustov, Magnetic moment of a two-particle bound state in quantum electrodynamics. Phys. Atom. Nucl. **65**, 271 (2002)
71. P.J. Mohr, G. Plunien, G. Soff, QED corrections in heavy atoms. Phys. Rep. **293**, 227 (1998)

72. P.J. Mohr, B.N. Taylor, CODATA recommended values of the fundamental physical constants: 2002. Rev. Mod. Phys. **77**, 1 (2005)
73. P.J. Mohr, B.N. Taylor, D.B. Newell, CODATA recommended values of the fundamental physical constants: 2006. Rev. Mod. Phys. **80**, 633 (2008)
74. P.J. Mohr, B.N. Taylor, D.B. Newell, CODATA recommended values of the fundamental physical constants: 2010. Rev. Mod. Phys. **84**, 1527 (2012)
75. A.V. Nefiodov, G. Plunien, G. Soff, Nuclear-polarization correction to the bound-electron g-factor in heavy hydrogenlike ions. Phys. Rev. Lett. **89**, 081802 (2002)
76. K. Pachucki, U.D. Jentschura, V.A. Yerokhin, Nonrelativistic QED approach to the bound-electron g factor. Phys. Rev. Lett. **94**, 229902(E) (2005)
77. K. Pachucki, U.D. Jentschura, V.A. Yerokhin, Nonrelativistic QED approach to the bound-electron g factor. Phys. Rev. Lett. **93**, 150401 (2004)
78. K. Pachucki, A. Czarnecki, U.D. Jentschura, V.A. Yerokhin, Complete two-loop correction to the bound-electron g factor. Phys. Rev. A **72**, 022108 (2005)
79. K. Pachucki, Nuclear mass correction to the magnetic interaction of atomic systems. Phys. Rev. A **78**, 012504 (2008)
80. H. Persson, S. Salomonson, P. Sunnergren, I. Lindgren, Radiative corrections to the electron g-factor in H-like ions. Phys. Rev. A **56**, R2499 (1997)
81. M. Phillips, The effect of nuclear motion on atomic magnetic moments. Phys. Rev. **76**, 1803 (1949)
82. W. Quint, J. Dilling, S. Djekic, H. Häffner, N. Hermanspahn, H.-J. Kluge, G. Marx, R. Moore, D. Rodriguez, J. Schönfelder, G. Sikler, T. Valenzuela, J. Verdú, C. Weber, G. Werth, HITRAP: a facility for experiments with trapped highly charged ions..Hyperfine Interactions **132**, 453 (2001)
83. J. Sapirstein, W.R. Johnson, The use of basis splines in theoretical atomic physics. J. Phys. B **29**, 5213 (1996)
84. J. Sapirstein, K.T. Cheng, Calculation of the lamb shift in neutral alkali metals. Phys. Rev. A **66**, 042501 (2002)
85. J. Schwinger, On quantum-electrodynamics and the magnetic moment of the electron. Phys. Rev. **73**, 416 (1948)
86. J. Schwinger, Quantum electrodynamics. III. The electromagnetic properties of the electron-radiative corrections to scattering. Phys. Rev. **76**, 790 (1949)
87. R. Serber, Linear modifications in the Maxwell field equations. Phys. Rev. **48**, 49 (1935)
88. V.M. Shabaev, Generalizations of the virial relations for the Dirac equation in a central field and their applications to the Coulomb field. J. Phys. B **24**, 4479 (1991)
89. V.M. Shabaev, QED theory of the nuclear recoil effect on the atomic g-factor. Phys. Rev. A **64**, 052104 (2001)
90. V.M. Shabaev, D.A. Glazov, M.B. Shabaeva, V.A. Yerokhin, G. Plunien, G. Soff, g-factor of high-Z lithiumlike ions. Phys. Rev. A **65**, 062104 (2002)
91. V.M. Shabaev, V.A. Yerokhin, Recoil correction to the bound-electron g factor in H-like atoms to all orders in αZ. Phys. Rev. Lett. **88**, 091801 (2002)
92. V.M. Shabaev, Two-time Green's function method in quantum electrodynamics of high-Z few-electron atoms. Phys. Rep. **356**, 119 (2002)
93. V.M. Shabaev, D.A. Glazov, M.B. Shabaeva, I.I. Tupitsyn, V.A. Yerokhin, T. Beier, G. Plunien, G. Soff, Theory of the g-factor of lithiumlike ions. Nucl. Instrum. Methods Phys. Res. B **205**, 20 (2003)
94. V.M. Shabaev, I.I. Tupitsyn, V.A. Yerokhin, G. Plunien, G. Soff, Dual kinetic balance approach to basis-set expansions for the Dirac equation. Phys. Rev. Lett. **93**, 130405 (2004)
95. V.M. Shabaev, D.A. Glazov, N.S. Oreshkina, A.V. Volotka, G. Plunien, H.-J. Kluge, W. Quint, g-factor of heavy ions: a new access to the fine structure constant. Phys. Rev. Lett. **96**, 253002 (2006)
96. N.J. Snyderman, Electron radiative self-energy of highly stripped heavy atoms. Ann. Phys. (N.Y.) **211**, 43 (1991)

97. G. Soff, P.J. Mohr, Vacuum polarization in a strong external field. Phys. Rev. A **38**, 5066 (1988)
98. R. Soria Orts, J. R. Crespo López-Urrutia, H. Bruhns, A. J. González Martínez, Z. Harman, U. D. Jentschura, C. H. Keitel, A. Lapierre, H. Tawara, I. I. Tupitsyn, J. Ullrich, and A. V. Volotka, Zeeman splitting and g factor of the $1s^2 2s^2 2p\ ^2P_{3/2}$ and $^2P_{1/2}$ levels in Ar^{13+}. Phys. Rev. A **76**, 052501 (2007)
99. S. Sturm, A. Wagner, B. Schabinger, J. Zatorski, Z. Harman, W. Quint, G. Werth, C.H. Keitel, K. Blaum, g-Factor of hydrogenlike $^{28}Si^{13+}$. Phys. Rev. Lett. **107**, 023002 (2011)
100. J. Sucher, S-matrix formalism for level-shift calculations. Phys. Rev. **107**, 1448 (1957)
101. I.I. Tupitsyn, A.V. Volotka, D.A. Glazov, V.M. Shabaev, G. Plunien, J.R. Crespo, Magnetic-dipole transition probabilities in B-like and Be-like ions. Phys. Rev. A **72**, 062503 (2005)
102. E.A. Uehling, Polarization effects in the positron theory. Phys. Rev. **48**, 55 (1935)
103. J.L. Verdú, T. Beier, S. Djekic, H. Häffner, H.-J. Kluge, W. Quint, T. Valenzuela, G. Werth, Measurement of the g_J-factor of a bound electron in hydrogen-like oxygen, $^{16}O^{7+}$. Can. J. Phys. **80**, 1233 (2002)
104. J.L. Verdú, S. Djekić, S. Stahl, T. Valenzuela, M. Vogel, G. Werth, T. Beier, H.-J. Kluge, W. Quint, Electronic g factor of hydrogenlike oxygen O^{7+}. Phys. Rev. Lett. **92**, 093002 (2004)
105. L. Veseth, Spin-extended hartree-fock calculations of atomic g_J factors. Phys. Rev. A **22**, 803 (1980)
106. M. Vogel, W. Quint, Trap-assisted precision spectroscopy of highly charged ions. Phys. Rep. **490**, 1 (2010)
107. A.V. Volotka, D.A. Glazov, G. Plunien, V.M. Shabaev, I.I. Tupitsyn, Radiative corrections to the magnetic-dipole transition amplitude in B-like ions. Eur. Phys. J. D **38**, 293 (2006)
108. A.V. Volotka, D.A. Glazov, I.I. Tupitsyn, N.S. Oreshkina, G. Plunien, V.M. Shabaev, Ground-state hyperfine structure of H-, Li-, and B-like ions in the intermediate-Z region. Phys. Rev. A **78**, 062507 (2008)
109. A.V. Volotka, D.A. Glazov, G. Plunien, V.M. Shabaev, I.I. Tupitsyn, Nuclear recoil effect on the magnetic-dipole decay rates of atomic levels. Eur. Phys. J. D **48**, 167 (2008)
110. A.V. Volotka, D.A. Glazov, V.M. Shabaev, I.I. Tupitsyn, G. Plunien, Screened QED corrections in lithiumlike heavy ions in the presence of magnetic fields. Phys. Rev. Lett. **103**, 033005 (2009)
111. A. Wagner, S. Sturm, F. Köhler, D.A. Glazov, A.V. Volotka, G. Plunien, W. Quint, G. Werth, V.M. Shabaev, K. Blaum, g-factor of lithiumlike silicon $^{28}Si^{11+}$. Phys. Rev. Lett. **110**, 033003 (2013)
112. G. Werth, H. Häffner, N. Hermanspahn, H.-J. Kluge, W. Quint, J. Verdú, in *The Hydrogen Atom*, ed. by S.G. Karshenboim, et al. (Springer, Berlin, 2001), p. 204
113. E.H. Wichmann, N.M. Kroll, Vacuum polarization in a strong Coulomb field. Phys. Rev. **101**, 843 (1956)
114. Z.-C. Yan, Calculations of magnetic moments for three-electron atomic systems. Phys. Rev. Lett. **86**, 5683 (2001)
115. Z.-C. Yan, Relativistic corrections to the zeeman effect of lithium and lithiumlike ions in the 2^2P_J and 3^2P_J states. Phys. Rev. A **66**, 022502 (2002)
116. Z.-C. Yan, Calculations of magnetic moments for lithium-like ions. J. Phys. B **35**, 1885 (2002)
117. A. Yelkhovsky, Recoil correction to the magnetic moment of a bound electron. E-print, hep-ph/0108091 (2001)
118. V.A. Yerokhin, V.M. Shabaev, First-order self-energy correction in hydrogenlike systems. Phys. Rev. A **60**, 800 (1999)
119. V.A. Yerokhin, P. Indelicato, V.M. Shabaev, Self-energy correction to the bound-electron g-factor in H-like ions. Phys. Rev. Lett. **89**, 143001 (2002)
120. V.A. Yerokhin, P. Indelicato, V.M. Shabaev, Evaluation of the self-energy correction to the g factor of S states in H-like ions. Phys. Rev. A **69**, 052503 (2004)
121. V.A. Yerokhin, U.D. Jentschura, Electron self-energy in the presence of a magnetic field: hyperfine splitting and g-factor. Phys. Rev. Lett. **100**, 163001 (2008)

122. V.A. Yerokhin, U.D. Jentschura, Self-energy correction to the hyperfine splitting and the electron g-factor in hydrogenlike ions. Phys. Rev. A **81**, 012502 (2010)
123. V.A. Yerokhin, K. Pachucki, Z. Harman, C.H. Keitel, QED theory of the nuclear magnetic shielding in hydrogenlike ions. Phys. Rev. Lett. **107**, 043004 (2011)

Chapter 5
The Magnetic Moments of the Proton and the Antiproton

Stefan Ulmer and Christian Smorra

Abstract A comparison of the magnetic moments of the proton and the antiproton provides a sensitive test of matter-antimatter symmetry. While the magnetic moment of the proton is known with a relative precision of 10^{-8}, that of the antiproton is only known with moderate accuracy. Important progress towards a high-precision measurement of the particle's magnetic moment was reported in 2011 by a group at Mainz when spin transitions of a single proton stored in a cryogenic Penning trap were observed. To resolve the single-proton spin flips, the so-called 'continuous Stern-Gerlach effect' was utilized. Using this technique, the proton magnetic moment was measured by two groups at Mainz and Harvard with relative precisions of 8.9×10^{-6} and 2.5×10^{-6}, respectively. Currently, two collaborations at the CERN antiproton decelerator (AD)—a part of ATRAP and BASE—are pushing their efforts to apply the methods developed for the proton to measure the magnetic moment of the antiproton. Very recently, DiSciacca et al. reported on a measurement of the antiproton's magnetic moment with a relative precision of 4.4 ppm, which is a improvement of the formerly best value by about a factor of 680. Using the so-called double Penning trap technique, both collaborations aim for a precision measurement at the level of at least 10^{-9} in future experiments, which would provide a highly sensitive test of the *CPT* symmetry using baryons.

5.1 Introduction

The Standard Model of elementary particle physics [1] is the pride and joy of modern physics. This locally gauge-invariant, unitary, relativistic SU(3)xSU(2)xU(1) quantum field theory describes many aspects observed in nature, and two of the

S. Ulmer (✉) · C. Smorra
RIKEN Ulmer Initiative Research Unit, Wako, Saitama, Japan
e-mail: Stefan.Ulmer@cern.ch

C. Smorra
e-mail: Christian.Smorra@cern.ch

W. Quint and M. Vogel (eds.), *Fundamental Physics in Particle Traps*,
Springer Tracts in Modern Physics 256, DOI: 10.1007/978-3-642-45201-7_5,

four known fundamental interactions, electromagnetism and weak interaction, were unified in its framework [2, 3]. Using the concept of spontaneous symmetry breaking, the SM explains the appearance of mass, mediated by the scalar Higgs field, which is propagated by a hypothetic massive boson, the so-called Higgs particle [4]. Very recently the ATLAS and the CMS collaborations at the large hadron collider at CERN announced the highly exciting 4.9 sigma significant discovery of a preliminarily unknown massive boson at an energy where the Higgs boson is expected to be, i.e. at 125 GeV [5, 6].

Although the SM witnesses a deep human understanding of some aspects of nature, even for massless neutrinos it still involves 18 free parameters, such as particle masses, CKM mixing angles, phases and coupling constants. Moreover, many physics mechanisms and experimental observations are introduced into the model 'by hand'. In addition, several physics phenomena are known which are not involved in the SM, such as dark matter and dark energy, gravitation, neutrino oscillations or the observed matter excess in the universe, the so-called baryon asymmetry. The investigation of these *physics beyond the standard model* is subject of numerous global physics research programs, which are dedicated to settle these questions and perhaps give guidance towards one of the major goals of modern fundamental physics-the development of a consistent grand unified theory (GUT).

Complementary to high-energy physics strategies, high-precision experiments at very low energies provide important contributions for the development of an even deeper understanding of nature. In these experiments, particle and antiparticle parameters such as for example charge, mass or magnetic moments are compared with very high precision, providing stringent test of the so-called *CPT*-symmetry. *C* stands for charge conjugation and transforms a particle to its antiparticle, *P* represents a parity transformation, and *T* a time reversal. It has been axiomatically proven, that this symmetry holds for all relativistic quantum field theories involved in the standard model [7, 8]. An important consequence of the *CPT* theorem is that particles and their antiparticles behave identical in the sense that masses and lifetimes are exactly the same, while the values of charges and magnetic moments are numerically identical but of opposite sign. Moreover, the *CPT* invariance also predicts that whenever a particle and its antimatter mirror image collide, the particles annihilate exactly, and so far no violation of this behaviour has been observed. On the other hand, this is in strong contradiction to our observations of the universe which is made out of matter, while antimatter can only be observed under highly exotic conditions, as in cosmic rays or particle accelerator facilities which are specifically dedicated to antimatter production [8]. This matter-antimatter asymmetry in the baryon sector can be expressed by the baryon asymmetry parameter [9]

$$\eta = \frac{n_B - n_{\bar{B}}}{n_\gamma} = 6 \cdot 10^{-10}, \tag{5.1}$$

where n_B is the baryon density, $n_{\bar{B}}$ the antibaryon density and n_γ the photon density in the universe. So far, no baryon number B violating processes have been observed, which indicates that the B-asymmetry is remanent from the early phase of the universe. However, an explanation of the absence of antimatter in our universe is lacking a satisfactory explanation. An intrinsic baryon asymmetry $B > 0$ from the Big Bang can most likely be excluded, since in thermal equilibrium the number of baryons and antibaryons evens out. The possibility that $B = 0$ and matter and antimatter dominated regions in the universe exist is also unlikely, since so far no annihilation radiation form the boundaries of these regions has ever been observed. Furthermore, large energy cosmic rays contain typically 10^4 protons per antiproton and no source of primary antiprotons or antinuclei has been found [10]. Thus, assuming that $B = 0$ at the beginning of the universe, the observed baryon asymmetry can only be explained by a process which generates an excess of baryons over time. According to Sakharov the necessary conditions required for such a 'baryogenesis' are:

1. The existence of a baryon number violating process.
2. A violation of the C and CP symmetry.
3. Thermodynamic non-equilibrium.

Within the Standard Model, B-violating processes and baryogenesis are plausible. However, theoretical considerations show that the CP-violating phase in the CKM-matrix is not sufficiently large to consistently explain the observed baryon excess. One proposed process consistent with the standard model is the so-called sphaleron process, which could have taken place during the electroweak phase transition. This process is a transition between two vacuum configurations. It conserves the $B - L$ quantum number, where L is the lepton number, and turns 3 leptons, one of each generation, into 9 quarks, three of each generation [11–13]. This requires at first a generation of a lepton asymmetry (leptogenesis), which is then converted into a baryon asymmetry. This is favored because a lepton asymmetry could arise from CP-violating phases in the neutrino sector, whereas the CP-violation in the quark sector is expected to be too weak to generate the observed baryon excess in time [14]. The sphaleron process is expected to occur at energies above 10 TeV, and thus cannot be probed with current accelerators. Other possibilities of baryogenesis incorporate physics beyond the Standard Model, such as in GUTs or supersymmetric theories. For example, a baryon-number and CPT-symmetry violating process can produce a baryon asymmetry independent of the C- and CP-violating process described by Sakharov, even in thermal equilibrium. This scenario is favoured by string theories [15]. Therefore, CPT-violation as manifestation of physics beyond the Standard Model and as a possible source for baryogenesis has been investigated in many experiments.

5.2 CPT Tests

In the search for a violation of the *CPT*-symmetry many different matter/antimatter systems have been compared so far (see Table 5.1). One of the most famous CPT-tests is the widely recognized comparison of neutral Kaon/Antikaon masses [16]

Table 5.1 A list of CPT test related to the proton antiproton g-factor comparison

	Measured quantity	Measurement precision	CPT test precision	Sensitivity
e^-/e^+ mass	ν_c	7.3×10^{-10} [47]	8×10^{-9} [48]	1.3×10^{-20}
e^+/e^- g-factor	ν_a	2.8×10^{-13} [49]	2.1×10^{-12} [21]	1.6×10^{-21} [50]
μ^+/μ^- g-factor	ν_a		3.7×10^{-9} [51]	8.7×10^{-24} [51]
$p/\overline{p}$ mass	ν_c and ν_{HFS}		2×10^{-9} [52]	
$p/\overline{p}$ charge/mass	ν_c	9×10^{-11} [46]	9×10^{-11} [46]	4×10^{-26} [46]
$p/\overline{p}$ g-factor	ν_L and ν_c	1.0×10^{-8} [53]	5×10^{-6} [54]	5×10^{-21}
$K_0/\overline{K}_0$ mass	Decay amplitudes		7×10^{-4} [55]	1.3×10^{-18} [55]
$H/\overline{H}$ HFS	ν_{HFS}	7×10^{-13} [26]	–	–
$H/\overline{H}$ 1S–2S	$\nu_{1S\rightarrow 2S}$	4.2×10^{-15} [27]	–	–

A recent complete overview of CPT tests can be found in Ref. [56]. The measured quantities for the CPT tests are given in the second column. ν_c denotes the cyclotron frequency, ν_L the Larmor frequency, ν_a the anomally frequency, ν_{HFS} hyperfine-structure transition frequencies, ν_{1S-2S} the 1^2S–2^2S transition frequency in hydrogen. The highest relative precision achieved for the respective quantities is listed in the third column. The fourth column lists the relative precision achieved in matter-antimatter comparisons. The last column called sensitivity lists the figure of merit for CPT tests as proposed in [57]. Further details are given in the text

$$\frac{m(K_0) - m(\bar{K}_0)}{m(avg)} < 6 \cdot 10^{-19}. \tag{5.2}$$

That figure of merit can be understood as a comparison of meson masses before and after a CPT transformation, and thus, is at the given level consistent with CPT conservation. In the framework of the standard model extension, which has been developed by Kostelecky and co-workers [17–20], bounds on CPT violation are derived which relate the achieved experimental precision to the intrinsic energy scale of the investigated systems. This approach predicts that the precision investigation of matter/antimatter systems with a low absolute intrinsic energy scale provides much sharper bounds on CPT-violation as the $K_0/\bar{K}_0$ mass comparison. In that context, comparisons of stable matter/antimatter systems as the electron and the positron, the proton and the antiproton or of bound systems as hydrogen and antihydrogen are specifically attractive. The stability of these systems allows for infinite observation times, and thus, such comparisons meet the requirements for experimental high precision studies. For instance, the $g-2$ values of the electron and the positron were compared with the fantastic precision of 2×10^{-12} [21]. These experiments were carried out by measuring anomaly frequencies $\nu_a = \nu_L - \nu_c$ with single particles stored in cryogenic Penning traps, where the continuous Stern-Gerlach effect [22] was applied to detect the spin direction of the investigated particles non-destructively. With an improved apparatus a group at Harvard has measured the $g-2$ value of the electron with even higher precision [23], and currently methods are being developed to improve $g-2$ for the positron as well. In other experiments, charges and masses of e^+ and e^- were compared at a level of 4×10^{-8} [24] and 8×10^{-9} [25], respectively.

For the near future, experiments are planned to perform highly sensitive CPT-tests by using antihydrogen, which is a positron e^+ bound to an antiproton $\bar{p}$. The electromagnetic spectrum of this anti-atom will be compared to that of hydrogen, which is one of the best understood and most precisely investigated simple systems in modern experimental physics [26, 27]. A precision measurement of the antihydrogen hyperfine structure with a relative precision of at least 5×10^{-7} [28] is planned by the ASACUSA collaboration [29] at the antiproton decelerator (AD) at CERN [30, 31]. In these experiments a polarized antihydrogen beam will be produced using a superconducting anti-Helmholtz cusp polarizer followed by a radio-frequency cavity, a sextupole analyzer and an antihydrogen detector. Using this system, Rabi-like microwave spectroscopy will be performed [32]. In later experiments the application of Ramsey beam spectroscopy is planned and in the far future even more elegant methods as atomic fountain spectroscopy [33] might be applied. Another experimental approach to access the hyperfine structure of antihydrogen is planned by the ALPHA collaboration [34] using low-field seeking antihydrogen atoms trapped in a magnetic minimum Ioffe trap. By inducing antihydrogen spin flips with magnetic radio-frequency drives, trapped low-field seekers are transformed into high-field seeking states, which escape from the magnetic bottle trap. Recording the antihydrogen annihilation signal as a function of the drive frequency gives as well access to the antihydrogen hyperfine structure. Although the typical depths of the magnetic minimum traps are only in the order of about 40 μeV, recently the first trapping of antihydrogen atoms, was reported [35, 36]. In these experiments, storage times of up to 1000 s were achieved. By applying the scheme described above, resonant quantum transitions in antihydrogen have been observed for the first time [37], which is an important milestone towards precision spectroscopy of antihydrogen.

In addition to these magnetic spectroscopy experiments, the ALPHA collaboration as well as the ATRAP collaboration [38] plan to measure the antihydrogen 1S–2S transition using two-photon spectroscopy. In companion experiments conducted with hydrogen, a high-precision measurement of the respective transition at the level of 10^{-15} was achieved [39]. To reach a comparable precision in magnetic minimum traps, the antihydrogen atoms have to be laser-cooled [40, 41]. Although it is challenging to produce coherent continuous-wave Lyman-α radiation with sufficient laser power, utilizing the non-linear electromagnetic susceptibility of mercury vapor in a so-called triple-resonant four-wave mixing approach, the cw-production of 121 nm wavelength with a power of 6 μW has been reported recently [42].

Other experimental efforts focus on microwave and optical spectroscopy of exotic atoms [43]. With a probability of about 3 % antiprotons passing a cryogenic helium target replace one of the two electrons and exotic $\bar{p}He^{+}$ is formed. Performing spectroscopy with these complex exotic atoms, in elegant experiments using frequency-comb-locked two-photon spectroscopy, the antiproton-to-electron mass ratio was determined with sub-ppb precision [43].

Also the proton/antiproton system was subject of high-accuracy studies. In pioneering experiments conducted by the TRAP collaboration, which demonstrated the first trapping of antiprotons from the Low Energy Antiproton Ring at CERN [44], first electron cooling of antiprotons [45], as well as the first observation of a single antiproton stored in a Penning trap, the charge-to-mass ratios of the two particles were compared with sub-ppb precision [46]. In a first approach TRAP aimed at a precision comparison at the level of 10^{-9} by measuring the cyclotron frequency of a single trapped proton, and then reversing the trapping potential to measure the same value for the antiproton. In using an H^- ion for these comparison experiments, systematic effects due to reversal of the trapping potential were suppressed, and finally a precision comparison at the level of 9×10^{-11} was achieved (Table 5.1).

5.3 The Magnetic Moments of the Proton and the Antiproton

One quantity of stable particle-antiparticle systems has not yet been compared with high precision, the magnetic moment

$$\mu_{p,\bar{p}} = g_{p,\bar{p}} \frac{e}{2m_{p,\bar{p}}} \cdot S \tag{5.3}$$

of the proton and the antiproton, where $e/m_{p,\bar{p}}$ is the charge-to-mass ratio, S the particle spin, and $g_{p,\bar{p}}$ the so-called g-factor. A summary of measurements conducted so far is shown in Fig. 5.1.

The first measurement of the magnetic moment of the proton goes back to Otto Stern. In using his famous ‘molecular ray method’ [58], sensationally, the magnetic moment of the proton was found to be $\mu_p = 2.5(3) \cdot \mu_N$, where $\mu_N = (e\hbar)/(2m_\mathrm{p})$ is the nuclear magneton. The measured value differed by a factor of 2.5 from the value expected from Dirac theory [59], and was the first experimental indication to baryon substructure. For this trailblazing discovery, Stern received the Nobel Prize

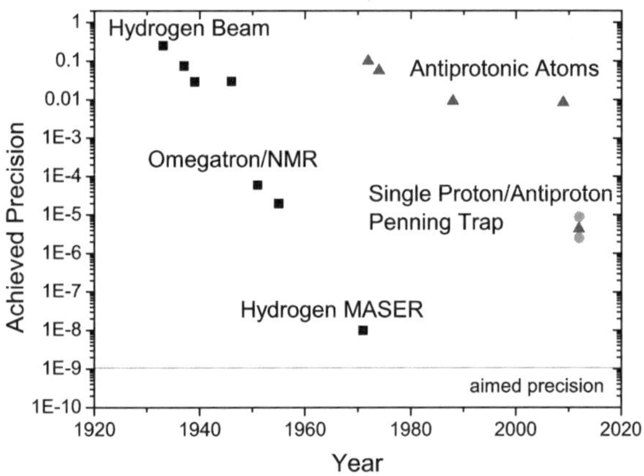

Fig. 5.1 Experimental precision achieved by different experiments within physic history. *Black squares* represent the measurements carried out with hydrogen. *Red dots* are results of precision experiments using a single trapped proton in a Penning trap. The *blue triangles* represent spectroscopy results with antiprotonic atoms. In 2012 the ATRAP collaboration determined the magnetic moment of a single trapped antiproton with a precision of 4.4 ppm

in 1943. This spectacular result inspired numerous experimentalists to perpetually increase the resolution of their techniques. Between the 1930s and the 1970s, Rabi [60] (molecular beam method), Bloch [61, 62] (nuclear magnetic resonance method), Sommer [63], and Collington [64] (omegatron method) continuously improved the value to a relative precision of $\Delta\mu_p/\mu_p = 1.4 \times 10^{-5}$. Presently, the most precise value for μ_p is extracted from an experiment conducted about forty years ago at the Massachusetts Institute of Technology [65]. The data are based on hyperfine spectroscopy of an atomic hydrogen maser in a magnetic field, yielding the magnetic moment ratio of the proton and the electron bound in atomic hydrogen $\mu_{\mathrm{p}}(\mathrm{H})/\mu_{\mathrm{e}}(\mathrm{H})$ at a level of 10 ppb. From this experiment the g-value of the free proton

$$g_{\mathrm{p}} = g_{\mathrm{e}} \cdot \frac{\mu_{\mathrm{p}}(\mathrm{H})}{\mu_{\mathrm{e}}(\mathrm{H})} \cdot \frac{g_{\mathrm{p}}}{g_{\mathrm{p}}(\mathrm{H})} \cdot \frac{g_{\mathrm{e}}(\mathrm{H})}{g_{\mathrm{e}}} \cdot \frac{m_{\mathrm{p}}}{m_{\mathrm{e}}} \tag{5.4}$$

is calculated, where the g-factor of the electron g_{e}, and the proton-to-electron mass ratio $m_{\mathrm{p}}/m_{\mathrm{e}}$ are known at a level of <0.001 ppb [23] and <1 ppb [66], respectively. The theoretical correction terms $g_{\mathrm{p}}/g_{\mathrm{p}}(\mathrm{H})$ and $g_{\mathrm{e}}(\mathrm{H})/g_{\mathrm{e}}$ are reviewed in [67, 68] and are also known at the sub-ppb level, resulting in [69]

$$g_{\mathrm{p}} = 5.585\,694\,713\,(46), \tag{5.5}$$

limited by the experimental precision achieved in the hydrogen-maser experiment [65].

In comparison, the magnetic moment of the antiproton $\mu_{\bar{p}}$ is only known with moderate precision. In the early 1970s a collaboration performing experiments at the Berkeley National Laboratory (BNL) have extracted the first $\mu_{\bar{p}}$-value from hyperfine measurements of antiprotonic lead and uranium atoms with percent precision [70]. Using a similar method, this value was improved by a factor of 2 in 1988 [71]. The ASACUSA collaboration [29] has extracted $\mu_{\bar{p}}$ from comparisons of super-hyperfine spectroscopy data of antiprotonic helium atoms ($\bar{p}\mathrm{He}^{+}$) and quantum electrodynamics (QED) calculations with a relative precision of 2.9×10^{-3} [72]. These exotic three-body systems have a hyperfine structure which is caused by the interaction of the electron magnetic moment with the antiproton angular momentum, as well as a super-hyperfine structure which is mainly due to electron/antiproton magnetic moment interaction. An energy level scheme is shown in Fig. 5.2. In these experiments, laser transitions between two specific quantum states (n, L) of the exotic atoms are driven, where n is the principal quantum number and L the angular momentum quantum number. One is the decay-dominated long-lived state with a lifetime of 1.5 μs (RD), and the other is the Auger-decay dominated short-lived state with 10 ns lifetime (AgD). For the determination of the magnetic moment of the antiproton, an annihilation signal is recorded as a function of a laser drive, which pumps the population from the RD state to the AgD state. Subsequently, microwave drives at $\nu_{\mathrm{HF},+}$ and $\nu_{\mathrm{HF},-}$ repopulate the F_{+} state, and the laser depopulation is repeated (compare Fig. 5.2). From the measured difference of $\nu_{\mathrm{HF},+}$ and $\nu_{\mathrm{HF},-}$, the antiproton magnetic moment is obtained [73]. The precision of these measurements is limited by statistics,

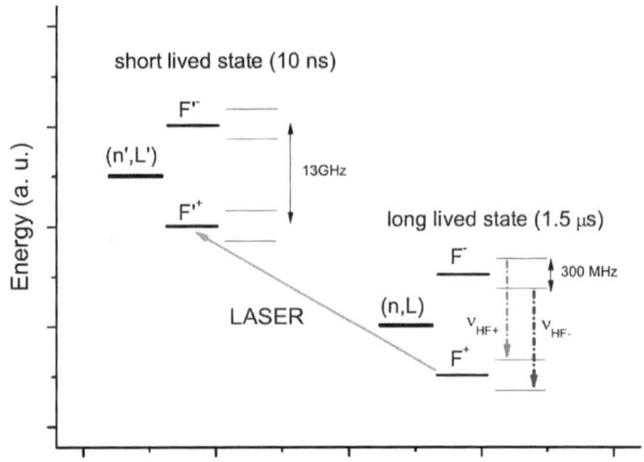

Fig. 5.2 Energy levels in exotic antiprotonic helium atoms and illustration of the laser/microwave spectroscopy scheme which was applied to measure the magnetic moment of the antiproton. For further details see text

Doppler and pressure broadening, as well as the three-body QED calculations which are based on the Breit-Pauli Hamiltonian. In the theoretical solution of the three-body problem, relative contributions of the order $\alpha^2 = 5 \times 10^{-5}$ are neglected. Assuming that all these limitations can be eliminated, the best possible precision will be ultimately limited by the natural linewidth of the longer-lived (n, L)-state to the order of about 5×10^{-7} [74].

However, important progress towards a high-precision determination of the magnetic moment of the antiproton was achieved [75]. Spin quantum transitions of a single proton stored in a cryogenic Penning trap were detected for the first time [76]. Based on these experiments, the magnetic moment of a single proton stored in a cryogenic Penning trap was measured with a relative precision of 8.9×10^{-6} by a group at Mainz [77]. Independently, a group at Harvard [78] reported a value with a relative precision of 2.5×10^{-6} [79]. This group (a part of the ATRAP collaboration) recently succeeded in measuring the magnetic moment of a single trapped antiproton with a precision of 4.4 ppm, which is a sensational improvement by a factor of about 680 compared to [72]. The experimental methods are very similar to the ones used in the $g-2$ comparisons of the electron and the positron [21]. However, in case of the proton/antiproton system, these experiments are more difficult compared to e^-/e^+, since the magnetic moment of the proton/antiproton is about 650 times smaller, which requires a much higher sensitivity of the experimental apparatus. In the future both groups aim for a precision measurement at the ppb level of accuracy by application of the so-called double Penning-trap technique [80], which will be explained later in this chapter.

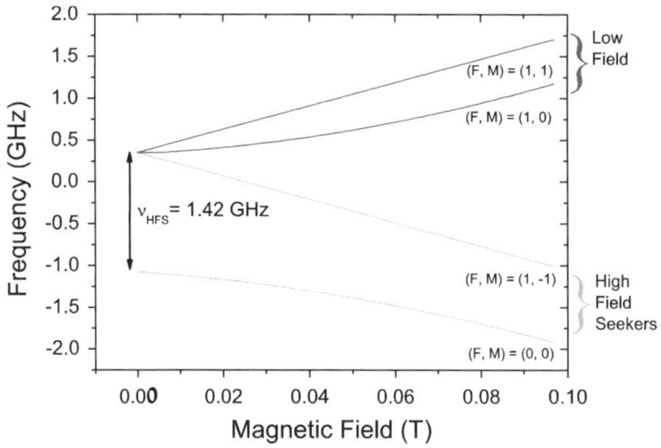

Fig. 5.3 Ground-state hyperfine structure of antihydrogen as a function of an externally applied magnetic field. For details see text

5.4 Antiproton Magnetic Moment and Antihydrogen Hyperfine Structure

As described above, the ASACUSA collaboration as well as the ALPHA collaboration plan to perform a measurement of the antihydrogen ground-state hyperfine structure (GS-HFS). While ASACUSA utilizes a beam-spectroscopy scheme [28], the experiments planned by ALPHA are based on trapped antihydrogen [37].

The GS-HFS, shown in Fig. 5.3, arises from the interaction of the magnetic moments of the positron μ_{e^+} and the antiproton $\mu_{\bar{p}}$. At vanishing magnetic field $B = 0$ the GS-HFS is split into two the energy levels $F = 0$ and $F = 1$ ($\nu_{\mathrm{HFS}} = \Delta E/h \approx 1.42$ GHz), where F is the total angular momentum quantum number. In the $F = 0$ state the $\mu_{e^+}/\mu_{\bar{p}}$ interaction is attractive and in $F = 1$ repulsive. For $B > 0$ the $F = 1$ levels split into three sub-levels $M = -1$, $M = 0$ and $M = 1$, where M is the magnetic quantum number. Low-field seeking states are characterized by positron magnetic moment antiparallel to the externally applied B-field, while in high-field seeking states the positron magnetic moment is parallel to B. At vanishing magnetic field the singlet ($F = 0$) to triplet ($F = 1$) transition frequency is given by

$$\nu_{\mathrm{HFS}} = \frac{16}{3}\left(\frac{m_{\bar{p}}}{m_{\bar{p}} + m_{e^+}}\right)^3 \frac{m_{e^+}}{m_{\bar{p}}}\frac{\mu_{e^+}}{\mu_{\mathrm{B}}}\frac{\mu_{\bar{p}}}{\mu_{\mathrm{N}}}\alpha^2 c R_\infty, \tag{5.6}$$

which is mainly a product of the positron magnetic moment μ_{e^+} and the antiproton magnetic moment $\mu_{\bar{p}}$. The constants involved in this expression are the positron and the antiproton mass, m_{e^+} and $m_{\bar{p}}$, respectively, the Bohr magneton μ_{B}, the nucleon magnetic moment μ_{N}, the fine structure constant α, the speed of light c, and the Rydberg constant R_∞. This energy splitting $\Delta E/h$ results from the point-like Fermi contact interaction of μ_{e^+} and $\mu_{\bar{p}}$. Equation (5.6) reproduces the experimental value at a level of about 5×10^{-5}. Contributions below that level are due to antiproton substructure and can be described theoretically by

$$\Delta\nu = \nu_{\text{HFS}} \frac{2Z\alpha m_e}{\pi^2} \int \frac{d^3q}{q^4} \left[\frac{G_E(-q^2)G_M(-q^2)}{1+\kappa} - 1 \right], \tag{5.7}$$

where $G_E(-q^2)$ and $G_M(-q)^2$ are the electric and magnetic form factors of the antiproton, and κ the particle's anomalous magnetic moment. Thus, together with an independent measurement of the magnetic moment of the antiproton, the precision measurement of the antihydrogen GS-HFS provides insight into the substructure of the antiproton.

5.5 *g*-Factor Measurements

In a magnetic field **B**, the determination of the *g*-factor[1] of a spin-carrying charged particle with charge-to-mass ratio e/m reduces to the measurement of two frequencies, namely the *free cyclotron frequency*

$$\nu_c = \frac{1}{2\pi} \frac{e}{m} B, \tag{5.8}$$

and the *Larmor frequency*

$$\nu_L = g \frac{1}{2\pi} \frac{e}{2m} B. \tag{5.9}$$

The Lorentz force $\mathbf{F} = e\mathbf{v} \times \mathbf{B}$ constrains the particle to a circular orbit, where it oscillates with the free cyclotron frequency ν_c. The Larmor frequency ν_L is the spin precession frequency. A measurement of the frequency ratio ν_L/ν_c corresponds to a direct measurement of the (anti)proton magnetic moment in units of the nuclear magneton μ_N since

$$\frac{g}{2} = \frac{\nu_L}{\nu_c} = \frac{\mu_{p,\bar{p}}}{\mu_N}. \tag{5.10}$$

5.6 The Penning Trap

To perform the measurement described above, a single particle stored in a cryogenic Penning trap [81] is used. A schematic of such a trap is shown in Fig. 5.4a. It is a superposition of a constant magnetic field $B\mathbf{e}_z$ in the axial direction and an electrostatic quadrupolar potential (see Fig. 5.4b)

---

[1] Since the charge-to-mass ratio of the proton/antiproton is known with a precision of about 0.1 ppb [46], we do not distinguish the terms *g-factor* and *magnetic moment* in the further text.

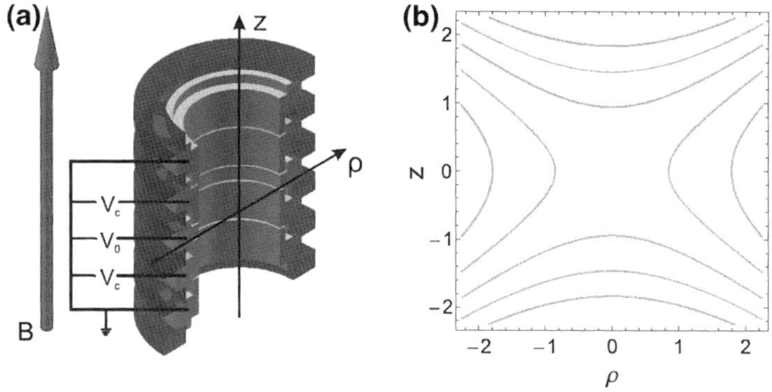

Fig. 5.4 **a** Schematic of of a cylindrical Penning trap. The magnetic field is indicated by the arrow in z-direction. Applying adequate voltages to the trap electrodes, the trapped particle sees an almost perfect quadrupolar potential. **b** shows a parametric plot of electrostatic equipotential lines

$$\Phi(z,\rho) = V_0 C_2 \left(z^2 - \frac{\rho^2}{2}\right), \tag{5.11}$$

where z and ρ are cylindrical coordinates. V_0 is the potential difference between the ring and the end cap electrodes of the trap and C_2 a coefficient which characterizes the trap geometry. The magnetic field constrains the particle to a circular orbit, while the electrostatic quadupolar potential confines it along the magnetic field lines. A typical trajectory of a particle trapped in such a superposition of static electric and magnetic fields is shown in Fig. 5.5. It can be described as a superposition of three independent harmonic oscillators,

- the axial oscillator at frequency

$$\nu_z = \frac{1}{2\pi}\sqrt{\frac{2C_2 e V_0}{m}}, \tag{5.12}$$

Fig. 5.5 Trajectory of a single particle in a Penning trap, which is a superposition of three uncoupled harmonic oscillators. For further details see text

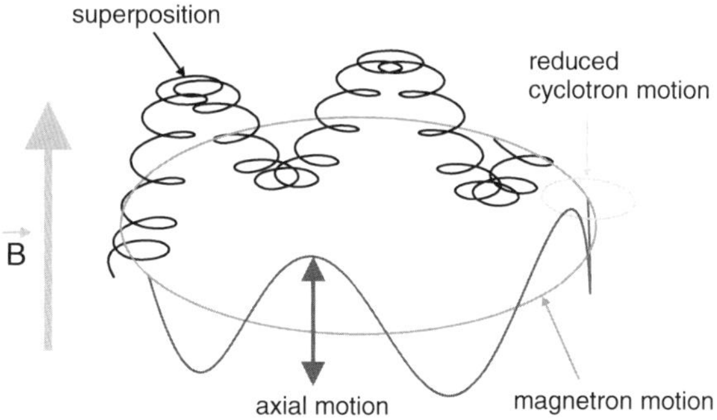

which is the oscillation of the particle along the magnetic field lines due to the electrostatic potential,

- the modified cyclotron oscillator at frequency

$$\nu_+ = \frac{1}{2}\left(\nu_c + \sqrt{\nu_c^2 - 2\nu_z^2}\right), \tag{5.13}$$

which is the free cyclotron oscillation, slightly modified by the radially pulling electrostatic potential,

- the magnetron oscillator at frequency

$$\nu_- = \frac{1}{2}\left(\nu_c - \sqrt{\nu_c^2 - 2\nu_z^2}\right), \tag{5.14}$$

which is due to a drift in the crossed static fields.

Typical precision Penning traps are operated at $\nu_+ \gg \nu_z \gg \nu_-$, where the stability condition $\nu_c^2 - 2\nu_z^2 > 0$ is fulfilled. The proton/antiproton cyclotron frequency is about 15.3 MHz/T, axial frequencies are of the order of 500 kHz to 1 MHz and thus, the magnetron frequency $\nu_- = \nu_z^2/(2\nu_+)$ is typically between 5 and 10 kHz. Note, that the magnetron energy E_- is of almost purely potential nature defined with respect to the ring electrode, and thus, maximal for a vanishing magnetron radius ρ_-. Consequently, the magnetron mode is meta-stable, however, typical radiative magnetron cooling-time constants are in the order of the age of the universe.

One of the most important relations for high-precision Penning trap physics is the so-called *Brown-Gabrielse invariance theorem* [81]

$$\nu_c = \sqrt{\nu_+^2 + \nu_z^2 + \nu_-^2}, \tag{5.15}$$

which directly relates the measurable trap frequencies ν_+, ν_- and ν_z to the free cyclotron frequency. This invariance theorem is robust against typical first-order trap errors such as a tilt between the magnetic trapping field and the electrostatic quadrupole potential or slight trap ellipticities. Thus, by measuring the trap frequencies, the charge-to-mass ratio of the trapped particle is directly accessible.

However, frequency shifts induced by trap imperfections as magnetic field inhomogeneities, anharmonicities of the trapping potential and relativistic effects affect the measured cyclotron frequency. To illustrate this, consider the axial electrostatic potential in presence of an octupolar perturbation $\Phi(z) = V_0(C_2z^2 + C_4z^4)$. The equation of motion is

$$\ddot{z} + \omega_z^2(1 + \frac{2C_4}{C_2}z^2)z = \ddot{z} + \omega_z^2(1 + \frac{2C_4}{C_2}\frac{2E_z}{m\omega_z^2}) = 0. \tag{5.16}$$

Since the perturbation scales as z^2, the solution of this equation of motion leads to a frequency shift which scales linearly with the axial energy

$$\frac{\Delta\omega_z}{\omega_z} = \frac{3}{4}\frac{C_4}{C_2^2}\frac{E_z}{qV_0}, \tag{5.17}$$

as obtained by application of harmonic balance [82]. Frequency shifts for other motions and different perturbations can be calculated in a similar way and are described in the chapter by Vogel and Quint.

5.7 Experimental Setup

High precision Penning trap experiments, such as one for the measurement of μ_p and $\mu_{\bar{p}}$, requires different ingredients. In the following, we discuss the general aspects of such an experiment, as for example implemented in the apparatus of BASE, which is shown in Fig. 5.6. The magnetic field is provided by a highly stable, extremely homogeneous, self-shielded superconducting magnet [83] with a typical homogeneity of $\Delta B/B \approx 10^{-6}$ in the central cubic centimeter, and an average magnetic field drift below 0.05 ppm/h. The cold stage of the apparatus, which includes the Penning trap, and sensitive single-particle detection electronics with cryogenic filter stages are connected to the 4 K heat exchangers of cryostats. This cold stage is surrounded by heat shields, which are connected to the liquid-nitrogen reservoirs of the cryostats. For the proton/antiproton g-factor measurement, five-electrode

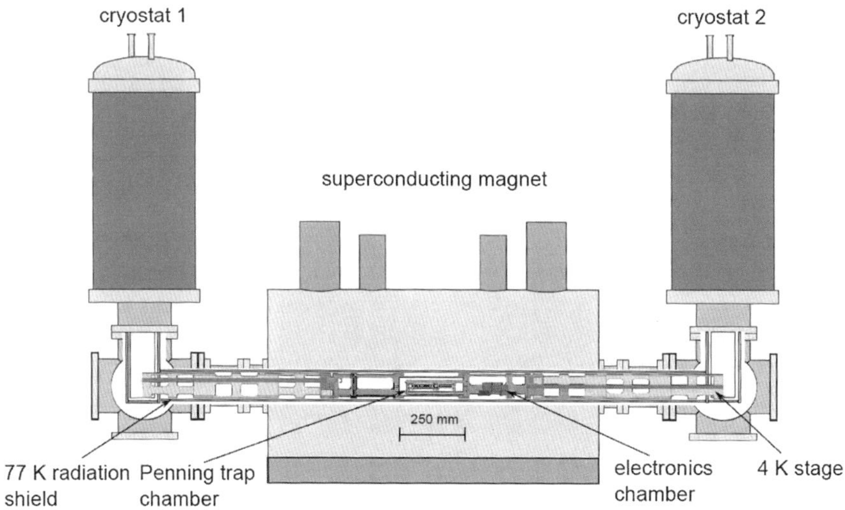

Fig. 5.6 Schematic of the BASE experiment. The trap is mounted in the center of the superconduction magnet. The cryogenic temperatures are provided by two cryostats placed at each end of the magnet. The cryogenic detection electronics are placed close to the trap

cylindrical Penning traps are used. The trap dimensions match the orthogonal, compensated design [84, 85]. The electrodes are made out of oxygen-free high-conductivity (OFHC) copper which has low magnetic susceptibility. The electrodes are gold-plated, hence preventing surface oxidation which would disturb the electric field due to surface potentials. Sapphire rings are used as spacers between the electrodes. In order to achieve long storage times, excellent vacuum conditions inside the Penning traps are required. Thus, the traps are placed in a hermetically sealed, cold-welded, cryo-pumped trap chamber, which is located in the center of the superconducting magnet. In this 'trap can' pressures of typically 10^{-16} mbar are achieved, which avoids particle loss due to charge exchange or annihilation with residual gas atoms. Under such conditions, the TRAP collaboration has stored a single antiproton for months [86]. Besides cryo-pumping, cryogenic operation of Penning-trap experiments has the advantage that electronic noise is reduced at low temperatures. In addition, superconducting resonators can be used for the particle-detection electronics, which increases detection efficiency and thus reduces the typical time scales required for the experimental cycles.

5.8 Measurement of the Eigenfrequencies

5.8.1 Peak Detection

Crucial parts of high-precision Penning trap experiments are highly sensitive detection systems with single-particle sensitivity. High performance enables fast frequency measurements at low particle energy, which avoids systematic energy-dependent shifts of the measured cyclotron frequency. Pioneering work on single-particle detection has been contributed by Wineland and Dehmelt who for the first time detected a single electron stored in a Penning trap in the early 1970s [87].

The basic principle of non-destructive detection is illustrated in Fig. 5.7. An oscillating trapped charged particle induces image currents I_p in the trap electrodes, which are typically of the order of a few fA. By connecting a large resistor to one of the trap electrodes, a voltage drop can be measured. In practice, the self capacitance C_T of the trap is compensated with an inductor L connected to the trap. L and C_T form a parallel

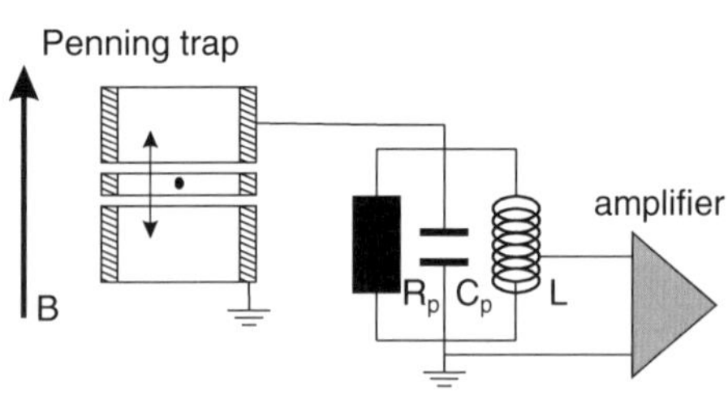

Fig. 5.7 Illustration of single particle detection. Image currents are induced in the trap electrodes which can be detected using resonant detection systems tuned to the particle's oscillation frequency

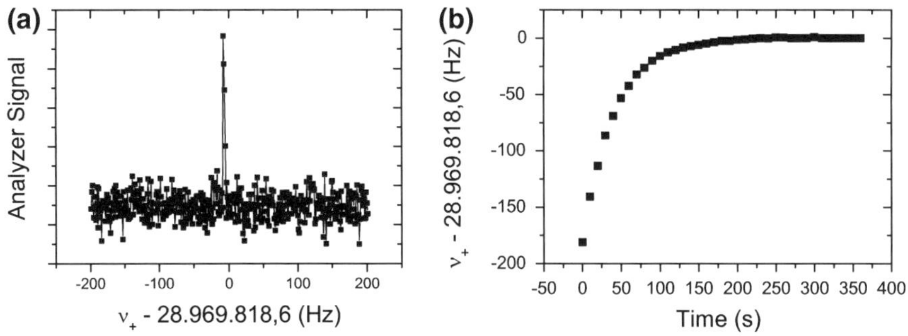

Fig. 5.8 **a** Signal of a single excited proton tuned to resonance with the detector for the modified cyclotron frequency. **b** Modified cyclotron frequency as a function of time. The change of ν_+ is due to resistive cooling in the presence of trap errors

tuned circuit [88–90] which acts as a resistor R_p, where ν_r it the resonance frequency $R_p = 2\pi\nu_r QL$. Q is the so-called 'quality factor', which characterizes the energy loss of the system per oscillation cycle. Tuning the particles's oscillation frequency (for example ν_z) to ν_r, the voltage signal $U = R_p \cdot I_p$ is obtained which is amplified with cryogenic low-noise amplifiers [91–93] and analyzed with a Fast-Fourier-Transform (FFT) spectrum analyzer. The signal of a single excited proton in resonance with the detector for the modified cyclotron frequency ν_+ is shown in Fig. 5.8a.The effective parallel resistance R_p of the detector cools the particle resistively. The cooling time constant

$$\tau = \frac{m}{R_p}\frac{D^2}{e^2} \tag{5.18}$$

is inversely proportional to R_p and proportional to the square of the parameter D, which is a trap-specific length. Figure 5.8b shows the measured modified cyclotron frequency as a function of time. The frequency change is due to magnetic field inhomogeneities. Fitting an exponential to these data, the modified cyclotron frequency can be determined with a precision of about 5 ppb.

In contrast to the modified cyclotron and the axial frequency, the magnetron frequency is not measured directly. Resistive detection of ν_- increases the magnetron radius ρ_- and eventually leads to particle loss. Thus, the magnetron frequency is measured by sideband coupling methods [94], which will be described below.

5.8.2 Dip Detection

A particle cooled to thermal equilibrium with the detection system involves some interesting aspects which can be utilized to increase the experimental precision. At its resonance frequency, a single particle behaves like an almost perfect conductor which shortcuts everything in parallel. This can be derived by considering the equations of motion. Equivalent inductances and capacitances can be defined to rewrite the equation of motion in the form of a differential equation which describes a series tuned circuit with an almost infinite quality factor [87]. As a consequence, a single particle cooled to thermal equilibrium appears as a dip in the thermal Johnson-noise spectrum $e_{th} = \sqrt{4k_B T \Delta B \mathrm{Re}[Z(\nu)]}$ of the detector [95, 96]. In this expression k_B is the Boltzmann constant, T the detector temperature, ΔB the measuring bandwidth and $\mathrm{Re}[Z(\nu)]$ the real part of the tuned circuit's impedance. Such a noise spectrum with a single particle tuned in parallel is shown in Fig. 5.9a. The linewidth $\Delta\nu_z$ of the dip is defined by the cooling resistor R_p and its coupling D to the particle

$$\Delta\nu_z = \frac{1}{2\pi\tau} = \frac{1}{2\pi}\frac{R_p e^2}{mD^2}. \tag{5.19}$$

The axial frequency is determined by a best fit to the data. For the parameters of the detection systems used at the proton g-factor experiment at the University of Mainz, the axial frequency can be determined with a precision of 10 ppb in an averaging time of 100 s. Cooled to thermal equilibrium with the detection system, the particle is typically located within a volume below 50 μm^3. In such a small volume, very homogeneous conditions can be produced and systematic frequency shifts are typically below the ppb level.

Since the linewidth of the dip scales as e^2, a trapped cloud of N uncorrelated particles produces a total linewidth $\Delta\nu_z(N) = N \cdot \Delta\nu_z$. Thus, for low particle numbers, the measurement of the linewidth directly corresponds to a determination of the number of trapped particles. This property is used to prepare a single particle in the trap. The linewidth as a function of N is shown in Fig. 5.9b. These data were produced by excitation of the trapped particles and subsequent lowering of the trapping

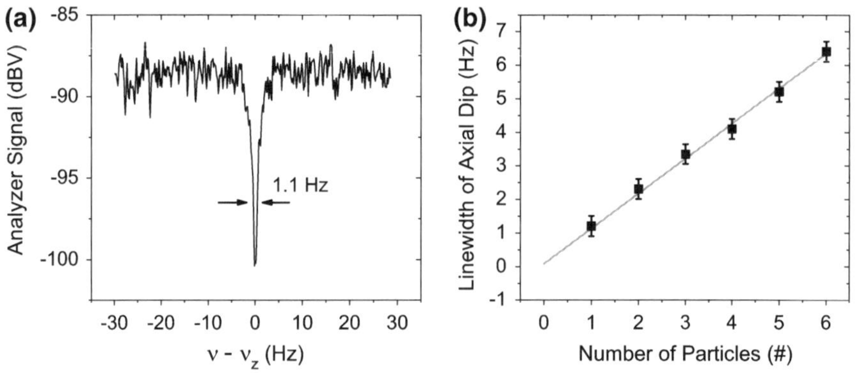

Fig. 5.9 **a** Noise spectrum of the axial detection system. The dip is due to a proton which is tuned in parallel to the detector and shorts the thermal noise of the resonator. **b** Line width of the 'noise dip' as a function of particle number

potential. After each heating step the particles were cooled to thermal equilibrium and the linewidth was measured.

5.8.3 Sideband Coupling

While the axial frequency is measured directly by means of dip detection, the radial frequencies are measured via peak detection or sideband coupling [94]. To this end, an rf-signal at a frequency $\nu_{\rm rf}$ is applied to a radially segmented trap electrode. This effectively produces an electrical rf-field

$$\mathbf{E}\,(\mathbf{r},t) = \mathrm{Re}\left(E_{\rm rf}\exp\left(2\pi\imath\,\nu_{\rm rf}t\right)\left(z\mathbf{e}_\rho - \rho\mathbf{e}_z\right)\right) \tag{5.20}$$

at the center of the Penning trap, where $E_{\rm rf}$ is the amplitude of the field. The term $\left(z\mathbf{e}_\rho - \rho\mathbf{e}_z\right)$ couples the eigenmotions. By adjusting the drive frequency $\nu_{\rm rf}$ to the sum of the axial and the magnetron frequencies $\nu_z + \nu_-$, energy is transferred between both modes. If the difference frequency $\nu_{\rm rf} = \nu_+ - \nu_z$ is irradiated, the axial and the modified cyclotron motion are coupled. The energy transfer induced by the coupling leads to an amplitude modulation of the axial motion

$$\begin{aligned} z(t) &= z_0\cos\left(\frac{\Omega_0}{2}t\right)\sin\left(2\pi\,\nu_z t\right) \\ &= z_0\left[\sin\left(2\pi\left(\nu_z + \frac{\Omega_0}{4\pi}\right)t\right) + \sin\left(2\pi\left(\nu_z - \frac{\Omega_0}{4\pi}\right)t\right)\right], \end{aligned} \tag{5.21}$$

where the resonant Rabi frequency Ω_0 is defined by the field amplitude $E_{\rm rf}$. The frequency spectrum of such a coupled motion shows two components, one at $\nu_l = \nu_z - \frac{\Omega_0}{4\pi}$ and the other at $\nu_r = \nu_z + \frac{\Omega_0}{4\pi}$. A frequency spectrum of a single proton tuned to resonance with the axial detector while a resonant coupling drive $\nu_{\rm rf} = \nu_z + \nu_-$ is applied is shown in Fig. 5.10a. If the drive frequency $\nu_{\rm rf}$ is detuned by $\delta_- = \nu_{\rm rf} - (\nu_z + \nu_-)$ or $\delta_+ = \nu_{\rm rf} - (\nu_+ - \nu_z)$, respectively, the resulting frequencies of such a *double-dip* are given by

$$\nu_l = \nu_z - \frac{\delta_\pm}{2} - \frac{\Omega_\pm}{4\pi} \tag{5.22}$$

$$\nu_r = \nu_z - \frac{\delta_\pm}{2} + \frac{\Omega_\pm}{4\pi}\,, \tag{5.23}$$

where $\Omega_\pm = \sqrt{\Omega_{0,\pm}^2 + \delta_\pm^2}$ is the off-resonant Rabi frequency. By recording ν_l and ν_r as a function of the drive frequency $\nu_{\rm rf}$, a 'classical avoided crossing' is observed (see Fig. 5.10b).

The system of Eqs. (5.22) and (5.23) includes three unknown quantities: the Rabi frequency $\Omega_\pm$, the frequency of the radial mode $\nu_\pm$, and the axial frequency ν_z.

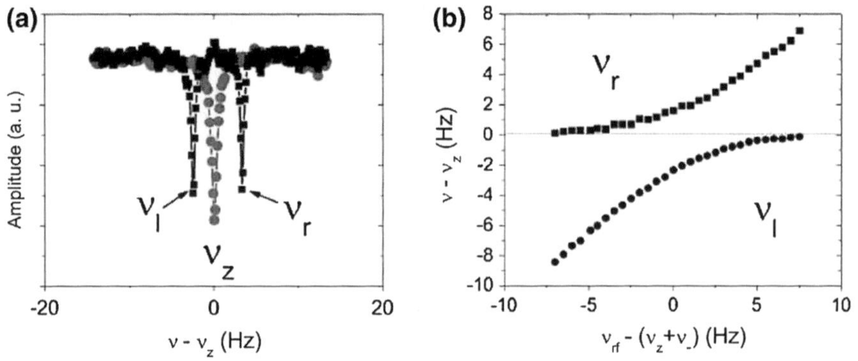

Fig. 5.10 **a** Single proton on resonance with the axial detection system. The 'double dip' spectrum (*filled squares*) was recorded while a sideband drive was irradiated. The single dip (*filled circles*) is recorded without sideband drive. **b** 'Classical Avoided Crossing'. The sideband frequencies as function of the detuning of the rf-signal is shown. For details see text

By recording such a sideband frequency spectrum (double dip) and, in an additional measurement, the axial frequency ν_z, the respective radial frequency can be extracted by solving the system of Eqs. (5.22) and (5.23).

Note, that by application of this sideband frequency measuring scheme, the frequencies are measured in thermal equilibrium with the cryogenic axial detection system. Thus, during such a frequency measurement, the particle amplitudes are typically below 100 μm and systematic shifts of the measured cyclotron frequency which arise form trap imperfections are low.

5.9 Advanced Frequency Measurements

In analogy to the AC Stark shift, the frequencies ν_l and ν_r, which are due to coupling of the axial motion to one of the radial modes, can be interpreted as 'classical dressed states' [97]. Applying an additional drive to the trap, these new states can be dressed with the remaining radial mode. Assume that a sideband drive $\nu_{\mathrm{rf}} = \nu_+ - \nu_z$ defines the new eigenstates $\nu_{l,+}$ and $\nu_{r,+}$ (cyclotron dressing). Once additional rf-drives at $\nu_{\mathrm{rf},1} = \nu_{l,+} + \nu_-$ and $\nu_{\mathrm{rf},2} = \nu_{r,+} + \nu_-$ are superimposed, the left and the right signal are split into another two modes

$$\nu_{l,l} = \nu_{l,+} - \frac{\delta_{l,-}}{4\pi} - \frac{\Omega_{l,-}}{4\pi} \tag{5.24}$$

$$\nu_{l,r} = \nu_{l,+} - \frac{\delta_{l,-}}{4\pi} + \frac{\Omega_{l,-}}{4\pi} \tag{5.25}$$

and

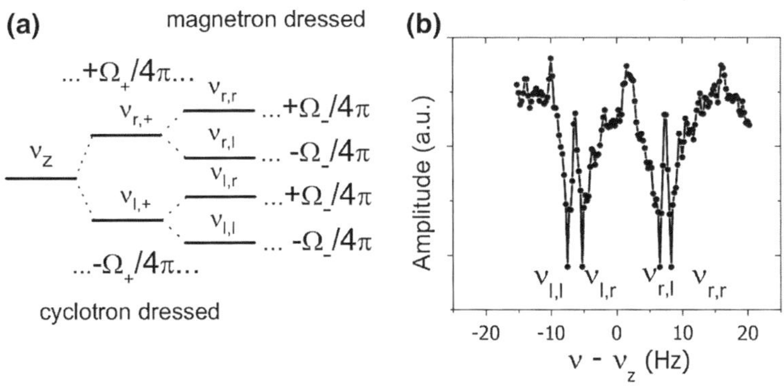

Fig. 5.11 **a** The principle for the simultaneous determination of all eigenfrequencies of a trapped proton is based on double-dressed states. The axial mode is first dressed with the cyclotron mode. Both cyclotron-dressed states are then coupled to the magnetron mode, which results in four double-dressed states. **b** Experimental result of a double-dressing. Applying three rf-drives to the trap, both radial modes are simultaneously coupled to the axial mode, which results in a quadruple-dip

$$\nu_{r,l} = \nu_{r,+} - \frac{\delta_{r,-}}{4\pi} - \frac{\Omega_{r,-}}{4\pi} \tag{5.26}$$

$$\nu_{r,r} = \nu_{r,+} - \frac{\delta_{r,-}}{4\pi} + \frac{\Omega_{r,-}}{4\pi}, \tag{5.27}$$

respectively, where $\Omega_{l,-} = \sqrt{\Omega_-^2 + \delta_{l,-}^2}$ and $\Omega_{r,-} = \sqrt{\Omega_-^2 + \delta_{r,-}^2}$ are off-resonant Rabi frequencies of the double-dressing, and $\delta_{l,-}/(4\pi) = \nu_{\mathrm{rf,l}} - (\nu_{l,+} + \nu_-)$ and $\delta_{r,-}/(4\pi) = \nu_{\mathrm{rf,2}} - (\nu_{r,+} + \nu_-)$ the detuning parameters. The result of such a double-dressing, a quadruple-dip, is shown in Fig. 5.11b. The Eqs. (5.24)–(5.27) relate four measured quantities $\nu_{l,l}$, $\nu_{l,r}$, $\nu_{r,l}$, and $\nu_{r,r}$ to five unknown quantities, the eigenfrequencies of the trapped proton ν_+, ν_z, and ν_-, and the resonant Rabi frequencies $\Omega_+/2\pi$ and $\Omega_-/2\pi$. To obtain an additional information, the rf drive is switches 'on' and 'off' while the spectrum is averaged, leading to an overlap of the fourfold dip and the unperturbed axial frequency ν_z. A quintuple-dip spectrum is obtained, which is shown in Fig. 5.12. This spectrum carries all eigenfrequency information of the trapped proton, and thus is a direct determination of the free cyclotron frequency ν_c of a single trapped particle [98]. From the spectrum shown in Fig. 5.12 a free cyclotron frequency of

$$\nu_c = 28\,974\,060.15(14)\ \mathrm{Hz} \tag{5.28}$$

is obtained, which corresponds to a precision of about 4.8 ppb.

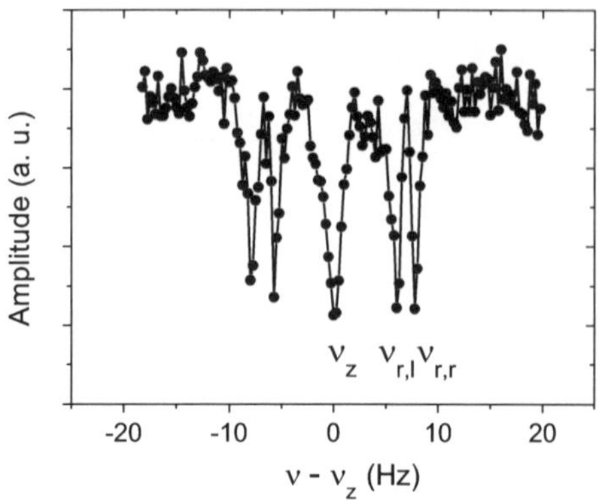

Fig. 5.12 This quintuple-dip spectrum was recorded in one measurement and carries all information of the eigenfrequencies of the trapped proton, and thus, the free cyclotron frequency ν_c

5.10 Continuous Stern Gerlach Effect

Compared to the detection of the free cyclotron frequency, the measurement of the Larmor frequency is much more challenging. The Larmor precession is not accompanied by a charge shift, and is thus not directly accessible by image current detection. To overcome this problem the spin direction is coupled to the axial oscillation frequency by superimposing a magnetic field inhomogeneity, a so-called magnetic bottle,

$$\Delta B = B_2 \left(\left(z^2 - \frac{\rho^2}{2} \right) \mathbf{e}_z + (z\rho)\, \mathbf{e}_\rho \right), \tag{5.29}$$

to the Penning trap. In practice such a ΔB is introduced by making the central ring electrode out of ferromagnetic material. The resulting magnetic field lines are shown in Fig. 5.13. Consider the energy of a magnetic moment μ in a magnetic field

$$\Phi_{\text{mag}} = -\left(\mu \mathbf{B}\right). \tag{5.30}$$

In the axial direction the magnetic bottle contributes a spin dependent quadratic potential

$$\Phi_{mag} = \pm \mu_z B_2 z^2 \tag{5.31}$$

which adds to the electrostatic potential,

$$\Phi = \Phi_{\text{mag}} + \Phi_{\text{el}} = \left(eC_2 V_0 \pm \mu_z B_2\right) z^2 \tag{5.32}$$

and thus, the axial oscillation frequency becomes a function of the spin eigenstate

$$\nu_z = \frac{1}{2\pi} \sqrt{2\left(eC_2 V_0 \pm \mu_z B_2\right)} \approx \nu_{z,0} \left(1 \pm \frac{\mu_z B_2}{4\pi^2 m \nu_{z,0}^2} \right). \tag{5.33}$$

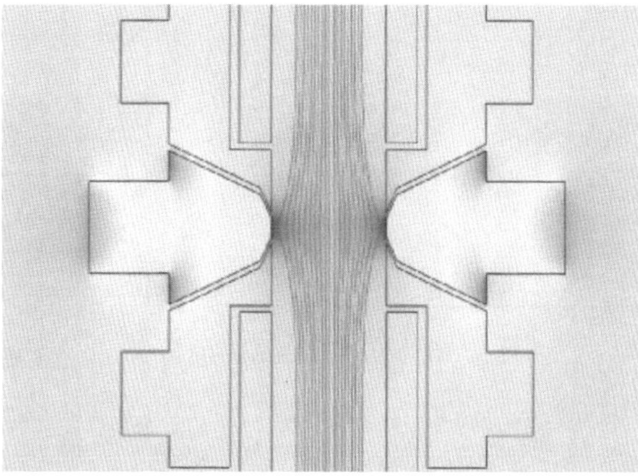

Fig. 5.13 Field lines produced by a magnetic bottle. The ferromagnetic central ring electrode distorts the field lines adding a magnetic field inhomogeneity B_2 to the homogeneous field B_0. The resulting magnetic field is described by: $\mathbf{B}(z, \rho) = B_0 \mathbf{e}_z + B_2 \left(\left(z^2 - \rho^2/2 \right) \mathbf{e}_z - z\rho \mathbf{e}_\rho \right)$

Consequently a change of the spin direction of the trapped particle changes the axial oscillation frequency by

$$\Delta \nu_{z,\,\mathrm{SF}} \approx \frac{\mu_z B_2}{2\pi^2 m \nu_{z,0}}, \tag{5.34}$$

which is proportional to the strength of the magnetic bottle B_2, proportional to the particle's magnetic moment μ and anti-proportional to its mass. This elegant measuring principle called continuous Stern-Gerlach effect, was invented by Hans Dehmelt [22], and has already been applied in the famous comparisons of the magnetic moments of the electron and the positron, where a precision of 2×10^{-12} has been achieved. However, the application of the same method to the proton/antiproton system is much more difficult, since the ratio μ_p/m_p is more than one million times smaller as in case of the electron/positron system. To construct an experimental apparatus which is sensitive enough to detect spin transitions of a proton or antiproton, a very strong magnetic inhomogeneity B_2 has to be used. At the Mainz proton g-factor apparatus, a Penning trap with an inner diameter of only 3.6 mm is used. The central ring electrode is made out of Co/Fe-alloy which has a saturation magnetization of 2.35 T. It produces a magnetic bottle of $B_2 = 300000\,\mathrm{T/m^2}$. When shifted along the trap axis by only about 1.5 mm, the average magnetic field seen by the particle changes already by 1 T. Even under these extreme magnetic conditions, a single proton spin flip changes the axial frequency of the particle by only about 200 mHz out of 680 kHz (Fig. 5.14).

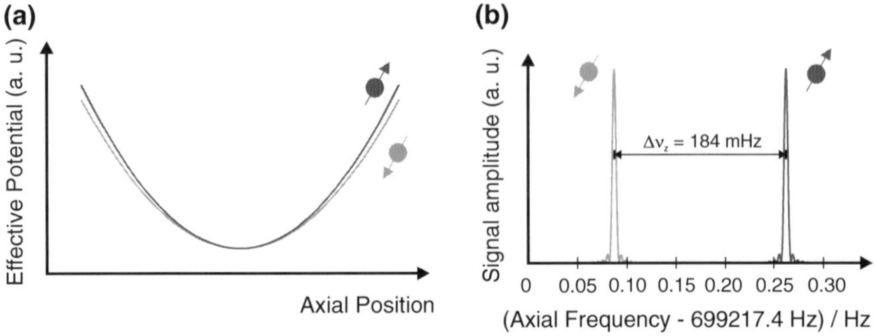

Fig. 5.14 **a** The effective potential for the axial motion in presence of the magnetic bottle. The potential depends on the spin state of the antiproton. **b** Simulation of the detection of the axial frequency of an antiproton in *spin down* state and in *spin up* state. The axial frequency difference due to the spin state of the antiproton is 184 mHz out of 699 kHz

5.11 Larmor Frequency Measurement

The continuous Stern-Gerlach effect is an elegant scheme to non-destructively detect the spin eigenstate of the trapped particle. This quantum non-demolition measurement can be applied to measure thc Larmor frequency of the trapped particle. Spin flips are driven by application of a simple Rabi resonance principle [99]. A magnetic radio-frequency field is irradiated to the trap with magnetic field vector perpendicular to the quantization axis which is defined by the axial magnetic field. On resonance and in its rest frame, the particle experiences a constant magnetic field component B_{rf} in ρ-direction, and in a classical picture, the spin starts to precess around the ρ axis. The quantum-mechanical treatment is similar to that of a simple two-level system. A measurement of the axial frequency collapses the wave function to one of the two possible spin eigenstates. The Larmor frequency is measured by applying the following idealized experimental sequence:

1. Measurement of the axial frequency ν_z in the trap with the superimposed magnetic bottle.
2. Irradiation of a spin flip drive with amplitude B_{rf} at a certain drive frequency ν_{rf}.
3. Repetition of the axial frequency measurement.

A change of the axial frequency indicates that the rf-drive induced a spin flip (see Fig. 5.15). Repeating this sequence for several times at different drive frequencies ν_{rf} gives the spin-flip probability P_{SF} as a function of the drive frequency. With a detailed understanding of the resulting lineshape $P_{\mathrm{SF}}(\nu_{\mathrm{rf}})$ the Larmor frequency is obtained from a best fit to the measured data.

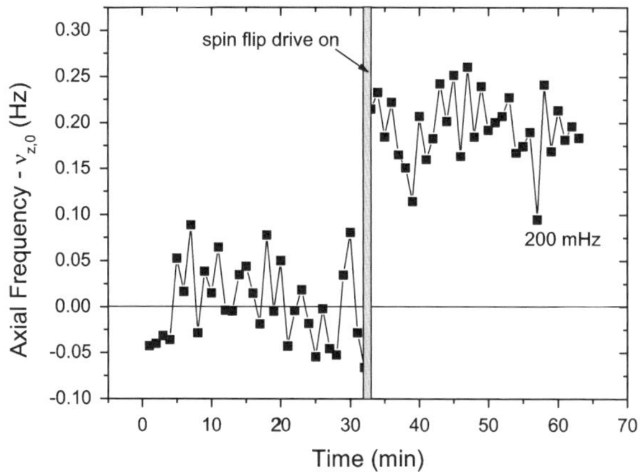

Fig. 5.15 Axial frequency as a function of time (simulated data). After 32 min a spin flip drive is irradiated and the axial frequency changes by about 200 mHz, which is due to a spin flip

5.12 Line Profile and Transition Rates

While a spin flip is driven, the particle is continuously thermalized by the axial detection system, which leads to a Brownian diffusion of the particle's axial energy E_z, and due to the presence of the magnetic bottle B_2, to fluctuations of the average Larmor frequency

$$\nu_L(E_z) = \nu_{L,0}\left(1 + \frac{B_2}{B_0}\frac{E_z}{2\pi^2 m\nu_z^2}\right). \tag{5.35}$$

The resulting line profile is thus a convolution of the approximately 'infinitesimally' sharp Lorentz line of the unperturbed Rabi resonance and the axial energy distribution $w(E_z) = \exp(-E_z/(k_B T_z))/(k_B T_z)$. The solution of the respective integral gives the line profile [100]

$$\chi_L(E_z) = \frac{\Theta(\nu - \nu_{L,0})}{\delta\nu_{LW}}\nu_{L,0}\exp\left(\frac{\Theta\,(\nu - \nu_L)}{\delta\nu_{LW}}\right). \tag{5.36}$$

where

$$\delta\nu_{LW} = \nu_{L,0}\frac{B_2}{B_0}\frac{k_B T_z}{4\pi^2 m\nu_z^2}\,, \tag{5.37}$$

is the linewidth parameter, which describes the average Larmor frequency shift at a given detector temperature T_z. The spin-transition rate is obtained by discussing the dynamics of the Hamiltonian

$$H = H_0 + \mu_i b_{\text{rf}} \tag{5.38}$$

in the interaction picture [101]. Here, H_0 is the unperturbed Penning trap Hamiltonian and b_{rf} characterizes the amplitude of the spin-flip drive. Finally the spin transition probability is obtained:

$$P_{\text{SF}} = \frac{1}{2}\left(1 - \exp\left(-\frac{1}{2}\Omega_R^2 t_0\, \chi(2\pi \nu_{\text{rf}}, B_2, T_z)\right)\right), \tag{5.39}$$

where t_0 is the irradiation time of the spin-flip drive and $\Omega_R^2 = (\mu b_{\text{rf}})/\hbar$ the Rabi frequency. Figure 5.16 shows theoretical results of Eq. (5.39) for different drive parameters. All curves show a sharp 'cut-off' arising from the convolution of the Rabi resonance with the Boltzmann-distributed axial energy. High drive amplitudes b_{rf} and irradiation times t_0 saturate the spin resonance, see parts a and c of Fig. 5.16. Part b of that figure shows P_{rf} for different temperatures of the axial detection system.

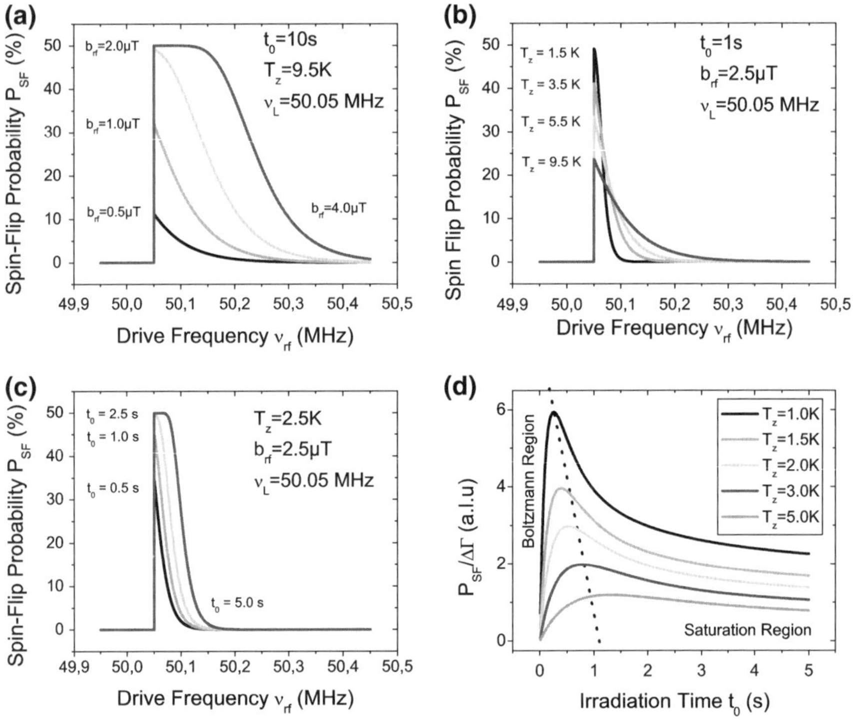

Fig. 5.16 Spin flip probability for different variable parameters. **a** spin-flip probability as a function of the drive frequency for variable drive amplitudes, **b** variable temperature, **c** variable drive time. **d** Ratio of spin flip probability and linewidth $\Delta\Gamma$

For higher temperatures T_z the line broadens since the particles traces an effectively higher magnetic field difference.

5.13 Statistical Detection of Spinflips

Above, an idealized measuring sequence for the determination of the Larmor frequency has been described. However, currently, in both experimental efforts aiming at a proton/antiproton g-factor measurement [76–79], the strong magnetic bottle avoids the application of this straight-forward measuring scheme. The main difficulty to resolve single proton/antiproton spin flips is the presence of the strong magnetic bottle, which also couples the radial modes' angular magnetic momenta to the axial oscillation frequency ν_z. In presence of the magnetic bottle, the axial frequency shift as a function of a shift in the modified cyclotron and the magnetron energies, $\mathrm{d}E_+$ and $\mathrm{d}E_-$ respectively, is given by

$$\Delta\nu_z = \frac{1}{4\pi^2 m\nu_z}\frac{B_2}{B_0}(dE_+ + dE_-)\ . \tag{5.40}$$

Under the conditions used at the proton g-factor experiment at Mainz, a shift in the radial energies dE_+ and dE_- of only 1 μeV causes an axial frequency shift of already ≈1 Hz. A single quantum jump between cyclotron states n_+ and $n_+ \pm 1$ changes the axial frequency already by 70 mHz.

In the following discussion only the modified cyclotron oscillator is considered. The transition rate between different cyclotron quantum states n_+ can be calculated by using Fermi's golden rule

$$\frac{dn_+}{dt} = \frac{2\pi}{\hbar}\Delta_+\rho_f(E_+)\Gamma_{i\to f}^2\ , \tag{5.41}$$

where Δ_+ is the linewidth parameter of the modified cyclotron resonance curve, $\rho_f(E_+)$ the density of cyclotron oscillator states, and

$$\Gamma_{i\to f} = eE_0\sqrt{\frac{\hbar}{2\pi\nu_z m}}\sqrt{\frac{n_+}{2}}\ , \tag{5.42}$$

the transition matrix element between different n_+ states. In this expression, E_0 is the electrical field noise density (which has the unit $\mathrm{Vm^{-1}Hz^{-1/2}}$). Since the transition rate scales with the quantum number n_+, it is crucial to cool the cyclotron mode to low quantum numbers. For example, at an effective modified cyclotron energy of only 86 μeV, which corresponds to a principal quantum number of about $n_+ \approx 1000$, a spurious white noise drive of only 20 $\mathrm{nVm^{-1}Hz^{-1/2}}$ is sufficient to drive one cyclotron quantum jump per second.

For further discussion, the 'axial frequency fluctuation' Ξ is introduced, which is defined as the root-mean-square (rms) value of the difference of two subsequent axial frequency measurements

$$\Xi_{\text{ax}} = \left((N-1)^{-1}\sum(\alpha_i - \bar{\alpha})^2\right)^{1/2}, \tag{5.43}$$

where $\alpha = \nu_z(t) - \nu_z(t + \Delta t)$. The best axial-frequency background fluctuation Ξ_{ref} so far achieved at the Mainz experiments is about 55 mHz. Using the formalism described above together with an absolute temperature calibration carried out in the magnetic bottle [102], this corresponds to an effective drive amplitude of only $E_0 = 7\,\text{nVm}^{-1}\cdot\text{Hz}^{-1/2}$. Although these background-noise conditions are remarkably low, the fluctuation is still too large to observe single proton/antiproton spin flips. However, using a statistical detection method [76], single-proton spin flips were observed for the first time.

For the statistical detection of single-proton spin flips the following measuring sequence is applied

1. measurement of the axial frequency,
2. irradiation of a resonant spin flip drive,
3. second measurement of the axial frequency,
4. irradiation of a reference drive below the Larmor frequency,
5. third measurement of the axial frequency.

The reference drive measurement is carried out to ensure that frequency generator background noise does not lead to additional heating. This scheme is repeated for several hundreds of times, and the background fluctuation Ξ_{ref} (axial frequency difference between measurements 3 and 5) is compared to the fluctuation which is obtained with spin-flip drive on Ξ_{SF} (axial frequency difference between measurements 1 and 3). The spin-flip-induced axial frequency shifts $\Delta\nu_{\text{z,SF}}$ add to the background fluctuation Ξ_{ref} in a statistical way

$$\Xi_{\text{SF}} = \sqrt{\Xi_{\text{ref}}^2 + P_{\text{SF}}\Delta\nu_{\text{z, SF}}^2}, \tag{5.44}$$

and from a comparison of Ξ_{SF} and Ξ_{ref} the spin flip probability P_{SF} is obtained. Results of such a toothed measurement are shown in Fig. 5.17a. The upper curve shows the evolution of Ξ_{SF} as a function of measurement cycles, while Ξ_{ref} is the measured background fluctuation. The clear separation of the two lines demonstrates the unambiguous detection of single-proton spin flips. An alternative analysis of the same data is shown in Fig. 5.17b. The entire data sequence was split into subsets and the values of Ξ_{SF} and Ξ_{ref} were binned to two histograms. Once this statistical detection method is repeated for different drive frequencies ν_{textrf} and $\Xi_{\text{SF}}^2 - \Xi_{\text{ref}}^2$ is evaluated, the resonance curve shown in Fig. 5.18 is obtained. From a best fit to these data, the Larmor frequency was determined with a precision of 5×10^{-5}.

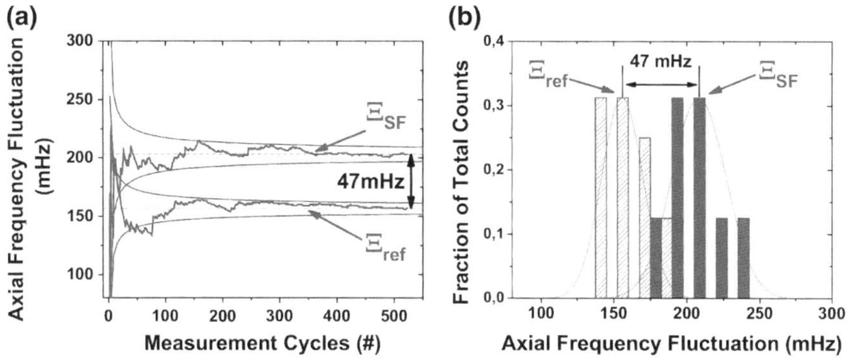

Fig. 5.17 In (**a**) the evolution of the frequency fluctuations Ξ_{SF} and Ξ_{ref} as a function of measurement cycles is shown. The clear separation of the two lines shows that spin flips are detected. An alternative analysis is shown in (**b**). The entire data sequence was split into subsets and the values of Ξ_{SF} and Ξ_{ref} were binned to two histograms

Fig. 5.18 Proton Larmor resonance obtained by statistical spin-flip detection in the analysis trap with its inhomogeneous magnetic field

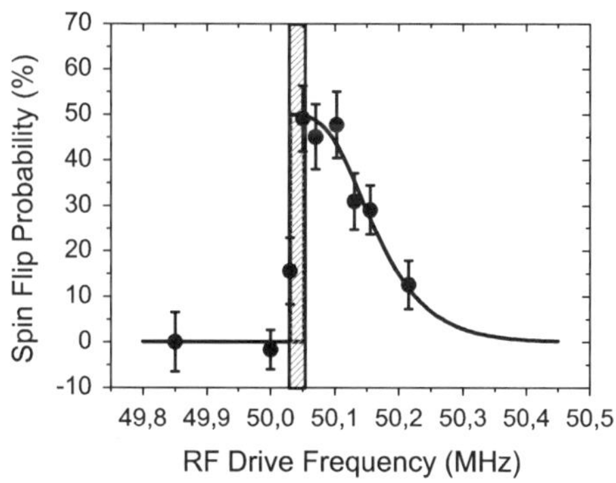

5.14 Feedback Cooling: Reduction of Linewidth

Although the first detection of single-proton spin flips described above was a major step to improve the magnetic moment of the antiproton, the Larmor-frequency resolution is only moderate. To further increase the experimental precision, active electronic feedback cooling was applied [103]. This reduces the temperature of the axial detection system T_z, and thus the width of the spin line. A schematic which illustrates the basic principle is shown in Fig. 5.19a. The single particle signal is picked up, amplified and split into two branches. One branch of the signal is phase-shifted and fed back to the axial detector. The detector temperature T_z can be adjusted by means of feedback phase and -gain. The ratio of T_z and the effective parallel resistance $R_{\mathrm{p,\,eff}}$ is a fluctuation-dissipation invariant [104]. Figure 5.19b shows noise spectra of a single proton tuned to resonance with the axial detection system with different feedback strengths applied. For positive feedback (feedback phase 0°) the signal-to-noise ratio as well as the linewidth of the single-particle dip become larger, which

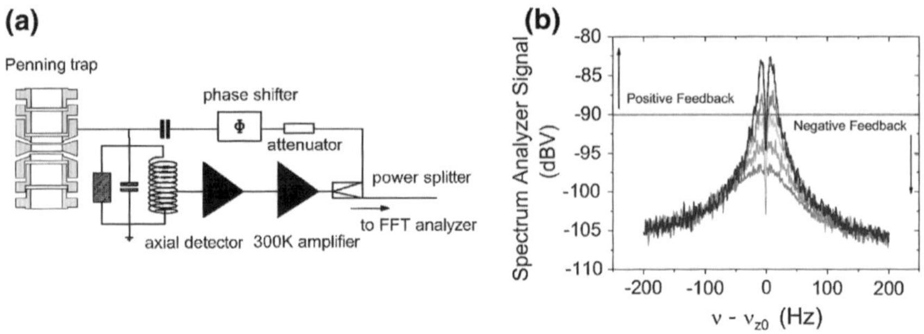

Fig. 5.19 **a** Principle of the feedback cooling circuit. **b** Axial frequency signal for different feedback temperatures

is due to an increase of $R_{\mathrm{p,\,eff}}$. For negative feedback (feedback phase 180°), the dip width and the signal-to-noise ratio decrease. The direct proportionality to the particle temperature can be proven by an independent temperature measurement using the magnetic bottle, as described in [102].

5.15 Determination of the *g*-Factor

For the *g*-factor determination, the Larmor frequency is measured as above while the axial detection system is feedback-cooled to about 2.5 K. The result is shown in Fig. 5.20a. A best fit of Eq. (5.39) to the data gives a Larmor frequency resolution of 1.8ppm. For a measurement of the cyclotron frequency, the same principle is applied as in case of the Larmor line, but with an electrical dipole drive scanning the modified cyclotron frequency. The drive amplitude is chosen weak enough that it does not affect significant heating of the modified cyclotron mode, but it increases the axial

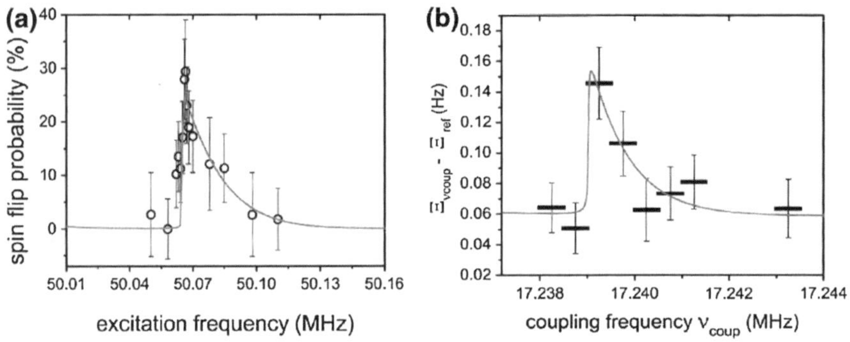

Fig. 5.20 **a** 1.8 ppm Larmor resonance **b** 8.9 ppm cyclotron resonance. Both data sets were measured with a single proton stored in a cryogenic Penning trap [77]

frequency fluctuation as a function of the drive frequency detectably. Figure 5.20b shows the corresponding data from which the free cyclotron frequency is obtained with 8.9 ppm precision. From these data the proton value

$$g/2 = 2.792\,849\,(25) \tag{5.45}$$

is extracted, limited by the precision of the cyclotron frequency determination [77]. However, at least another order of magnitude in the determination of the cyclotron frequency can be obtained easily just by more accurate sampling of the cyclotron resonance line. This strategy was applied by the Harvard group, which recently published the currently best-known single-particle-based free-proton magnetic moment [79]

$$g/2 = 2.792\,846\,(7), \tag{5.46}$$

while the Mainz group focussed on establishing a more advanced experimental method.

5.16 Double Penning Trap Technique

For further improvement of precision, the so-called 'double Penning trap method' will be applied. This technique was developed at the University of Mainz by a group which measured the magnetic moment of the electron bound to highly charged ions [80] for stringent tests of bound-state quantum electrodynamics. In experiments carried out with C^{5+} [105, 106], O^{7+} [107], and Si^{13+} [108] precisions at the sub-ppb level were achieved. A schematic of such a double Penning trap is shown in Fig. 5.21. The basic idea of this scheme is the spacial separation of the spin state analysis and the precision frequency measurements to two traps:

1. An *analysis trap* with the superimposed magnetic bottle.
2. A *precision trap* in which the magnetic field is about 10^5 times more homogeneous than in the analysis trap.

A g-factor measuring sequence in the double trap is as follows (Fig. 5.22): First, the spin state is analyzed in the analysis trap, and afterwards, the particle is transported to the precision trap, where the cyclotron frequency ν_c is measured and a magnetic radio-frequency drive is applied to flip the spin. Subsequently, the particle is transported back to the analysis trap and the spin state is analyzed. As in case of a g-factor measurement in the magnetic bottle, this sequence is repeated for several times at different drive frequencies ν_{rf}, and the spin-flip probability $P_{\mathrm{SF}}(\nu_{\mathrm{rf}})$ is obtained. The difference to a g-factor measurement in the magnetic bottle is, that the frequency determination is carried out in a magnetic field which is much more homogeneous than in the analysis trap. This reduces the line width of the g-factor resonance significantly, and increases the experimental precision. However, to decide whether a

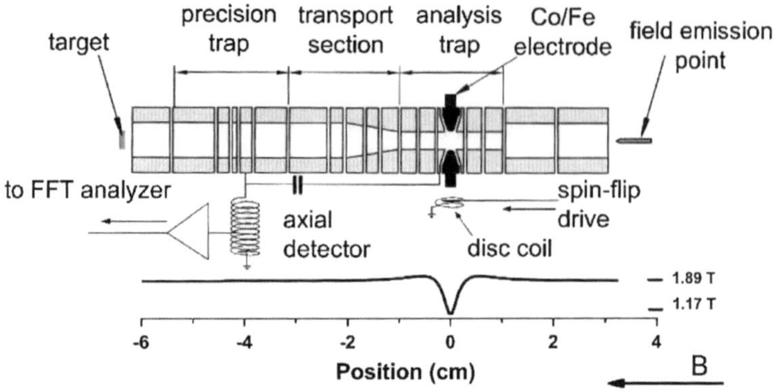

Fig. 5.21 Double Penning trap setup. It consists of an analysis trap with a the superimposed magnetic bottle and a precision trap in which the magnetic field is about 10^5 times more homogeneous than in the analysis trap. For further details see text

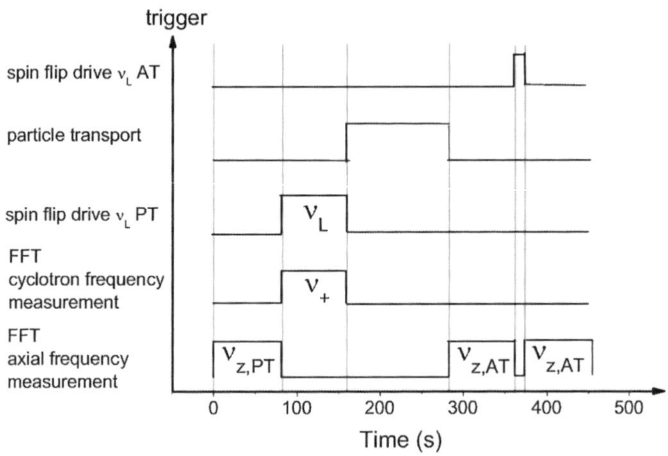

Fig. 5.22 Measuring sequence for the double Penning-trap based g-factor determination, where the spin state analysis and the precision measurements of ν_L and ν_c are separated to two traps. This sequence is repeated for several hundred times at different spin-flip drive frequencies. Since the spin is flipped in the by a factor of 100,000 more homogeneous precision trap, the spin resonance is much narrower

spin flip happened in the precision trap, single spin-flip resolution in the analysis trap is required. At the BASE experiment at Mainz encouraging progress towards the application of this method was reported recently. Single spin flips were observed with a fidelity of up to about 90 %[109], which allowed the first demonstration of the double-trap technique using a single proton[110]. With these encouraging results a measurement of μ_p and $\mu_{\bar{p}}$ is in close reach.

5.17 Towards a High Precision Measurement of the Antiproton Magnetic Moment

All methods described above for the proton can be directly transferred to measure the magnetic moment of the antiproton. Two collaborations at CERN AD, a part of ATRAP (Harvard) and BASE (RIKEN, Mainz) [111, 112], are aiming at this high-precision CPT test. Applying the methods described above, in 2012 diSciacca et al. (ATRAP) reported on a first Penning-trap-based measurement of the magnetic moment of a single antiproton. In these experiments, a relative precision of 4.4 ppm was achieved, which is a sensational 680-fold improvement of the value measured in [72]. In future steps, the double-trap scheme will be applied to eventually improve this value by another factor of 1,000.

However, to conduct this measurement with antimatter, several modifications of the apparatus are necessary. Trapping of antiprotons from AD has to be established. This requires a beamline to guide the 5.3 MeV antiprotons to the experiment, as well as a degrader structure, which typically consists of a sequence of several thin metallic foils, to bridge the energy range from MeV to about 10 keV which can be handled by Penning trap setups. Passing through the degrader, antiprotons lose energy due to electromagnetic interaction with the degrader material. The transmission and energy distribution of the antiprotons is sensitive to the degrader thickness [113, 114]. Thus, for sensitive tuning a rotatable degrader with azimutally increasing thickness will be used. Typically, a fraction of 10^{-4} moderated antiprotons passes the structure and is trapped by high voltages applied to specifically designed electrodes. The degrader technique has been pioneered by the TRAP collaboration [46, 115] and is meanwhile routinely applied in several antimatter experiments [29, 34, 116].

Figure 5.23 shows a section view of the Penning trap used by the BASE collaboration. It is an assembly of cylindrical stacked electrodes in carefully designed geometry. At the entrance of the electrode stack, a beryllium degrader is placed. Between two high-voltage electrodes which are used to catch the keV antiprotons, two traps are located. The first trap is called reservoir trap with the purpose of catching, cooling and storing a cloud of antiprotons. Several AD-pulses are accumulated and single antiprotons are suspended into the second trap, which is the precision trap, used for the frequency measurements in the double-trap measurement scheme. The capability of storing antiprotons ensures the experimental operation of BASE independent of accelerator cycles. Downstream to the second high voltage electrode the *analysis trap* and a *monitor trap* (MT) are located. The MT has an inner diameter of only about 3 mm and will be used for efficient cooling of the modified cyclotron motion during the g-factor measurement cycle. In later experiments in which the collaboration plans an improved comparison of the proton/antiproton charge-to-mass ratio, a single particle stored into this trap may serve as a probe to monitor and correct for magnetic field fluctuations.

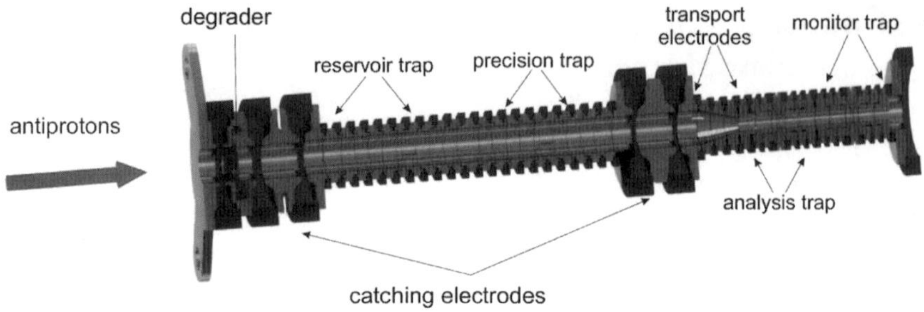

Fig. 5.23 Penning trap stack which is used by the BASE collaboration to measure the magnetic moment of the antiproton

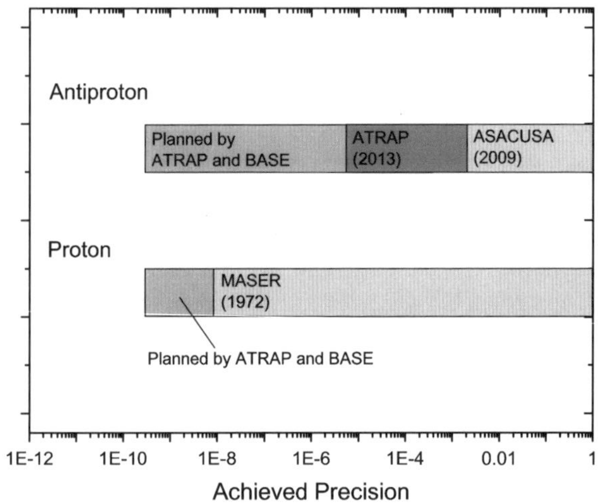

Fig. 5.24 Summary of the current proton and antiproton magnetic moment precision achieved

5.18 Summary

In this chapter, experiments to perform high-precision measurements of the magnetic moments of the proton and the antiproton were presented. Starting with an overview on precision tests of matter-antimatter asymmetry, the efforts of two independent collaborations, BASE and a part of ATRAP, which aim at such high-precision measurements, were placed into context. The applied experimental methods have been described in detail, starting form an introduction of the Penning trap and followed by an explanation of the key technique which employs the continuous Stern-Gerlach effect. The current status of the experiments has been presented and described in detail. While ATRAP has succeeded in measuring the antiprotonic magnetic moment at the level of 4.4 ppm in 2012, the BASE collaboration is preparing the apparatus for the application of the double-trap technique at the Antiproton Decelerator of CERN.

To close the chapter, the current status of precisions regarding the proton/antiproton magnetic moments is shown in Fig. 5.24.

References

1. S.L. Glashow, Partial-symmetries of weak interactions. Nucl. Phys. **22**, 579 (1961)
2. S. Weinberg, A model of leptons. Phys. Rev. Lett. **19**, 1264 (1967)
3. A. Salam, in Elementary Particle Physics: Relativistic Groups and Analyticity, Almquvist and Wiksell, Stockholm, 367 (1968).
4. P.W. Higgs, Broken symmetries and the masses of gauge bosons. Phys. Rev. Lett. **13**, 508 (1964)
5. ATLAS Collaboration, Observation of a new particle in the search for the Standard Model Higgs boson with the ATLAS detector at the LHC. Phys. Lett. B **716**, 1–29 (2012)
6. C.M.S. Collaboration, Observation of a new boson at a mass of 125 GeV with the CMS experiment at the LHC. Phys. Lett. B **716**, 30 (2012)
7. G. Luders, Proof of the TCP theorem. Ann. Phys. **2**, 1 (1957)
8. V.T. Cocconi et al., Mass analysis of the secondary particles produced by the 25-Gev proton beam of the cern proton synchrotron. Phys. Rev. Lett. **5**, 19 (1960)
9. D.N. Spergel et al., First-year wilkinson microwave anisotropy probe (WMAP) observations: determination of cosmological parameters. Astrophys. J. Suppl. Ser. **148**, 175 (2003) (WMAP Collaboration)
10. G. Steigman, Observational tests of antimatter cosmologies. Ann. Rev. Astr. Astrophys. **14**, 339 (1976)
11. F.R. Klinkhammer, N.S. Manton, A saddle-point solution in the Weinberg-Salam theory. Phys. Rev. D **30**, 2212 (1984)
12. V.A. Kuzmin, V.A. Rubakov, M.E. Shaposhnikov, On anomalous electroweak baryon-number non-conservation in the early universe. Phys. Lett. B **155**, 36 (1985)
13. M. Sher, Electroweak Higgs potential and vacuum stability. Phys. Rep. **179**, 273 (1989)
14. K. Zuber, *Neutrino Physics* (Taylor and Francis Group, Boca Raton, 2010)
15. V.A. Kostelecky, R. Potting, CPT and strings. Nucl. Phys. B **359**, 545 (1991)
16. J. Beringer et al., Review of Particle Physics. Phys. Rev. D **86**, 010001 (2012) (Particle Data Group)
17. D. Colladay, V.A. Kostelecky, CPT violation and the standard model. Phys. Rev. D **55**, 6760 (1997)
18. R. Bluhm, V.A. Kostelecky, N. Russell, CPT and Lorentz tests in Penning traps. Phys. Rev. D **57**, 3932 (1998)
19. V.A. Kostelecky, Sensitivity of CPT Tests with neutral mesons. Phys. Rev. Lett. **80**, 1818 (1998)
20. R. Bluhm, V.A. Kostelecky, N. Russell, CPT and Lorentz Tests in hydrogen and antihydrogen. Phys. Rev. Lett. **82**, 2254 (1999)
21. R.S. VanDyck, P.B. Schwinberg, H.G. Dehmelt, New high-precision comparison of electron and positron g factors. Phys. Rev. Lett. **59**, 26 (1987)
22. H. Dehmelt, P. Ekström, Proposed g-2/dν_z experiment on stored single electron or positron. Bull. Am. Phys. Soc. **18**, 727 (1973)
23. D. Hanneke, S. Fogwell, G. Gabrielse, New measurement of the electron magnetic moment and the fine structure constant. Phys. Rev. Lett. **100**, 120801 (2008)
24. R.J. Hughes, B.I. Deutch, Electric charges of positrons and antiprotons. Phys. Rev. Lett. **69**, 578 (1992)
25. M.S. Fee et al., Measurement of the positronium $1^3S_1 2^3S_1$ interval by continuous-wave two-photon excitation. Phys. Rev. A **48**, 192 (1993)

26. L. Essen, R. Donaldson, M. Bangham, E Hope, Frequency of the hydrogen maser. Nature **229**, 110 (1971)
27. C.J. Partley, Improved measurement of the hydrogen 1S2S transition frequency. Phys. Rev. Lett. **107**, 203001 (2011)
28. Y. Enomoto et al., Synthesis of cold antihydrogen in a cusp trap. Phys. Rev. Lett. **105**, 243401 (2010)
29. ASACUSA Collaboration, http://asacusa.web.cern.ch/ASACUSA/
30. S. Mauri, The antiproton decelerator. Hyperfine Interact **109**, 43 (1997)
31. P. Belochitskii, T. Eriksson, S. Maury, The CERN antiproton decelerator (AD) in 2002: status, progress and machine development results. Nucl. Instr. Meth. B **214**, 176 (2004)
32. E. Widmann, M. Diermaier, B. Juhasz, C. Malbrunot, O. Massiczek, C. Sauerzopf, K. Suzuki, B. Wünschek, J. Zmeskal, S. Federmann, N. Kuroda, S. Ulmer, and Y. Yamazaki, Measurement of the hyperfine structure of antihydrogen in a beam, arXiv:1301.4670 (2013)
33. A.M. Kasevich, E. Riis, R.G. DeVoe, S. Chu, Rf spectroscopy in an atomic fountain. Phys. Rev. Lett. **63**, 612 (1989)
34. ALPHA Collaboration, http://alpha-new.web.cern.ch/
35. ALPHA Collaboration, Trapped antihydrogen. Nature **468**, 673 (2010)
36. G. Gabrielse et al., Trapped antihydrogen in its ground state. Phys. Rev. Lett. **108**, 113002 (2012)
37. ALPHA Collaboration, Resonant quantum transitions in trapped antihydrogen atoms. Nature **483**, 439 (2012)
38. ATRAP Collaboration, http://gabrielse.physics.harvard.edu/
39. Th Udem, A. Huber, B. Gross, J. Reichert, M. Prevedelli, M. Weitz, T.W. Hänsch, Phase-coherent measurement of the hydrogen 1S–2S transition frequency with an optical frequency interval divider chain. Phys. Rev. Lett. **79**, 2646 (1997)
40. K.S.E. Eikema, J. Walz, T.W. Hänsch, Continuous wave coherent Lyman-α radiation. Phys. Rev. Lett. **83**, 3828 (1999)
41. P.H. Donnan, M. Fujiwara, F. Robicheaux, A proposal for laser cooling antihydrogen atoms. J. Phys. B: At. Mol. Opt. Phys. **46**, 025302 (2013)
42. D. Kolbe, M. Scheid, J. Walz, Triple resonant four-wave mixing boosts the yield of continuous coherent vacuum ultraviolet generation. Phys. Rev. Lett. **109**, 063901 (2012)
43. M. Hori et al., Two-photon laser spectroscopy of antiprotonic helium and the antiproton-to-electron mass ratio. Nature **475**, 484 (2011)
44. G. Gabrielse et al., First capture of antiprotons in a penning trap: a kiloelectronvolt source. Phys. Rev. Lett. **57**, 2504 (1986)
45. G. Gabrielse et al., Cooling and slowing of trapped antiprotons below 100 meV. Phys. Rev. Lett. **63**, 1360 (1989)
46. G. Gabrielse, A. Khabbaz, D.S. Hall, C. Heimann, H. Kalinowsky, W. Jhe, Precision mass spectroscopy of the antiproton and proton using simultaneously trapped particles. Phys. Rev. Lett. **82**, 3198 (1999)
47. T. Beier et al., New determination of the electrons mass. Phys. Rev. Lett. **88**, 011603 (2002)
48. M.S. Fee et al., Measurement of the positronium $1^3S_1 2^3S_1$ interval by continuous-wave two-photon excitation. Phys. Rev. A **48**, 192–220 (1993)
49. G. Gabrielse et al., New measurement of the electron magnetic moment and the fine structure constant. Phys. Rev. Lett. **100**, 120801 (2008)
50. R.K. Mittleman et al., Bound on CPT and Lorentz symmetry with a trapped electron. Phys. Rev. Lett. **83**, 2116–2119 (1999)
51. G.W. Bennett et al., Search for Lorentz and CPT violation effects in muon spin precession. Phys. Rev. Lett. **100**, 091602 (2008)
52. R.S. Hayano et al., Antiprotonic helium and CPT invariance. Rep. Prog. Phys. **70**, 1995–2065 (2007)
53. P.F. Winkler, D. Kleppner, T. Myint, F.G. Walther, Magnetic moment of the proton in Bohr magnetons. Phys. Rev. A **5**, 83 (1972)
54. J. DiSciacca et al., Phys. Rev. Lett. **110**, 130801 (2013)

55. B. Schwingenheuer et al., CPT tests in the neutral kaon system. Phys. Rev. Lett. **74**, 4376–4379 (1995)
56. V.A. Kostelecky, N. Russell, Data tables for Lorentz and CPT violation. Rev. Mod. Phys. **83**, 11–31 (2011)
57. R. Bluhm, V.A. Kostelecky, N. Russell, Testing CPT with anomalous magnetic moments. Phys. Rev. Lett. **79**, 1432–1435 (1997)
58. R. Frisch, O. Stern, Über die magnetische Ablenkung von Wasserstoff-Molekülen und das magnetische moment des protons. Z. Phys. A **85**, 4 (1933)
59. P.A.M. Dirac, The quantum theory of the electron. Proc. Roy. Soc. Lond. A. Mat. **117**, 610 (1928)
60. I.I. Rabi, J.M.B. Kellogg, J.R. Zacharias, The magnetic moment of the proton. Phys. Rev. **46**, 157 (1934)
61. F. Bloch, W.W. Hansen, M. Packard, The nuclear induction experiment. Phys. Rev. **70**, 474 (1946)
62. F. Bloch, C. Jeffries, A direct determination of the magnetic moment of the proton in nuclear magnetons. Phys. Rev. **80**, 305 (1950)
63. H. Sommer, H.A. Thomas, J.A. Hipple, The measurement of e/M by cyclotron resonance. Phys. Rev. **82**, 697–702 (1951)
64. D. Collington, A. Dellis, J. Sanders, K. Turberfield, Magnetic moment of the proton. Phys. Rev. **99**, 1622 (1955)
65. P. F Winkler, D. Kleppner, T. Myint, F. G. Walther, Magnetic moment of the proton in Bohr magnetons. Phys. Rev. A **5**, 83 (1972)
66. T. Beier, H. Häffner, N. Hermanspahn, S.G. Karshenboim, H.-J. Kluge, W. Quint, S. Stahl, J. Verdù, G. Werth, New determination of the electrons mass. Phys. Rev. Lett. **88**, 011603 (2001)
67. P. Mohr, B. Taylor, D. Newell, CODATA recommended values of the fundamental physical constants: 2006. Rev. Mod. Phys. **80**, 633 (2008)
68. S.G. Karshenboim, V.G. Ivanov, The g factor in a light two-body atomic system: a determination of fundamental constants to test QED. Can. J. Phys. **80**, 1305 (2002)
69. S.G. Karshenboim, V.G. Ivanov, The g factor of the proton. Phys. Lett. B **566**, 27 (2003)
70. B.L. Roberts, Measurement of the antiproton magnetic moment and mass. Phys. Rev. D **17**, 358 (1978)
71. A. Kreissl, A.D. Hancock, H. Koch, T. Kohler, H. Poth, U. Raich, D. Rohmann, A. Wolf, L. Tauscher, A. Nilsson, Remeasurement of the magnetic moment of the antiproton. Z. Phys. C **37**, 557 (1988)
72. T. Pask, D. Barna, A. Dax, R.S. Hayano, M. Hori, D. Horvath, S. Friedreich, B. Juhasz, O. Massiczek, N. Ono, Antiproton magnetic moment determined from the HFS of antiprotonic helium. Phys. Lett. **B678**, 55 (2009)
73. E. Widmann, J. Eades, T. Ishikawa, J. Sakaguchi, T. Tasaki, H. Yamaguchi, R.S. Hayano, M. Hori, H.A. Torii, B. Juhasz, D. Horvath, T. Yamazaki, Hyperfine structure of antiprotonic helium revealed by a Laser-Microwave-Laser resonance method. Phys. Rev. Lett. **89**, 243402 (2002)
74. E. Widmann, private communication (2013)
75. R. Blatt, Quantum physics: Spin flips with a single proton. Nature **475**, 298 (2011)
76. S. Ulmer, C.C. Rodegheri, K. Blaum, H. Kracke, A. Mooser, W. Quint, J. Walz, Observation of spin flips with a single trapped proton. Phys. Rev. Lett. **106**, 253001 (2011)
77. C.C. Rodegheri, K. Blaum, H. Kracke, S. Kreim, A. Mooser, W. Quint, S. Ulmer, J. Walz, An experiment for the direct determination of the g-factor of a single proton in a Penning trap. New J. Phys. **14**, 063011 (2012)
78. N. Guise, J. DiSciacca, G. Gabrielse, Self-excitation and feedback cooling of an isolated proton. Phys. Rev. Lett. **104**, 143001 (2010)
79. J. DiSciacca, G. Gabrielse, Direct measurement of the proton magnetic moment. Phys. Rev. Lett. **108**, 153001 (2012)

80. H. Häffner, T. Beier, S. Djekic, N. Hermanspahn, H.-J. Kluge, W. Quint, S. Stahl, J. Verdù, T. Valenzuela, G. Werth, Double Penning trap technique for precise g factor determinations in highly charged ions. Eur. Phys. J. D **22**, 163 (2003)
81. L.S. Brown, G. Gabrielse, Geonium theory: physics of a single electron or ion in a Penning trap. Rev. Mod. Phys. **58**, 233 (1986)
82. M. Kretzschmar, Single particle motion in a Penning trap: description in the classical canonical formalism. Phys. Scr. **46**, 544 (1992)
83. G. Gabrielse, J. Tan, Self-shielding superconducting solenoid systems. J. Appl. Phys. **63**, 5143 (2011)
84. G. Gabrielse, L. Haarsma, S.L. Rolston, Open-endcap Penning traps for high precision experiments. Int. J. Mass Spec. **88**, 319 (1989)
85. J. Verdu et al., Calculation of electrostatic fields using quasi-Green's functions: application to the hybrid Penning trap. New J. Phys. **10**, 103009 (2008)
86. G. Gabrielse, D.F. Phillips, W. Quint, H. Kalinowsky, G. Rouleau, W. Jhe, Special relativity and the single antiproton: fortyfold improved comparison of antiproton and proton charge-to-mass ratios. Phys. Rev. Lett. **74**, 3544 (1995)
87. D.J. Wineland, H.G. Dehmelt, Principles of the stored ion calorimeter. J. Appl. Phys. **46**, 919 (1975)
88. X. Feng, M. Charlton, M. Holzscheiter, R. Lewis, Y. Yamazaki, Tank circuit model applied to particles in a Penning trap. J. Appl. Phys. **79**, 8 (2009)
89. S. Ulmer, H. Kracke, K. Blaum, S. Kreim, A. Mooser, W. Quint, C.C. Rodegheri, J. Walz, The quality factor of a superconducting rf resonator in a magnetic field. Rev. Sci. Inst. **80**, 123302 (2009)
90. A. Mooser, K. Blaum, H. Kracke, W. Quint, C.C. Rodegheri, S. Ulmer, J. Walz, Towards a direct measurement of the g-factor of a single isolated proton. Can. J. Phys. **89**, 165 (2011)
91. S. Ulmer, H. Kracke, K. Blaum, A. Mooser, W. Quint, C. C. Rodegheri, J. Walz, A cryogenic detection system at 28.9 MHZ for the non-destructive observation of a single proton at low particle energy. Nucl. Inst. Meth. A **705**, 55 (2013).
92. S.R. Jefferts, T. Heavner, P. Hayes, G.H. Dunn, Superconducting resonator and a cryogenic GaAs field-effect transistor amplifier as a single-ion detection system. Rev. Sci. Instrum. **64**, 737 (1993)
93. J. Ketelaer et al., Recent developments in ion detection techniques for Penning trap mass spectrometry at TRIGA-TRAP. Eur. Phys. J. A **42**, 311 (2009)
94. E.A. Cornell, R.M. Weisskoff, K.R. Boyce, D.E. Pritchard, Mode coupling in a Penning trap: π pulses and a classical avoided crossing. Phys. Rev. A **41**, 312 (1990)
95. J. Johnson, Thermal agitation of electricity in conductors. Phys. Rev. **32**, 97 (1928)
96. H. Nyquist, Thermal agitation of electric charge in conductors. Phys. Rev. **32**, 110 (1928)
97. R.J. Glauber, Coherent and incoherent states of the radiation field. Phys. Rev. **131**, 2766 (1963)
98. S. Ulmer, K. Blaum, H. Kracke, A. Mooser, C.C. Rodegheri, W. Quint, J. Walz, Direct measurement of the free cyclotron frequency of a single particle in a Penning trap. Phys. Rev. Lett. **107**, 103002 (2011)
99. I. Rabi, J. Zacharias, S. Millman, P. Kusch, A new method of measuring nuclear magnetic moment. Phys. Rev. **53**, 318 (1938)
100. L.S. Brown, Geonium lineshape. Ann. Phys. **159**, 62 (1985)
101. J. J. Sakurai, Modern Quantum Mechanics, 1st edn. (Addison Wesley, Redwood City, 1993)
102. S. Djekic, J. Alonso, H.-J. Kluge, W. Quint, S. Stahl, T. Valenzuela, J. Verdu, M. Vogel, G. Werth, Temperature measurement of a single Ion in a Penning trap. Eur. Phys. J. D **31**, 451 (2004)
103. H. Dehmelt, W. Nagourney, J. Sandberg, Self-excited mono-ion oscillator. Proc. Nat. Acad. Sci. **83**, 5761 (1986)
104. B. d'Urso, B. Odom, G. Gabrielse, Feedback cooling of a one-electron oscillator. Phys. Rev. Lett **90**, 43001 (2003)
105. N. Hermanspahn, H. Häffner, H. Kluge, W. Quint, S. Stahl, J. Verdu, G. Werth, Observation of the continuous Stern-Gerlach effect on an electron bound in an atomic ion. Phys. Rev. Lett. **84**, 427 (2000)

106. H. Häffner, T. Beier, N. Hermanspahn, H. Kluge, W. Quint, S. Stahl, J. Verdu, G. Werth, High-accuracy measurement of the magnetic moment anomaly of the electron bound in hydrogenlike carbon. Phys. Rev. Lett. **85**, 5308 (2000)
107. J. Verdù, S. Djekic, S. Stahl, T. Valenzuela, M. Vogel, G. Werth, T. Beier, H. Kluge, W. Quint, Electronic g-factor of hydrogenlike oxygen $^{16}O^{7+}$. Phys. Rev. Lett. **92**, 093002 (2004)
108. S. Sturm, K. Blaum, B. Schabinger, A. Wagner, W. Quint, G. Werth, On g-factor experiments with individual ions. J. Phys. B **43**, 074016 (2010)
109. A. Mooser, H. Kracke, K. Blaum, C.C. Rodegheri, W. Quint, S. Ulmer, J. Walz, Resolution of single spin flips of a single proton. Phys. Rev. Lett. **110**, 140405 (2013)
110. A. Mooser et al., Phys. Lett. B **723**, 78–81 (2013)
111. S. Ulmer et al., ICPEAC Proceedings, submitted (2013)
112. C. Smorra et al., LEAP Proceedings, submitted (2013)
113. G. Gabrielse, X. Fei, L.A. Orozco, R.L. Tjoelker, Cooling and slowing of trapped antiprotons below 100 meV. Phys. Rev. Lett. **63**, 1360 (1989)
114. L.H. Andersen, P. Hvelplund, H. Knudsen, et al., Further studies of double ionization of He, Ne, and Ar by fast and slow antiprotons. Phys. Rev. A **40**, 7366 (1989)
115. G. Gabrielse, X. Fei, L.A. Orozco, et al., Barkas effect with use of antiprotons and protons. Phys. Rev. A **40**, 481 (1989)
116. A. Kellerbauer et al., Proposed antimatter gravity measurement with an antihydrogen beam. Nucl. Instrum. Meth. in Phys. Res. B **266**, 351 (2008)

Chapter 6
Fundamental Physics with Antihydrogen

J. S. Hangst

Abstract Antihydrogen—the antimatter equivalent of the hydrogen atom—is of fundamental interest as a test bed for universal symmetries—such as CPT and the Weak Equivalence Principle for gravitation. Invariance under CPT requires that hydrogen and antihydrogen have the same spectrum. Antimatter is of course intriguing because of the observed baryon asymmetry in the universe—currently unexplained by the Standard Model. At the CERN Antiproton Decelerator (AD) [1], several groups have been working diligently since 1999 to produce, trap, and study the structure and behaviour of the antihydrogen atom. One of the main thrusts of the AD experimental program is to apply precision techniques from atomic physics to the study of antimatter. Such experiments complement the high-energy searches for physics beyond the Standard Model. Antihydrogen is the only atom of antimatter to be produced in the laboratory. This is not so unfortunate, as its matter equivalent, hydrogen, is one of the most well-understood and accurately measured systems in all of physics. It is thus very compelling to undertake experimental examinations of the structure of antihydrogen. As experimental spectroscopy of antihydrogen has yet to begin in earnest, I will give here a brief introduction to some of the ion and atom trap developments necessary for synthesizing and trapping antihydrogen, so that it can be studied.

6.1 Some History

Antihydrogen was initially produced and observed in in-beam experiments at CERN and Fermilab [2, 3], but these experiments had little potential for future measurements of the anithydrogen spectrum, and were quickly abandoned. During the operation

J. S. Hangst (✉)
Department of Physics and Astronomy, Aarhus University, DK-8000 Aarhus C Denmark
and CERN Division PH, CH-1211 Geneve 23, Switzerland
e-mail: jeffrey.hangst@cern.ch

W. Quint and M. Vogel (eds.), *Fundamental Physics in Particle Traps*,
Springer Tracts in Modern Physics 256, DOI: 10.1007/978-3-642-45201-7_6,

of the LEAR facility at CERN, the TRAP collaboration developed the necessary techniques for slowing, cooling, and trapping antiprotons [4, 5]. In parallel, Surko and colleagues developed the technology for accumulating and storing positrons [6] emitted from a radioactive source. The Surko technique was adapted for antihydrogen production by the ATHENA collaboration [7] at the AD. ATHENA successfully demonstrated synthesis of antihydrogen atoms from trapped plasmas of antiprotons and positrons in 2002 [8]. Following the 1-year shutdown of the AD in 2005, the ATRAP and ALPHA (successor to ATHENA) collaborations embarked on efforts to magnetically trap neutral antihydrogen atoms. The ALPHA collaboration succeeded in trapping atoms of antihydrogen in 2010 [9]. Progress with trapped antihydrogen has been brisk in recent years. ALPHA has shown that it is possible to hold trapped anti-atoms for up to 1,000 s [10], and to drive resonant quantum microwave transitions (positron spin flip) in the trapped atoms [11]. More recently, ALPHA has performed the first systematic study of antihydrogen atoms in gravitational free fall [12]. ATRAP has reported evidence for trapped antihydrogen atoms [13], although the experiment appears to suffer from a lack of reproducible conditions. Using a very different approach, the ASACUSA collaboration hopes to study the hyperfine spectrum of antihydrogen atoms in flight [14]. They have recently demonstrated progress on generating a beam of antihydrogen atoms produced in their novel cusp trap [15]. Two new experiments at the AD hope to study the effect of the Earths gravitational field on antihydrogen atoms. The AEgIS experiment [16] began operation in 2012, and the Gbar experiment [17] should begin in a few years. It is fair to say that there has never been more activity in low-energy antihydrogen physics than at the present time. In the following I will concentrate on the experimental techniques that have been developed to produce trappable antihydrogen atoms. The point of reference will of course be the authors ALPHA and ATHENA experiments, but differences between these approaches and those of other groups will be noted along the way. This chapter is intended as an overview, with the technical details to be found in the referenced literature.

6.2 Producing Antihydrogen: ATHENA

The basic idea for producing antihydrogen atoms is deceptively simple. Clouds of positrons and antiprotons, stored in Penning traps, are mixed, i.e., allowed to spatially overlap and interact (we will not review all of the techniques for catching and accumulating antiprotons and positrons here; see the above-referenced literature). The basic workhorse potential used for mixing is the so-called nested potential [18], shown in Fig. 6.1. The positrons are trapped in the center well of the nested potential. In the ATHENA solenoidal field of 3 T, the positrons would cool by cyclotron radiation and attempt to come into equilibrium with the surrounding trap structure—which was at about 15 K; however, no absolute measure of temperature was available in ATHENA. The antiprotons were injected from a side well (Fig. 6.1) and would interact in the positron cloud to form antihydrogen. The dominant formation mechanism

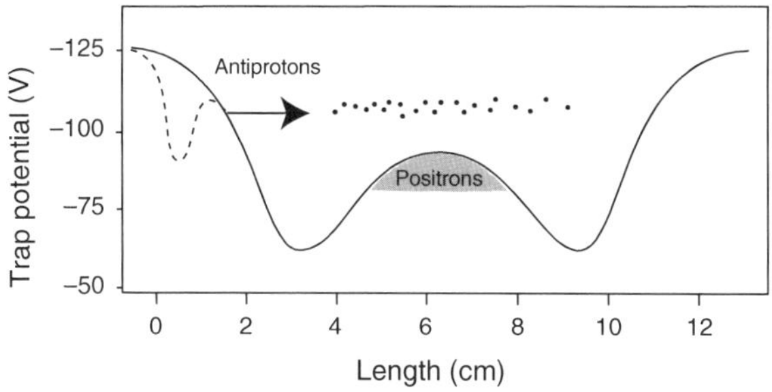

Fig. 6.1 The potential well configuration used in the first demonstration of low energy antihydrogen production by ATHENA. Antiprotons were injected from the well at the left of the figure (*dashed line*) into the nested potential well (reproduced from [8])

is a three-body event involving two positrons and an antiproton—the extra positron carries away the binding energy of the atom. Three-body formation typically results in loosely-bound Rydberg atoms. In ATHENA, a typical mixing cycle involved a few tens of thousands of antiprotons and up to 10^8 positrons. Several thousand antihydrogen atoms could be produced per cycle, with a peak rate of a few hundred per second [19].

6.3 Detecting Antihydrogen: ATHENA

The charged particles used to make antihydrogen are confined by the fields of the Penning trap. When a neutral antihydrogen atom forms, it is insensitive to the trapping fields and escapes, to annihilate on the inner surface of the trap electrodes. ATHENA pioneered the method of annihilation detection to identify the lost atoms when they hit the wall [7]. The ATHENA detector is shown schematically in Fig. 6.2. The two-layer silicon detector could identify the tracks of charged pions from the antiproton annihilation, and the CsI crystals detected the back-to-back, 511 keV gamma rays from positron annihilation. Spatial and temporal coincidence of these two signatures on the wall of the Penning trap provided the first confirmation for production of low energy antihydrogen [7]. Spatial characterization of the antihydrogen annihilation distribution proved to be a powerful tool for studying and optimising antihydrogen production in ATHENA [20]. For example, it is possible to distinguish antihydrogen annihilation from the annihilation of bare antiprotons lost from the trap (or resulting from field ionisation of anti-atoms) without relying on the positron detection. A position sensitive annihilation detector is also a key feature of the ALPHA antihydrogen trapping apparatus; see below. Note that the ATRAP collaboration has relied heavily on field-ionisation detection of Rydberg antihydrogen for their production experiments [21]. In this technique, a drifting Rydberg anti-atom is stripped by a spatially

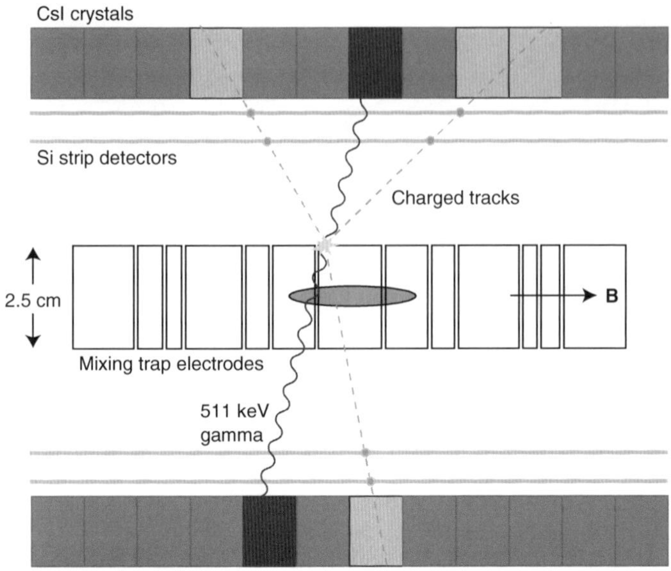

Fig. 6.2 Schematic diagram of the ATHENA antihydrogen production trap and annihilation detector. The *dashed green lines* represent charged pion tracks from an antiproton annihilation; the *wavy blue lines* are 511 keV gammas from a positron annihilation. The location of the positron plasma is indicated by the *blue shape* at the center of the Penning trap, which employed a 3 T axial field (reproduced from [8])

localized electric field, and the freed antiproton is re-trapped, to be released, detected and counted at a later time. This technique cannot be applied to antihydrogen in the ground state—which is the ultimate system of interest.

6.4 Antihydrogen and Ion Trap Physics

Production of antihydrogen requires careful control and manipulation of one component plasmas of electrons, antiprotons and positrons. The Penning traps employed comprise solenoidal magnetic fields of typically a few T, and stacks of hollow, cylindrical electrodes that can be used to create and dynamically manipulate longitudinal electric fields. (see Figs. 6.1 and 6.2.) The traps are typically placed in contact with a liquid helium cryostat for thermal and vacuum considerations. Electrons are used to cool the trapped antiprotons—which have initial energies of up to several keV—down to cryogenic temperatures [5]. Electrons pre-loaded into the catching well cool through cyclotron radiation in the solenoidal magnetic field. A bunch of antiprotons from the AD (5.3 MeV kinetic) can be dynamically trapped using a pulsed, high voltage electrode. The antiprotons are either slowed by passing through a thin foil (ATRAP, ATHENA, ALPHA) or are first decelerated to 100 keV by a radiofrequency quadrupole before the final degrading stage (ASACUSA). The AD cycle is typically about 100 s in duration and results in about 3×10^7 antiprotons delivered to the experiment in each cycle.

Many complex manipulations of the plasmas are necessary to produce antihydrogen reproducibly. The electrons must first be removed from the combined electron-antiproton plasma before antihydrogen formation can proceed. In ALPHA, this plasma would typically comprise several tens of millions of electrons and about 50,000 antiprotons. The electrons are removed by pulsing open the confining potential for intervals short enough that the faster electrons can escape while the antiprotons remain trapped. This process inevitably heats the antiprotons—so even if the initial mixed plasma was close to the cryogenic wall temperature, the antiprotons will typically end up at a few hundred K. We will consider further cooling of antiprotons later.

Luckily, just producing antihydrogen doesnt require terribly cold antiprotons. The antiprotons are typically injected into a much colder positron plasma and can be cooled by Coulomb collisions inside the positron cloud. Indeed, in the initial ATHENA experiments, antiprotons were injected into the positron plasma with tens of eV of longitudinal energy (note that 1 eV is equivalent to about 12,000 K). Temperature plays a much more important role when we discuss trapping of antihydrogen later.

Plasma radii and densities must also be controlled. The rotating wall technique [22] plays a crucial role in production of antihydrogen cold enough to trap, and is also typically used for tailoring electron and positron densities in any production experiment. In ALPHA, the rotating wall compression technique is used extensively: on electrons, on positrons, and on combined electron/antiproton plasmas [23]. It is obviously important to have reliable diagnostics for measuring the transverse sizes and density distributions of the trapped plasmas. In ALPHA we have relied heavily on microchannelplate/phosphour screen detectors [24]. The trapped cloud is extracted longitudinally from the trap and dumped onto the imaging detector. A rather extreme example of an unstable antiproton cloud extracted from the ALPHA-2 device is shown in Fig. 6.3. Note that linear tracks from the annihilation products can be seen traversing the face of the detector. ASACUSA has used similar detectors to study rotating wall compression of plasmas in their traps [25]. Monitoring of plasma vibrational modes has also proved to be a useful technique for diagnosing the behaviour of lepton plasmas for antihydrogen production and related experiments. In particular, monitoring changes in temperature through changes in the quadrupole frequency [26, 27] was very useful in ATHENA [28, 29]. In ALPHA we have recently used similar electron plasma mode diagnostics to study the effect of injected microwaves, resonant at the cyclotron frequency, on stored electrons, and used this information to help characterize the microwave field profile [30] for the antihydrogen spin flip experiment [11].

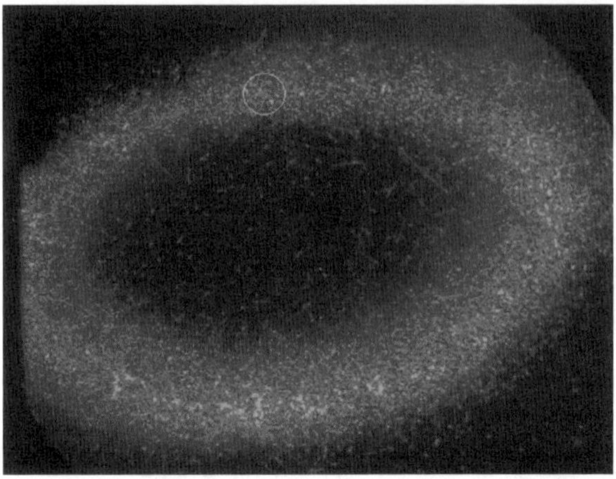

Fig. 6.3 A false color image of an antiproton plasma extracted from the ALPHA-2 device. The antiprotons strike a multichannel plate/phosphor screen detector; the phosphor is imaged by a CCD camera. This image is of an unstable antiproton cloud that was initially captured off-axis in the Penning trap. The hollow profile results from diochotron oscillations of the plasma. Tracks from annihilation products can be seen in the plane of the detector

6.5 Trapping Antihdyrogen for Spectroscopy: ALPHA

The ALPHA device was purpose-built to trap antihydrogen atoms so that their properties can be studied. The basic idea is to produce antihydrogen atoms at the field minimum of a magnetic gradient trap. The $\mu \cdot B$ interaction of the atoms magnetic dipole moment with the external field creates a potential well; if the atom is born with a small enough energy, it cannot escape the well and is trapped. Unfortunately, the interaction is very weak compared to what can be obtained with charged particles. For ground state antihydrogen, one obtains about 0.7 K of confinement depth for every 1 T of field change. The ALPHA trap has a well depth of about 0.5 K [31]. Thus atoms must be created with energies corresponding to less than 0.5 K in order to be trapped. When one compares this to the typical energy scales of the charged particle plasmas used to produce the neutral anti-atoms, it is clear that the task is rather daunting. The antiprotons—whose momentum determines the antihydrogen atoms momentum at production—must have meV energies when they produce anti-atoms. Recall that they start by being trapped at keV energies, and that even very careful removal of electrons leaves them at order of 100 K—to be compared to the 0.5 K trapping depth. Again, one can rely on interactions with a cyclotron-radiation cooled positron plasma to cool the antiprotons. However, in practice, the positrons dont often reach equilibrium with the cryogenic walls of the trap. The reasons for this arent yet understood in any quantitative way, but black body radiation from warm areas of the apparatus plays a role, as does electrical noise on the Penning trap electrodes. In ALPHA we generally find that plasmas with fewer numbers of positrons equilibrate at lower temperatures; the plasmas used in the first demonstration of trapping were typically 70–80 K, at equilibrium, and then evaporatively cooled (see discussion below) to about 40 K before mixing. These temperatures can be compared to the trap electrodes, which were at about 10 K.

The bottom line is that the charged particle temperatures achieved in the experiment to date are still quite high compared to the neutral trapping well depth. Thus one can only hope to catch a fraction of the antihydrogen atoms produced.

6.5.1 ALPHA Configuration

A schematic of the ALPHA central trapping region is shown in Fig. 6.4. The Penning trap electrodes are immediately inside the inner wall of the crystostat for the superconducting magnets that make up the neutral atom trap. An external solenoid (not pictured) provides a 1 T uniform field for the Penning trap. The atom trap magnets comprise an octupole and two solenoidal mirror coils. The resulting field has a minimum at the center of the Penning trap, where the antihydrogen is produced. The ALPHA device employs a transverse octupole—not the quadrupole generally used in Ioffe-Pritchard traps for atoms of matter. The motivation here is to try to minimize perturbations from transverse magnetic fields on the charged plasmas that are needed to form antihydrogen. Recall that most Penning traps use a very uniform axial field in order to maintain rotational symmetry. The transverse fields of the atom

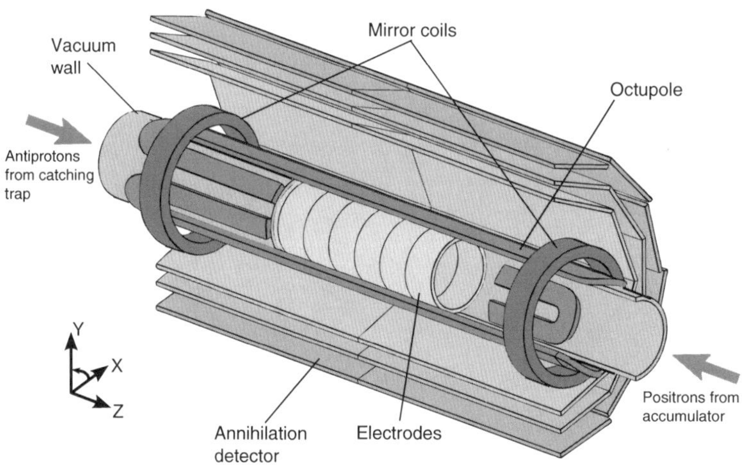

Fig. 6.4 Schematic representation of the ALPHA antihydrogen production and trapping region. The Penning trap electrodes (*yellow*) have an inner diameter of 44.5 mm. The inner cryostat wall and vacuum chamber wall is shown in *grey*. The atom trap coils are shown in *green* (axial confinement) and *red* (transverse confinement). The effective length of the atom trap is 274 mm. The modular annihilation detector is shown in a *cutaway view*; it covers the full azimuthal region. An external solenoid (not pictured) provides a 1 T axial field for the Penning trap (Reproduced from [9])

trap strongly break this symmetry. In the design stages of ALPHA, we were very concerned that these transverse fields could have a serious negative effect on stability of the positron and antiproton clouds, and experiments at Berkeley demonstrated the that strong quadrupole fields had serious effects on the stability of stored electron plasmas [32]. The particles can be directly lost if their axial excursions are so long that they follow the transverse field lines into the trap wall—so called ballistic loss. This was a big concern for antihydrogen production—as the plasmas typically have to be moved together over distances that would cause such loss [32].

Fajans and Schmidt had earlier pointed out that, by using a higher-order multipole magnet, it should be possible to achieve the same total atom trap well depth while having a much flatter field profile at the axis of the Penning trap [33]. This is illustrated in Fig. 6.5 for the octupole versus quadrupole case. (An octupole—as opposed to a higher order multipole—was chosen due to practical tradeoffs in the engineering and construction of the magnets.) The operational goal then is to confine the charged particle clouds to small radii where the transverse fields are small. One of the first important results of ALPHA was to demonstrate that, in the presence of the atom trap fields, both antiprotons and positrons could indeed be trapped for times long enough to allow for antihydrogen synthesis [34]. ATRAP demonstrated antihydrogen production in a quadrupole atom trap in 2008 [35].

The ALPHA atom trap magnets are of a special construction developed by Brookhaven National Laboratory (BNL) [31]. The superconductor is wound directly onto the cryostat wall and secured with a tensioned fiberglass composite. There are no metal collars to counter the magnetic forces. This is very important for ALPHA, as the material in the magnet structure scatters the charged pions that are used to determine the antihydrogen annihilation position—the vertex. Denser material in the magnet structure leads to degraded vertex position resolution. Position sensitive

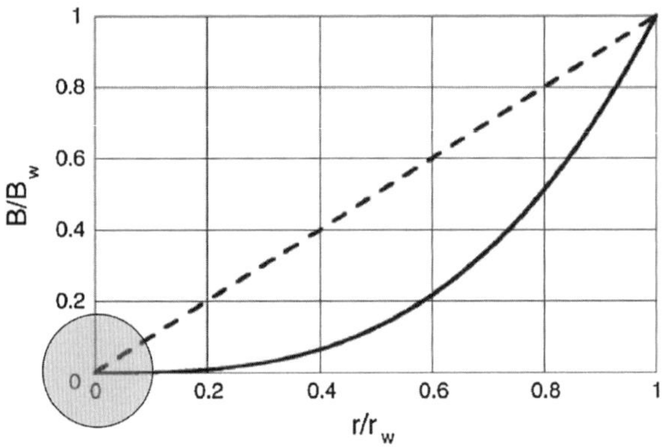

Fig. 6.5 Plot illustrating the magnetic field strength profiles for an octupole (*solid curve*) and a quadupole (*dashed curve*). The scalar field magnitudes are normalized to the maximum field B_w at the inner wall radius r_w of the Penning trap. The maximum obtainable field is assumed to be the same for both magnet types. The *shaded region* roughly indicates the maximum radius at which plasmas are typically stored in ALPHA

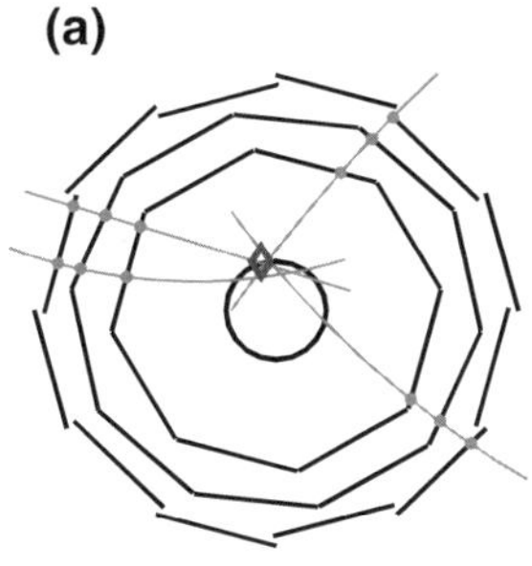

Fig. 6.6 Schematic representation (in axial projection) of event topologies in the ALPHA detector. **a** an antiproton from an antihydrogen atom annihilates on the inner wall of the Penning trap electrodes and yields four charged pions. The pion tracks (*red curves*) are reconstructed from the hit positions (*red dots*) in the three-layer silicon detector. Analysis of the tracks determines the vertex position (*blue diamond*). **b** a typical track from a cosmic ray arriving from overhead (Reproduced from [9])

detection of the antiproton annihilation has proven to be critical in all of the recent results involving trapped antihydrogen. The octupole is wound in eight distinct layers of 1 mm diameter superconductor. The effective atom trapping region is a cylindrical volume of 274 mm length and 44.5 mm diameter.

The ALPHA production and trapping region is surrounded by a three-layer, imaging vertex detector [36] comprised of segmented silicon. The detector is used to identify antiproton annihilations from lost antihydrogen, and to distinguish these from cosmic rays—which are the source of the dominant background for this experiment. The typical topologies for antiproton and cosmic ray events are shown in Fig. 6.6. Data samples of bare antiproton annihilations and cosmic ray annihilations can be collected independently of any of our antihydrogen experiments and used to develop the criteria for distinguishing signal from background in an unbiased manner. In the latest published results, we were able to reject cosmic rays to a background level of $1.7 \times 10^{-3}\,\mathrm{s}^{-1}$ [11]. Note that, unlike ATHENA, ALPHA does not have a detector for identifying the gamma rays from the positron annihilation. Absorption by the material in the cryostat and atom trap magnets precludes efficient detection of these photons. Thus we rely mainly on the antiproton annihilation detector to analyse what is happening in our experiments.

6.5.2 Detecting Trapped Antihydrogen

When the ALPHA machine was being designed in 2004–2005, it seemed unlikely that conventional ways of detecting trapped atoms of matter (e.g., by laser fluorescence) could be utilised—due to the small number of atoms expected to be trapped and the restricted geometry of the experiment itself. We thus settled on an alternative:

detection of trapped anti-atoms by controlled release from the trap and observation of the resulting annihilation. A key to this technique is rapid shutdown of the superconducting magnets in the atom trap—in order to minimise the probability that a cosmic ray arrives during the release interval. The ALPHA magnets can be de-energised with a time constant of about 9 ms. Note that this is extreme for a superconducting magnet and the experiment was not undertaken without some risk. The magnets are shut down by diverting their current (about 900 A for the octupole and 700 A for the mirror coils) to a resistor network by means of an IGBT (isolated-gate bipolar transistor) switch. The magnets quench during this shutdown, but the BNL produced magnets have survived many thousands of cycles of this treatment.

6.5.3 Antihydrogen Trapping

Ignoring many of the subtle details for now, the sequence for trapping antihydrogen involves preparing plasmas of antiprotons and positrons in adjacent wells, with the positrons centered in the atom trap. The atom trap magnets are then energized, and the plasmas are mixed to form antihydrogen. After the mixing the Penning potentials are removed and pulsed electric fields are used to remove any un-reacted charged particles that may be mirror-trapped by the atom trap magnetic fields [37].

Then the magnets are shut down, and we analyze the annihilation detector output for events resulting from antihydrogen annihilation during the release interval (typically 30 ms). To be certain that any released particles are neutral antihydrogen and not bare antiprotons, we apply an axial electric field to the trap volume while the trap is being shut down and look for evidence of displacement of the annihilation vertex positions under influence of the fields. Data sets are accumulated with fields in one axial direction left, the opposite direction right, and with no field. Figure 6.7 shows the results of the first published experiment [9]. We observed 38 events consistent with the release of trapped antihydrogen, out of 335 attempts. The expected background for the total sample was 1.4 ± 1.4 events. The antihydrogen would have been held for at least 172 ms, which is the time necessary to perform the field manipulations to remove any remaining trapped particles. The axial distribution of the annihilation positions is consistent with computer simulations of the release process for neutral antihydrogen and completely inconsistent with simulations for release of mirror trapped antiprotons [9, 37]. Refinements during the 2010 and 2011 AD running periods led to an improvement in the initial trapping rate (about one in nine attempts) to about one trapped atom per attempt [10]. An attempt takes about 20 min of real time.

6.5.4 Holding Antihydrogen

An obvious next question is How long can one hold onto trapped antihydrogen? We investigated this by simply delaying the shutdown of the magnets in our usual

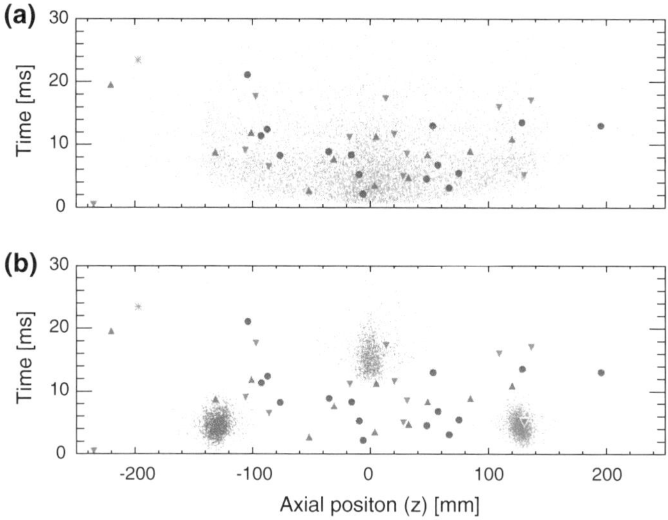

Fig. 6.7 Data and simulation of annihilation events from particles released from the ALPHA antihydrogen trap. The axial position of an annihilation is plotted versus the time from the initiation of the shutdown of the trap magnets. The measured data are the *green circles* (no axial deflection field), *red triangles* (axial field to deflect antiprotons to the right), and *blue triangles* (axial field to deflect antiprotons to the left). The measured data set is compared to (**a**) computer simulations of the release of trapped antihydrogen (*grey dots*) and (**b**) computer simulations of the release of mirror trapped antiprotons (coloured dots corresponding to the bias fields described above). One background point (*purple star*) was obtained with conditions (*heated positrons*) under which no trappable antihydrogen should have been produced (Reproduced from [9])

trapping sequence to study the survival rate at various times. The short answer is that there is clear evidence for trapped antihydrogen after 1,000 s of hold time [10], Fig. 6.8. About half of the atoms survive for this time. We have yet to investigate any loss mechanisms in detail (there may be multiple mechanisms with different time scales) or to carefully study longer times—these measurements are quite frankly rather tedious. But 1,000 s is a very long time on an atomic scale and is sufficient to allow one to contemplate spectroscopic measurements and laser cooling—even for just one atom trapped at time. The long hold times also guarantee that the trapped antihydrogen—which may have been initially trapped in a positronic excited state—has decayed to the ground state [10]. This is an extremely important point, as ground state antihydrogen is the object we wish to study precisely—for example using the 1s–2s transition. This transition in material hydrogen has been measured very precisely (a fractional frequency uncertainty of 4.2×10^{-15}) and referenced to a frequency standard by the Hänsch group [38]. Comparison of the frequencies of this transition for hydrogen and antihydrogen is one of the longest-standing goals of the AD physics program.

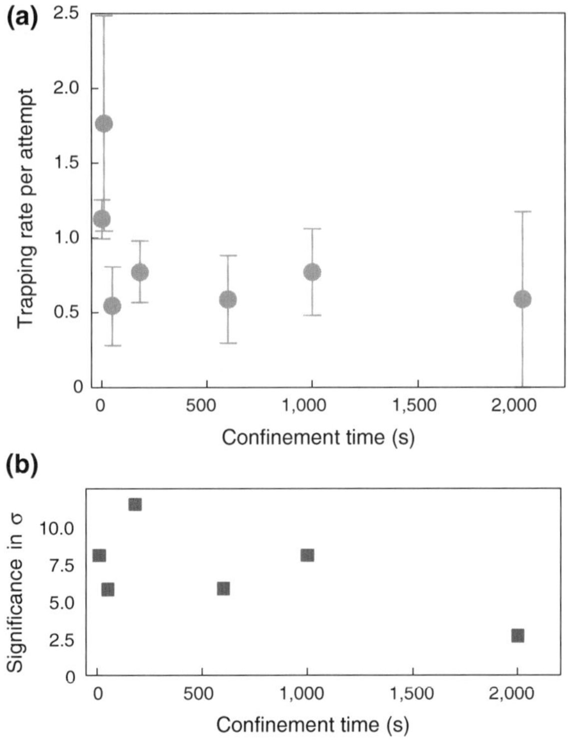

Fig. 6.8 Determination of the storage lifetime of trapped antihydrogen in the ALPHA trap. **a** The average number of antihydrogen atoms released per attempt is plotted versus the holding time. The error bars are due to counting statistics. **b** statistical significance of the data points in the graph above, expressed in σ (Reproduced from [10])

6.5.5 Measuring Trapped Antihydrogen

The original ALPHA device was not equipped with windows to allow laser access to the trapping volume, but we were able, in the shutdown between the 2010 and 2011 AD runs, to modify the apparatus to allow for axial injection of microwaves into the anti-atom trapping volume. The goal of this modification was to try to observe resonant quantum transitions between the ground-state hyperfine levels in trapped antihydrogen. The microwaves were introduced using a vacuum mounted horn antenna that could be manipulated onto the experiments access. Of the four ground state hyperfine levels, two are low-field seeking (the energy level increases with B-field) and can be trapped in the minimum-B trap. The other two states are high field seeking and untrappable. The idea behind this experiment was to resonantly drive microwave transitions between these two level manifolds [11]. The anti-atoms that make a transition go from being trapped to being untrappable and are thus lost. The transition corresponds to a positron spin flip. The experiment is conducted in the same way as the storage time experiment. Atoms are trapped and held, and while being held they are illuminated with microwaves for some time (180 s in this case). After the irradiation, the trap is released to see if any atoms remain trapped.

Three types of data sets are accumulated: microwaves on resonance, microwaves off resonance, and no microwaves present. In this case on resonance" means resonant with the minimum field in the highly inhomogeneous trap—obviously the field varies greatly over the trapping volume. (Note that both of the low field seeking states were addressed by alternating the microwave frequency back and forth between the two during the irradiation interval.) It turns out that we could clearly see the effects of the microwaves in the release data: the resonant microwaves effectively empty the trap; using no microwaves or off-resonant microwaves does not [11]. We were also able to directly detect the annihilation of the atoms when their spin-flip occurred. This was a rather simple on-off measurement—the off-resonant microwaves were detuned 100 MHz from the roughly 29 GHz resonant field, and we did not attempt to scan the frequency to measure a line shape. Nevertheless, it represents the first-ever resonant interaction with an anti-matter atom, and demonstrates that it is possible to do some interesting physics with just a few anti-atoms (the total sample here was about 100 atoms detected).

6.5.6 Trapped Antihydrogen and Ion Trap Physics

As mentioned above, producing antihydrogen in the first place requires the successful and reliable implementation of many manipulation and diagnostic tools for Penning traps. In addition to the ones summarized above, a few innovations that are unique to ALPHA are worth mentioning, as they are directly related to producing cold antihdyrogen. We attempted for many years to trap antihydrogen using ATHENA-type mixing. These attempts failed to produce any antihydrogen cold enough to be trapped—at least at the level of our sensitivity and patience. Two techniques have been particularly important to our success: autoresonant injection of antiprotons into a positron plasma and evaporative cooling of charged particle plasmas. These are described briefly below; again the details can be found in the references.

6.6 Autoresonant Injection of Antiprotons into a Positron Plasma

As described above, in ATHENA the antiprotons were injected into the positron plasma with relative energies of many electron volts. There was evidence, both from ATHENA [20] and ATRAP [39], that this type of mixing produced hot antihydrogen of perhaps hundreds of degrees K. Simulations suggested that the antihydrogen forms before the antiprotons cool into thermal equilibrium in the positron plasma [40].

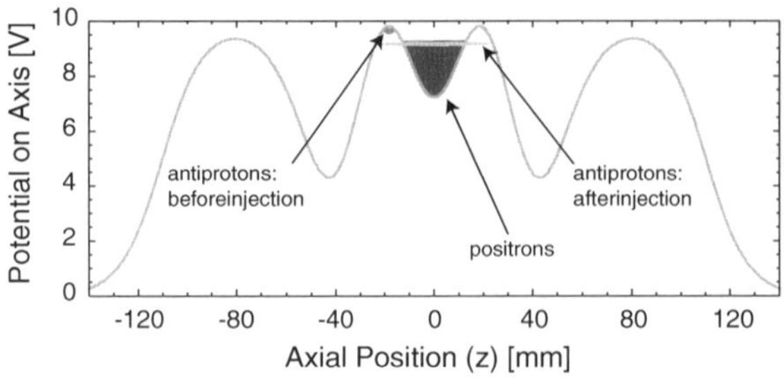

Fig. 6.9 Representation of the axial potential used for autoresonant injection of antiprotons in ALPHA. The externally applied potential is shown in *red*. The positron space charge significantly flattens the total potential in the center of the nested well

As mentioned above, the inability of ALPHA to trap antihdyrogen atoms produced a la ATHENA tends to confirm this expectation. The scheme for autoresonant (see [41]) injection can be illustrated with help of Fig. 6.9. The positron and antiproton plasmas are prepared in adjacent wells. Both plasmas can be cooled in-situ by evaporative cooling—see discussion below. The axial motion of the antiprotons in the potential well is driven by applying a periodic signal to one of the Penning electrodes. Under the correct conditions of drive strength and plasma density and temperature, the antiproton cloud will behave as a coherent macroparticle [42]. The nested well shape for the antiprotons results in nonlinear behaviour: the frequency at the bottom of the antiproton well (the antiproton starting position in Fig. 6.9) is higher than that at the energy level at which the antiprotons would enter the positron plasma. The cold antiproton cloud is captured by the drive in the linear part of the well and the frequency is then swept from high to low. The antiproton cloud autoresonantly follows the drive (downward in Fig. 6.9)—matching its axial oscillation amplitude to the corresponding frequency determined by the well shape. By careful control of this frequency sweep, the antiprotons can be injected into the positron plasma with very small relative velocity—allowing for production of cold antihydrogen. All of the antihydrogen trapped in ALPHA has been produced with this kind of mixing, but this technique is still relatively new and has yet to be fully exploited. A theoretical study involving simulations of the plasma behaviour under autoresonant injection was recently published [43] and gives some guidance for optimising this technique in the future.

6.7 Evaporative Cooling of Charged Antimatter Plasmas

The technique of evaporative cooling is a mainstay of cold atom research [44]. In ALPHA, we have applied this technique to both antiprotons [45] and positrons in our efforts to create trappable antihydrogen (see [46] for a description of the experimental situation before evaporative cooling was introduced).

As noted above, the positron plasmas in our machine do not reach thermal equilibrium with the cryogenic trap walls (10 K). The antiproton plasmas are usually left at a temperature of a few hundred K after electron removal. In order to improve this situation we can evaporatively cool both species by simply reducing the Penning trap well depth to allow warmer particles to escape. We have earlier cooled a sample of antiprotons to about 9 K using this technique [45]. This of course involves throwing away precious antimatter particles, but the idea is to end up with more particles that can produce trappable antihydrogen.

For positrons, the situation is not very dire; we can easily trap close to 10^8 positrons, but we use only about 2 million positrons in the successful antihydrogen trapping experiments. It hurts more to throw away antiprotons—we only have about 40,000 to start with, but current thinking suggests that the positron temperature still plays the dominant role, as the antiprotons quickly equilibrate when they are gently introduced by the autoresonant technique.

Indeed, our first hint of antihydrogen trapping came in 2009, when we saw six candidate events in 212 attempts with positrons at about 70 K [46]. These measurements were taken without the electrical bias fields that ensure that the events were not due to mirror-trapped antiprotons, but simulations indicate that it is very unlikely that these were charged events. The most substantial change between the 2009 and 2010 runs was the introduction of evaporative cooling to the positron plasma. This resulted in positrons of about 40 K, and the measured trapping rate of one atom in about nine attempts.

While the antiproton temperature may only have a higher order effect on the efficiency of the current mixing technique, evaporative cooling can in principle lead to better control and reproducibility of the autoresonance injection. In other mixing techniques in which the antiproton plasma is stationary—for example, interaction of an antiproton cloud with positronium atoms [47], evaporative cooling could be a very important tool for increasing the number of trapped atoms. Both ATRAP and AEGIS use variations of the positronium technique. Efforts in ALPHA will continue to reduce the positron plasma temperature, as well as optimization of the whole production cycle to increase the number of trapped atoms per attempt. Madsen [48] is pursuing the use of sympathetic cooling of positrons by laser cooled trapped ions, as demonstrated by the Bollinger group [49], to reduce the positron temperature. This project is funded and just beginning to be implemented for ALPHA.

6.8 Towards Antihydrogen Spectroscopy

At the time of the writing of this chapter (Summer 2013), the AD is currently shut down in connection with the upgrade of the LHC machine, and we will not see antiproton beam again before mid-2014 at the earliest. To prepare for the next generation of experiments with antihydrogen, we have (in 2012) constructed a completely new experimental apparatus, known as ALPHA-2. A schematic diagram is shown in Fig. 6.10. As mentioned above, the original ALPHA machine did not allow for

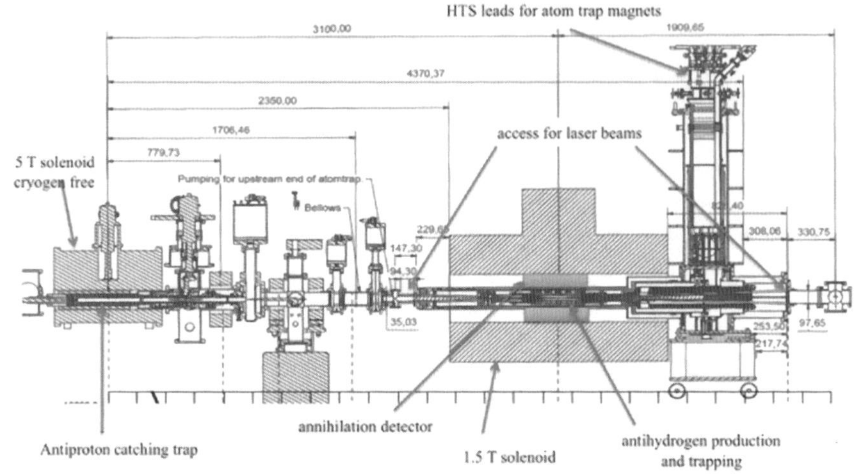

Fig. 6.10 Graphic depiction of the ALPHA-2 device. The antiprotons from the AD (arriving from the left of the figure) are caught, cooled and accumulated in a separate trap with its own solenoid magnet. The atom trap features a similar construction to that of the original ALPHA machine, but with added solenoidal coils for tailoring the longitudinal field shape. Current is introduced to the magnets (eight individual coils) through high temperature superconducting leads developed for the LHC

access of laser beams to the anti-atom trapping volume. ALPHA-2 remedies this by including access and egress windows for up to four laser beams. One of the laser paths features a Fabry-Perot buildup cavity within the cryogenic UHV system to allow for power enhancement of the 243 nm laser light needed for the two-photon excitation of the 1s–2s transition. Investigation of this transition has been one of the goals of the collaboration since its inception. We hope to begin initial investigations of this line with ALPHA-2 when the AD beam returns.

Some ALPHA colleagues have recently considered the feasibility of laser cooling of trapped antihydrogen atoms using pulsed Lyman-α light [50]. The potential for this is also included in the design of ALPHA-2, and the necessary laser is being developed within the collaboration. We will also continue our microwave investigations. ALPHA-2 features extra magnetic coils that will allow us to flatten the longitudinal field profile in the trap center, yielding a larger resonant volume for the microwave transitions. We eventually hope to probe the antihydrogen NMR (nuclear magnetic resonance or antiproton spin flip) transitions using trapped antihydrogen, but some technical development remains to allow injection of the required wavelength into the trapping volume. Meanwhile, ASACUSA is actively pursuing their Ramsey-type drift experiment [14] to measure the hyperfine spectrum. I will refrain from making predictions about how the development of antihydrogen spectroscopy will proceed—these always ultimately lead to embarrassment. It is in many ways remarkable that we are now in a position to design real measurements using trapped antimatter atoms—the landscape was somewhat bleaker a few short years ago. Clearly we would like to have more trapped atoms per attempt, and the truly precision experiments may require vastly different experimental setups than those we have now. I remain confident that this inexhaustible and creative community of physicists will overcome the necessary obstacles to answer the very fundamental and intriguing question: Do atoms of matter and antimatter obey the same laws of physics?

To aid us in this quest, CERN has recently approved a major upgrade to the AD complex. The ELENA ring will further decelerate antiprotons from the AD down to 100 keV. This will allow researches to capture a much higher fraction of the delivered antiprotons and allow us to make more rapid progress in our research. The entire community is looking at a very bright future for low energy antiproton physics.

6.8.1 Dropping Antihydrogen

Although this is a volume about spectroscopy, we note in passing that ALPHA has recently published the first experimental study to directly address gravitational effects on neutral antimatter [12]. The concept is simple—we looked at the position distribution of the annihilation vertices due to antihydrogen atoms that were intentionally released from the trap. This was not an experiment per se, but a retrospective analysis of data taken in the course of other experiments. The antihydrogen in ALPHA is typically too warm (recall the well depth is up to about 0.5 K) that we would expect to see the effect of gravitational free fall when the atoms are released in the horizontally oriented atom trap. However, there is a distribution of energies, and the coldest atoms can be expected to emerge later in the release process, as the confining potential decays. So while we cannot directly observe free fall (or upward acceleration, for antigravity), we can compare the measurements to computer simulations of the release of trapped atoms. In the simulations, we can assume that the gravitational mass of antihydrogen is different from its inertial mass by some ratio F. By statistically comparing the data with simulations of varying F, we can use the data to place limits on F. The answer is that F lies between -65 and 110 (negative signs imply antigravity). This is not a very interesting band yet, but the experiment shows how one might approach such measurements with trapped antihydrogen, and we have examined the improvements that could be made to this technique in the future. It is not unreasonable to expect to be able to rule out $F = -1$ (antigravity) with this technique in the not too distant future. Both AEGIS and Gbar will aim for more precise measurements of the value of g for antiatoms. These are both very technically challenging experiments.

Acknowledgments The author would like to thank the editors, Professors Quint and Vogel, for taking the initiative to prepare this volume and for the hard work of editing it. My many colleagues in PS200, ATHENA, and ALPHA are gratefully acknowledged for outstanding collaboration over the years; their names are to be found in the references. I would also like to thank the CERN AD and injector staff for delivering reliable beam over the years, and the members of the other AD and LEAR experiments, past and present, for creating an extremely stimulating working environment at CERN. The authors work has been supported by the Danish National Research Council (SNF, FNU), the Carlsberg Foundation, and the European Research Council.

References

1. S. Maury, The antiproton decelerator: AD. Hyp. Int. **109**, 4352 (1997)
2. G. Baur et al., Production of antihydrogen, Phys. Lett. B **368**(3), 251 (1996). Bibcode:1996PhLB.368.251B. doi:10.1016/0370-2693(96)00005-6
3. G. Blanford et al., Observation of antihydrogen, Phys. Rev. Lett. **80**(14), 3037 (1998). Bibcode:1998PhRvL.80.3037B. doi:10.1103/PhysRevLett. 80.3037
4. G. Gabrielse et al., First capture of antiprotons in a penning trap: a kiloelectronvolt source. Phys. Rev. Lett. **57**, 2504–2507 (1986)
5. G. Gabrielse et al., Cooling and slowing of trapped antiprotons below 100 meV. Phys. Rev. Lett. **63**, 13601363 (1989)
6. C.M. Surko, R.G. Greaves, Emerging science and technology of antimatter plasmas and trap-based beams. Phys. Plasmas **11**, 2333–2348 (2004)
7. L.V. Jorgensen et al., New source of dense cryogenic positron plasmas. Phys. Rev. Lett. **95**, 025002 (2005)
8. M. Amoretti et al., Production and detection of cold antihydrogen atoms. Nature **419**, 456459 (2002)
9. G.B. Andresen et al., Trapped antihydrogen. Nature **468**, 673–676 (2010)
10. C. Amole et al., ALPHA collaboration, confinement of antihydrogen for 1000 seconds. Nat. Phys. **7**, 558 (2011)
11. C. Amole et al., Resonant quantum transitions in trapped antihydrogen atoms. Nature **483**, 439 (2012)
12. C. Amole et al., Description and first application of a new technique to measure the gravitational mass of antihydrogen. Nat. Commun. **4**, 1785 (2012)
13. G. Gabrielse et al., Trapped antihydrogen in its ground state. Phys. Rev. Lett. **108**, 113002 (2012)
14. E. Widmann et al., Measurement of the hyperfine structure of antihydrogen in a beam, Hyp. Int. **215**, 1 (2013) (http://dx.doi.org/10.1007/s10751-013-0809-6)
15. Y. Enomoto et al., Synthesis of cold antihydrogen in a cusp trap, Phys. Rev. Lett. **105**, 243401 (2010) (http://link.aps.org/doi/10.1103/PhysRevLett. 105.243401)
16. A. Kellerbauer et al., (AEgIS collaboration), proposed antimatter gravity measurement with an antihydrogen beam. Nucl. Inst. Meth. B **266**, 351 (2008). doi:10.1016/j.nimb.2007.12.010
17. P. Perez, Y. Sacquin, The GBAR experiment: gravitational behaviour of antihydrogen at rest. Class. Quantum Grav. **29**, 184008 (2012)
18. G. Gabrielse, S.L. Rolston, L. Haarsma, W. Kells, Antihydrogen production using trapped plasmas. Phys. Lett. A **129**, 38 (1988)
19. M. Amoretti et al., High rate production of antihydrogen. Phys. Lett. B **578**, 23–32 (2004)
20. N. Madsen et al., Spatial distribution of cold antihydrogen formation. Phys. Rev. Lett. **94**, 033403 (2005)
21. G. Gabrielse et al., Background-free observation of cold antihydrogen with field ionization analysis of its states. Phys. Rev. Lett. **89**, 213401 (2002)
22. X.-P. Huang, F. Anderegg, E.M. Hollmann, C.F. Driscoll, T.M. ONeil, Steady-state confinement of non-neutral plasmas by rotating electric fields, Phys. Rev. Lett. **78**, 875 (1997)
23. G.B. Andresen et al., Compression of antiproton clouds for antihydrogen trapping. Phys. Rev. Lett. **100**, 203401 (2008)
24. G.B. Andresen et al., Antiproton, positron, and electron imaging with a microchannel plate/phosphor detector. Rev. Sci. Inst. **80**, 123701 (2009)
25. N. Kuroda et al., Radial compression of an antiproton cloud for production of intense antiproton beams. Phys. Rev. Lett. **100**, 203402 (2008)
26. M.D. Tinkle et al., Low-order modes as diagnostics of spheroidal non-neutral plasmas. Phys. Rev. Lett. **72**, 352 (1994)
27. M.D. Tinkle, R.G. Greave, C.M. Surko, Low-order longitudinal modes of single-component plasmas. Phys. Plasmas **2**, 2880 (1995)

28. M. Amoretti et al., Complete nondestructive diagnostic of nonneutral plasmas based on the detection of electrostatic modes. Phys. Plasmas **10**, 3056 (2003)
29. M. Amoretti et al., Positron plasma diagnostic and temperature control for antihydrogen production. Phys. Rev. Lett. **91**, 055001 (2003)
30. C. Amole et al., In-situ electromagnetic field diagnostics with an electron plasma in a Penning-Malmberg trap. New J. Phys. (Submitted) (2013)
31. W. Bertsche et al., A magnetic trap for antihydrogen confinement. Nucl. Inst. Meth. A **566**, 746 (2006)
32. J. Fajans, W. Bertsche, K. Burke, S.F. Chapman, D.P van der Werf. Phys. Rev. Lett. **95**, 15501 (2005)
33. J. Fajans, A. Schmidt, Malmberg-penning and minimum-B trap compatibility: the advantages of higher-order multipole traps. Nucl. Inst. Meth. A **521**, 318 (2004)
34. G. Andresen et al., Antimatter plasmas in a multipole trap for antihydrogen. Phys. Rev. Lett. **98**, 023402 (2007)
35. G. Gabrielse et al., Antihydrogen production within a penning-ioffe trap. Phys. Rev. Lett. **100**, 113001 (2008)
36. G. Andresen et al., The ALPHA detector: module production and assembly. JINST **7**, C01051 (2012). doi:10.1088/1748-0221/7/01/C01051
37. C. Amole et al., Discriminating between antihydrogen and mirror-trapped antiprotons in a minimum-B trap. New J. Phys. **14**, 105010 (2012)
38. C.G. Parthey et al., Improved measurement of the hydrogen 1S2S transition frequency. Phys. Rev. Lett. **107**, 203001 (2011)
39. G. Gabrielse et al., First measurement of the velocity of slow antihydrogen atoms. Phys. Rev. Lett. **93**, (2004)
40. F. Robicheaux, Simulations of antihydrogen formation. Phys. Rev. A **70**, 022510 (2004)
41. J. Fajans, L. Friedland, Autoresonant (nonstationary) excitation of pendulums, plutinos, plasmas, and other nonlinear oscillators. Am. J. Phys. **69**, 1096 (2001)
42. I. Barth, L. Friedland, E. Sarid, A.G. Shagalov, Autoresonant transition in the presence of noise and self-fields. Phys. Rev. Lett. **103**, 155001 (2009)
43. C. Amole et al., Experimental and computational study of the injection of antiprotons into a positron plasma for antihydrogen production. Phys. Plasmas **20**, 043510 (2013). doi:10.1063/1.4801067
44. H.F. Hess, Evaporative cooling of magnetically trapped and compressed spin-polarized hydrogen. Phys. Rev. B **34**, 3476 (1986)
45. G. Andresen et al., Evaporative cooling of antiprotons to cryogenic temperatures. Phys. Rev. Lett. **105**, 013003 (2010)
46. G. Andresen et al., Search for trapped antihdyrogen. Phys. Lett. B **695**, 95 (2011)
47. B.I. Deutch et al., Antihydrogen by positronium-antiproton collisions. Hyp. Int. **44**(1–4), 271–286 (1989)
48. N. Madsen, Private Commun. (2011)
49. B.M. Jelenkovic et al., Sympathetically laser-cooled positrons. Nucl. Inst. Meth. B **192**, 117127 (2002)
50. P.H. Donnan et al., A proposal for laser cooling antihydrogen atoms. J. Phys. B **46**, 025302 (2013). doi:10.1088/0953-4075/46/2/025302

Chapter 7
High-Precision Mass Measurements of Radionuclides with Penning Traps

Michael Block

Abstract The mass of an atom is directly related to the binding energy of all its constituents. Thus, it provides information about all the interactions inside the atom. High-precision mass measurements hence allow studies of fundamental interactions and are of great importance in many different fields in physics. The masses of radionuclides provide information on their stability and their structure and are therefore of particular interest for nuclear structure investigations and as input for nucleosynthesis models in nuclear astrophysics. Penning trap mass spectrometry provides masses of radionuclides with unprecedented accuracies on the order of 10^{-8} and can nowadays be applied even to nuclides with short half-lives and low production rates. Utilizing advanced ion manipulation techniques radionuclides from essentially all elements produced in a broad range of nuclear reactions can be accessed. In this chapter the standard procedures of on-line Penning trap mass spectrometry are introduced and some representative examples of recent mass measurements are given.

7.1 Importance of Masses of Radionuclides

The mass of an atom is one of its fundamental properties and contains information about all the interactions inside the atom. The atomic mass m of the atom is smaller than the sum of the masses of all its constituents. This difference is due to the binding energy $B(N, Z)$ of all its constituents that can be obtained from the relation

$$B(N, Z) = [Nm_n + Zm_p + Zm_e - m(Z, N)]c^2; \qquad (7.1)$$

where N is the neutron number, Z is the atomic number, m_n is the mass of a neutron, m_p is the proton mass, and m_e is the electron mass. Since the binding energy of the

M. Block (✉)
Gesellschaft für Schwerionenforschung mbH, Planckstrasse 1, 64291 Darmstadt, Germany
e-mail: m.block@gsi.de

W. Quint and M. Vogel (eds.), *Fundamental Physics in Particle Traps*,
Springer Tracts in Modern Physics 256, DOI: 10.1007/978-3-642-45201-7_7,

electrons is much lower than the nuclear binding energy it does not play a significant role for nuclidic masses at the present uncertainty level that can be reached for radionuclides so that $B_{nucl} \approx B(N, Z)$.

Masses play an important role in several fields of physics ranging from astrophysics to tests of the Standard Model [1, 2]. Mass measurements on stable nuclides can nowadays be performed with relative uncertainties on the order of about 10^{-11} [3–5], an accuracy that is required for the determination of fundamental constants and for very sensitive tests of fundamental symmetries. For example, a high-precision comparison of the proton-to-antiproton-mass ratio [6] provides a sensitive test of the charge, parity, and time reversal (CPT) symmetry. High-precision mass measurements also support precise tests of the Standard Model via the unitarity of the Cabbibo-Kobayashi-Maskawa quark mixing matrix by studies of superallowed β decays [7].

Mass measurements of radionuclides are a powerful tool for nuclear physics since they provide information on the stability and the nuclear structure of the investigated nuclides. Structural effects are revealed by several indicators such as the nucleon separation energies [1, 2]. In this context masses are often expressed by the mass excess ME defined as

$$ME = [m(\text{in u}) - A]; \tag{7.2}$$

where $A = N + Z$ is the mass number and the atomic mass unit u is defined as 1/12th of the mass of ^{12}C. A moderate precision on the order of 100 keV already allows identifying global structure phenomena such as major shell closures. For the investigation of nucleon pairing or deformation a higher precision on the order of 10 keV is often required. For tests of relations such as the isobaric multiplet mass equation (IMME) [8] even a higher precision on the level of about 1 keV is desirable. The IMME relation describes the masses of nuclides within an isobaric multiplet. It is based on the fact the strong interaction is approximately charge independent, a symmetry of this interaction that was already realized by Heisenberg [9] and Wigner [10] leading to the introduction of the isospin concept.

For nuclear structure studies it is important to reach radionuclides far away from stability whose proton-to-neutron ratio is very different from nuclides close to stability. Such extreme ratios often lead to changes in the structure and to new discoveries. It has been observed for example that the shell closure at $N = 28$ established in nuclides close to stability erodes in more neutron-rich nuclides [11]. The shell structure evolution can be tracked by masses.

A prominent nuclear structure phenomenon observed in light nuclides are halo structures. In such cases one or two nucleons are very loosely bound giving rise to exceptional large radii. For example, the two-neutron halo nucleus ^{11}Li [12] has a similar size as the much heavier nucleus ^{208}Pb illustrated also by mean square charge radii of ^{11}Li obtained by laser spectroscopy [13]. Halo candidate nuclides can be identified by mass measurements that reveal the neutron binding energy.

Another peculiar example are 'superheavy' elements ($Z \geq 104$) that are predicted to inhabit an 'island of stability' [14]. Their nuclei owe their very existence to nuclear shell effects that stabilize them against the disintegration by spontaneous fission due to the strong Coulomb repulsion. The strength of these shell effects can be experimentally determined by direct mass measurements with Penning traps [15].

In general the proton and neutron separation (binding) energies determine the limits of stability given by the drip lines ($S_p(N, Z), S_n(N, Z) = 0$) [16]. Since mass differences also determine the energy available for radioactive decays mass measurements allow identifying nuclides that exhibit exotic decay modes as for instance two-proton decay [17] or cluster radioactivity [18].

Masses and mass differences (Q values) also play an important role in nuclear astrophysics for the modelling of the formation of elements in stellar and explosive burning scenarios [19]. They determine the pathway of the different nucleosynthesis processes such as the rapid proton capture, the slow neutron capture and the rapid neutron capture process [20–22].

Accurate Q values are moreover relevant for neutrino physics. The search for rare processes such as neutrino-less double β decay and neutrino-less double electron capture needs very accurate Q values [23, 24]. The observation of one of these rare neutrino-less decay modes would unambiguously establish the neutrino as a Majorana particle and allow us to to determine the (anti)neutrino mass. The determination of the neutrino mass from the endpoint of the β decay spectrum of tritium as planned in the KATRIN experiment [25] for example, requires the very accurate knowledge of the ^{3}H-^{3}He mass difference.

The present knowledge on the masses of all about 3,500 nuclides known to date is summarized in the latest Atomic-Mass Evaluation (AME 2012) [26]. It contains all experimental data related to nuclidic masses obtained by different mass spectrometry techniques.

7.2 Mass Measurements at On-line Facilities

The development of mass spectrometry goes along with some important discoveries in physics. The 'birth' of mass spectrometry dates back to the time around 1900, a very productive period in physics with several ground breaking discoveries. A brief history on mass measurements and their evaluation can be found in an article by Audi [27].

An important milestone was the discovery of the electron by J. J. Thompson in 1897 that was based on a measurement of its mass-to-charge ratio [28], the same principle still applied in Penning trap mass spectrometry (PTMS) to date. Another important discovery was made 100 years ago by the observation of different neon isotopes using a mass spectrograph with electric and magnetic fields by the very same J. J. Thompson [29–32] who was awarded the Nobel prize in physics in 1906. F. W. Aston, J. J. Thompson's student, improved the mass spectrograph, where the different ions were registered on a photo plate, and discovered several new nuclides [33]. He was awarded the Nobel prize in chemistry in 1922. Further people involved in the development and improvement of mass spectrometers with higher resolution include for example Dempster [34], Mattauch and Herzog [35], Bainbridge, and Jordan [33].

Such devices reached a relative mass uncertainty of up to 10^{-5} and were used for a number of mass measurements of stable nuclides. Besides the discovery of multiple nuclides it led to the experimental observation of the mass defect, i.e. the difference between the atomic mass and the sum of the masses of its constituents as discussed in the introduction.

Mass measurements of radionuclides had to be carried out on-line at radioactive beam facilities. This endeavour started in the 1970s [36] at ISOLDE [37], the isotope separator on-line at CERN Geneva, and were initially performed by conventional mass spectrometers. However, such spectrometers reach their limits in applications to radionuclides due to the limited yield of such particles since in conventional mass spectrometers, for example of Mattauch-Herzog type, a high resolution often requires the use of narrow slits that reduces the transmission.

The advantage of a mass determination based on a cyclotron frequency measurement was realized by L. Smith who constructed an radiofrequency (rf) mass spectrometer in the 1960s [38] that could reach a higher precision. The 'Mistral' spectrometer installed at ISOLDE was a mass spectrometer of Smith type that was used in particular for measurements of light halo nuclides [39, 40]. However, its operation was rather complicated and the efficiency was limited so that this spectrometer was later decommissioned.

High-precision mass measurements of radionuclides in a Penning trap were pioneered almost thirty years ago, again at ISOLDE. The first mass measurement in a Penning trap was performed a couple of years earlier [41]. The installation of the Penning trap mass spectrometer ISOLTRAP [42, 43] marked the beginning of a new era in mass spectrometry.

Penning traps provide masses of radionuclides with unprecedented accuracies based on a direct cyclotron frequency measurement. Relative uncertainties on the order of 10^{-8} are nowadays feasible [1] and even radionuclides with short half-life [44] and low production rates [15] are accessible. For certain radionuclides even an accuracy of a few parts in 10^9 was already reported [45]. Since the installation of ISOLTRAP the landscape at radioactive beam facilities has changed dramatically. Today many Penning trap mass spectrometers are operated or planned at facilities around the world [43, 46–53]. The different devices show a large complementarity utilizing different production and measurement schemes in order to access practically the full nuclear chart.

A different technique for mass measurements that is also well suited for the application to radionuclides is time-of-flight mass spectrometry (ToFMS). First devices were used by A.E. Cameron and and D.F. Eggers around 1950 [54]. ToFMS can reach radionuclides far away from stability. However, the precision using conventional devices was typically only on the order of 10^{-6} for a reasonable flight path. This is often not sufficient to resolve nuclear isomers, and even nuclear isobars is some cases, and renders an accurate mass determination difficult. Recently new devices have been developed for ToFMS that allow a longer flight path of the ions of interest to achieve a higher resolving power maintaining a compact design. In so-called multi-reflection time-of-flight mass spectrometers [55–57] ion bunches are reflected multiple times between electrostatic mirrors. Such devices can reach a mass resolving power of up to about 200,000 for a several turns within a few milliseconds cycle time. They are thus attractive as isobar separators [58] and may be used for mass measurements of very short-lived radionuclides with reasonable precision.

Another workaround in ToFMs is the use of a high-energy storage ring such as the experiment storage ring (ESR) at GSI Darmstadt [59]. An ion circulates in the ring with about 100 m circumference with a typical frequency on the order of MHz. In the so-called isochronous mode the mass of short-lived nuclides with microsecond half-lives can be determined by ToFMS with a precision of about 10^{-6} [60, 61]. Alternatively the Schottky mode, where even single ions can be detected electronically by the signal they induce into pick up electrodes while they orbit, mass measurements with a precision of about 10^{-7} can be obtained. However, the latter technique requires (electron) cooling of the radioactive ions so that only nuclides with half-lives of more than about one second can be tackled [60].

7.3 Penning-Trap Mass Spectrometry

Penning-trap mass spectrometry (PTMS) has in recent years developed into an elite method for high-precision mass measurements not only of stable nuclides but also for radionuclides [1]. This has opened up many new possibilities for studies related to various fields in physics as discussed in Sect. 7.1. With respect to nuclear structure investigations and to nuclear astrophysics it is necessary to extend the reach to exotic nuclides far away from stability. To this end, the introduction of buffer-gas stopping of ion beams along with advanced ion-manipulation techniques has been crucial. Nowadays PTMS can be applied to radioisotopes of essentially all elements, even of the elements above fermium, are meanwhile accessible to PTMS (c.f. Sect. 7.5.4). In the following the basic principles and some of the standard methods in PTMS are briefly introduced. Many of the general aspects discussed here also apply to PTMS of stable nuclides, however the focus here is on techniques relevant for experiments with radionuclides, where one has to deal with the challenges related to short half-lives and low production rates.

7.4 Production of Radionuclides at On-line Facilities

For the production and separation of radioactive ion beams a variety of techniques has been developed. The most important methods at accelerator facilities are presently the so-called ISOL method [62] and the in-flight technique. In ISOL facilities radionuclides are produced by fission and spallation reactions, typically induced by high-energy protons or light ions impinging on a thick target of a few mg/cm^2, for example made of uranium carbide. Alternatively neutron-induced fission or photo-fission are utilized. The radioactive ions are stopped in the target, diffuse out of the hot target cavity to the ion source where the radioactive atoms or molecules are ionized by different methods and extracted as an ion beam at energies of typically 30–60 keV.

In addition to surface ionization nowadays more selective ionization methods such as laser resonant ionization are more and more often used to provide radioactive ion beams of higher purity [63]. The ion beam is mass separated in a magnetic separator, typically with a mass resoling power of about $m/\delta m \approx 3{,}000$ sufficient to separate isotopes, and delivered to the experiments. Since the diffusion out of the hot cavity takes time the access to very short-lived nuclides with half-lives below a few milliseconds is not feasible. Moreover, certain elements are not available due to their chemical properties. For example refractory elements cannot be produced at ISOL facilities.

The first ISOL facility, the the LISOL facility in Louvain-la-neuve [64], came into operation in 1989. The major ISOL facilities operated today are ISOLDE at CERN, Geneva [37], the ISAC facility at TRIUMF, Vancouver [65], and SPIRAL at GANIL, Caen [66]. The ion guide isotope separator on line (IGISOL) method [67] is a variant of the ISOL technique that overcomes some of its limitations. In the IGISOL technique the reaction target is installed in a buffer gas-filled chamber. Ions that recoil out of the target are slowed down in the gas and swept out by the gas flow through an extraction hole. The ion source system is kept at high voltage, typically about 30 keV, and the extracted ion beam is also mass separated by a magnetic separator prior to the distribution to the experiments. The IGISOL facility at Jyvaeskylae [68] is a user facility where this approach is applied.

In order to get access to all elements and to increase the reach to more exotic, short-lived radionuclides other production schemes have to be employed. The fragmentation of relativistic projectiles in combination with the in-flight separation by electromagnetic fragment separators gives access to all elements lighter than the projectile, i.e. in the case of uranium fragmentation isotopes of all elements up to uranium can be produced. This method requires projectile energies of about 100 *A*MeV to about 1 *A*GeV and can use thick targets of a few mg/cm^2. The beam quality (emittance) is worse than at ISOL facilities but rather pure beams can be obtained by two-stage separators. In addition to fragmentation also in-flight fission of fragments is utilized, an approach that is favourable for the production of neutron-rich nuclides. Facilities based on projectile fragmentation in operation comprise the radioactive isotope beam factory (RIBF) [69] at RIKEN, Wako-shi, Japan, the fragment separator (FRS) [70] at GSI in Darmstadt, Germany, and the A1900 [71] at the NSCL at Michigan State University, East Lansing, USA.

Another approach for the production of radionuclides is via fusion-evaporation reactions. Light to heavy-ion beams with energies around the Coulomb barrier, i.e. about 5 *A*MeV, are reacted with thin targets of about 0.5 mg/cm^2 to form a compound nucleus. The compound nucleus de-excites by the evaporation of protons, neutrons, or α particles depending on its excitation energy. The evaporation residue recoils out of the target and is separated from the primary beam by kinematic separators in flight. Either vacuum separators such as the velocity filter SHIP at GSI Darmstadt [72], Vassilissa at Dubna [73], or gas-filled separators such as the DGFRS in Dubna, Russia [74], the BGS at LBNL Berkeley, USA [75], GARIS at RIKEN

[76], RITU at Jyvaeskylae [77], or TASCA [78] at GSI Darmstadt are employed. These separators are optimized for high transmission and a high primary beam rejection but do generally not contain a separation based on the mass-to-charge ratio of the reaction products. This production scheme is presently the only way to access the heaviest elements. Using beams of stable nuclides and various targets including radioactive actinides targets this approach has been used for production of elements up to $Z = 118$ [74]. However, in general neutron-deficient nuclides are produced by this method.

7.4.1 Typical Layout of a Penning Trap Mass Spectrometer

The typical layout of an on-line Penning trap mass spectrometer setup for high-precision mass measurements of radionuclides is schematically depicted in Fig. 7.1. The initial stage has to be adapted to the properties of the radioactive ion beams delivered from a radioactive beam facility. The ion beams delivered by isotope separator online (ISOL) facilities have a moderate beam energy of 30–60 keV and a relatively good beam quality. Radioactive beams from in-flight separators have a much higher energy of tens of MeV up to hundreds of MeV per nucleon and a rather high emittance. Then buffer gas cells have to be employed for the conversion into a low-energy ion beam.

Independent of the production scheme one has to deal with low production rates that require highly efficient setups and sensitive diagnostics. The first beam preparation stage that is nowadays included in practically all state-of-the-art Penning trap mass spectrometers is a buffer-gas filled radiofrequency quadrupole (RFQ) cooler and buncher. In such buffer gas-filled linear Paul traps ions can be cooled by an inert buffer gas in a few milliseconds, accumulated over several cycles, and bunched. Thereby low emittance bunched ions beams can be prepared for an efficient injection into Penning trap systems. In this way also the properties of the ion beam are decoupled from the production stage.

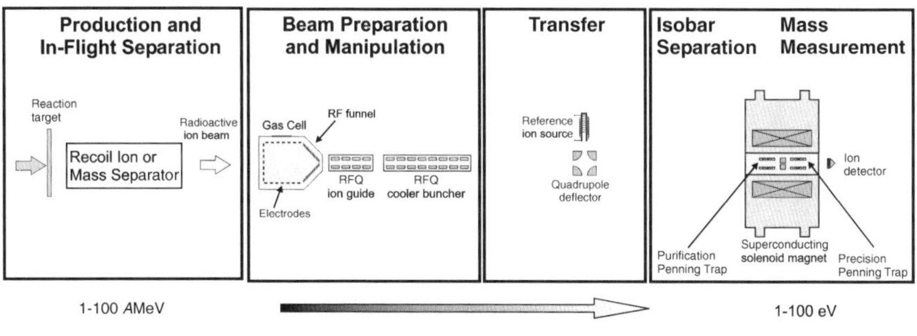

Fig. 7.1 Typical Layout of a Penning Trap Mass Spectrometer at an accelerator facility

In nuclear reactions often several nuclides are produced at the same time. In addition, contaminant ions may be created in the different beam preparation stages, for example by charge exchange reactions in the buffer gas, due the formation of molecular ions, or by background ions arising from nuclear decay products and unwanted ionization. The presence of unwanted ions that may even be much more abundant than the ion of interest has to be handled. The preparation of pure samples is crucial for high-precision mass measurements. To this end, in many setups a second Penning trap is utilized as a purification trap. In this trap a mass-selective buffer gas cooling and centering technique [79] is employed. This universal technique is suitable for radionuclides and provides a high mass resolving power of about 100,000 so that the separation of nuclear isobars can be accomplished. Alternatively other isobar separators such as magnetic separators [80] or multi-reflection time-of-flight mass spectrometer can be utilized [58]. The actual cyclotron frequency measurement is performed in the measurement trap, a precision Penning trap where the ions of interest are stored and manipulated in ultrahigh vacuum conditions. In the following the production of radionuclides is briefly addressed and the individual stages of an on-line Penning trap mass spectrometer are discussed in more detail.

7.5 Beam Preparation

The kinetic energy of the radionuclides produced by either of the techniques discussed above is too high for ion trapping. Thus, the ions have to be slowed down and prepared for the injection into a Penning trap first. In the case of an ISOL facility this can be accomplished utilizing an RFQ cooler buncher operated at high voltage [81]. The development of RFQ cooler-buncher devices was crucial to boost the performance of on-line Penning trap mass spectrometers and has made them a standard component for the preparation for low-energy radioactive beams. They are addressed in more detail in the next section.

Due to their much higher kinetic energy, about 50 MeV for fusion products up to several 100 *A*MeV for projectile fragments, the reaction products from in-flight separation cannot be slowed down electrostatically. An additional beam preparation stage in this case is provided by so-called buffer gas cells or ion-catcher devices [82–86] that are based on the IGISOL approach. In such gas cells the ions are slowed down in an inert buffer gas, typically helium, after passing a solid degrader that takes away most of the initial kinetic energy. The gas cells are operated at pressures on the order of 100 mbar for fusion products to about 1 bar for fragment beams. The ions remain singly or doubly charged in this process depending on their ionization potential and the cleanliness of the system. The ions are extracted by a combination of electric fields and buffer gas flow through an extraction hole, often a de Laval nozzle, within a few milliseconds at low pressure up to some hundred milliseconds at high pressure. In addition, cone-shaped electrode stacks [82, 85], so-called funnels, or ‘rf carpets’ [83, 87, 88] are utilized to facilitate an efficient extraction. In these devices a repelling force is created by an appropriate rf field to minimize ion losses during the focussing on the extraction hole. The typical efficiency of gas stopping cells is presently about 10 % [82–85]. Since for every incoming ion up to 10^6 helium ion electron pairs are created a very high cleanliness and high gas purity are essential to avoid losses by charge exchange and molecule formation. Thus, cryogenic gas cells have recently been developed [88, 89] that will be operated at temperatures of 40–80 K so that most impurities freeze out. They are expected to boost the efficiency of gas stopping systems and provide purer beams. A so-called 'cyclotron gas stopper' [90, 91] has been suggested to handle radioactive beams with high rates. In this device the ions are also slowed down in a buffer gas in a weak focussing field of a cyclotron-type magnet that results in a spiraling motion that extends the ions' pathway. Such a device can be operated at a lower pressure and may be beneficial for the stopping of light particles. A prototype of a cyclotron gas stopper is being constructed at the NSCL [92].

7.5.1 Beam Preparation with an RFQ Cooler and Buncher

RFQ ion-beam cooler and buncher devices allow the preparation of low-energy rare isotope beams with high beam quality [46, 93–95]. Their main benefits are

- a reduction of the transversal emittance and the energy spread of the injected beam
- the beam properties of the extracted beam are decoupled from the beam properties of the injected beam
- ion accumulation over several accelerator cycles becomes possible
- bunched beams with variable repetition rate can be provided.

This is a prerequisite for many precision experiments on radionuclides. It allows an efficient injection of rare isotope beams into ion traps and improves the sensitivity of laser spectroscopy experiments. In this section the basic principle of an RFQ cooler-buncher is briefly summarized. More detailed information about RFQs and Paul traps can be found in various textbooks and monographs [96–99].

An RFQ cooler-buncher is a buffer-gas filled linear Paul trap [100] where the ions are confined in a quadrupolar time varying potential of the form

$$\phi(x, y, t) = \frac{V_0 \cos \Omega t}{r_0^2}(x^2 - y^2) \tag{7.3}$$

that is created by an rf voltage

$$U(t) = V_0 \cos \Omega t \tag{7.4}$$

applied to pairs of opposite electrodes of a four-rod structure as shown in Fig. 7.2. In contrast to the original mass filter design by Paul [100] nowadays mostly cylindrical electrodes are used instead of hyperbolic-shaped ones as they are easier to machine and the potential sensed by the ions close to the symmetry axis of the RFQ is similar.

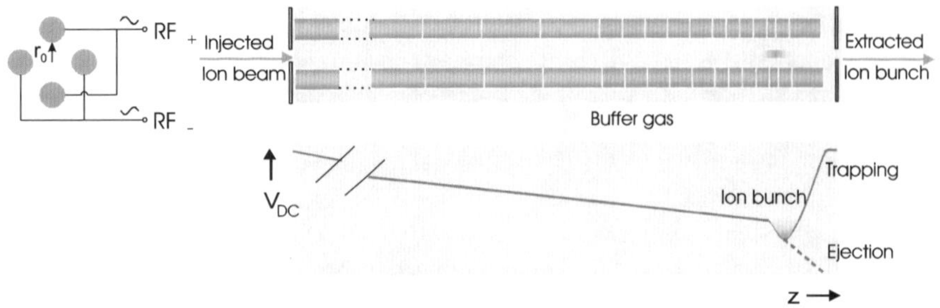

Fig. 7.2 Schematic view of an RFQ ion beam cooler and buncher. (*Top, left*) Cut in the transversal plane showing the connection of the rf voltage to the electrode rods. (*Top, right*) Cut of along the beam axis. The axially segmented buffer gas-filled electrode structure is shown. (*Bottom*) Axial DC potential in bunch mode

The ion motion can be described by Mathieu-type equations of motion of the form

$$\frac{\delta^2 u}{\delta^2 \xi} = (a - 2q\cos(2\Omega t))u = 0. \tag{7.5}$$

with generalized coordinates u and a parameterized time ξ. The Mathieu parameters a and q represent the DC and rf amplitudes of the trapping potential, respectively.

$$a = \frac{4eU_0}{mr_0^2\Omega^2} \tag{7.6}$$

$$q = \frac{2eU_0}{mr_0^2\Omega^2}, \tag{7.7}$$

Stable solutions of the Mathieu equation are obtained for certain values of a and q. The stability criterion can be depicted in a diagram as shown in Fig. 7.3 where a stable confinement is achieved choosing operating parameters within the shaded region. For an RFQ ion-beam cooler-buncher typically $a = 0$ is chosen leading to the stability condition $q < 0.908$ [98]. From this condition the required rf amplitude for a selected operation frequency Ω and a given charge-to-mass ratio can be derived. The motion in the rf potential can also be described in terms of a time-averaged pseudo-potential as introduced by Dehmelt [101]. The ion motion

$$x(t) = \{x_0\cos(\omega t)\}\cos\Omega t \tag{7.8}$$

is a composition of a forced oscillation following the driving rf field resulting in a so-called micro-motion at frequency Ω and a harmonic oscillation, the macro-motion, at a frequency

$$\omega = \frac{q}{\sqrt{8}}\Omega. \tag{7.9}$$

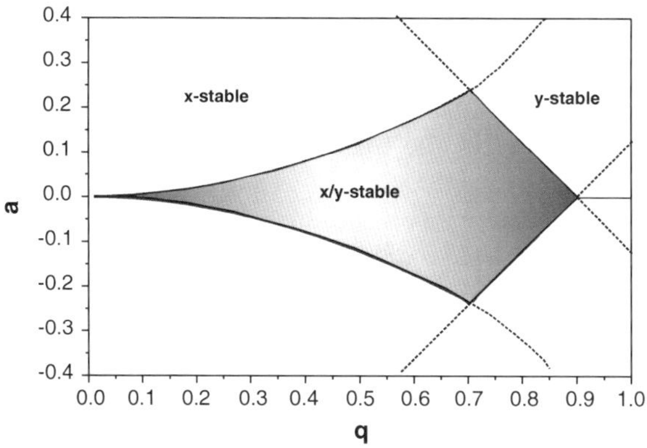

Fig. 7.3 Stability diagram of an rf trap. A stable confinement is achieved for operating points (a_o, q_o) within the *shaded* region

For typical parameters of an RFQ ion-beam cooler of $a = 0$, $q \approx 0.4$, and $\Omega = 1$ MHz an oscillation frequency of $\omega = 2\pi \cdot 76$ kHz is obtained for an ion with mass $A = 100$.

While the radial confinement in the RFQ is provided by an rf field as discussed above a static electric field gradient is superimposed for the axial confinement and to drag the ions through the buffer gas into a potential well at the end of the RFQ. To this end the electrode rods are segmented so that the desired axial field gradient can be created as indicated in Fig. 7.2. The RFQ is filled with a light inert buffer gas, typically high-purity helium, at a pressure of about 10^{-3} mbar so that the ions are thermalized in collisions with the buffer gas atoms within a few milliseconds. If the RFQ is operated as ion guide in transmission mode a higher pressure is typically used since the ions have to be cooled in a single pass, whereas a lower pressure is favourable for bunching. This buffer gas cooling technique is universal and thus suitable for radioactive ions. It can be applied as long as the mass of the buffer gas atoms is small compared to the mass of the ion to be cooled. The application of buffer gas cooling results in an emittance reduction of an injected ion beam to about $10\,\pi\,\text{mm}\times\text{mrad}$ or less and results in a longitudinal energy spread of less than 1 eV. The cooled ions can be accumulated in a potential well at the end of the rod structure where they form an ion bunch. An ion bunch is extracted by switching the voltages applied to the end electrodes of the RFQ. A typical bunch width of about 100 ns–1 μs is achieved. The bunch properties can be influenced by the trapping potential and the extraction voltages. There are several designs in use for RFQ cooler bunchers, in some of which the electrodes for the DC potential are separated from the rf electrodes [95].

7.5.2 Penning Traps

Penning traps are well described in the literature in several review articles, monographs, and textbooks [96, 97, 99, 102, 103]. The application of Penning traps for high-precision mass spectrometry has also been addressed in various works [1, 104, 105]. In this section the basics are briefly summarized and the relevant methods with respect to PTMS of radionuclides are introduced.

The concept of a Penning trap goes back to the works of F. Penning, who actually worked on more accurate pressure gauges but was credited with the name nonetheless [106], and Pierce [107]. However, the real breakthrough using Penning traps for high-precision experiments was related to Dehmelt's famous experiment on the anomalous magnetic moment of a free electron [101]. This ingenious experiment earned him the the Nobel prize in physics in 1989 together with Wolfgang Paul, another ion trap pioneer introducing the RF trap nowadays referred to as Paul trap, and Norman Ramsey, whose method of separate oscillatory fields has meanwhile also found its way into PTMS [108].

In a Penning trap a charged particle is confined by a strong homogeneous magnetic field $\mathbf{B} = B_0\mathbf{z}$ and a superimposed electrostatic quadrupolar field $\mathbf{E}$ created by a set of electrodes as shown in Fig. 7.4. A quadrupolar potential of the form

$$U(\rho, z) = \frac{U_0}{\rho_0^2 + 2z_0^2}(\rho^2 - 2z^2), \tag{7.10}$$

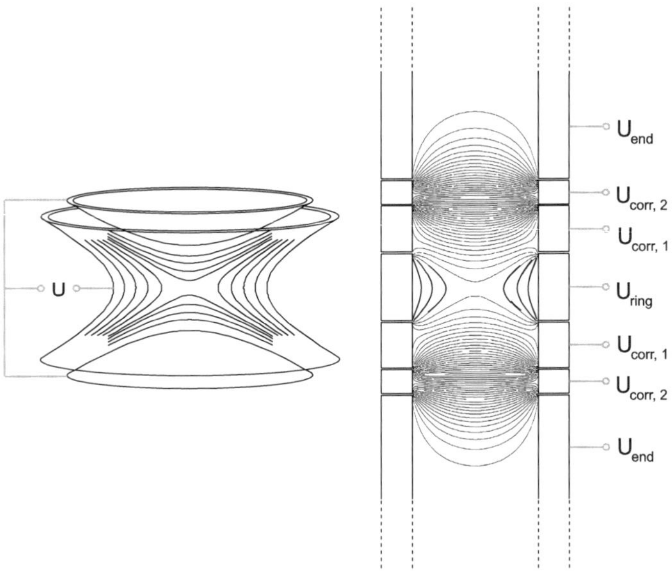

Fig. 7.4 Ideal Penning trap with hyperbolical electrodes (*left*) and the motion of a single ion in the trap (*right*)

where ρ_0 and z_0 are the trap dimensions as shown in Fig. 7.4, is created applying a voltage U between a ring electrode and two endcap electrodes of the trap. In practice traps with hyperbolic electrodes as well as with cylindrical electrodes are in use. With the characteristic dimension d

$$\rho_0^2 + 2z_0^2 = 4d^2 \tag{7.11}$$

the potential can be rewritten as

$$U(\rho, z) = \frac{U_0}{4d^2}(\rho^2 - 2z^2). \tag{7.12}$$

Solving the equations of motion for a single trapped ion leads to three independent eigenmotions: A harmonic oscillation in axial direction at frequency

$$\omega_z = \sqrt{\frac{qU}{md^2}} \tag{7.13}$$

and two radial motions, the cyclotron motion at the modified cyclotron frequency

$$\omega_+ = \frac{\omega_c}{2} + \sqrt{\frac{\omega_c^2}{4} - \frac{\omega_z^2}{2}} \tag{7.14}$$

and the magnetron motion at a frequency

$$\omega_- = \frac{\omega_c}{2} - \sqrt{\frac{\omega_c^2}{4} - \frac{\omega_z^2}{2}}. \tag{7.15}$$

The magnetron frequency is to first order independent of the mass of the stored ion. For typical parameters the hierarchy of the eigenfrequencies is $\omega_- < \omega_z < \omega_+$. In a Penning trap even single ions can be confined under well-defined, almost perturbation-free conditions in ultrahigh vacuum. This results in long observation times, in the case of radionuclides ultimately limited by the half-life, that lay the ground for experiments of unprecedented precision.

The mass m of an ions with charge q can be obtained from its free cyclotron frequency ω_c at which it orbits in a magnetic field of strength B.

$$\omega_c = 2\pi\nu_c = \frac{qB}{m}. \tag{7.16}$$

For PTMS in particular the following relations between the free cyclotron frequency and the eigenfrequencies in the trap are important:

$$\omega_c = \omega_+ + \omega_-; \tag{7.17}$$

$$\omega_z^2 = 2\omega_+\omega_-; \tag{7.18}$$

$$\omega_c = \omega_+\omega_- = \sqrt{\omega_+^2 + \omega_-^2 + \omega_z^2}; \tag{7.19}$$

where the latter relation is also known as the Brown-Gabrielse invariance theorem [102]. It is often used for the cyclotron frequency determination of stable nuclides. Equation 7.17 is typically employed for PTMS of radionuclides since ν_c can be measured directly using the so-called time-of-flight ion-cyclotron resonance (ToF-ICR) technique [109, 110]. This method is universal and fast and has become the method of choice for the cyclotron frequency measurement in on-line PTMS applications. A discussion on its applicability and limitations for high-precision PTMS in real Penning traps was recently presented by Gabrielse [111].

The basic principle of the ToF-ICR method is explained in the following. The ion's magnetron motion is excited by a dipolar rf field in the azimuthal plane of the Penning trap at the frequency ν_-. Thereby the ion, initially injected into the trap on axis, is prepared in a pure magnetron motion on a radius R_m. Subsequently a quadrupolar rf field at the frequency ν_{rf} is applied in the azimuthal plane. The frequency of this rf field is varied around the expected cyclotron frequency. If $\nu_{rf} = \nu_c$ the two radial motions are coupled and energy is transferred converting the magnetron motion into cyclotron motion. If the length and the amplitude of the rf pulse are chosen properly then the ion is prepared in a pure cyclotron motion with radius $R_c = R_m$ after the second excitation. This conversion is accompanied by an increase of the radial kinetic energy of the ion, i.e. the energy in the plane perpendicular to the magnetic field axis, giving rise to a resonance. This can be detected via the orbital magnetic moment $\boldsymbol{\mu}_\mathbf{B}$ of the ion that increases on resonance. This change can be detected via the time of flight of the ion through the magnetic field gradient of the superconducting solenoid of the Penning trap to an ion detector outside the magnet field [109, 110]. In the magnetic field gradient the ion experiences an accelerating axial force

$$\mathbf{F} = \nabla\boldsymbol{\mu}_\mathbf{B} \cdot \mathbf{B};\,, \tag{7.20}$$

so that resonantly excited ions reach the detector earlier than those off resonance. If the time of flight is recorded as a function of the excitation frequency ν_{rf} a resonance as shown in Fig. 7.5 is obtained. From a fit of the theoretically expected line shape [103, 110] to the resonance curve the cyclotron frequency is derived. In an ideal case the linewidth of such a time-of-flight cyclotron resonance is Fourier limited and thus approximately given as [104] $\delta\nu = 0.8 \cdot T_{RF}^{-1}$.

The resolving power $\Re$ of the ToF-ICR method scales according to

$$\Re = \frac{\nu}{\Delta\nu} = \frac{m}{\Delta m} \propto \nu_c T_{rf}. \tag{7.21}$$

It is proportional to the product of the excitation time T_{RF} and the cyclotron frequency. For radionuclides the excitation time is ultimately limited by the half-life

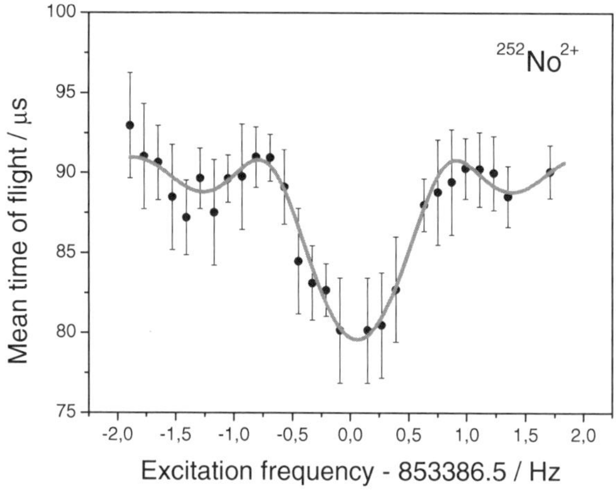

Fig. 7.5 Cyclotron resonance of ^{252}No adapted from [112]. The *line* represents a fit of the theoretically expected lineshape to the data points

of the investigated particle. An optimum is then reached for $T_{rf} \approx 3T_{1/2}$ [113]. The statistical uncertainty of the frequency determination is given by the relation [114]

$$\frac{\delta m}{m} \propto \frac{1}{\Re\sqrt{N_{det}}} \propto \frac{m}{qBT_{rf}\sqrt{N_{det}}} \tag{7.22}$$

where N_{det} is the number of detected ions after excitation and extraction from the trap. Typically at least about 30 ions have to be accumulated to obtain a resonance curve [15, 112]. In on-line PTMS the accuracy of cyclotron frequency measurements is often limited by statistics due to the low production rates of the radionuclides of interest. For an excitation time of 1 s a resolving power of about $R \approx 1{,}000{,}000$ can be reached for an ion of mass $A = 100$ in a magnetic field of $B = 7\ T$. The corresponding statistical uncertainty that can be reached for a nuclide with a half-life of about one second for a number of 1,000 detected ions is then $(\frac{\delta m}{m})_{stat} = 3 \times 10^{-8}$.

A higher accuracy can be obtained applying different excitation schemes to the ToF-ICR method such as an octupolar excitation [115–117] or a Ramsey-type excitation [108]. Alternatively highly charged ions can be utilized as recently demonstrated with the TITAN spectrometer at the TRIUMF facility in Vancouver [118]. Another approach that has recently been developed is the so-called phase sensitive PI-ICR method [119] where the ion motion is imaged onto a spatially resolving microchannel plate detector. Then the (radial) eigenfrequencies can be measured via a phase measurement. In this way one can overcome the Fourier limit of the ToF-ICR method. With SHIPTRAP a gain in resolving power by a factor of forty and a gain in precision by a factor of about five has already been demonstrated with this novel method [119]. This allows measurements with a high accuracy in rather short time and thus extends the reach of PTMS to shorter-lived nuclides.

7.5.3 Contributions to the Systematic Uncertainty in PTMS

For high-precision mass measurements in addition to the statistical uncertainty several systematic uncertainties have to be considered. The most relevant systematic effects for PTMS of radionuclides have been discussed for example by Bollen et al. [104] following an analysis for electrons by Brown and Gabrielse [102]. For a high-precision cyclotron frequency measurement the following effects can lead to cyclotron frequency shifts

- a deviation of the electric trapping field from a pure quadrupolar field
- spatial magnetic field inhomogeneities
- temporal fluctuations of the confining fields
- a tilt of the electrical field axis relative to the magnetic field axis
- ion-ion interaction if more than one ion is trapped at a time.

7.5.3.1 Temporal Magnetic Field Changes

The magnetic field strength B has to be known precisely in order to derive the mass with high precision from the measured cyclotron frequency according to Eq. 7.16. However, the magnetic field also changes in time. An intrinsic decay due to flux creep that is approximately linear over time on the order of 10^{-8}/h is accompanied by short-term fluctuations. These can arise for example from changes of parameters like the ambient temperature that lead to changes in the magnetic susceptibility of the materials inside the magnet bore and from changes of the pressure in the liquid helium cryostat of the solenoid, and thus the boiling point of the liquid helium. Moreover, external magnetic stray fields that are often unavoidable at accelerator facilities can play a role.

An example of the magnetic field stability is presented in Fig. 7.6 where the magnetic field evolution of the solenoid of the SHIPTRAP mass spectrometer is shown. The general trend shows a linear decay of the magnetic field over time with

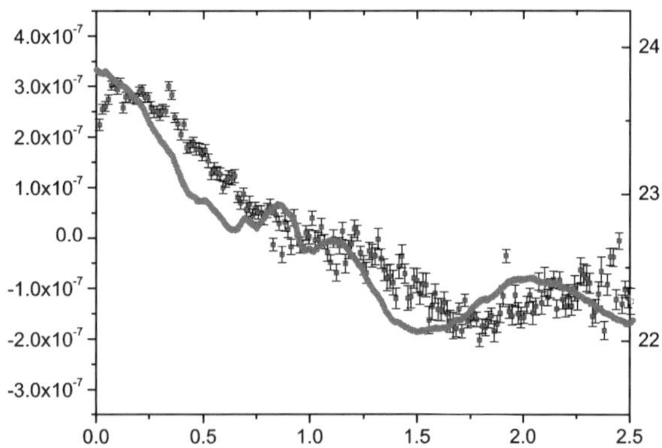

Fig. 7.6 Relative magnetic field changes in the SHIPTRAP solenoid over a period of 2.5 days. The magnetic field was monitored by cyclotron frequency measurements of $^{85}Rb^{+}$ ions. The *red line* represents the temperature in the bore of the superconducting solenoid

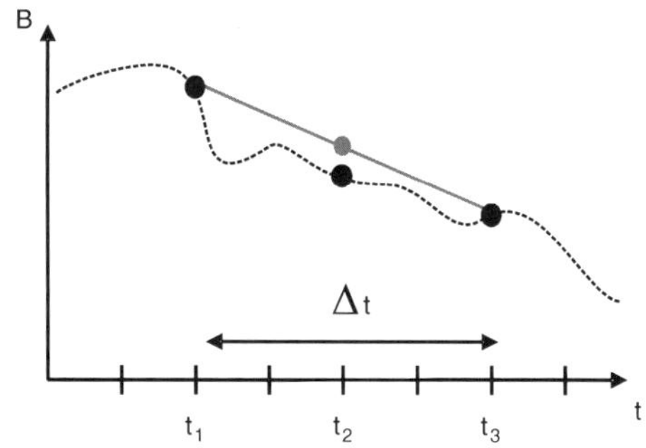

Fig. 7.7 Principle of the magnetic field calibration using a linear interpolation. The actual magnetic field is represented by the *dashed line*. The *black dots* marked the time where a calibration measurements is performed. The interpolation is indicated by the *full line*

superimposed fluctuations on shorter time scales. These are mainly correlated to changes of the temperature measured inside the bore of the solenoid.

The temporal fluctuations of the magnetic field are accounted for in PTMS by calibrating the magnetic field via a cyclotron frequency measurement of a reference ion with well-known mass that is chosen such that its mass-to-charge ratio is similar to that of the ion of interest. Then both ions are trapped at the same positions in the trap and sense the same field. Calibration measurements are then performed before (t_1) and after (t_3) the frequency measurement of the ion of interest at time t_2. The scheme is qualitatively illustrated in Fig. 7.7. From the measured cyclotron frequencies a linear extrapolation to the time of the actual measurement t_2 is performed providing the frequency $\nu_c^{ref}(t_2)$ according to the relation

$$\nu_c^{ref}(t_2) = \nu_c^{ref}(t_1) + \frac{\nu_c^{ref}(t_3) - \nu_c^{ref}(t_1)}{t_3 - t_1}(t_2 - t_1), \tag{7.23}$$

Thereby the linear decay of the magnetic field is corrected. The effect of nonlinear magnetic field fluctuations is considered by a systematic uncertainty that depends on the measurement time needed to acquire a single cyclotron resonance. This uncertainty due the difference between the actual magnetic field value and the interpolated values can be estimated experimentally by an analysis as described in [120]. In the example of SHIPTRAP mentioned above a time dependent uncertainty due to nonlinear magnetic field fluctuations was determined [121] to be on the order of

$$\sigma_B \approx 2 \times 10^{-9}/min \times \Delta t, \tag{7.24}$$

where Δt is the time interval between consecutive calibration measurements in minutes. This uncertainty is quadratically added to the statistical uncertainty of the frequency measurements. Since part of the temporal magnetic field fluctuations are due to changes of the temperature in the bore of the solenoid and the pressure in the liquid helium dewar of the solenoid an active stabilization system for both parameters can improve the long time stability of the magnetic field. At SHIPTRAP such a system has recently been implemented and reduced the uncertainty related to temporal fluctuation of magnetic field to [122].

$$\sigma_{\omega B} \approx 2 \times 10^{-11}/min \times \Delta t. \tag{7.25}$$

The primary result of a PTMS experiment following this calibration procedure is thus the frequency ratio r between the cyclotron frequencies of the ion of interest and the reference ion

$$r = \frac{\nu_c(t_2)}{\nu_{c,ref}(t_2)}. \tag{7.26}$$

The uncertainty δr of a measured frequency ratio r is comprised by the statistical uncertainty σ_{stat} and a systematic uncertainty σ_{syst} that includes for example the uncertainty due to nonlinear fluctuations of the magnetic field σ_B as discussed above

$$\delta r = \sqrt{\sigma_{stat}^2 + \sigma_{syst}^2} = \sqrt{\sigma_B^2 + \sigma_{stat}^2 + \sigma_{res}^2}, \tag{7.27}$$

and σ_{res} contains all other systematic uncertainties.

The atomic mass is obtained from the measured frequency ratio according to the following relation using the most recent and hence most accurate mass value of the reference ion m_{ref}

$$m = \frac{q}{q_{ref}} r(m_{ref} - q_{ref} m_e) + q m_e, \tag{7.28}$$

where m_e is the electron mass and q and q_{ref} are the charge states of the ion of interest and the reference ion. For the typical uncertainties achieved for radionuclides the binding energy of the electrons can be neglected. If carbon cluster ions are used as mass reference [120] absolute mass measurements can be performed since the atomic mass is defined via the mass of ^{12}C.

7.5.3.2 Spatial Homogeneity of the Confining Fields

The electric trapping potential of a real trap deviates from a pure quadrupole potential even for a hyperbolic electrode geometry due to several reasons, for example, machining imperfections, misalignments, the truncation of electrodes, and additional holes required for the injection and extraction of particles. The impact of such perturbations can be compensated to some extent by additional correction electrodes. In a similar way a harmonic trapping potential can be realized using a cylindrical trap geometry with open endcaps [123]. Cylindrical traps are simpler to machine and can be pumped more efficiently than traps with hyperbolic electrodes. In such a cylindrical trap a careful tuning of the trapping voltages (applied to the correction electrodes) is particularly important. In addition, patch potentials on the electrode surfaces can play a role. Those are generally avoided by gold plating of the electrodes.

In general a perfect harmonic potential is only realized in a small region around the trap center whereas for larger distances from the trap center higher order multipoles of the trapping field become significant. The deviation from the ideal quadrupole field thus also depends on the amplitude of the ion motion. Therefore the application of ion cooling to constrain the amplitudes of the ion motion to a region close to trap center is crucial for high precision measurements.

The deviation of the real trapping potential from a pure quadrupole is expressed by the coefficients C_i of a multipole expansion of the real potential ϕ_{real}:

$$\phi_{real} = \frac{U_0}{2} \sum_i C_i (\frac{r}{d})^i P_i(cos\theta) \tag{7.29}$$

For an ideal quadrupole potential $C_2 = -2$, while all other coefficients vanish. In real traps typically the next two higher order terms, the octupole (C_4) and dodecapole (C_6) term, play a role. As long as the radial symmetry is not broken all odd expansion coefficients vanish. The cyclotron frequency shift in case of non-vanishing C_4 and C_6 terms is given [102] by

$$\Delta\omega_c^{electr} = C_4 \frac{3}{4}(\rho_+^2 - \rho_-^2)\frac{\omega_z^2}{\omega_c^2} + \frac{15}{4}\frac{C_6}{d^2}[\rho_z^2(\rho_-^2 - \rho_+^2) - (\rho_-^4 - \rho_+^4)]. \tag{7.30}$$

It depends on the amplitudes of the ion motion but is approximately mass independent. The size of the different multipole coefficients can be estimated for a given electrode geometry by fitting Eq. 7.29 to the numerically calculated potential in a certain region around the trap center.

Besides electric field anharmonicities also spatial magnetic field inhomogeneities lead to frequency shifts. The intrinsic homogeneity of a commercially available state-of-the-art superconducting solenoid magnet is on the order of $\Delta B/B = 0.1 - 1$ ppm/cm^3 if no special measures are taken. The actual magnetic field B sensed by the trapped ions can be additionally modified by materials introduced into the magnetic field such as the trap electrodes or the trap vacuum tube due to a non-zero magnetic susceptibility. The modification of the magnetic field around the trap center can be described by

$$\frac{\delta B}{B_0} = \beta_2[(z^2 - \rho^2/2)z - z\rho], \tag{7.31}$$

where B_0 is the magnetic field at the trap center, and z and ρ are the motional amplitudes, since a linear term vanishes. The resulting cyclotron frequency shift is then given by [102]

$$\Delta\omega_c^{magn} = \omega_c\beta_2[z^2 - \frac{\rho_+^2}{4}(1 + \frac{\omega_c}{\omega_+ - \omega_-}) - \frac{\rho_-^2}{4}(1 + \frac{\omega_c}{\omega_+ - \omega_-})] \tag{7.32}$$

This shift is proportional to the cyclotron frequency and hence, depends on the mass of the ion. This effect can be minimized by using materials with low magnetic susceptibility for the trap construction and reducing the amount of material introduced. Moreover, the design should be such that the radial symmetry is maintained.

7.5.3.3 Trap Misalignments

Misalignments of the individual trap electrodes can be minimized by a careful mechanical trap design and thus do not play a significant role in particular for cylindrical traps. However, a misalignment of the electrical field axis with respect to the magnetic field axis is much harder to avoid in practice. A tilt of the axis by an angle θ results in a shift of the eigenfrequencies [104] by

$$\Delta\omega_c{}^{tilt} \approx \frac{9}{4}\omega_- \sin^2\theta, \tag{7.33}$$

where ω_- is the magnetron frequency. However, all eigenfrequencies shift in the same way. This shift is to first order mass independent. By a careful alignment of the trap in the magnet bore this tilt angle is minimized. Typical values that can be achieved are on the order of 1 mrad.

7.5.3.4 Ion-Ion Interactions

In experiments with radioactive particles contaminant ions can be created either in the production stage by unwanted nuclear reactions, in the ion preparation process, or by the radioactive decay of the investigated particle itself. The presence of of unwanted ions in the trap can result in a shift the cyclotron frequency due to the Coulomb interaction between different ions. In general this effect is avoided if a measurement is performed with only a single trapped ion at a time, however this is not always possible in on-line experiments. If the different ions in the trap have the same mass-to-charge ratio then the field acts on the center of mass and no frequency shift occurs. In the case of contaminant ions with different mass-to-charge ratio one resonance shifted in frequency is observed if the two different cyclotron frequencies cannot be resolved. If the mass difference of the ions is large enough that the individual resonances are resolved both resonances shift to lower frequencies and the frequency shift is proportional to the number of trapped ions [124]. Therefore, in a precision measurement possible frequency shifts are revealed by an analysis as described in [120]. In this method the cyclotron frequency is determined as a function of the number of ions in the trap at a time by dividing the data into subsets with similar statistics. Then the cyclotron frequency is linearly extrapolated to one ion in the trap taking into account the ion detection efficiency. Frequency shifts due to image charge effects do not play a role for radionuclides yet as they are typically two or three orders of magnitude smaller than the statistical uncertainty.

7.5.4 *Applications of High-Precision Mass Measurements*

In this section some recent examples of mass measurements of radionuclides performed by PTMS are given. They illustrate the versatility and the present performance of PTMS for radionuclides. The first example is related to mass measurements in the context of nuclear astrophysics. The second example addresses experiments in the region of the heaviest elements.

In the stellar nucleosynthesis of heavy elements in the universe various scenarios and processes are discussed. The slow and rapid neutron capture process (s and r process) are expected to be responsible for the creation of most of the heavy elements beyond iron [21, 22]. The rapid proton-capture process (rp process) is believed to be responsible for so-called X-ray bursts [20] and may also take place in a neutrino driven wind in core collapse supernovae [125]. Type I x-ray bursts can occur when a neutron star accretes matter from an expanded companion star and last for a few seconds. For a detailed astrophysical modelling of the various processes under different conditions several nuclear properties such as half-lives and masses are required input parameters. In order to eliminate or at least minimize the impact of nuclear physics-related uncertainties on the astrophysical conclusions accurate data of the relevant properties are needed for many, often very exotic, radionuclides.

The predicted pathway of the r process [21] proceeds in a region of very neutron-rich nuclides some of which will only be accessible at next generation radioactive beam facilities such as FAIR in Darmstadt [126], FRIB at Michigan State university [127], or SPIRAL 2 at GANIL [128].

The rp process consists of a sequence of proton capture reactions and subsequent beta decays in competition with photodisintegration. The predicted rp process pathway [20] reaches out to exotic nuclides at the proton drip line that can already be accessed for practically all nuclides relevant to the rp process at present radioactive beam facilities. However, some of them are short-lived and can only be produced in minute quantities. The impact of masses on the r and rp process was discussed in several articles [19, 129, 130]. In these investigations it was found that masses are required with an uncertainty on the order of 10 keV or below. For so-called ‘waiting points’ a higher precision is often needed.

Waiting points in the rp process are rather long-lived nuclides encountered when the proton capture reaches the drip line so that the process only proceeds towards heavier nuclei after the β decay of the waiting point nuclide (or if it can be bypassed via two-proton capture). Thus waiting point nuclides have a large impact on the characteristics of an x-ray burst, for example the corresponding light curve. Depending on the density and temperature conditions in the stellar environment the effective half-life of a waiting point, i.e. the time until the rp process proceeds, can be altered significantly. An example is shown in Fig. 7.8 for the waiting point ^{68}Se to illustrate the impact of precise mass values. The effective lifetime of ^{68}Se as a function of temperature was determined by a local network calculation based on mass measurements with LEBIT [131]. The effective lifetime previous to this experiment was 10.7(6.8) s so that ^{68}Se would not be a strong waiting point within the

Fig. 7.8 Impact on the mass on the effective half-life of the rp process waiting point ^{68}Se (adapted from [131]) for different temperatures

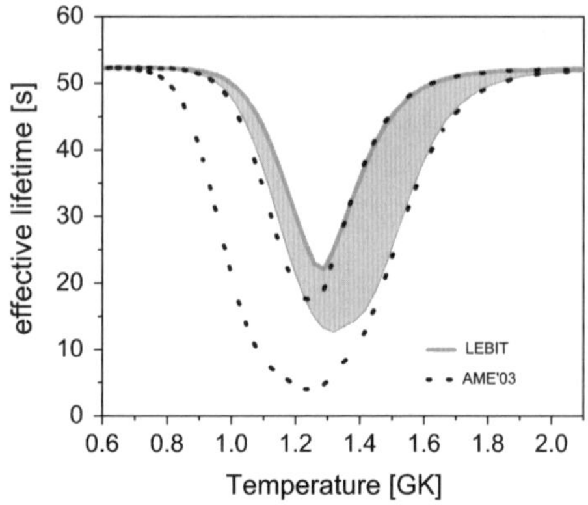

experimental uncertainty. Based on the new accurate mass values a lifetime of 17.4(4.5) s was obtained indicating that ^{68}Se is indeed a strong waiting point for all relevant conditions.

In recent years the masses of many radionuclides relevant for nuclear astrophysics have have been measured with high accuracy by direct mass measurements using PTMS at various radioactive beam facilities [132–139]. In certain areas, for example around ^{84}Mo, the estimated masses have been found to deviate from the new data by as much as 1 MeV [135]. Based on the new data improved calculations were performed that contribute to a better understanding of nucleosynthesis.

An example for mass measurements in the context of nuclear structure studies is related to direct mass measurements of the heaviest elements. Superheavy nuclides owe their very existence to nuclear shell effects that stabilize them against the disintegration by spontaneous fission due to the strong Coulomb repulsion in nuclides with high-Z. The fission barrier calculated according to macroscopic nuclear models [14] vanishes for nuclides with $Z \approx 104$. However, a finite fission barrier is obtained for superheavy elements when including nuclear shell effects. Superheavy elements are predicted to inhabit an 'island of stability' by different theoretical models that, however, still disagree on the exact location an extension of this island. The predicted center is located at $Z = 114$, 120, or 126 and $N = 172$ or 184 depending on the parametrization [14, 140–142]. Experimentally this concept is supported by several new elements that have been discovered in recent years in different laboratories worldwide with the latest claims reaching up to element $Z = 118$ [72, 74, 143–147].

However, binding energies that are at the heart of superheavy nuclides have experimentally been out of reach for a long time. The transuranium nuclides have not been accessible by PTMS either due to their low production rates of due to to their production scheme that requires additional preparation steps. Isotopes of the elements above fermium ($Z = 100$) can only be produced by fusion-evaporation reactions with heavy-ion beams typically from ^{40}Ar to ^{70}Zn and either lead and bismuth targets or actinide targets from uranium to californium. However, the production rates are very low even for primary beam intensities of almost 10^{13} particles per second.

About four ions per second can be delivered to a PTMS setup for ^{254}No ($Z = 102$), but the rate drops steeply to about one atom per three days for element flerovium ($Z = 114$) [144]. The actual number of ions available for PTMS is even lower due to the losses during the slowing down and beam preparation procedures resulting in overall efficiencies on the order of a few percent. Such low rates are a major challenge for PTMS since they result in very long measurement times of a few days over which all relevant parameters such as the trapping fields have to be stable.

Nonetheless, recently direct mass measurements of several nobelium and lawrencium isotopes have been performed with the Penning trap mass spectrometer SHIPTRAP at GSI in Darmstadt, Germany [15, 112]. In case of the nuclide ^{256}Lr the yield was on the order of two particles per minute. For a cyclotron resonance about 50 ions were detected statistically distributed over a period of about four days. The masses have been determined with an accuracy of down to about 10 keV and now provide reliable anchor points in the region of the heaviest elements.

New elements in this region have so far been identified by the observation of α decay chains that connect new nuclides to well known nuclides. Prior to the SHIPTRAP experiments also the masses of nuclides in this region were determined in this way, i.e. indirectly from the measured α decay energy and the known mass of the daughter nuclide. However, this method works only well if the α decay occurs between nuclear ground states as for even-even nuclides, else it requires a detailed knowledge about the nuclear level scheme. In addition, the mass uncertainties accumulate going along the chain.

In addition to providing anchor points direct mass measurements allow mapping the strength of nuclear shell effects as demonstrated by SHIPTRAP [15]. From accurate experimental mass values the strength of the shell effects can be derived according to Eq. 7.34. The difference of several nuclidic masses provides the so-called shell gap parameter $\delta_{2n}(N, Z)$ that amplifies the visibility of a shell closure.

$$\delta_{2n}(N, Z) = S_{2n}(N, Z) - S_{2n}(N + 2, Z) \tag{7.34}$$

$$= 2ME(N, Z) + ME(N - 2, Z) + ME(N + 2, Z), \tag{7.35}$$

where $S_{2n}(N, Z)$ is the two-neutron separation energy. The binding energy of neighboring nuclides changes most significantly when a closed shell is reached since the nuclides are most strongly bound for closed shells. Such differences can also be visualized by the other mass differences such as the two-nucleon separation energy $S_{2n/2p}(N, Z)$. Since the pairing between nucleon results in a change of the binding energy that would result in an odd even staggering of one-nucleon separation energies two nucleon separation energies are used to visualize trends in the nuclear structure evolution.

Figure 7.9 shows the shell gap parameter for nobelium isotopes around neutron number $N = 152$. The effect of a closed neutron shell is clearly visible by the peak in $\delta_{2n}(N, Z)$. The strength of this shell closure in a region of deformed nuclei is much weaker though than for example in the spherical doubly magic nucleus ^{208}Pb where the shell gap is about 8 MeV large. The accurate experimental data provide a sensitive

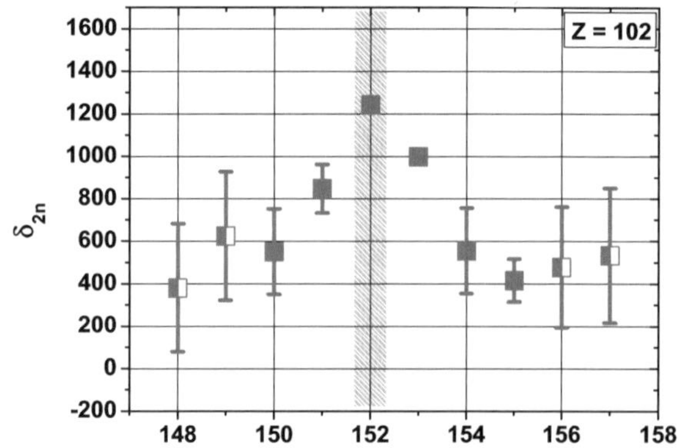

Fig. 7.9 Neutron shell gap for neutron numbers around $N = 152$ (adapted from [15]). Experimental data are given by the *red points*. Filled symbols are exclusively based on SHIPTRAP data [15, 112]

benchmark for nuclear models that are used to describe superheavy elements and can thus be used to improve predictions of the island of stability.

7.6 Conclusions

PTMS has proven to be a versatile method that provides mass values of unprecedented accuracy even for radionuclides with short half-lives and with production rates as low as a few particles per second. Since the installation of the first Penning trap at a radioactive beam facility almost 30 years ago a new era in precision mass spectrometry has begun and resulted in many exciting scientific results contributing to great discoveries in various disciplines. Penning traps dominate the landscape in mass spectrometry to date and with ongoing and future devolvements they will advance further towards higher precision, higher selectivity, and higher sensitivity. This will pave the way to extend the reach to more exotic nuclides and lead to new breakthroughs in physics.

References

1. K. Blaum, High-accuracy mass spectrometry with stored ions. Phys. Rep. **425**, 1 (2006)
2. D. Lunney, J.M. Pearson, C. Thibault, Recent trends in the determination of nuclear masses. Rev. Mod. Phys. **75**, 1021 (2003)
3. S. Rainville, J.K. Thompson, D.E. Pritchard, An ion balance for ultra-high-precision atomic mass measurements. Science **303**, 334 (2004)
4. R.S. Van Dyck Jr., S.L. Zafonte, S. Van Liew, D.B. Pinegar, P.B. Schwinberg, Ultraprecise atomic mass measurement of the α particle and^4 He. Phys. Rev. Lett. **92**, 220802 (2004)
5. W. Shi, M. Redshaw, E.G. Myers, Atomic masses of ^{32,33}S, ^{84,86}Kr, and ^{129,132}Xe with uncertainties below 0.1 ppb. Phys. Rev. A **72**, 022510 (2005)
6. G. Gabrielse et al., Precision mass spectroscopy of the antiproton and proton using simultaneously trapped particles. Phys. Rev. Lett. **82**, 3198 (1999)

7. I.S. Towner, J. Hardy, The evaluation of V_{ud} and its impact on the unitarity of the Cabibbo-Kobayash-i-Maskawa quark-mixing matrix. Rep. Prog. Phys. **73**, 046301 (2010)
8. J. Britz, A. Pape, M.S. Antony, Coefficients of the isobaric mass equation and their correlations with various nuclear parameters. At. Data Nucl. Data Tables **69**, 125 (1998)
9. W. Heisenberg, Über den Bau der Atomkerne. Z. Phys. **77**, 1 (1932)
10. E.P. Wigner, in *Proceedings of the Robert A. Welch Conferences on Chemical Research*, vol 1 (Robert A. Welch Foundation, Houston, 1957), p. 67
11. O. Sorlin, M.-G. Porquet, Evolution of the N = 28 shell closure: a test bench for nuclear forces. Physica Scripta **T152**, 014003 (2013)
12. I. Tanihata, H. Savajols, R. Kanungo, Recent experimental progress in nuclear halo structure studies. Prog. Part. Nucl. Phys. **68**, 215 (2013)
13. W. Noertershaeuser, T. Neff, R. Sanchez, I. Sick, Charge radii and ground state structure of lithium isotopes: Experiment and theory reexamined. Phys. Rev. C **84**, 024307 (2011)
14. Z. Patyk, A. Sobiczewski, Ground-state properties of the heaviest nuclei analyzed in a multidimensional deformation space. Nucl. Phys. A **533**, 132 (1991)
15. E. Minaya Ramirez et al., Direct mapping of nuclear shell effects in the heaviest elements. Science **337**, 1207 (2012)
16. M. Thoennessen, Reaching the limits of nuclear stability. Rep. Prog. Phys. **67**, 1187 (2004)
17. M. Pfützner, M. Karny, L.V. Grigorenko, K. Riisager, Radioactive decays at limits of nuclear stability. Rev. Mod. Phys. **84**, 567 (2012)
18. R.G. Lovas, R.J. Liotta, A. Insolia, K. Varga, D.S. Delion, Microscopic theory of cluster radioactivity. Phys. Rep. **294**, 265 (1998)
19. H. Schatz, The importance of nuclear masses in the astrophysical rp-process. Int. J. Mass Spectrom. **251**, 293 (2006)
20. H. Schatz et al., rp-process nucleosynthesis at extreme temperature and density conditions. Phys. Rep. **294**, 167 (1998)
21. F.-K. Thielemann et al., Operation of the r-process and cosmochronology. Phys. Rep. **227**, 269 (1993)
22. F. Käppeler, The origin of the heavy elements: The s process. Prog. Part. Nucl. Phys. **43**, 419 (1999)
23. S.A. Eliseev, YuN Novikov, K. Blaum, Search for resonant enhancement of neutrinoless double-electron capture by high-precision Penning-trap mass spectrometry. J. Phys. G **39**, 124003 (2012)
24. J.J. Gomez-Cadenas, J. Martin-Albo, M. Mezzetto, F. Monrabal, M. Sore, The search for neutrinoless double beta decay. Riv. Nuovo Cim. **35**, 29 (2012)
25. J. Wolf for the KATRIN collaboratio, The KATRIN neutrino mass experiment. Nucl. Instrum. Meth. A **623**, 442 (2010)
26. G. Audi et al., The AME 2012 atomic mass evaluation. Chin. Phys. C **36**(12), 1603–2014 (2012)
27. G. Audi, The history of nuclidic masses and of their evaluation. Int. J. Mass Spectrom. **251**, 85 (2006)
28. J.J. Thomson, Cathode rays. Phil. Mag. **3**, 293 (1897)
29. J.J. Thomson, On the masses of the ions in gases at low pressures. Phil. Mag. **5**, 547 (1899)
30. J.J. Thomson, On rays of positive electricity. Phil. Mag. **13**, 561 (1907)
31. J.J. Thomson, Further experiments on positive rays. Phil. Mag. **24**, 209 (1912)
32. J.J. Thomson, Multiply-Charged atoms. Phil. Mag. **24**, 668 (1912)
33. F.W. Aston, *Mass Spectra and Isotopes*, 2nd edn. (Edward Arnold and Co., London, 1942)
34. A.J. Dempster, The energy content of the heavy nuclei. Phys. Rev. **53**, 869 (1938)
35. J. Mattauch, A double-focusing mass spectrograph and the masses of N15 and O18. Phys. Rev. **50**, 617 (1936)
36. R. Klapisch, R. Prieels, C. Thibault, A.M. Poskanzer, C. Rigaud, E. Roeckl, On-line mass-spectrometric measurement of the masses of neutron-rich sodium isotopes. Phys. Rev. Lett. **31**, 118 (1973)
37. E. Kugler, The ISOLDE facility. Hyperfine Int. **129**, 23 (2000)

38. L.G. Smith, C.C. Damm, Mass synchrometer. Rev. Sci. Instr. **27**, 638 (1956)
39. D. Lunney, D. Vieira, G. Audi et al., Mass measurements of the shortest-lived nuclides a la MISTRAL. Int. J. Mass Spectrom. **251**, 286 (2006)
40. C. Gaulard, C. Bachelet, G. Audi et al., Mass measurements of the exotic nuclides ^{11}Li and ^{11,12}Be performed with the Mistral spectrometer. Nucl. Phys. A **826**, 1 (2009)
41. G. Gärtner, E. Klempt, A direct determination of the proton-electron mass ratio. Z. Physik A **287**, 1 (1978)
42. G. Bollen, P. Dabkiewicz, P. Egelhof, T. Hilberath, H. Kalinowsky, F. Kern, H. Schnatz, L. Schweikhard, H. Stolzenberg, R.B. Moore, H.-J. Kluge, G.M. Temmer, G. Ulm,and the ISOLDE Collaboration, First absolute mass measurements of short-lived isotopes. Hyperfine Interact. **38**, 793 (1987)
43. M. Mukherjee, D. Beck, K. Blaum, G. Bollen, J. Dilling, S. George, F. Herfurth, A. Herlert, A. Kellerbauer, H.-J. Kluge, S. Schwarz, L. Schweikhard, C. Yazidjian, ISOLTRAP: An on-line Penning trap for mass spectrometry on short-lived nuclides. Eur. Phys. J. A **35**, 1 (2008)
44. M. Smith et al., First Penning-trap mass measurement of the exotic Halo nucleus ^{11}Li. Phys. Rev. Lett. **101**, 202501 (2008)
45. R. Ringle et al., High-precision Penning trap mass measurements of ^{37,38}Ca and their contributions to conserved vector current and isobaric mass multiplet equation. Phys. Rev. C **75**, 055503 (2007)
46. J. Clark et al., Improvements in the injection system of the Canadian Penning trap mass spectrometer. Nucl. Instrum. Methods B **204**, 487 (2003)
47. T. Eronen et al., JYFLTRAP: a Penning trap for precision mass spectroscopy and isobaric purification. Eur. Phys. J. A **48**, 46 (2012)
48. M. Block et al., Towards direct mass measurements of nobelium at SHIPTRAP. Eur. Phys. J. D **45**, 39 (2007)
49. G. Bollen et al., Beam cooling at the low-energy-beam and ion-trap facility at NSCL/MSU. Nucl. Instrum. Methods A **532**, 203 (2004)
50. J. Dilling et al., Mass measurements on highly charged radioactive ions, a new approach to high precision with TITAN. Int. J. Mass. Spectrom. **251**, 198 (2006)
51. J. Ketelaer et al., TRIGA-SPEC: A setup for mass spectrometry and laser spectroscopy at the research reactor TRIGA Mainz. Nucl. Instrum. Meth. A **594**, 162 (2008)
52. V.S. Kolhinen et al., MLLTRAP: A Penning trap facility for high-accuracy mass measurements. Nucl. Instrum. Meth. B **266**, 4547 (2008)
53. D. Rodriguez et al., MATS and LaSpec: High-precision experiments using ion traps and lasers at FAIR. Eur. Phys. J. ST **183**, 1 (2010)
54. A.E. Cameron, D.F. Eggers Jr, An ion 'velocitron'. Rev. Sci. Instr. **19**, 605 (1948)
55. W.R. Plass et al., Isobar separation by time-of-flight mass spectrometry for low-energy radioactive ion beam facilities. Nucl. Instrum. Meth. B **266**, 4560 (2008)
56. P. Schury et al., Multi-reflection time-of-flight mass spectrograph for short-lived radioactive ions. Eur. Phys. J. A **42**, 343 (2009)
57. R. Wolf et al., A multi-reflection time-of-flight mass separator for isobaric purification of radioactive ion beams. Hyperfine Int. **199**, 115 (2011)
58. R. Wolf et al., On-line separation of short-lived nuclei by a multi-reflection time-of-flight device. Nucl. Instrum. Meth. A **686**, 82 (2012)
59. B. Franzke, The heavy ion storage and cooler ring project ESR at GSI. Nucl. Instrum. Meth. B **24/25**, 18–25 (1987)
60. B. Franzke, H. Geissel, G. Münzenberg, Mass and lifetime measurements of exotic nuclei in storage rings. Mass Spectrom. Rev. **27**, 428 (2008)
61. YuA Litvinov, H. Geissel, R. Knöbel, B. Sun, H. Xu, Direct Mass Measurements of Exotic Nuclei in Storage Rings. Acta Physica Polonica B **41**, 511 (2010)
62. U. Koester, Intense radioactive-ion beams produced with the ISOL method. Eur. Phys. J. A **15**, 255 (2002)
63. V.N. Fedoseyev, G. Huber, U. Koster, J. Lettry, V.I. Mishin, H. Ravn, V. Sebastian, The ISOLDE laser ion source for exotic nuclei. Hyperfine Int. **127**, 409 (2000)

64. M. Huyse, R. Raabe, Radioactive ion beam physics at the cyclotron research centre Louvain-la-Neuve. J. Phys. G **38**, 024001 (2011)
65. P.G. Bricault, M. Dombsky, P.W. Schmor, G. Stanford, Radioactive ion beams facility at TRIUMF. Nucl. Instrum. Meth. B **126**, 231 (1997)
66. A. Navin, F. de Oliveira Santos, P. Roussel-Chomaz, O. Sorlin, Nuclear structure and reaction studies at SPIRAL. J. Phys. G **38**, 024004 (2011)
67. J. Arje et al., The ion guide isotope separator on-line, IGISOL. Nucl. Instrum. Meth. A **247**, 431 (1986)
68. J. Äystö, T. Eronen, A. Jokinen, A. Kankainen, I.D. Moore, H. Penttilä(eds.), An IGISOL Portrait—selected contributions. Eur. Phys. J. A **48** (2012)
69. H. Sakurai, RI beam factory project at RIKEN. Nucl. Phys. A **805**, 526c (2008)
70. H. Geissel et al., The GSI projectile fragment separator (FRS): a versatile magnetic system for relativistic heavy ions. Nucl. Instrum. Meth. B **70**, 286 (1992)
71. D.J. Morrissey, B.M. Sherrill, M. Steiner, A. Stolz, I. Wiedenhöver, Commissioning the A1900 projectile fragment separator. Nucl. Instrum. Meth. B **204**, 90 (2003)
72. S. Hofmann, G. Münzenberg, The discovery of the heaviest elements. Rev. Mod. Phys. **72**, 733 (2000)
73. A.V. Yeremin, The kinematic separator VASSILISSA performance and experimental results. Nucl. Instrum. Meth. A **350**, 608 (1994)
74. Yu. Ts, Oganessian, Heaviest nuclei from ^{48}Ca-induced reactions. J. Phys. G **34**, R165 (2007)
75. V. Ninov, K.E. Gregorich, C.A. McGrath, The Berkeley Gas-filled separator. AIP Conf. Proc. **455**, 704 (1998)
76. K. Morita et al., RIKEN isotope separator on-line GARIS/IGISOL. Nucl. Instrum. Meth. B **70**, 220 (1992)
77. M. Leino et al., Gas-filled recoil separator for studies of heavy elements. Nucl. Instrum. Meth. B **99**, 653 (1995)
78. A. Semchenkov et al., The TransActinide separator and chemistry apparatus (TASCA) at GSI optimization of ion-optical structures and magnet designs. Nucl. Instrum. Meth. B **266**, 4153 (2008)
79. G. Savard, St Becker, G. Bollen, H.-J. Kluge, R.B. Moore, L. Schweikhard, H. Stolzenberg, U. Wiess, A new cooling technique for heavy ions in a Penning trap. Phys. Lett. A **158**, 247–252 (1991)
80. C.N. Davids, D. Peterson, A compact high-resolution isobar separator for the CARIBU project. Nucl. Instrum. Meth. B **266**, 4449 (2008)
81. F. Herfurth, Segmented linear RFQ traps for nuclear physics. Nucl. Instrum. Meth. B **204**, 587 (2008)
82. G. Savard et al., Large radio-frequency gas catchers and the production of radioactive nuclear beams. J. Phys. Conf. Ser. **312**, 052004 (2011)
83. M. Wada et al., Slow RI-beams from projectile fragment separators. Nucl. Instrum. Meth. B **204**, 570 (2003)
84. L. Weissman et al., First extraction tests of the NSCL gas cell. Nucl. Phys. A **746**, 655 (2004)
85. J.B. Neumayr et al., The ion-catcher device for SHIPTRAP. Nucl. Inst. Meth. B **244**, 489 (2005)
86. Yu. Kudryavtsev et al., Dual chamber laser ion source at LISOL. Nucl. Inst. Meth. B **267**, 2908 (2009)
87. S. Schwarz, RF ion carpets: the electric field, the effective potential, operational parameters and an analysis of stability. Int. J. Mass Spectrom. **299**, 71 (2011)
88. M. Ranjan et al., New stopping cell capabilities: RF carpet performance at high gas density and cryogenic operation. Europhys. Lett. **96**, 52001 (2011)
89. C. Droese et al., GSI Scientific Report 2012
90. I. Katayama et al., Cyclotron ion guide for energetic radioactive nuclear ions. Hyperfine Int. **115**, 165 (1998)
91. G. Bollen, D.J. Morrissey, S. Schwarz, A study of gas-stopping of intense energetic rare isotope beams. Nucl. Inst. Meth. A **550**, 27 (2005)

92. G. Bollen et al., Efficient and fast thermalization of Rare Isotope Beam from projectile fragmentation. The cyclotron gas stopper project at the NSCL. Eur. Phys. J. Spec. Top. **150**, 265 (2007)
93. F. Herfurth, A linear radiofrequency ion trap for accumulation, bunching, and emittance improvement of radioactive ion beams. Nucl. Instrum. Meth. A **469**, 254 (2001)
94. A. Nieminen et al., Beam cooler for low-energy radioactive ions. Nucl. Instrum. Meth. A **469**, 244 (2001)
95. S. Schwarz et al., A second-generation ion beam buncher and cooler. Nucl. Instrum. Meth. B **204**, 474 (2003)
96. F. Major, V.N. Gheorghe, G. Werth, *Charged Particle Traps— Physics and Techniques of Charged Particle Field Confinement*, 1st edn. (Springer, Heidelberg, 2005)
97. G. Werth, V.N. Gheorghe, F.G. Major, *Charged Particle Traps II* (Springer, Heidelberg, 2009)
98. P.H. Dawson, *Quadrupole Mass Spectrometry and Its Applications* (Elsevier, Amsterdam, 1995)
99. P. Ghosh, *Ion Traps* (Oxford University Press, Oxford, 1995)
100. W. Paul, Electromagnetic traps for charged and neutral particles. Rev. Mod. Phys. **62**, 531 (1990)
101. H.G. Dehmelt, Experiments with an isolated subatomic particle at rest. Rev. Mod. Phys. **62**, 525 (1990)
102. L.S. Brown, G. Gabrielse, Geonium theory: Physics of a single electron or ion in a Penning trap. Rev. Mod. Phys. **58**, 233 (1986)
103. M. Kretzschmar, Int. J. Mass Spectrom., submitted (2013)
104. G. Bollen et al., The accuracy of heavy-ion mass measurements using time of flight-ion cyclotron resonance in a Penning trap. J. Appl. Phys. **68**, 4355 (1990)
105. L. Schweikhard, G. Bollen (eds.). Int. J. Mass Spectrom. 251 (2006)
106. F.M. Penning, Verzögerungen bei der Zündung von gasgefüllten Photozellen im Dunkeln. Physica **3**, 563 (1936)
107. J.R. Pierce, *Theory and Design of Electron Beams* (D. van Nostrand Co., NewYork, 1949). (Chapter 3)
108. S. George et al., The Ramsey method in high-precision mass spectrometry with Penning traps: experimental results. Int. J. Mass Spectrom. **264**, 110 (2007)
109. G. Gräff, H. Kalinowsky, J. Traut, A direct determination of the proton electron mass ratio. Z. Physik A **297**, 35 (1980)
110. M. König, G. Bollen, H.-J. Kluge, T. Otto, J. Szerypo, Quadrupole excitation of stored ion motion at the true cyclotron frequency. Int. J. Mass Spec. Ion Proc. **142**, 95 (1995)
111. G. Gabrielse, The true cyclotron frequency for particles and ions in a Penning trap. Int. J. Mass Spectrom. **279**, 107 (2009)
112. M. Block, Direct mass measurements above uranium bridge the gap to the island of stability. Nature **463**, 785 (2010)
113. A. Kellerbauer, Recent improvements of ISOLTRAP: absolute mass measurements of exotic nuclides at 10^{-8} precision. Int. J. Mass Spectr. **229**, 107 (2003)
114. G. Bollen, Mass measurements of short-lived nuclides with ion traps. Nucl. Phys. A **693**, 3 (2001)
115. S. Eliseev et al., Octupolar excitation of ions stored in a Penning trap mass spectrometerA study performed at SHIPTRAP. Int. J. Mass Spectrom. **262**, 45 (2007)
116. R. Ringle et al., Octupolar excitation of ion motion zin a Penning trapA study performed at LEBIT. Int. J. Mass Spectrom. **262**, 33 (2007)
117. S. Eliseev et al., Octupolar-excitation Penning-trap mass spectrometry for Q-value measurement of double-electron capture in 164Er. Phys. Rev. Lett. **107**, 152501 (2011)
118. S. Ettenauer et al., First use of high charge states for mass measurements of short-lived letN=Y nuclides in a Penning trap. Phys. Rev. Lett. **107**, 272501 (2012)
119. S. Eliseev et al., Phase-Imaging Ion-Cyclotron-Resonance measurements for short-lived nuclides. Phys. Rev. Lett. **110**, 082501 (2013)

120. A. Kellerbauer et al., From direct to absolute mass measurements: a study of the accuracy of ISOLTRAP. Eur. Phys. J. D **22**, 53 (2003)
121. M. Block et al., Towards direct mass measurements of nobelium at SHIPTRAP. Eur. Phys J. D **45**, 39 (2007)
122. C. Droese et al., Investigation of the magnetic field fluctuation and implementation of a temperature and pressure stabilization at SHIPTRAP. Nucl. Instrum. Meth. A **632**, 157 (2011)
123. G. Gabrielse, F.C. Mackintosh, Cylindrical Penning traps with orthogonalized anharmonicity compensation. Int. J. Mass Spec. Ion Proc. **57**, 1 (1984)
124. G. Bollen et al., Resolution of nuclear ground and isomeric states by a Penning trap mass spectrometer. Phys. Rev. C **46**, R2140 (1992)
125. C. Fröhlich et al., Composition of the innermost core-collapse Supernova Ejecta. Ap. J. **637**, 415 (2006)
126. T. Beier, The status of FAIR. Hyperfine Int. **194**, 99 (2009)
127. G. Bollen, FRIBFacility for rare isotope beams. AIP conf. proc. **1224**, 432 (2010)
128. . M. Lewitowicz, Status of the SPIRAL2 Project. Acta Physica Polonica B **42**, 877 (2011)
129. S. Wanajo, S. Goriely, M. Samyn, N. Itoh, The r-Process in Supernovae: impact of new microscopic mass formulae. Ap. J. **606**, 1057 (2004)
130. A. Arcones, G.F. Bertsch, Nuclear Correlations and the r Process. Phys. Rev. Lett. **108**, 151101 (2012)
131. J. Savory et al., Nuclear Scission and Quantum Localization. Phys. Rev. Lett. **107**, 132501 (2009)
132. A. Martín et al., Mass measurements of neutron-deficient radionuclides near the end-point of the rp-process with SHIPTRAP. Eur. Phys. A **34**, 341 (2007)
133. P. Schury et al., Precision mass measurements of rare isotopes near N=Z=33 produced by fast beam fragmentation. Phys. Rev. C **75**, 055801 (2007)
134. C. Weber et al., Mass measurements in the vicinity of the r p-process and the p-process paths with the Penning trap facilities JYFLTRAP and SHIPTRAP. Phys. Rev. C **78**, 054310 (2008)
135. E. Haettner et al., Mass measurements of very Neutron-Deficient Mo and Tc isotopes and their impact on rp Process Nucleosynthesis. Phys. Rev. Lett. **106**, 122501 (2011)
136. J. Fallis et al., Mass measurements of isotopes of Nb, Mo, Tc, Ru, and Rh along the vp- and rp-process paths using the Canadian Penning trap mass spectrometer. Phys. Rev. C **84**, 045807 (2011)
137. A. Kankainen et al., Isomer and decay studies for the rp process at IGISOL. Eur. Phys. J. A **48**, 49 (2012)
138. J. Van Schelt et al., Mass measurements near the r-process path using the Canadian Penning Trap mass spectrometer. Phys. Rev. C **85**, 045805 (2012)
139. R. Wolf et al., Plumbing neutron stars to new depths with the binding energy of the exotic nuclide ^{82}Zn. Phys. Rev. Lett. **110**, 041101 (2013)
140. M. Bender et al., Shell structure of superheavy nuclei in self-consistent mean-field models. Phys. Rev. C **60**, 034304 (1999)
141. S. Cwiok et al., Shape coexistence and triaxiality in the superheavy nuclei. Nature **433**, 705 (2005)
142. P. Møller, J.R. Nix, Nuclear ground-state masses and deformations. At. Data Nucl. Data Tab. **59**, 185 (1995)
143. L. Stavsetra et al., Independent verification of element 114 production in the ^{48}Ca + ^{242}Pu reaction. Phys. Rev. Lett. **103**, 132502 (2009)
144. ChE Düllmann et al., Production and decay of element 114: High cross sections and the new nucleus ^{277}Hs. Phys. Rev. Lett. **104**, 252701 (2010)
145. S. Hofmann et al., The reaction ^{48}Ca + ^{248}Cm $\rightarrow$ 296,116* studied at the GSI-SHIP. Eur. Phys. J. A **48**, 62 (2012)
146. K. Morita et al., New result in the production and decay of an isotope, ^{278}113, of the 113th element. J. Phys. Soc. Jpn. **81**, 103201 (2012)
147. Yu.Ts. Oganessian et al., Synthesis of the heaviest elements in ^{48}Ca-induced reactions, Radiochim. Acta **99**, 429 (2011)

Chapter 8
Quantum Information Processing with Trapped Ions

Christian Roos

Abstract Trapped ions constitute a well-isolated small quantum system that offers low decoherence rates and excellent opportunities for quantum control and measurement by laser-induced manipulation of the ions. These properties make trapped ions an attractive system for experimental investigations of quantum information processing. In the following, the basics of storing, manipulating and measuring quantum information encoded in a string of trapped ions will be discussed. Based on these techniques, entanglement can be created and simple quantum protocols like quantum teleportation be realized. This chapter concludes with a discussion of the use of entangling laser-ion interactions for quantum simulations and quantum logic spectroscopy.

8.1 Introduction

Information processing by computers is based on transistors and other fundamental elements whose functioning can only be understood within the framework of quantum theory. However, the theory describing the processing of information by computers is entirely based on classical terms: the basic unit of information, the bit, is encoded as one of two classical physical states of a memory element. Similarly, the processing of information by logical gates that change the state of a bit depending on the state of other bits is based on Boolean logic that does not use quantum physics.

In contrast to traditional computer science, quantum information science is a research field where the notions of information processing are investigated within the framework of quantum theory. Its origins go back to studies that tried to generalize theories of classical information processing to the quantum domain. In the 1980s and

C. Roos (✉)
Institute for Quantum Optics and Quantum Information,
Technikerstrasse 21a, 6020 Innsbruck, Austria
e-mail: christian.roos@uibk.ac.at

W. Quint and M. Vogel (eds.), *Fundamental Physics in Particle Traps*,
Springer Tracts in Modern Physics 256, DOI: 10.1007/978-3-642-45201-7_8,

1990s, it was shown that by making explicit use of the laws of quantum physics certain information processing parts could be carried out more efficiently than by classical information processing [1]. The best-known example of such a quantum speed-up was discovered by Peter Shor who showed that the time required for factoring large numbers increases only polynomially with the size of the number whereas all known classical algorithms show an exponential scaling and are thus inefficient for the factoring of large numbers.

In a quantum computer, bits are replaced by quantum bits or qubits, i.e. two-level quantum systems storing the most basic unit of information in superposition states $|\psi\rangle = \alpha|\uparrow\rangle + \beta|\downarrow\rangle$. Processing of quantum information can be effected by a sequence of quantum gates acting on a register of quantum bits initially prepared in a suitable pure quantum state. Here, quantum gates replace the logic gates of a conventional information processor and they are realized by unitary operations acting on a small number of quantum bits. Importantly, quantum gates acting on more than a single qubit have the capability of entangling qubits with each other. In addition to unitary transformations of the processor's quantum state, measurements play a key role in quantum information processing. The speed-up of certain information processing tasks by running them on a quantum processor is often attributed to the possibility of preparing the quantum register at the beginning of the processing task in a giant superposition state and carrying out computations in parallel on all input states at once. However, this parallelism does not carry over to the stage where the result of the computation is read out by a quantum measurement. Nevertheless, algorithms like the one invented by Peter Shor, exploit quantum interference effect by clever quantum measurements that extract the vital information from the register's quantum state.

Quantum computation might be the best-known branch of quantum information processing (QIP). But the term encompasses also many other fields like quantum communication, the study of properties of entangled states, the enhancement of measurements by protocols processing quantum information, and quantum simulation.

While the study of QIP started as a purely theoretical endeavour, it turned into a major research field when proposals were put forward which discussed actual quantum information processing in experiments and showed ways of dealing with errors arising from unwanted interactions between the qubits and their physical environment as well as imperfect quantum gate operations. The discovery of quantum error correction came as a big surprise as it is not obvious that qubits whose state space is continuous and not discrete can be protected against errors by measurements that tend to destroy the coherence of a quantum state.

The first practical proposal for QIP was made for trapped ions in 1995 and soon followed by other proposals for systems like ultracold neutral atoms, photons, quantum dots or superconducting devices. Any physical system considered for QIP needs to fulfil a set of basic requirements formulated by David DiVincenzo [2] as the following five criteria:

1. A scalable physical system with well characterized qubits.
2. The ability to initialize the state of the qubits to a pure state.

3. Long relevant decoherence times, much longer than the gate operation time.
4. A universal set of quantum gates into which the computation can be broken down.
5. A qubit-specific measurement capability.

All of these requirements are met to an excellent degree by experiments with trapped ions. Paul and Penning traps enable the trapping of single or few ions for days or even weeks. When placed in ultra-high vacuum environments with a pressure of about 10^{-11} mbar or below, ions will be perturbed by collisions with background gas atoms and molecules only very infrequently. The availability of lasers for exciting atomic transitions constitutes a versatile tool which can be used for trapping and cooling, quantum state manipulation and state detection. The ions' motional state can be localized to a region of about 10 nm by laser cooling them to the ground state of the confining potential.

A long time before the DiVincenco criteria were written down, the technology for meeting the criteria (excluding the fourth one) in trapped ion experiments had already been developed. These technical developments were driven by the goal of using single trapped ions for precision spectroscopy and in particular for the development of (optical) frequency standards. Stable or metastable states connected by a narrow atomic transition as used in frequency standards are also well-suited for encoding a quantum bit as a superposition state of those two electronic levels. To achieve long coherence times, the electromagnetic field at the location of the ions has to be controlled very well and a pair of electronic states should be chosen for the qubit that is not much affected by changes of the field. Under such conditions, an ion can become a quantum object with coherence times in the range of milliseconds to minutes [3]. Linear ion strings can be used to create a multi-qubit system. The initialization of the qubit states is accomplished by optical pumping using techniques. Once a quantum bit is prepared in a pure inital quantum state, the qubit can be manipulated by pulses of coherent laser light similar to the ones for probing an ion in a frequency standard. The qubit-specific measurement capability by which the state of the qubit can be detected at the end of the computation refers to the ability to discriminate between the ion being in either one or the other of two long-lived states. An efficient state read-out based on fluorescence detection is possible if one of the two states can be excited to a third short-lived state which decays back to the original state [4–6] which is nothing else than Dehmelt's well-known electron shelving technique.

The availability of these techniques developed for precision spectroscopy have made trapped ions an interesting system for quantum optics experiments, fundamental tests of quantum physics and for quantum information processing. For the latter, the challenge to be overcome consists in realizing entangling interactions between the quantum bits that allows the system to be scaled up—at least in principle—to a large number of quantum bits. The problem is rooted in the difficulty of engineering interactions between different ions in the string that depend on the internal states of the ions even though the distances between the ions are on the order of micrometers, i.e. many orders of magnitude bigger than the characteristic atomic length scale.

The solution, proposed by Ignacio Cirac and Peter Zoller in 1995 [7], is to use a laser-ion interaction that couples the qubit states to the vibrational modes of the ion string. Due to the Coulomb interaction between the ions, the vibrational modes are shared between all ions in the string. This proposal and further ideas [8, 9] inspired by it have been successfully implemented in the last couple of years and have led to the demonstration of different two-qubit quantum gates [10–13] and the realization of simple quantum algorithms. In other experiments, the same interactions have been used for creating entangled states and studying their entanglement.

8.2 Storing Quantum Information in Trapped Ions

In linear radio-frequency ion traps, the trapping frequency along the axial direction is typically much weaker than along the radial directions, i. e. $\omega_z \ll \omega_x, \omega_y$. For a collection of N ions, the equilibrium configuration minimizing their energy will be a linear ion string along the axial direction (with uneven ion spacings) as long as the anisotropy of the confining potential is sufficiently large [14]. The typical distance between neighbouring ions d is of the order of a few micrometers for axial frequencies of about 1 MHz.

The magnitude of d is interesting for several reasons. First of all, d is considerably bigger than optical wavelengths which makes it possible to address a single ion within an ion string by a strongly focussed laser beam [15] and to spatially resolve the fluorescence emitted by the ions on a CCD camera. Secondly, the inter-ion distance d is much bigger than the Bohr radius. For this reason, state-dependent ion-ion interactions are extremely weak and the energy of an ion hardly depends on the state of a neighbouring ion. Therefore, trapped ions are well suited for storing quantum information in their internal states.

The storage of a quantum bit of information in a single ion requires the use of long-lived atomic states. Otherwise, the information will be quickly lost by spontaneous decay. As shown in Fig. 8.1, there are different possibilities for encoding a qubit: it can be encoded as a *hyperfine qubit* in a pair of two hyperfine ground states if the ion has non-zero nuclear spin or as a *Zeeman qubit* in a superposition of two Zeeman ground states if the total angular momentum of the ground state is different from

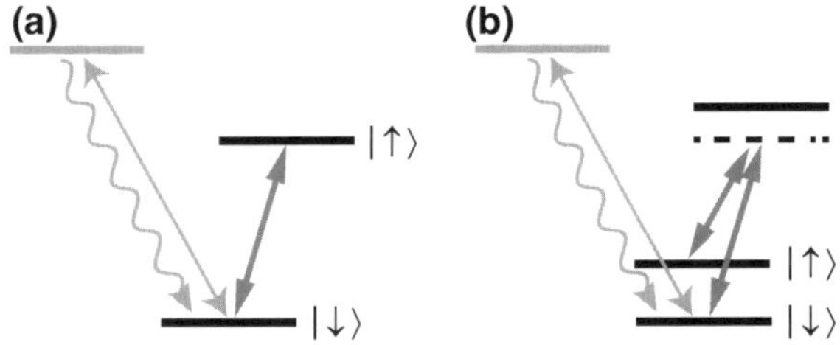

Fig. 8.1 Quantum information may be encoded in an optical qubit (**a**) or in a hyperfine or Zeeman qubit (**b**). The qubit is manipulated by laser pulses exciting either a single-photon or a Raman transition. For fluorescence state detection, one of the qubit levels is coupled to a short-lived state

zero. Alternatively, a ground state and a metastable state can be used for encoding the qubit as an *optical qubit*, the name referring to the energy difference between the two qubit states that corresponds to the one of a photon in the visible or near-infrared range of wavelengths. In this case, the lifetime of the metastable state has to be considerably longer than the time it takes to carry out the quantum information processing task.

The availability of states suitable for encoding a qubit is a first requirement that limits the choice of ion species. Further constraints arise from the requirements of having a qubit that can be efficiently detected and conveniently manipulated by laser light. For these reasons, experiments have focussed on singly-charged alkali-earth ions and some further ion species. Ions that have been used for encoding hyperfine qubits are $^{9}Be^{+}$, $^{25}Mg^{+}$, $^{43}Ca^{+}$, $^{111}Cd^{+}$, $^{137}Ba^{+}$ and $^{171}Yb^{+}$. Optical qubits have been encoded in $^{40}Ca^{+}$ and $^{88}Sr^{+}$.

Hyperfine qubits have qubit states with transition frequencies in the microwave range and the transition frequency of a Zeeman qubit may even lie in the radio-frequency domain. Therefore, single-qubit operations can in principle be realized by coupling the states with microwave or radio-frequency fields. However, due to their long wavelengths, this approach is unable to address a single qubit within an ion string unless ion-qubit s are mad unequal by spatially inhomogeneous external fields in order to achieve addressing in the frequency domain.

For this reason, the method of choice has been to use lasers for qubit manipulation. In the case of hyperfine or Zeeman qubits, two-photon Raman transitions are used for coupling the qubit states by using a laser field that off-resonantly couples the qubit states to a short-lived excited state [16]. Decoherence caused by spontaneous decay of the excited state can be sufficiently reduced by choosing large detunings. Coupling the qubit states by a two-photon Raman process has the benefit of relaxing the frequency stability requirements for the laser source which are much higher in the case of optical qubits. Optical qubit states are directly coupled on a single-photon dipole-forbidden transition using a laser with ultra-stable frequency [17]. In the following, procedures for qubit preparation, manipulation and measurement will be discussed for case of optical qubits.

8.3 Preparation and Detection of a Qubit Encoded in a Single Ion

Most experiments processing quantum information using optical qubits have been carried out with $^{40}Ca^{+}$ ions (see Fig. 8.2) which will serve us to illustrate the procedures for preparing, manipulating and detecting the qubit. In $^{40}Ca^{+}$, we make the identifications $|\downarrow\rangle = |S_{1/2}, m\rangle$ and $|\uparrow\rangle = |D_{5/2}, m'\rangle$ where $|\uparrow\rangle$ ($|\downarrow\rangle$) is an eigen-state of the operator σ_z with eigenvalue $+1$ (-1) and m, m' denote different Zeeman states of the $S_{1/2}$ ground and $D_{5/2}$ metastable state. The metastable $D_{5/2}$ state has a lifetime of about 1.15 s which is long compared with the typical duration of coherent manipulation and state detection.

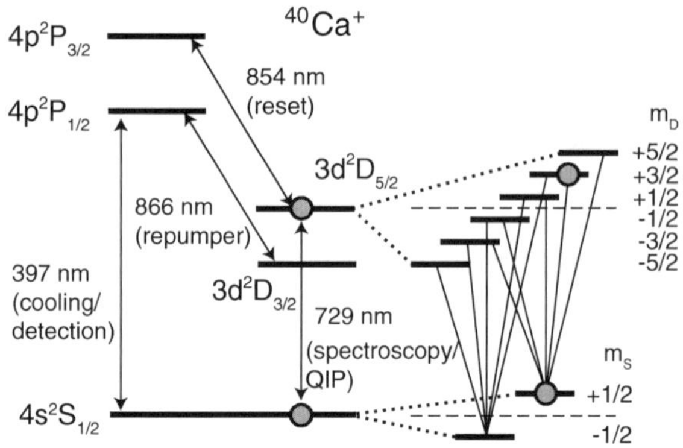

Fig. 8.2 Example of an optical qubit. Energy levels and transitions used in the experiments with $^{40}Ca^{+}$. The optical qubit is encoded in a suitable pair of Zeeman states of the $S_{1/2}$ and the $D_{5/2}$ metastable state. The *right-hand side* shows the Zeeman structure of the $S_{1/2} \leftrightarrow D_{5/2}$ quadrupole transition after lifting the degeneracy by a small magnetic field which shifts the Zeeman states by about 10 MHz

Laser cooling allows for preparing the vibrational state of a single trapped ion or an ion crystal in a pure quantum state. For quantum information processing, also the ion's internal state needs to be initialized in a pure state which is achieved by optical pumping. To illustrate this process, we consider the problem of initializing a $^{40}Ca^{+}$ ion in the $S_{1/2}, m = 1/2$ state. The metastable state $D_{3/2}$ and $D_{5/2}$ are easily pumped to the $S_{1/2}$ by coupling them to the $P_{1/2}$ or $P_{3/2}$ state with lasers running at 866 and 854 nm. To initialize the ion in the $m = 1/2$ Zeeman ground state, an ion placed in a weak magnetic field which defines the quantization axis can be excited with σ_+-circularly polarized light on the $S_{1/2} \leftrightarrow P_{1/2}$ transition. In this way, the polarization of the laser beam ensures that only the $m = -1/2$ state couples to the excited state. In practice, about 99 % of the population can be transferred to the qubit state by optical pumping.

The optical qubit can be conveniently detected by coupling the $S_{1/2}$ state to the short-lived excited state $P_{1/2}$ by a laser with a wavelength close to 397 nm as shown in Fig. 8.3. As the $P_{1/2}$ state decays after about 8 ns back to the $S_{1/2}$ state, tens of millions of photons per second can be scattered on this transition by near-resonant laser light provided that a second laser at 866 nm is used to prevent optical pumping to the $D_{3/2}$ state. By suitably setting the frequencies of these lasers, the ion can be Doppler-cooled at the same time. If the qubit is originally in the $|{\downarrow}\rangle$ state, the ion will scatter laser light that is detected by either a photomultiplier tube or a CCD camera. If on the other hand, the qubit is in the $|{\uparrow}\rangle$ state, the ion does not couple to the laser fields and will not emit fluorescence photons. In this case, the only light that reaches the detectors is due to scattering of stray light from the trap electrodes. To the detect the state of a single qubit, a photo-multiplier tube can be used. If several qubits in a string of ions are to be measured, the ions' fluorescence has to be spatially resolved by imaging it onto a low-noise CCD camera.

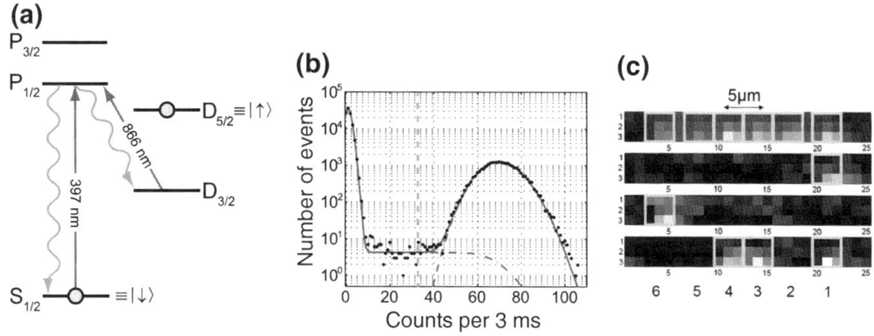

Fig. 8.3 Quantum state detection by electron shelving. **a** Lasers used for exciting the $S_{1/2}$-$D_{5/2}$ optical qubit in $^{40}Ca^{+}$. *Fluorescence* indicates that the ion resides in state $S_{1/2} \equiv |\downarrow\rangle$, absence of fluorescence indicates that the ion is in the $D_{5/2} \equiv |\uparrow\rangle$ state. **b** The fluorescence histogram measured by a photo-multiplier shows a bimodal distribution, the *left peak* caused by stray light scattered off trap electrodes, the *right peak* due to atomic fluorescence (and *stray light*). If the count rate is below the threshold indicated by the *dashed line*, the ion is detected in state $|\uparrow\rangle$. **c** A low-noise CCD camera can be used for spatially resolving the fluorescence of a string of ions. In this way, correlations between the qubit states can be measured

The detection process constitutes a projective measurement which projects the qubit state onto either $|\downarrow\rangle$ or $|\uparrow\rangle$. Detection errors can arise if the qubit changes its state during the detection interval or if photons scattered off trap electrodes are mistaken for fluorescence photons. In a recent experiment, the Oxford ion trapping group demonstrated qubit detection errors as small as 10^{-4} [18].

8.4 Coherent Manipulation of a Qubit

We will start by discussing how an ion-qubit interacts with a monochromatic travelling-wave laser beam.

8.4.1 Laser-Ion Interactions

The qubit can be coherently manipulated by coherently exciting it with a monochromatic travelling-wave laser beam. This process is described by a Hamiltonian consisting of three parts,

$$H = H^{(e)} + H^{(m)} + H^{(i)}$$

where $H^{(m)}$ describes the ion's motional state, $H^{(e)}$ the electronic structure of the ion, and $H^{(i)}$ the interaction of the ion with the laser beam. For a laser-cooled ion in a harmonic potential, $H^{(m)}$ is the sum of three quantum harmonic oscillators. Here,

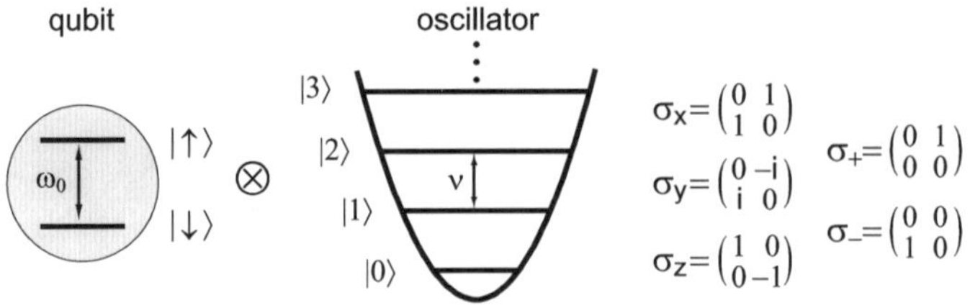

Fig. 8.4 Simplest quantum system for modelling the interactions of a trapped ions with a laser beam. The ion is described as a two-level system trapped in a harmonic potential. The quantum state is described by two quantum number, respectively specifying the internal and the motional state of the ion. The spin matrices to the *right* are used for describing coherent operations changing the qubit state

we will assume that the laser beam is aligned with the direction of one of these normal modes with angular frequency ν. This simplifies the description and reduces the Hamiltonian to

$$H^{(m)} = \hbar\nu(a^\dagger a + 1/2)$$

where $a^\dagger$ and a are creation and annihilation operators of vibrational quanta inducing changes between the vibrational quantum states $|0\rangle, |1\rangle, |2\rangle \dots$ and so on. The electronic part of the Hamiltonian, $H^{(e)}$ can often be approximated by describing the ion as a two-level system as in Fig. 8.4 with excited state $|\uparrow\rangle$ and ground state $|\downarrow\rangle$. In this approximation,

$$H^{(e)} = \frac{\hbar\omega_0}{2}\sigma_z,$$

where σ_z is a Pauli spin matrix, accounts for the energy difference $\hbar\omega_0$ between the two states. The Hamiltonian describing the laser-ion interaction is then given by

$$H^{(i)} = \hbar\Omega(\sigma_+ + \sigma_-)\cos(k\hat{x} - \omega_L t - \phi). \tag{8.1}$$

Here, ω_L and ϕ_L denote the frequency and phase of the laser field at the location of the ion which is described by the position operator $\hat{x}$. k is the magnitude of the laser beam's wave vector and the Rabi frequency Ω characterizes the strength of the interaction. The jump operator σ_+ induces change of the qubit state from $|\downarrow\rangle$ to $|\uparrow\rangle$, σ_- induces the inverse process.

The laser-ion interaction can be considerably simplified if the motional energy of the ion is so low that its wave packet is localized on a length scale much smaller than the wavelength of the laser. In this regime, which is called the Lamb-Dicke regime, the ion is sensitive to the spatial gradient of the electric field of the laser beam but not to any higher derivative. For an ion cooled to the ground state of motion, Lamb-Dicke parameter η, defined as $\eta = kx_0$, compares the width $x_0 = \sqrt{\frac{\hbar}{2m\nu}}$ of its wave packet with the laser wavelength. With the help of this definition, we can express in

Eq. (8.1) the coupling of the laser to the ion motion $k\hat{x} = \eta(a + a^\dagger)$ by the creation and annihilation operators.

If in addition the laser resonantly or near-resonantly couples a pair of quantum states as shown in Fig. 8.5b, a standard procedure of quantum optics can be employed. It consists in going into an interaction picture defined with respect to $H^{(e)} + H^{(m)}$ and carrying out the rotating wave approximation which amounts to dropping rapidly time-varying terms from the resulting interaction Hamiltonian. Mathematical details regarding this procedure can be found in Ref. [19].

As a result, one obtains three simple Hamiltonians that describe the situations where the laser-ion detuning $\Delta = \omega_L - \omega_0$ is either zero or equal to $\pm\nu$. These Hamiltonians resonantly couple the quantum state $|\downarrow\rangle|n\rangle$ with the states $|\uparrow\rangle|m\rangle$, where $m = n$ or $m = n \pm 1$. There are three resonances in the spectrum shown in Fig. 8.5a which correspond to the following situations:

1. If the laser is resonant with the atomic transition, $\omega_L = \omega_0$, the Hamiltonian

$$H_{carr} = \frac{\hbar\Omega}{2}(\sigma_+ e^{i\phi} + \sigma_- e^{-i\phi}) \tag{8.2}$$

   couples the ground state to the excited state without affecting the motional state. This excitation is called *carrier excitation* and it corresponds to the strong center resonance in Fig. 8.5a
2. If $\omega_L = \omega_0 - \nu$, the laser is resonant with the *red sideband* of the transition. The Hamiltonian

$$H_{RSB} = i\hbar\frac{\eta\Omega}{2}(\sigma_+ a e^{i\phi} - \sigma_- a^\dagger e^{-i\phi}). \tag{8.3}$$

   describes processes where upon excitation to the upper state the vibrational quantum number is reduced by one unit. This process corresponds to the resonance to the left of the carrier transition in Fig. 8.5a. The resonance is weaker because the coupling strength between the states $|\downarrow\rangle|n\rangle \leftrightarrow |\uparrow\rangle|n-1\rangle$ is reduced by $\eta\sqrt{n}$ with respect to the carrier transition. This interaction is known in quantum optics as

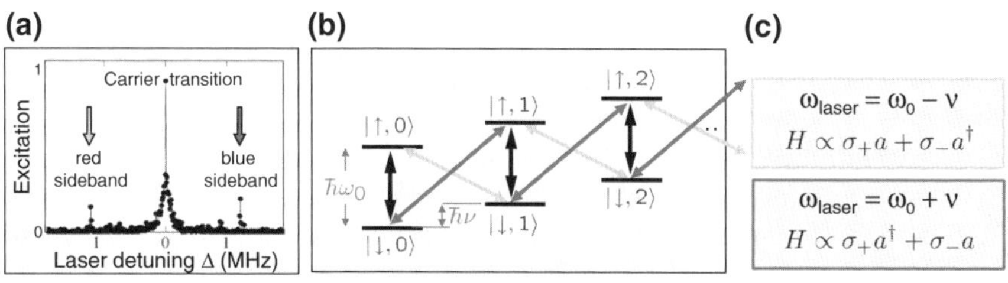

Fig. 8.5 Spin-motion coupling. **a** Coherent excitation on vibrational sidebands is used for coupling the qubit to the motional state. **b**, **c** By exciting the *red* (*blue*) sideband, a qubit state change is accompanied by a decrease (increase) of the vibrational quantum number, realizing an (anti-) Jaynes-Cummings Hamiltonian

a Jaynes-Cummings model which describes the interaction between a two-level atom and mode of the electromagnetic field.

3. Similarly, if $\omega_L = \omega_0 + \nu$, the laser is resonant with the *blue sideband* of the transition. The Hamiltonian

$$H_{BSB} = i\hbar\frac{\eta\Omega}{2}(\sigma_+ a^\dagger e^{i\phi} - \sigma_- a e^{-i\phi}). \tag{8.4}$$

describes processes where upon excitation to the upper state the vibrational quantum number is increased by one unit, giving rise to the resonance to the right of the carrier transition. In this case, the coupling strength between the states $|\downarrow\rangle|n\rangle \leftrightarrow |\uparrow\rangle|n+1\rangle$ is reduced by $\eta\sqrt{n+1}$ with respect to the strength of the carrier transition.

The laser-ion interactions described above can easily be generalized to cover also situations where multiple ions interacting with the laser. In most cases, the laser will be resonant with no more than a single vibrational mode of the ion string so that all the other modes don't need to be taken into account for the description of the process. If in addition the laser interacts with just a single ion, the only difference with the single-ion case described above concerns the ion's position operator which needs to be expressed in terms of the joint vibrational modes of the ion string. More details on this topic and on the case of laser-ion interactions involving Raman processes are provided by Ref. [20].

8.4.2 Laser Cooling of Single Ions

Different methods have been developed to reduce the kinetic energy of trapped ions. The motion of the ions can be damped by coupling the trap electrodes to an electric circuit that dissipates energy, by collisions with background gas atoms or by inelastic scattering of laser light. Out of these methods only laser cooling offers the prospect of cooling the ion into the quantum regime. Besides, it is flexible as it allows one to switch the interaction on and off at will. Laser cooling works by near-resonant excitation of an atomic transition. In the rest frame of the oscillating ion, the laser frequency appears frequency-modulated with the trap frequency. The strength of the sidebands depends on the amplitude of the ionic motion. Two limiting cases will be considered:

$\nu \ll \Gamma$: If the trap frequency ν is much lower than the decay rate Γ of the transition used for cooling, then the spacing of the sidebands is much smaller than the absorption width of the transition. The velocity of the ion due to the confining potential changes on a longer timescale than the time it takes the ion to absorb or emit a photon. So these processes can be assumed to change the momentum of the ion instantaneously. The ion behaves like a free particle seeing a time-dependent Doppler-shifted laser frequency. A velocity-dependent radiation pressure can provide cooling [21, 22], which has been termed *Doppler cooling*.

$\nu \gg \Gamma$: In the opposite case, the sidebands are resolved so that the laser can be tuned to a specific sideband. If the energy of the absorbed photons is smaller than the mean energy of spontaneously emitted photons, the kinetic energy of the ion shrinks.

In the following, the cooling mechanisms will be explained by means of simple models for the case of a two-level atom. A more elaborate and mathematically stringent treatment of Doppler cooling can be found in [23, 24].

The complementary aspects of both cooling techniques can be combined in a cooling scheme that allows the cooling of a trapped ion from room temperature to the quantum mechanical ground state of the trapping potential.

Doppler cooling: Consider a two-level atom moving with velocity v interacting with a travelling wave laser field. The laser is characterised by its frequency ω_L and the Rabi frequency Ω, ω_0 denotes the atomic transition frequency and $\Delta = \omega_L - \omega_0$ the detuning of the laser. The laser exerts a radiation pressure force on the atom, given by

$$F = \hbar k \Gamma \rho_{ee},$$

where the excited state probability is

$$\rho_{ee} = \frac{\Omega^2}{\Gamma^2 + 4(\Delta - kv)^2}$$

in the limit of low saturation. The radiation pressure force can be linearised around $v = 0$ if the velocity is already small:

$$F = F_0 + \left.\frac{dF}{dv}\right|_{v=0} v, \tag{8.5}$$

with

$$F_0 = \hbar k \Gamma \frac{\Omega^2}{\Gamma^2 + 4\Delta^2}$$

and

$$\left.\frac{dF}{dv}\right|_{v=0} = F_0 \frac{8k\Delta}{\Gamma^2 + 4\Delta^2}.$$

F_0 is the time-averaged radiation pressure which displaces the ion slightly from the centre of the trap. The velocity-dependent part of Eq. (8.5) provides a viscous drag, if the detuning Δ is negative. The random nature of the absorption and emission processes counteracts this cooling force. The cooling rate is given by

$$\dot{E}_{cool} = \langle F v \rangle = \langle (F_0 + \left.\frac{dF}{dv}\right|_{v=0} v) v \rangle = \left.\frac{dF}{dv}\right|_{v=0} \langle v^2 \rangle.$$

Spontaneously emitted photons lead to a diffusive spreading of the mean squared momentum. Assuming uncorrelated spontaneous emission events, a heating rate results, given by

$$\dot{E}^{em}_{heat} = \frac{1}{2m}\frac{d}{dt}\langle p^2\rangle = \frac{1}{2m}(\hbar k)^2\Gamma\langle\rho_{ee}(v)\rangle \approx \frac{1}{2m}(\hbar k)^2\Gamma\rho_{ee}(v=0),$$

provided that the atomic velocity is small. A second contribution to the heating rate, similar in size to $\dot{E}^{em}_{heat}$, results from the fluctuations of the cooling force due to the discreteness of the absorption process. In the steady state, the cooling rate $\dot{E}_{cool}$ equals the total heating rate $\dot{E}_{heat} = \dot{E}^{em}_{heat} + \dot{E}^{abs}_{heat}$ leading to

$$\dot{E}_{cool} + \dot{E}_{heat} = 0 \iff \left.\frac{dF}{dv}\right|_{v=0}\langle v^2\rangle = \frac{1}{m}(\hbar k)^2\Gamma\rho_{ee}(v=0).$$

Using the relation $m\langle v^2\rangle = k_B T$ the cooling limit can be formulated as

$$k_B T = \frac{\hbar\Gamma}{4}\left(\frac{\Gamma}{-2\Delta} + \frac{-2\Delta}{\Gamma}\right).$$

The lowest temperatures $k_B T_{min} = \frac{\hbar\Gamma}{2}$ are obtained by setting $\Delta = -\frac{\Gamma}{2}$. Expressing the energy in terms of the average vibrational quantum number, $k_B T = \langle n\rangle\hbar\nu$ yields the equation

$$\langle n\rangle = \frac{\Gamma}{2\nu}. \tag{8.6}$$

In the derivation it was assumed that the wave vector of spontaneously emitted photons points in or against the direction of the laser beam. Cooling of the harmonic motion in all three oscillator directions $\mathbf{e}_i$ is accomplished by choosing a wave vector $\mathbf{k}$ that has a non-vanishing overlap with every direction $\mathbf{e}_i$.

Sideband cooling: Sideband cooling on an atomic transition requires the trap frequency to be large compared to the laser linewidth and the decay rate of the excited state, allowing the laser to be tuned to the lower motional sideband of the transition. Then each excitation to the upper level is accompanied by a reduction in the vibrational quantum number. In the Lamb-Dicke regime, spontaneous decay does not affect the ion's motional state most of the time. So every cycle of absorption followed by spontaneous emission takes out a motional quantum until the ion reaches the ground state of motion that is decoupled from the laser excitation. Figure 8.6a illustrates the basic process.

The cooling process has been described in great detail [23]. For a vibrational mode that is aligned with the direction of the laser beam, a simple model provides estimates for the cooling rate

$$R_n = \Gamma\frac{(\eta\sqrt{n}\Omega)^2}{2(\eta\sqrt{n}\Omega)^2 + \Gamma^2}$$

for a particle residing in the nth vibrational state, and for the average quantum number in steady state

$$\langle n \rangle \approx \frac{\Gamma^2}{4\omega^2}. \tag{8.7}$$

In the resolved sideband limit, $\Gamma \ll \omega$, the particle is cooled to the zero-point energy of motion with high probability. Non-resonant excitation from the $|n = 0\rangle$ state (see Fig. 8.6b) prevents the motional ground state from being a perfect 'dark state' and imposes a limit upon the minimum energy achievable. The cooling limit has been calculated under the assumption that $\langle n \rangle \ll 1$ and that the sideband Rabi frequency $\eta\Omega$ is small compared to the decay rate Γ. Therefore, the magnitude of the decay rate Γ constrains the cooling rate maximally attainable. If the natural lifetime of the excited level is very long, it can be advantageous to shorten it by coupling the level to an auxiliary level having a shorter lifetime [25]. In this way, the cooling rate can be tailored to the desired value by adjusting the laser power coupling the metastable state to the auxiliary state.

To cool an ion to the vibrational ground state, a two-step process is used. In a first step, Doppler cooling provides high cooling rate required to cooling an ion having a temperature in the range of 100–1000 K into the mK regime. Next, starting with an energy of a few to a few tens of vibrational quanta, sideband cooling takes the ion to the ground state. Sideband cooling cannot only be carried out on narrow single-photon transitions [25] but also between hyperfine or Zeeman ground states coupled by Raman processes [16]. Figure 8.7 shows the typical signature of an ion having reached the ground state $|n = 0\rangle$, excitation on the red sideband is no longer possible. By comparing the excitation probabilities p_r and p_b on the red and the blue vibrational sidebands, the average motional energy can be calculated via $\langle n \rangle = p_r/(p_b - p_r)$. As the coupling strength on the vibrational sidebands depends on the vibrational quantum number (cf. Eqs. (8.3), (8.4)), it is even possible to infer the probabilities of occupying the nth vibrational level from the population dynamics on vibrational sidebands [26].

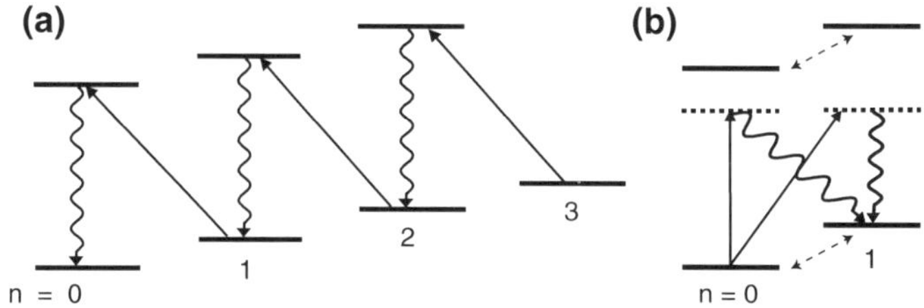

Fig. 8.6 **a** Principle of sideband cooling: each excitation to the upper state lowers the number of vibrational quanta by one while spontaneous emission does not change the motional state (to first order). **b** Heating processes depleting the vibrational ground state are off-resonant excitation of the ion involving *blue* sideband processes and motional heating due to fluctuating electric fields by noise at the trap frequency (indicated as *dashed lines* coupling $n = 0$ to $n = 1$)

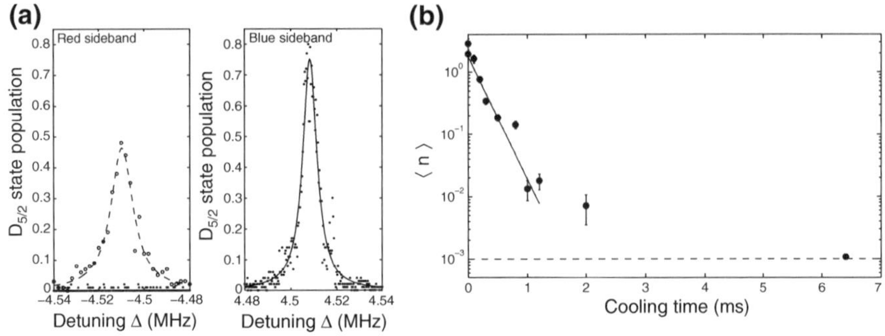

Fig. 8.7 **a** Cooling to the ground state is detected by the vanishing of the *red* vibrational sideband in the excitation spectrum. The average vibrational quantum number $\langle n \rangle$ can be extracted by a comparison of the excitation probabilities on the *red* and the *blue* vibrational sidebands. **b** The plot shows the rate at which $\langle n \rangle$ decreases as a function of the length of the sideband cooling pulse (Adapted from [17])

8.4.3 Single-Qubit Gates

Single-qubit rotations can be carried out with by resonantly exciting the qubit with laser pulses on the carrier transition as described by Eq. (8.2). This interaction will act only on the qubit state without affecting the ion's motional state. The resulting Rabi oscillations are shown in Fig. 8.8. By applying the interaction for a duration τ, qubit state transformations

$$U(\theta, \phi) = \exp(-\frac{i}{\hbar} H_{carr} \tau) \tag{8.8}$$

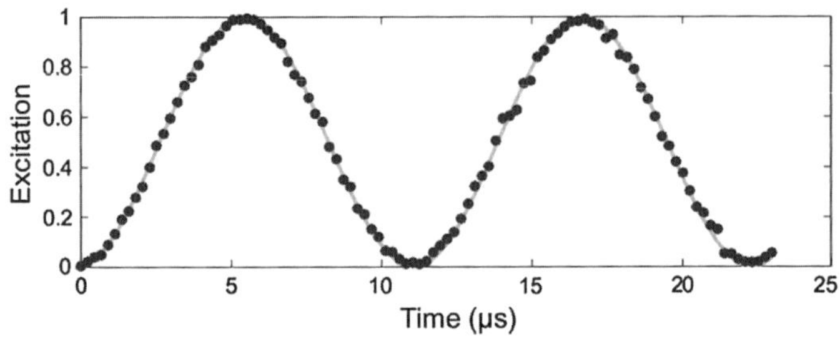

Fig. 8.8 Rabi oscillations of an ion initially prepared in the $|\downarrow\rangle$ state and resonantly excited on the qubit transition. The figure shows the probability of finding the ion in the upper state $|\uparrow\rangle$ as a function of the duration of the laser pulse. Each point is an average of 1,000 individual experiments consisting of qubit preparation, manipulation and measurement. A $\pi/2$ pulse, corresponding to an excitation of about 3 μs creates an equal superposition of the $|\downarrow\rangle$ and $|\uparrow\rangle$ state. A π pulse, corresponding to a 6 μs pulse, exchanges the population of the qubit states

are induced that correspond to rotations of the Bloch sphere around a rotation axis located in the equatorial plane of the Bloch sphere with the rotation angle θ given by $\theta = \Omega\tau$. In particular, rotations around the x- and the y-axis are realized by setting the laser phase ϕ either to 0 or to $\pi/2$.

To illustrate this in some more detail, let us assume that $\phi = 0$. Then, $H_{carr} = \frac{\hbar\Omega}{2}(\sigma_+ + \sigma_-) = \frac{\hbar\Omega}{2}\sigma_x$ induces a rotation around the x-axis. If the duration τ of the pulse is chosen such that $\theta = \Omega\tau = \pi/2$, the laser pulse induces the coherent operation described by $U(\pi/2, 0) = \exp(-i\frac{\pi}{4}\sigma_x) = \frac{1}{\sqrt{2}}(I - i\sigma_x)$. A qubit initially prepared by optical pumping in $|\downarrow\rangle$ is mapped by this operation to the superposition state

$$|\downarrow\rangle \longrightarrow \psi_f = \frac{1}{\sqrt{2}}(|\downarrow\rangle - i|\uparrow\rangle). \tag{8.9}$$

At this stage, the assignment of a particular relative phase to this state is somewhat arbitrary. If instead of the basis states $\{|\uparrow\rangle, |\downarrow\rangle\}$ one had chosen to introduce the states $|\tilde{\downarrow}\rangle = |\downarrow\rangle$, $|\tilde{\uparrow}\rangle = -i|\uparrow\rangle$, the resulting state would have been $\psi_f \propto |\tilde{\downarrow}\rangle + |\tilde{\uparrow}\rangle$. However, once a basis has been chosen, it has to be kept for all subsequent operations.

For example, in the Ramsey experiment shown in Fig. 8.9, a $\pi/2$ pulse around the x-axis applied to $|\downarrow\rangle$ prepares the superposition state (8.9) which is the $+1$ eigenstate of the σ_y Pauli matrix. This state is mapped by a second pulse (having the same laser phase as the first pulse) to the $|\uparrow\rangle$ state. If, in the time between the pulses, the phase of the superposition state is changed to $\frac{1}{\sqrt{2}}(|\downarrow\rangle - i|\uparrow\rangle)$ which is the -1 eigenstate of σ_y, the second pulse maps it back to the initial $|\downarrow\rangle$ state. This experiment can also be viewed in the following way: The time before the start of the second pulse can be interpreted as a state preparation procedure. The second pulse in combination with the fluorescence measurement probes the coherence of the prepared state by

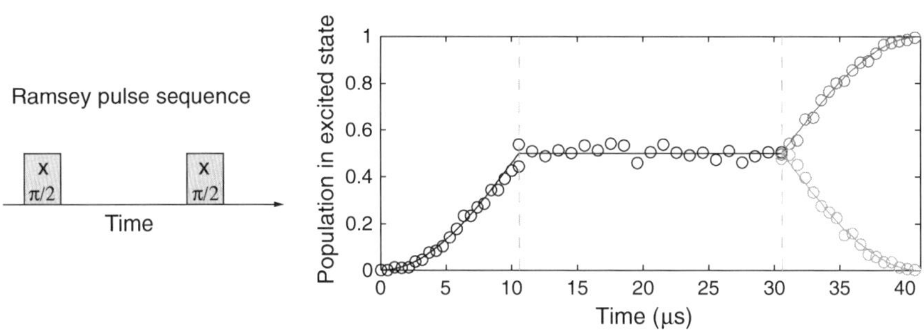

Fig. 8.9 The easiest non-trivial single-qubit manipulation is a Ramsey experiment. The first Ramsey pulse prepares the qubit in an superposition of the $|\uparrow\rangle$ and $|\downarrow\rangle$ states. The second Ramsey pulse probes this superposition. If the relative phase of the superposition state did not change in the time between the pulses, the two $\pi/2$ pulses act as a single π pulse which flips the initial state from $|\downarrow\rangle$ to $|\uparrow\rangle$. This is shown in the data set as the time evolution marked by *black* and *blue* data points where each point corresponds to the average excitation probability measured in 400 experiments. If during the waiting time, the relative phase changed by π, the second Ramsey pulse would reverse the time evolution induced by the first pulse (*red* data points)

projecting the state onto the y-axis of the Bloch sphere describing the state space of the qubit. Such measurements are the basis for a complete characterization of multi-qubit states by state tomography described in Sect. 8.6.

In the usual Schrödinger picture, the relative phase of the superposition state ψ_f of (8.9) would evolve in time as $\frac{1}{\sqrt{2}}(|\downarrow\rangle - e^{-i\omega_0 t} i|\uparrow\rangle)$. For optical qubits where $\omega_0/(2\pi) > 10^{14}$ Hz, the only way to track the phase evolution of such a state is to compare its phase with the phase of the laser field coupling the qubit states which evolves at the same rate. This is exactly what one does by going into an interaction picture in which the laser-ion interaction is described by the Hamiltonian of Eq. (8.2). It also explains why high-fidelity single-qubit operations on an optical qubit require a very frequency-stable laser. If the laser frequency fluctuates, its phase at a time t after the start of a gate operation is no longer precisely defined with respect to its initial phase. Therefore the phase of the qubit state can no longer be precisely determined which is equivalent to a dephasing of the qubit state. For a hyperfine or Zeeman qubit, the frequency stability requirements of the source inducing single-qubit gates are much less stringent provided that the same source is used for creating the two laser frequencies of the Raman process coupling the qubit states.

Arbitrary single qubit operations (which realize rotations of the Bloch sphere around any desired rotation axes and for any rotation angle) are realized by combining single-qubit operations of type (8.8) around the x- and the y-axis in a way which is similar to the description of the rotation of a rigid body by the Euler angles.

Additionally, rotations around the z-axis can be accomplished by illuminating the ion with a far-detuned laser field coupling either to the quadrupole or to dipole transitions that induce a differential Stark shift of the qubit states. In comparison to the rotations induced by the resonant interaction (8.2), the light-shift-based interaction has the advantage of being independent of path-length fluctuation of the optical beam path that shift the phase of the laser light.

Single-qubit operations become more challenging to achieve if they are to be carried out on a single qubit of a qubit register. There are several possible approaches. The most direct way consists in strongly focussing a laser beam so that it interacts with only a single ion in the ion string. An alternative approach is based on using segmented micro-fabricated ion traps that provide flexible trapping potentials which enable splitting ion strings into substrings so that the qubit to me manipulated can be spatially separated from the other qubits. Finally, one can also try to spectrally separate the transition frequency of the qubit to be manipulated from the other qubits by using inhomogeneous fields.

8.5 Entangling Quantum Gates

The big challenge in trapped-ion quantum information processing is the generation of entanglement between different ion-qubits. As the distance between ions is much bigger than the Bohr radius, no state-dependent ion–ion interactions exist in the absence of additional external electromagnetic fields. Fortunately, the Coulomb force not only pushes the ions apart but also leads to a coupling of the ion motion which provides a means for mediating entanglement between the qubits. For this, laser fields are used that lead to correlated changes of the state of the qubits and the motional degrees of the ion crystal and thus entangle the two subsystems. The laser-ion interactions can be arranged in such a way that in the end an entangling operation has been carried out between two or more qubits and no entanglement persists between the qubits and the vibrational degrees of freedom.

Another possibility for creating entanglement between ion qubits is the use of qubit-state-dependent external potentials in addition to the trap potential which would enable entanglement generation by microwave fields [27]. This approach leads to a coupling of the qubits with the motional degrees of freedom of the ion string which is similar to the one induced by coupling on the vibrational sidebands of a qubit transition. A third strategy, developed to entangle ions over macroscopic distances, consists in entangling ion qubits with photons that are subsequently projected onto an entangled state by a measurement [28]. Currently, however, it is not very efficient as the entanglement generation process is non-deterministic and succeeds only in about 1 out of 10^8 trials.

All current quantum gate realizations that deterministically entangle ions rely on a transient coupling of the internal states to the vibrational modes of the ion string by either laser light or inhomogeneous microwave fields. In the following, we will focus on laser-induced entangling operations. The basic processes underlying them are described by Eqs. (8.3) and (8.4).

These gate operations can be broadly classified according to the way the lasers interact with the ions:

1. Quantum gates induced by a laser beam that interacts with a single ion at a time as originally proposed in the seminal chapter by Cirac and Zoller [7]. In these gates, a single ion is entangled with a vibrational mode of the ion string and the entanglement is subsequently transferred from the vibrational mode to the internal state of a second ion.
2. Quantum gates induced by a bi-chromatic laser that collectively interacts with two or more ions. Here, a vibrational mode becomes transiently entangled with the qubits before getting disentangled at the end of the gate operation, resulting in an effective interaction between the qubits capable of entangling them [8, 9, 29, 30].

Even though both classes of gates are applicable to hyperfine qubits as well as optical qubits, for a long time experiments with optical qubits have relied on the former [12] and experiments with hyperfine qubits on the latter type of interaction [11]. In any case, the main goal consists in demonstrating fast operations creating entanglement with high fidelity.

8.5.1 Cirac–Zoller-Type Gate Interactions

An interaction on the red or the blue sideband leads to entanglement between the ion's internal and vibrational states. For experiments aiming at entangling internal states of several ions, the goal consists in finding pulse sequences that transiently entangle internal and vibrational states in such a way that in the end the internal states are entangled, with no entanglement remaining between the internal states and the motion. The simplest example of such a pulse sequence is the following: initially, two ions are prepared in the state $|\downarrow\rangle|\downarrow\rangle|0\rangle$ by optical pumping, sideband cooling and a carrier π-pulse to flip the state of one of the ions. Next, a $\pi/2$ pulse is applied to the other ion on the blue sideband. These pulses induce the mapping

$$|\downarrow\rangle|\downarrow\rangle|0\rangle \rightarrow |\downarrow\rangle|\uparrow\rangle|0\rangle \rightarrow |\downarrow\rangle|\uparrow\rangle|0\rangle + |\uparrow\rangle|\uparrow\rangle|1\rangle$$

and entangle one of the ions with the motion (here, for simplicity, the normalization factor was dropped). In the last step, the entanglement is transferred from the vibrational mode to the second ion by a π-pulse on the blue sideband, this time applied to yet unentangled ion, in order to map the state to

$$|\downarrow\rangle|\uparrow\rangle|0\rangle + |\uparrow\rangle|\uparrow\rangle|1\rangle \rightarrow |\downarrow\rangle|\uparrow\rangle|0\rangle + |\uparrow\rangle|\downarrow\rangle|0\rangle.$$

While the sequence given above maps a product state onto a maximally entangled state, it does not constitute a quantum gate operation since it performs the mapping only for a specific input state while it transforms all other input states into two-qubit states that are entangled with the vibrational state.

The best-known entangling two-qubit gate is the controlled-NOT gate which is acts on the two-qubit basis states as follows:

$$|\uparrow\rangle|\uparrow\rangle \longrightarrow |\uparrow\rangle|\uparrow\rangle \tag{8.10}$$
$$|\uparrow\rangle|\downarrow\rangle \longrightarrow |\uparrow\rangle|\downarrow\rangle \tag{8.11}$$
$$|\downarrow\rangle|\uparrow\rangle \longrightarrow |\downarrow\rangle|\downarrow\rangle \tag{8.12}$$
$$|\downarrow\rangle|\downarrow\rangle \longrightarrow |\downarrow\rangle|\uparrow\rangle. \tag{8.13}$$

The state of the second qubit (the target qubit) is flipped if the state of the first qubit (the control qubit) is the $|\downarrow\rangle$ state. This is an entangling operation. If the control qubit is prepared in a superposition state, the gate creates a Bell state:

$$(|\uparrow\rangle + |\downarrow\rangle)|\uparrow\rangle \longrightarrow |\uparrow\rangle|\uparrow\rangle + |\downarrow\rangle|\downarrow\rangle$$

The controlled-NOT (CNOT) gate derives its importance from the fact that any quantum computation on a set of qubits can be broken down into a circuit of elementary quantum gates consisting of CNOT gates and single-qubit gates.

Slightly more complicated pulse sequences than the ones mentioned above for creating a two-ion Bell state can indeed realize a CNOT gate operation as was first shown by I. Cirac and P. Zoller. In their proposal [7], the state $\psi = \alpha|\uparrow\rangle + \beta|\downarrow\rangle$ of the control qubit is swapped by a π-pulse on the blue sideband onto a vibrational mode initially prepared in the ground state. After this pulse, the vibrational mode is in a superposition $\alpha|0\rangle + \beta|1\rangle$ of the two lowest vibrational states and acts as an additional qubit. This approach reduces the problem to carrying out a controlled-NOT gate between the qubit encoded in the vibrational mode and the target qubit. The gate operation is completed by mapping the quantum information from the vibrational mode back to the control qubit using again a π-pulse on the blue motional sideband as shown in Fig. 8.10a.

The CNOT operation between the vibrational mode and the target qubit can be further broken down into a conditional phase gate enclosed in a Ramsey experiment consisting of two $\pi/2$-carrier pulses on the target qubit. Here, the conditional phase is an entangling operation that realizes the mapping

$$|\uparrow\rangle|\uparrow\rangle \longrightarrow |\uparrow\rangle|\uparrow\rangle \tag{8.14}$$
$$|\uparrow\rangle|\downarrow\rangle \longrightarrow |\uparrow\rangle|\downarrow\rangle \tag{8.15}$$
$$|\downarrow\rangle|\uparrow\rangle \longrightarrow |\downarrow\rangle|\uparrow\rangle \tag{8.16}$$
$$|\downarrow\rangle|\downarrow\rangle \longrightarrow -|\downarrow\rangle|\downarrow\rangle \tag{8.17}$$

which multiplies the state by -1 if both qubits are in the spin-down state. To realize such a gate with trapped ions, in Ref. [7] it was suggested to use a 2π-pulse on the red sideband of a transition connecting one of the target qubit states to an auxiliary level. Such a pulse does not change the state of the target qubit if the vibrational state is in $n = 0$ or if the target qubit is in the state not coupling to the auxiliary state. If on the other hand the vibrational state is in $n = 1$ and the target qubit is in the state coupling to the auxiliary state, then a 2π rotation multiplies the initial state by -1 as required for the conditional phase gate.

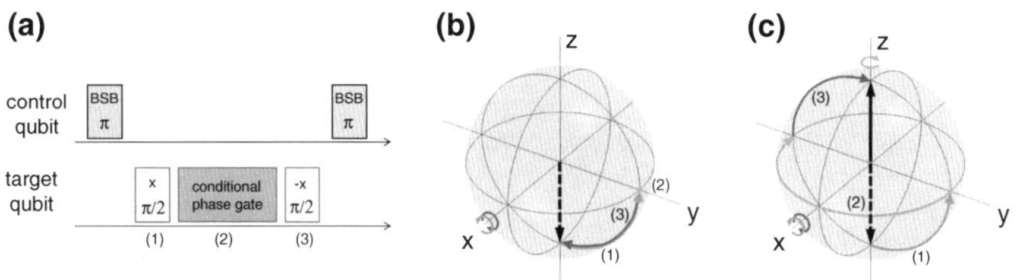

Fig. 8.10 Cirac–Zoller CNOT gate operation. **a** The state of the control qubit is transiently mapped on the vibrational mode by *blue* sideband pulses. A composite gate operation realizes a CNOT between the vibrational mode qubit controlling the target qubit. **b**, **c** Bloch sphere dynamics induced on the target qubit by the pulses (*1*–*3*). Depending on the state of the control qubit, the target qubit is either flipped or remains in its initial state

The equivalence between the CNOT gate and the conditional phase gate can be understood in terms of Ramsey spectroscopy experiments (see Fig. 8.10b, c). If the control qubit (i.e the qubit encoded in the vibrational states) is in state $|\uparrow\rangle$, the conditional phase gate has no effect. Therefore, a sequence of two Ramsey $\pi/2$ pulses enclosing the phase gate, the second pulse being the inverse of the first one, has no effect on the state in this case. If, however, the control qubit is in state $|\downarrow\rangle$, the conditional phase gate shifts the phase of the target qubit's $|\downarrow\rangle$ state by π. As a result, now the two Ramsey pulses flip the state of the target qubit as required for the operation of a CNOT gate. Experimental data illustrating this process were already presented in Fig. 8.9.

The experimental realization of a controlled-NOT gate operation is important for trapped-ion quantum computing as the gate allows for realizing arbitrary unitary operations when combined with the single-qubit operations described in Sect. 8.4. A CNOT gate on two qubits encoded in the vibrational state and the internal state of a single ion was already demonstrated in 1995 [31]. Since the first demonstration of a two-ion CNOT gate in 2003 [12], the gate fidelity has been raised from the initial 70 % to about 90 % [32], and the gate has been employed as a building block in experiments demonstrating elementary quantum protocols.

8.5.2 *Quantum Gates Based on Bichromatic Light Fields*

An alternative route towards the realization of a universal two-qubit gate consists in coupling both ions simultaneously to the vibrational mode. A number of different entangling gate operations has been demonstrated that is based on this approach [10, 11]. Interestingly, there is a unified description for both types of gates even though at a first glance the physical mechanisms at work seem to be fairly different [33]. This description, which is based on the driven quantum harmonic oscillator, will be reviewed in the following:

The Hamiltonian $\tilde{H} = \hbar\nu a^\dagger a + \hbar\gamma i(a^\dagger e^{i\omega t} - ae^{-i\omega t})$ describes a harmonic oscillator oscillating at frequency ν and driven by a periodic force with frequency ω and coupling strength γ. Going into an interaction picture defined by $H_0 = \hbar\nu a^\dagger a$ yields the Hamiltonian

$$H = \hbar\gamma i(a^\dagger e^{i\delta t} - ae^{-i\delta t}), \tag{8.18}$$

where $\delta = \omega - \nu$. Under the action of the driving force, an oscillator that is initially in a coherent state remains in a coherent state. For a force that is slightly detuned from resonance, the coherent state maps out a circle in phase space and returns to the initial state after a period $\tau = 2\pi/\delta$. This operation multiplies the oscillator state by a phase factor whose magnitude is given by the ratio of the strength of the force and the detuning [11, 29]. After time t, the propagator is $U(t) = \hat{D}(\alpha(t))\exp(i\Phi(t))$ with $\alpha(t) = i\left(\frac{\gamma}{\delta}\right)(1 - e^{i\delta t})$ and $\Phi = \left(\frac{\gamma}{\delta}\right)^2(\delta t - \sin\delta t)$. Here, $\hat{D}(\alpha) = e^{\alpha a^\dagger - \alpha^* a}$ denotes the displacement operator in the motional state space. After a time $\tau_N = 2\pi N/|\delta|$, $N = 1, 2, \ldots$, the motional state returns to its initial state in phase space with its phase changed by an amount $\Phi(\tau_N) = 2\pi N\left(\frac{\gamma}{\delta}\right)^2 \text{sign}(\delta)$.

By making this phase change depend on the states two qubits are prepared in, an entangling gate operation can be achieved. For

$$H = \hbar\gamma i(a^\dagger e^{i\delta t} - a e^{-i\delta t})\mathcal{O} \tag{8.19}$$

where $\mathcal{O}$ is an operator acting on the qubit states, the propagator U is replaced by

$$U(t) = \hat{D}(\alpha(t)\mathcal{O}) \exp(i\Phi(t)\mathcal{O}^2). \tag{8.20}$$

If the interaction time τ is chosen such that $\alpha(\tau) = 0$, the displacement operator $\hat{D}$ vanishes from the equation. Therefore, the vibrational state is not altered and thus a gate operation $U(\tau) = \exp(i\Phi(\tau)\mathcal{O}^2)$ is obtained which depends on $\mathcal{O}$ in a non-linear way. Various entangling gates can be realized by different choices of $\mathcal{O}$ where this operator is the sum of single-qubit operators acting on a pair of qubits. In the following, two widely used entangling gates will be discussed where either $\mathcal{O} = \sigma_z^{(1)} + \sigma_z^{(2)}$ or $\mathcal{O} = \sigma_y^{(1)} + \sigma_y^{(2)}$.

8.5.3 Conditional Phase Gates

Conditional phase gates are obtained by setting $\mathcal{O} = \sigma_z^{(1)} + \sigma_z^{(2)}$. With this choice and for a suitable coupling strength, the gate operation becomes

$$U = \exp\left(i\frac{\pi}{4}\sigma_z^{(1)} \otimes \sigma_z^{(2)}\right).$$

In the eigenstate basis of the σ_z operator, the gate multiplies the basis states by the phase factors shown in Eqs. (8.14)–(8.17) that cannot be obtained by single-qubit gate operations. As already discussed, the resulting conditional phase gate can be converted into a controlled-NOT gate operation by additional single-qubit operations.

This type of gate was first implemented in Ref. [11] on hyperfine qubits encoded in two $^9Be^+$ ions. To realize the required gate Hamiltonian, the ions were placed into a standing wave created by two counter-propagating beams off-resonantly exciting the qubit states to an excited state. As the qubit states coupled differently to this state, they experienced spatially varying ac-Stark shifts of opposite sign as indicated in Fig. 8.11. When placed in the node of the standing wave, an ion experiences a state-dependent force described by the Hamiltonian $H \propto (a + a^\dagger)\sigma_z$. The force is made time-varying by detuning the frequency of one of the beams forming the standing wave by an amount close to a vibrational mode frequency of the two-ion crystal. This turns the standing wave into a walking wave. In the experiment by Leibfried et al. [11], the two counter-propagating Raman beams had a frequency difference

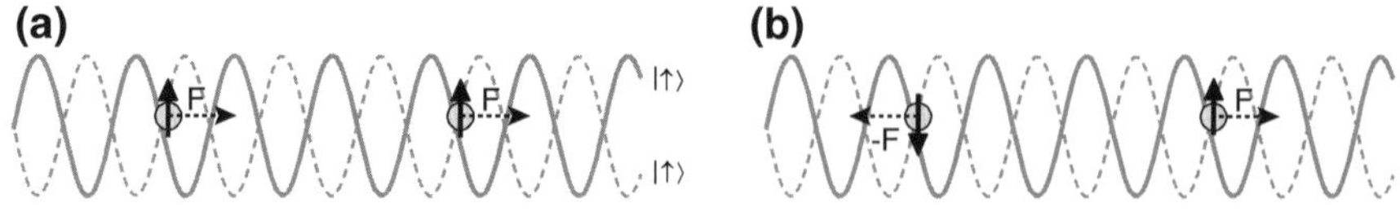

Fig. 8.11 Realization of a conditional phase gate. Two ions are placed into a qubit-state-dependent potential created by a spatially varying ac-Stark shift that is temporally modulated with a frequency close to one of the vibrational modes. The *solid* (*dashed*) *line* indicates the potential experienced by an ion in the $|\uparrow\rangle$ ($|\downarrow\rangle$) state, **a** If both ions are in the same state, the resulting force couples to the center-of-mass mode but not to the stretch mode. **b** If the ions are in different states, a coupling to the stretch mode exists but not to the center of mass mode

$\delta\omega = \sqrt{3}\nu + \delta$ which coupled the ions to the stretch mode (oscillating at frequency $\sqrt{3}\nu$) if the ions were in different qubit states.[1] In this way, a state-dependent displacement force was created based on the strong field gradients created by the Raman light field which realized the Hamiltonian (8.19). An gate was realized that entangled the ions with a fidelity of 97 % in less than 40 μs.

8.5.4 Mølmer-Sørensen Gates

Another entangling gate, which was first investigated by Sørensen, Mølmer [8] and others [9], is obtained by setting $\mathcal{O} = \sigma_y^{(1)} + \sigma_y^{(2)}$. This choice leads to the gate operation

$$U = \exp\left(i\frac{\pi}{4}\sigma_y^{(1)} \otimes \sigma_y^{(2)}\right).$$

In contrast to $\sigma_z \otimes \sigma_z$ gates that do not change the internal states of the ions, the Mølmer-Sørensen gate operation induces collective spin flips $|\downarrow\downarrow\rangle \leftrightarrow |\uparrow\uparrow\rangle$, $|\downarrow\uparrow\rangle \leftrightarrow |\downarrow\uparrow\rangle$ by processes coupling to the lower and upper motional sidebands as illustrated in Fig. 8.12. The gate operation U maps the product state basis $\{|\uparrow\uparrow\rangle, |\uparrow\downarrow\rangle, |\downarrow\uparrow\rangle, |\downarrow\downarrow\rangle\}$ onto a basis of entangled states.

In order to see that the simultaneous application of a red and a blue sideband excitation indeed produces the desired interaction, the Hamiltonians (8.3) and (8.4) have to be summed up. The phases appearing in the red and the blue sideband Hamiltonian become time-dependent when the laser is not perfectly resonant with the sideband transition. If the sum of the frequencies exciting the red and the blue sideband equals twice the qubit frequency, an interesting situation arises. For $\omega_r = \omega_0 - (\nu + \varepsilon)$ and $\omega_b = \omega_0 + (\nu + \varepsilon)$, the phases become $\phi_r = \varepsilon t$ and $\phi_b = -\varepsilon t$, and

$$H_{RSB} + H_{BSB} = -\hbar\frac{\eta\Omega}{2}\sigma_y(a^\dagger e^{-i\varepsilon t} + ae^{i\varepsilon t}). \tag{8.21}$$

[1] The actual experiment was slightly more involved as the ac-Stark shifts experienced by the qubit states had not the same magnitude.

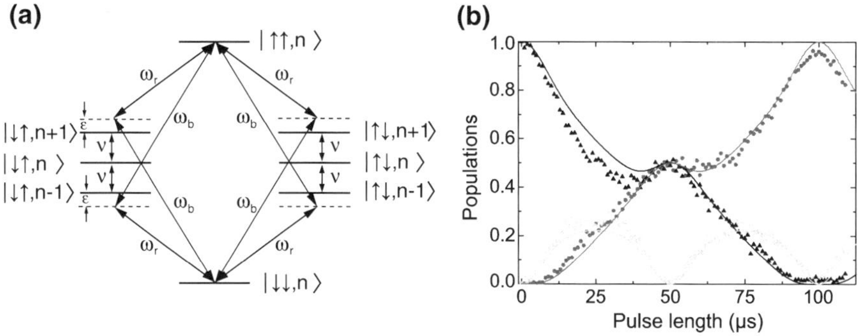

Fig. 8.12 **a** Mølmer-Sørensen gate. A bichromatic laser field with frequencies ω_b, ω_r satisfying $2\omega_0 = \omega_b + \omega_r$ is tuned close to the upper and lower motional sideband of the qubit transition. The field couples the qubit states $|\downarrow\downarrow\rangle \leftrightarrow |\uparrow\uparrow\rangle$ via the four interfering paths shown in the figure. Similar processes couple the states $|\uparrow\downarrow\rangle \leftrightarrow |\downarrow\uparrow\rangle$ with the same strength provided that the Rabi frequencies of the light fields ω_b, ω_r are equal. **b** Dynamics of the populations $p_{\downarrow\downarrow}$ (*black symbols*), $p_{\uparrow\uparrow}$ (*blue symbols*) and $p_{\downarrow\uparrow} + p_{\uparrow\downarrow}$ (*red symbols*) induced by the bichromatic laser field. After $\tau = 50\ \mu$s, the $|\downarrow\downarrow\rangle$ has been transformed into the entangled state $|\downarrow\downarrow\rangle + i|\uparrow\uparrow\rangle$

This Hamiltonian is valid for a single ion. If two ions are simultaneously illuminated by the bichromatic light field, a similar Hamiltonian is obtained that corresponds to setting $\mathscr{O} = \sigma_y^{(1)} + \sigma_y^{(2)}$ in (8.19). A Mølmer-Sørensen gate is produced by a laser pulse of duration $\tau = \frac{2\pi}{|\varepsilon|}$ for a coupling strength $\eta\Omega = \frac{\varepsilon}{2}$. Figure 8.12 shows the population evolution under the action of a Mølmer-Sørensen gate in an experiment [13] demonstrating the creation of Bell states with a fidelity above 99 %. To prove that Bell states of high fidelity were indeed produced, additional measurements have to be carried out to test the coherence of the superposition states produced by the gate.

Other gates like the $\sigma_x \otimes \sigma_x$ gate are generated by shifting the overall phase of the bichromatic light field. When applied to more than two ions, the Mølmer-Sørensen gate produces a pairwise coupling of the spin-flip operators of the different qubits which can be used for generating maximally entangled states.

8.6 Quantum State Tomography

All information about physical systems is inferred from measurements. In quantum physics, sometimes a measurement on only a single quantum system is performed, the outcome of which is one of the eigenvalues of the observable that is measured. More often, however, the expectation value $\langle A\rangle = \mathrm{Tr}(\rho A)$ of an observable A is measured using an ensemble of systems all prepared in the same quantum state ρ where Tr denotes the trace operation. Here, the measurement is either carried out in parallel on many quantum systems of the same type or it is carried out sequentially on an individual quantum system that is repeatedly prepared and measured to gather the required information.

In the field of quantum information, quantum measurements are a key element. Among others, measurement of observables called *entanglement witnesses* can provide information about whether a quantum state is entangled. There are, however, also questions that cannot be answered by the measurement of a single observable because the quantity to be measured cannot be written as a linear function of the density operator. A simple example of this kind is the task to determine the purity of a single-qubit state. Another example is the characterization of the entanglement of a two-qubit entangled state by means of an entanglement measure.

In these cases, the brute-force approach is to determine the whole density matrix that describes the quantum state. In this way, the complete characterization of the quantum state by means of its density matrix provides the answer to any question one might want to pose. The complete reconstruction of a state has been termed quantum state tomography. The following discussion of quantum state tomography will be restricted to the case of finite-dimensional Hilbert spaces. The assumption that the description of the quantum state to be measured requires only a finite number of dimensions is well satisfied in the experiments described in this article. Formally, the basic principle of quantum state tomography can be summarized as follows:

In a d-dimensional quantum system, it is possible to expand the density matrix ρ using a basis of d^2 hermitian matrices A_i that are mutually orthogonal. Mathematically, this condition is expressed as

$$\rho = \sum_{i=1}^{d^2} \lambda_i A_i \tag{8.22}$$

with $\mathrm{Tr}(A_i A_j) = \mu_i \delta_{ij}$ where $\mu_i \neq 0$. Since the matrices A_i are assumed to be hermitian, they are observables with mean value

$$\langle A_j \rangle = \mathrm{Tr}(\rho A_j) = \sum_{i=1}^{d^2} \lambda_i \mathrm{Tr}(A_i A_j) = \lambda_i \mu_i, \tag{8.23}$$

and the coefficients that completely characterize the state are given by

$$\lambda_i = \mu_i^{-1} \langle A_i \rangle. \tag{8.24}$$

Therefore, a measurement of the set of observables $\mathscr{S} = \{A_i\}$ is sufficient for reconstructing the density operator ρ. The most basic example for a measurement of this kind is the reconstruction of the density matrix of a spin-1/2 system or any other two-level quantum system. In this case, a possible set of observables is given by $\mathscr{S} = \{\mathbb{I}, \sigma_x, \sigma_y, \sigma_z\}$ where $\mathbb{I}$ is the identity and σ_j are the Pauli matrices. Because of $\mathrm{Tr}(\sigma_i \sigma_j) = 2\delta_{ij}$ and $\mathrm{Tr}(\sigma_i) = 0$, the observables are orthogonal and the density matrix can be written as

$$\rho = \frac{1}{2}(\mathbb{I} + n_x\sigma_x + n_y\sigma_y + n_z\sigma_z) \text{ with } n_k = \langle\sigma_k\rangle. \tag{8.25}$$

A measurement of the density matrix is accomplished by measuring the components n_j of the Bloch vector $\mathbf{n} = (n_x, n_y, n_z)$. The observable $\mathbb{I}$ does not need to be measured since $\langle\mathbb{I}\rangle = 1$ for any quantum state.

The generalization of this procedure to the reconstruction of an N-qubit system is straightforward. In this case, it is sufficient to measure observables $A_{\mathbf{j}}$ that are tensor products of Pauli matrices, i. e. $A_{\mathbf{j}} = A_{j_1} \otimes A_{j_2} \otimes \cdots$ with $A_{j_k} \in \{I, \sigma_x, \sigma_y, \sigma_z\}$.

For a qubit encoded in a trapped ion, the following measurements need to be carried out: the observable σ_z is determined by a fluorescence measurement as described in Sect. 8.3. The other two observables cannot be directly measured. Instead, for measuring σ_x, we apply first the single-qubit rotation (8.8) that maps the eigenvector corresponding to the eigenvalue $+1$ of σ_x onto the $+1$-eigenvector of σ_z before measuring σ_z on the transformed state. A similar procedure is carried out for measurements of σ_y. In case of an N-qubit system encoded in a string of ions, this procedure requires the possibility of addressing single ions and also measuring the fluorescence of a single ion. To measure a spin correlation like $\sigma_x^{(1)} \otimes \sigma_z^{(2)}$, first the Bloch sphere of qubit 1 is rotated around the y-axis by a $\pi/2$ pulse. Then, a spatially resolved measurement of the fluorescence of both ions is performed in order to determine $\sigma_z^{(1)} \otimes \sigma_z^{(2)}$. This requires correlating the measurement results (either $+1$ or -1) found for ion 1 and ion 2 by multiplying them with each other. The procedure is easily generalized to N qubits where it requires measurements in 3^N different bases that are reduced to measurements of $\sigma_z^{(j_1)}$, $\sigma_z^{(j_1)} \otimes \sigma_z^{(j_2)}$, $\sigma_z^{(j_1)} \otimes \sigma_z^{(j_2)} \otimes \sigma_z^{(j_3)}$ where $j_m \in \{1, 2, \ldots N\}$ and so on. Figure 8.13 shows the experimentally reconstructed density matrix of a two-ion Bell state.

Although the general strategy for measuring the density matrix ρ is very straightforward, its practical implementation becomes non-trivial once the inevitable condition is taken into account that in any experiment only a finite number of copies of ρ is available for carrying out the measurements. For this reason, it is impossible to measure the expectation values $\langle A_i\rangle$ of Eq. (8.24) to arbitrary precision. Any value entering this equation can therefore only be an estimate $\tilde{\lambda}_i$ of the observable's expectation value λ_i that will only converge to the true value in the limit of an infinite number of measurements.

This little difference has important consequences that are illustrated in Fig. 8.14 for the case of a two-level system. For simplicity, the Bloch vector describing the state of the system is assumed to have a vanishing y-component. Any valid Bloch vector has to lie within the unit circle. However, errors in the measurement of $\langle\sigma_x\rangle$ and $\langle\sigma_z\rangle$ can give rise to a state reconstruction with a vector whose tip lies outside the 'Bloch' circle. Then, the corresponding density matrix has one eigenvalue bigger than one and one eigenvalue smaller than zero; thus, it is no longer positive semi-definite and does not describe a valid physical state. This problem occurs in particular for states of high purity that lie close to the boundary of the Bloch sphere. However, the likelihood of encountering this problem increases the higher the dimension of the state space becomes. In practice, these problems can be overcome by state

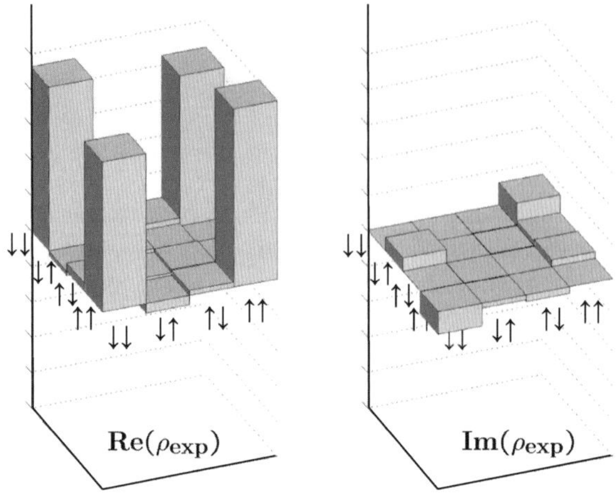

Fig. 8.13 Density matrix reconstruction of the two-ion Bell state $|\psi\rangle = \frac{1}{\sqrt{2}}(|\downarrow\rangle|\downarrow\rangle + |\uparrow\rangle|\uparrow\rangle)$. On the *left*, the real part of the experimentally reconstructed density matrix ρ is shown. As expected, the $\downarrow\downarrow$ and $\uparrow\uparrow$ states have populations close to 0.5. In addition, the existence of off-diagonal elements close to 0.5 demonstrates the coherence between the $\downarrow\downarrow$ and $\uparrow\uparrow$ parts and proves that the state is indeed entangled and not just a mixture of $\downarrow\downarrow$ and $\uparrow\uparrow$

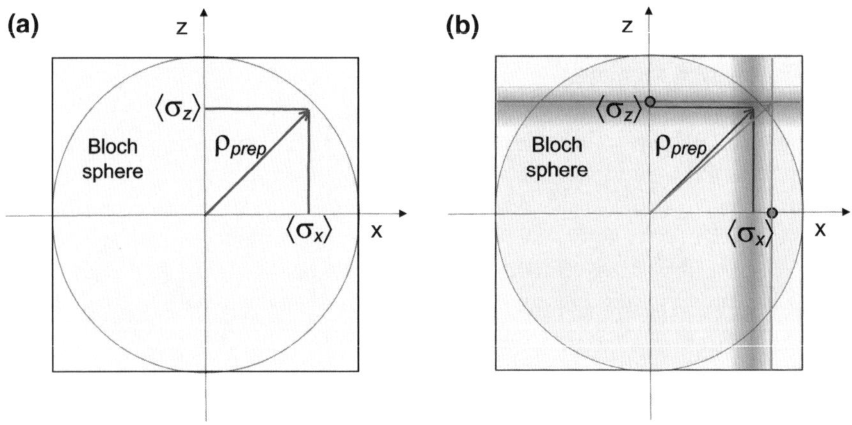

Fig. 8.14 Quantum state tomography of a two-level system. **a** The Bloch vector describing the system is indicated by an *arrow*. Its components are determined by measurements of the corresponding observables. **b** Errors in the determination of Bloch vector components λ_x and λ_z can give rise to a reconstructed vector whose tip (indicated by the '+' sign) lies outside the unit circle containing the physically allowed states

reconstruction schemes based on maximum-likelihood or Bayesian estimation that go beyond the simple linear reconstruction scheme sketched in Eqs. (8.22)–(8.24). By these schemes physically valid density matrices such as the one shown in Fig. 8.13 can be reconstructed from the same set of measurements used for the linear reconstruction technique.

While the principles of quantum state tomography were already considered 50 years ago, its extension to the complete characterization of quantum processes [1] is more recent. In quantum process tomography, the characterization of a physical transformation of a state by means of unitary or decohering processes and measurements is achieved by preparing a set of input states that are subjected to the process in question and reconstructing the quantum state at the output. Process tomography is one possible way of characterizing experimental realizations of entangling gate operations [32].

8.7 Entangled States and Elementary Quantum Protocols

In current experiments, entangling gate operations have been applied to ion strings of up to 16 ions. Experiments investigating high-fidelity operations have shown that single qubit gates can be performed with error rates as low a one part in 10^5 whereas for two-qubit gates error rates as low as 1 % have been observed. While these achievements are still far from the requirements of large-scale quantum computation which would require millions of gates acting on thousands of qubits, they have enabled interesting entanglement studies in few-ion strings and the demonstration of a number of elementary quantum protocols.

Various experiments have, for example, investigated Greenberger-Horne-Zeilinger (GHZ)-states

$$\psi = \frac{1}{\sqrt{2}}(|\uparrow\uparrow\cdots\uparrow\rangle + |\downarrow\downarrow\cdots\downarrow\rangle)$$

of up to 14 ions [10, 34]. Figure 8.15a shows an example of a three-ion GHZ state. The off-diagonal elements of the reconstructed density matrix prove the coherence of the state. GHZ states are maximally entangled in the sense that a measurement of single ion can reveal the full information about the state of the other ions. This property makes GHZ states fragile with respect to interactions with a decohering environment. It does not, however, signify that any measurement will completely destroy the quantum character of the state. If, for example, one ion of the three-qubit state $\psi = \frac{1}{\sqrt{2}}(|\uparrow\uparrow\uparrow\rangle + |\downarrow\downarrow\downarrow\rangle)$ is measured in the σ_z-basis $\{|\uparrow\rangle, |\downarrow\rangle\}$, the other two will be projected into either the product state $|\downarrow\downarrow\rangle$ or $|\uparrow\uparrow\rangle$. If, on the other hand, a measurement onto the eigenstates of σ_x ($|\uparrow\rangle \pm |\downarrow\rangle$) is carried out, the remaining ions are projected into one of the Bell state states $|\uparrow\downarrow\rangle - |\downarrow\uparrow\rangle$ as demonstrated in Ref. [35]. Figure 8.15b, c shows the state after projecting the first qubit into the x-basis (onto the states $|\uparrow\rangle + |\downarrow\rangle$ and $|\uparrow\rangle - |\downarrow\rangle$) and after a correction operation that deterministically turns the state into a Bell state.

Elementary quantum algorithms with trapped ions include demonstrations of quantum teleportation, quantum error correction, entanglement purification and entanglement swapping [36–40]. More details about some of these experiments can

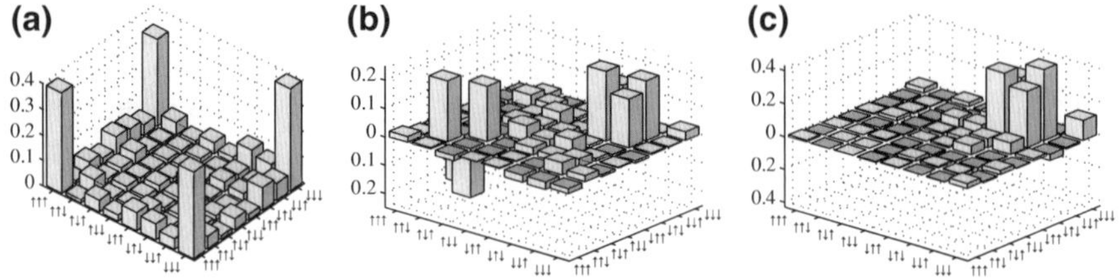

Fig. 8.15 Quantum state tomography of a GHZ state. **a** The figure shows the absolute value of the density matrix. **b** After measuring the second qubit in the x-basis, the resulting state is a mixture of two Bell states if the measurement result is disregarded. **c** If, depending on the measurement result, a correction operation is applied, the GHZ-state can be deterministically transformed into a Bell state

be found in references [41, 42]. Efforts to advance experiments along these lines focus on reducing error rates in the coherent operations, speeding them up and handling larger number of ions by using microfabricated ion traps.

8.7.1 Deterministic Quantum Teleportation

Quantum teleportation is an simple example of an important quantum communication protocol. It solves the problem of sending a quantum bit prepared in an unknown state from one party, often called Alice, to another party, usually named Bob, in the absence of a 'quantum channel' by measurements and local quantum operations. The key to accomplishing this task is the existence of a quantum resource shared between Alice and Bob in the form of a (maximally entangled) Bell state $|\Psi_-\rangle = \frac{1}{\sqrt{2}}(|\uparrow\downarrow\rangle - |\downarrow\uparrow\rangle)$. The teleportation protocol and its implementation with a string of three ions is sketched in Fig. 8.16. In addition to the unknown quantum state $|\phi\rangle = \alpha|\uparrow\rangle + \beta|\downarrow\rangle$, Alice is in the possession of the first qubit of the Bell state $|\Psi_-\rangle$ she shares with Bob (who has the second qubit of $|\Psi\rangle$). By means of the four Bell states $|\Psi_\pm\rangle = \frac{1}{\sqrt{2}}(|\uparrow\downarrow\rangle \pm |\downarrow\uparrow\rangle)$, $|\Phi_\pm\rangle = \frac{1}{\sqrt{2}}(|\uparrow\uparrow\rangle \pm |\downarrow\downarrow\rangle)$, the initial three-qubit state of Alice and Bob can be rewritten as follows

$$
\begin{aligned}
(\alpha|\uparrow\rangle_1 + &\beta|\downarrow\rangle_1)((|\uparrow\downarrow\rangle_{23} - |\downarrow\uparrow\rangle_{23}) \\
&= \frac{1}{2}\{|\Psi_-\rangle_{12}(-\alpha|\uparrow\rangle_3 - \beta|\downarrow\rangle_3) + |\Psi_+\rangle_{12}(-\alpha|\uparrow\rangle_3 + \beta|\downarrow\rangle_3) \\
&\quad + |\Phi_-\rangle_{12}(\beta|\uparrow\rangle_3 + \alpha|\downarrow\rangle_3) + |\Phi_+\rangle_{12}(-\beta|\uparrow\rangle_3 + \alpha|\downarrow\rangle_3)\} \\
&= \frac{1}{2}\{|\Psi_-\rangle_{12}(-I|\phi\rangle_3) + |\Psi_+\rangle_{12}(ie^{i\frac{\pi}{2}\sigma_z}|\phi\rangle_3) + |\Phi_-\rangle_{12}(-ie^{i\frac{\pi}{2}\sigma_x}|\phi\rangle_3) \\
&\quad + |\Phi_+\rangle_{12}(-e^{i\frac{\pi}{2}\sigma_y}|\phi\rangle_3), \}
\end{aligned}
$$

where the index 1 refers to qubit 1 and 23 to the two qubits on Bob's side. The last line of the equations contains the key to recovering $|\phi\rangle$ in qubit 3 on Bob's side. If

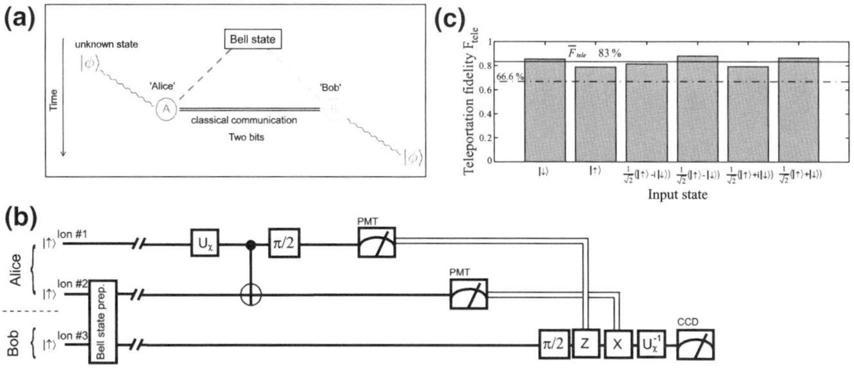

Fig. 8.16 **a** Quantum teleportation protocol: an unknown quantum state $|\phi\rangle$ is teleported from Alice to Bob by means of measurements on the original state and a Bell state shared between the two parties. **b** Quantum teleportation protocol with trapped ions. Alice prepares the initial state by the unitary U_χ in ion 1 and carries out a two-qubit measurement (using an entangling gate) followed by state measurements. The measurement results are communicated to Bob who carries out additional single-qubit gates, which depend on the measurement outcomes, in order to restore the state on his side

qubits 1 and 2 are projected onto the Bell state basis, qubit 3 is equal to $|\phi\rangle$ up to single qubit rotations around the x, y, and z-axis. If Bob applies the inverse of these single qubit operations to qubit 3 after having received the measurement outcome telling him on which Bell state qubits 1 and 2 had been projected, he will reconstitute the initial state $|\phi\rangle$ on his side.

Figure 8.16b shows protocol in some more detail using a graphical representation of the quantum circuit which widely used in quantum information processing. Here, lines represent qubits, boxes are gate operations acting on them as well as measurements. The double lines after measurements of qubits 1 and 2 indicate classical information (instead of quantum information encoded in a qubit). First a Bell state is prepared in ions 2 and 3. Next ion 1 is prepared in $|\phi\rangle$ by the unitary operation U_χ. The symbol connecting qubit 1 and 2 to the right of U_χ represents a CNOT gate operation. The CNOT together with the $\pi/2$ pulse on ion 1 is used for mapping the Bell state basis onto the product state basis, $\uparrow\uparrow$, $\uparrow\downarrow$, $\downarrow\uparrow$, $\downarrow\downarrow$ which can be measured by flourescence measurements. Conditioned on the results of these two measurements, additional single-qubit gates are applied to qubit 3 to recover the initial state.

Deterministic quantum teleportation was first demonstrated [36, 37] by groups in Boulder and Innsbruck in 2004. Figure 8.16c shows the measured teleportation fidelity of different input states [43], i.e. the probability of successfully recovering the state on Bob's side which on average was 83 %. By purely classical communication, this result would have been 66.6 % at most which demonstrates that genuine quantum effects were at work in these experiments.

8.8 Quantum Simulation

Quantum simulations are another research direction that has recently come into the focus of trapped-ion experiments. Quantum simulation aims at simulating a quantum system of interest with a controllable laboratory system described by the same mathematical model. In this way, it might be possible to simulate quantum systems that can neither be efficiently simulated on a classical computer [44] nor easily accessed experimentally. For this, the laboratory system needs to be very well understood in terms of the Hamiltonian describing it. To turn it into a useful quantum simulator, it should allow for parameter tunability and for the measurement of observables that provide important insights into the physics of the system to be simulated. There are two types of quantum simulators currently discussed in the literature. Digital quantum simulators [45] try to translate the unitaries describing the system dynamics into quantum circuits consisting of elementary gate operations. In this approach, a universal quantum computer could be used for efficiently simulating all quantum systems with local interactions. The second class, that one might call analog quantum simulators, builds on the principle of engineering a system having exactly the Hamiltonian one is interested in. The main motivation behind these approaches is to find solutions of problems in quantum-many body physics that cannot be efficiently simulated on classical computers. First experiments demonstrating basic elements of analog and digital quantum simulators with trapped ions have been carried out in the last couple of years (for an overview, see Refs. [46, 47]).

One class of Hamiltonians that seem to be particularly well-suited for being simulated by trapped ions, are models describing systems of interacting spins. Experiments simulating the Ising Hamiltonian with a transverse field, $H = \sum_{i,j} J_{ij}\sigma_z^i\sigma_z^j + B\sum_i \sigma_x^i$ started with 2–3 ions [48, 49] demonstrating adiabatic state changes and spin frustration. In the mean time, experiments involving up 16 ions [50] have been carried out in order to show the onset of a quantum phase transition. In all these experiments, each ion-qubit encodes a spin-1/2 system that interacts with the other spins via laser-mediated effective spin-spin interactions similar to the ones used for the entangling quantum gates discussed previously. The action of the transverse field can be fairly straight-forwardly implemented by further interactions coupling the two qubit states.

If the present experiments using the analog simulation approach could be scaled up to a few tens of ions, trapped-ion experiments might enter the regime where numerical simulation aiming at predicting experimental outcomes becomes infeasible due to the big size of the Hilbert space required to describe the quantum dynamics. If these conditions are met, trapped-ion quantum experiments would no longer be only demonstration experiments but also become tools for predicting quantum behaviour in a new regime.

8.9 Quantum Information for Precision Measurements

Techniques for entangling ions with each other are not only of interest in the context of quantum computation and simulation but also for spectroscopic precision measurements. There are different ways of enhancing spectroscopic measurements with trapped ions by entangling interactions:

1. **Improved signal-to-noise ratio.** In a Ramsey experiment (see Fig. 8.9), an ion initially prepared in a superposition state $\psi(0) = |\alpha\rangle + |\beta\rangle$ is used to measure the difference Δ between the laser (or the microwave) frequency and the atomic transition frequency. At the end of the Ramsey probe time τ, a non-vanishing Δ manifests itself as a phase factor in the quantum state $\psi(\tau) = |\alpha\rangle + e^{i\Delta\tau}|\beta\rangle$. The second Ramsey pulse maps the phase $\phi = \Delta\tau$ into a population imbalance of the upper and lower state which is subsequently measured by a fluorescence measurement. If this experiment is carried out on N ions instead of a single one, the signal-to-noise ratio is improved by a factor $\sqrt{N}$.
   Instead of preparing N ions in such a product state, on which experiments are performed in parallel, they can also be prepared in the entangled state $\Psi(0) = |\alpha\rangle|\alpha\rangle \cdots |\alpha\rangle + |\beta\rangle|\beta\rangle \cdots |\beta\rangle$ by one of the interactions discussed in Sect. 8.5. At the end of the probe time, this state has evolved into $\Psi(t) = |\alpha\rangle|\alpha\rangle \cdots |\alpha\rangle + e^{iN\Delta\tau}|\beta\rangle|\beta\rangle \cdots |\beta\rangle$ and the phase factor can be mapped into changes of the state populations by another laser pulse. The phase of this entangled state, $\phi = N\Delta\tau$, is more sensitive to changes in Δ than a single-ion superposition state. The signal-to-noise ratio is improved by a factor N compared to the single-ion experiment provided that the state remains fully coherent over the time of the experiment. Thus entanglement enhances the sensitivity by a factor of $\sqrt{N}$. Use of entangled states for enhancing the signal-to-noise ratio of spectroscopic measurements was first proposed by Bollinger and colleagues [51] and experimentally implemented in a demonstration experiment a few year later [52]. However, when evaluating the advantage brought by entangled states, the effects of decoherence need to be considered with great care [53].
2. **Longer coherence times.** Entanglement can also be used to achieve longer coherence times by preparing a quantum state that is not susceptible to the dominant source of decoherence. In Ref. [54], a maximally entangled state of two ions was prepared for measuring the atomic quadrupole moment of the $D_{5/2}$ state in $^{40}Ca^{+}$. The entangled state was made insensitive to fluctuations of the strengths of the magnetic field which limited the coherence time of single-ion superposition states to few milliseconds. In contrast, the entanglement of the two-ion state persisted for hundreds of milliseconds which greatly enhanced the spectroscopic resolution.
3. **Quantum logic spectroscopy.** Undoubtedly, the practically most relevant use of entanglement for spectroscopic purposes is the development of quantum logic spectroscopy [55]. This technique enables the spectroscopic investigations of narrow transitions in ions that cannot be easily measured by the electron shelving technique (see Fig. 8.3). The key idea is to transfer the information about whether

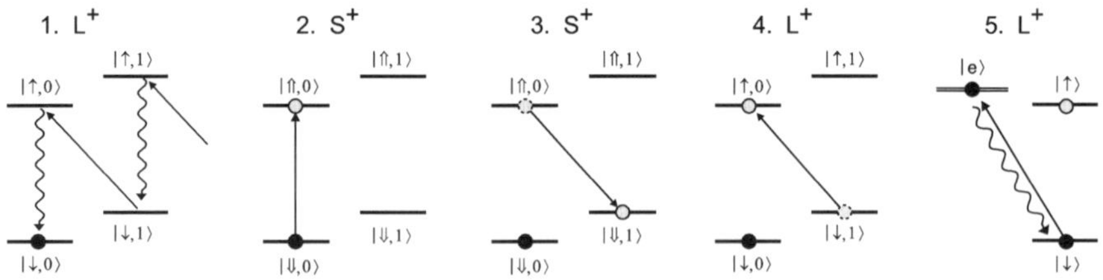

Fig. 8.17 Quantum logic spectroscopy protocol. After ground state cooling and initialization of the logic ion (L^+), the spectroscopy ion (S^+) is probed. In steps 3 and 4, a *red* sideband π pulse on S^+ mapping the upper state population to the motional mode is followed by another *red* sideband π pulse on L^+ which transfers the information from the mode into the qubit encoded in L^+. Finally, the qubit is read out by a fluorescence measurement using the electron shelving technique

the excitation of the "spectroscopy" ion was successful or not to an second "logic" ion of a different species that is co-trapped with the other ion. The logic ion is an ion species that can be easily detected. The information transfer proceeds by mapping the information via red sideband excitation from the spectroscopy ion to a joint vibrational mode of the two-ion crystal that was initially cooled to the ground state. In a second step, the information is mapped from the vibrational mode to the internal states of the logic ion where it is read out. This protocol is illustrated in Fig. 8.17.

Quantum logic spectroscopy was developed for the construction of atomic clocks based on a single $^{27}Al^+$. This isotope has a clock transition in the ultraviolet spectral region with very favourable properties but cannot be measured and laser-cooled easily. The Boulder ion trapping group has demonstrated the operation of Al^+ single ion clocks with fractional frequency uncertainties in the range of 10^{-17} by transferring the spectroscopic information to a co-trapped $^9Be^+$ or a $^{27}Mg^+$ ion [56, 57].

Other variants of quantum logic spectroscopy are likely to find applications in spectroscopic investigations of atomic dipole transitions that are of astro-physical interest and in experiments with molecular ions.

8.10 Decoherence and Scalability Issues

The coherent quantum operations described so far will fail if the experimental control is not good enough or if the ions couple to strongly to fluctuating electromagnetic fields created by the ions' environment. Also, some of the most important quantum computation applications require control over a very large number of qubits which makes scalability of the approach an important issue. These two topics will be briefly discussed in the following.

8.10.1 Decoherence in Trapped-Ion Experiments

The coherence of multi-qubit entangled states can be affected by decoherence in many different ways. In the absence of control fields inducing gate operations, the most relevant type of decoherence is dephasing of the qubit states by electromagnetic fields leading to shifts in the energy difference of the qubit states. Such shifts are often caused by fluctuating magnetic fields causing Zeeman shifts. For this reason, it is favourable to store quantum information in qubits whose transition frequency does not depend in first order on the Zeeman effect. In this way, single-qubit coherence times of greater than 10 s can be achieved [58].

Spontaneous decay is only relevant for optical qubits where one of the qubit states is metastable. However, it is only a minor error source as the lifetime of these states usually is on the order of a second and therefore much longer than the typical duration of most of the current experiments.

Since all current deterministic entangling gates make use of qubit-qubit couplings mediated by the ion motion, we also need to consider the influence of an environment on the ion's motional state. Trapped ions couple via their charge to fluctuating electric fields. Field fluctuations at frequencies close to the vibrational frequency heat up the ion motion [59]. Motional heating reduces the fidelity of entangling gates. In case of the Cirac-Zoller gate, the ion motion has to be cooled to the vibrational ground state. The conditional phase gate and the Mølmer-Sørensen require the ion motion to be only in the Lamb-Dicke regime, but they are affected by heating occurring during the gate operation.

The physical origin of the fluctuating electric fields are still under discussion. Thermal electric noise and noise generated by processes involving the surfaces of the trap are being considered. For smaller traps, the problem becomes more and more severe due to the much stronger electric fields that voltage fluctuations on a trap electrode create. While for millimeter-sized ion traps and oscillation frequencies in the MHz-range, heating rates are typically on the order of 1 phonon per second, for microfabricated traps with ion-electrode distances below 100 μm, heating rates can be orders of magnitude higher. Operating ion traps at cryogenic temperatures is one approach for reducing motional heating rates [60]. Recently, it was found that motional heating could be strongly improved by in-situ cleaning of an ion trap with an argon-ion beam [61].

Limited classical control is yet another source of errors in trapped ion quantum information experiments. Laser intensity noise leads to errors in single- and multi-qubit rotations. Laser frequency noise is an important source of errors in experiments with optical qubits as the laser frequency serves as a local oscillator for measuring the phase of qubit superposition states. In experiments with hyperfine qubits, temporal fluctuations of the length of the laser path of the Raman beam setup can give rise to phase errors as well. In multi-qubit experiments, qubit cross-talk needs to be avoided. Finally, real ions have more stable states than the two energy levels the quantum information is encoded in. Absorption or emission of photons on transitions connecting the qubit states to other states leads to a loss of information from the computational subspace.

8.10.2 Increasing the Number of Qubits

The prospect of building a universal quantum computer started the field of trapped-ion quantum information processing. Building such a device with imperfect components is in principle possible if quantum error correction protocols are used. However, the implementation of fault-tolerant quantum computing comes with a huge overhead in terms of the number of qubits that are required for carrying out even simple gate operations.

The simple approach of manipulating quantum information stored in a linear string of ions by coupling to the ion motion becomes impractical for ion numbers beyond a few tens of particles. There are two obvious problems with this approach:

- There is a structural phase transition from a linear ion string to a zig-zag configuration when more and more ions are added to an anisotropic harmonic potential characterized by the axial and radial frequencies ν_{ax} and ν_{tr} with $\nu_{ax} \ll \nu_{tr}$. To keep the crystal linear, the trap anisotropy ν_{ax}/ν_{tr} has to be increased. While raising ν_{tr} is difficult, lowering ν_{ax} renders laser cooling more difficult and comes at the price of increased motional heating and reduced speed of two-qubit gate operations.
- In longer strings, the speed of two-qubit gate operations is reduced for an additional reason. Because of the higher mass of longer ion strings, the momentum transfer of $\hbar k$ from photon absorption or emission causes smaller changes to the motional state of the string. In addition, coupling to a single vibrational mode becomes more challenging due to the increased number of vibrational mode frequencies.

Different strategies are being explored to overcome these problems. One approach is based on storing smaller number of ions in different traps and using photons as a quantum interface for entangling ions held in different traps. While it appears to be challenging to achieve deterministic entangling gates with this approach, entanglement distribution is feasible. In this approach, ion qubits are entangled with a degree of freedom of an emitted photon. When photons emitted from different ions are interfered on a beam splitter and subsequently measured, entanglement between the ions can be created probabilistically but heralded by the photon measurement results [28].

Another strategy is to split a long ion string into smaller substrings that are dynamically merged, split and moved around in order to carry out entangling gates on a small number of ions. This approach, which was pioneered by D. Wineland's ion trapping group in Boulder, requires the ability to create anharmonic trapping potentials that can be dynamically changed. The efficient generation of anharmonic potentials necessitates in turn a reduction of the ion-electrode distance which can be achieved by making use of microfabrication technologies for building microstructured segmented ion traps. This approach is currently pursued by many different research groups.

By segmenting the electrode opposing the rf-electrodes in a linear trap as shown in Fig. 8.18, flexible axial potentials are generated by application of suitable dc-voltages. An even more radical approach consists in arranging all electrodes in a

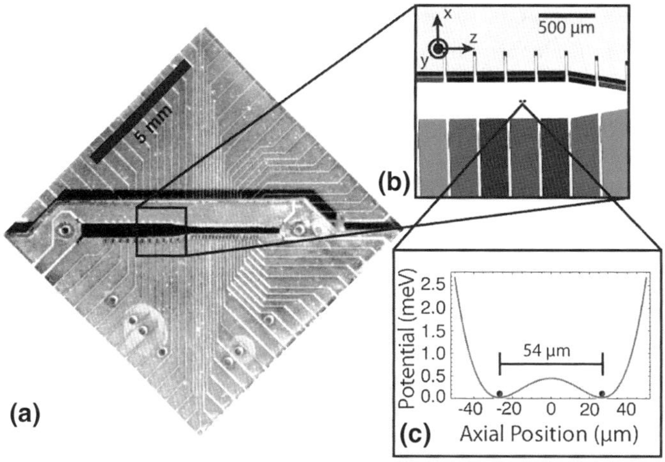

Fig. 8.18 A Two-layer segmented linear trap. By segmenting the electrode opposite to the radio-frequency electrode, anharmonic potentials along the axis of the trap are produced by applying dc-voltages to the different trap segments (figure reproduced from Ref. [63])

single plane in such a way that a two-dimensional rf-quadrupole field providing transverse confinement is created a few tens of micrometers above the trap surface and axial potentials are produced by dc-voltages applied to the segmented electrodes. In segmented microtraps, entanglement between ions in different trapping wells was demonstrated [62, 63]. With very low-noise programmable voltage sources, it is possible to transport ions fast and non-adiabatically over macroscopic distances [64, 65] and to split ions strings on timescales of a few tens of microseconds without significantly heating up the ion motion. A second ion species co-trapped with the quantum logic ions can be used for re-cooling the ion string after splitting operations. Microstructured ion traps are also being considered for achieving quantum simulations with two-dimensional arrays of ions each held in a separate trap.

8.11 Outlook

Over the last decade, techniques have been developed for quantum information processing using small crystals with about 2–20 ions. Qubits encoded in ions can be prepared, manipulated and measured with high fidelity. While the experimental demonstrations are still far from coming close to what is needed for full-scale quantum computation, they have enabled experiments demonstrating fundamental principles of QIP and have helped to shed new light on fundamental properties of quantum physics like entanglement between particles and the quantum measurement process.

In addition to the multi-ion gates discussed in this chapter, new techniques are currently being developed. These include the development of new entangling gates based on microwaves and magnetic field gradients [66, 67], the use of Rydberg instead of Coulomb interactions [68] and the development of fast gates based on ultrashort laser pulses [69]. Other experimental efforts are directed at interfacing trapped ions with the modes of microfabricated optical resonators and with other quantum systems capable of storing quantum information.

The biggest impact QIP techniques with trapped ions will have in the short term is most likely to be found in the field of quantum simulation where the gap between the experimental achievements and the theoretical requirements is much smaller than for quantum computation. Also, QIP techniques will certainly continue to gain even more importance in experiments where quantum logic techniques are used for turning ions into sensitive sensors and measurement devices for quantum logic spectroscopy.

References

1. M.A. Nielsen, I.L. Chuang, *Quantum Computation and Quantum Information* (Cambridge University Press, Cambridge, 2000)
2. D.P. DiVincenzo, The physical implementation of quantum computation. Fortschr. Phys. **48**, 771 (2000)
3. J.J. Bollinger, D.J. Heinzen, W.M. Itano, S.L. Gilbert, D.J. Wineland, A 303 MHz frequency standard based on trapped Be^+ Ions. IEEE Trans. Instr. Meas. **40**, 126 (1991)
4. W. Nagourney, J. Sandberg, H. Dehmelt, Shelved optical electron amplifier: observation of quantum jumps. Phys. Rev. Lett. **56**, 2797 (1986)
5. J.C. Bergquist, R.G. Hulet, W.M. Itano, D.J. Wineland, Observation of quantum jumps in a single atom. Phys. Rev. Lett. **57**, 1699 (1986)
6. T. Sauter, W. Neuhauser, R. Blatt, P.E. Toschek, Observation of quantum jumps. Phys. Rev. Lett. **57**, 1696 (1986)
7. J.I. Cirac, P. Zoller, Quantum computations with cold trapped ions. Phys. Rev. Lett. **74**, 4091 (1995)
8. A. Sørensen, K. Mølmer, Quantum computation with ions in thermal motion. Phys. Rev. Lett. **82**, 1971 (1999)
9. E. Solano, R.L. de Matos Filho, N. Zagury, Deterministic bell states and measurement of the motional state of two trapped ions. Phys. Rev. A **59**, R2539 (1999)
10. C.A. Sackett, D. Kielpinski, B.E. King, C. Langer, V. Meyer, C.J. Myatt, M. Rowe, Q.A. Turchette, W.M. Itano, D.J. Wineland, C. Monroe, Experimental entanglement of four particles. Nature **404**, 256 (2000)
11. D. Leibfried, B. DeMarco, V. Meyer, D. Lucas, M. Barrett, J. Britton, W.M. Itano, B. Jelenković, C. Langer, T. Rosenband, D.J. Wineland, Experimental demonstration of a robust, high-fidelity geometric two ion-qubit phase gate. Nature **422**, 412 (2003)
12. F. Schmidt-Kaler, H. Häffner, M. Riebe, S. Gulde, G.P.T. Lancaster, T. Deuschle, C. Becher, C.F. Roos, J. Eschner, R. Blatt, Realization of the Cirac-Zoller controlled-NOT quantum gate. Nature **422**, 408 (2003)
13. J. Benhelm, G. Kirchmair, C.F. Roos, R. Blatt, Towards fault-tolerant quantum computing with trapped ions. Nat. Phys. **4**, 463 (2008)
14. J.P. Schiffer, Phase transitions in anisotropically confined ionic crystals. Phys. Rev. Lett. **70**, 818 (1993)
15. H.C. Nägerl, D. Leibfried, H. Rohde, G. Thalhammer, J. Eschner, F. Schmidt-Kaler, R. Blatt, Laser addressing of individual ions in a linear ion trap. Phys. Rev. A **60**, 145 (1999)

16. C. Monroe, D.M. Meekhof, B.E. King, S.R. Jefferts, W.M. Itano, D.J. Wineland, P.L. Gould, Resolved-sideband Raman xooling of a bound atom to the 3d zero-point energy. Phys. Rev. Lett. **75**, 4011 (1995)
17. C. Roos, T. Zeiger, H. Rohde, H. Nägerl, J. Eschner, D. Leibfried, F. Schmidt-Kaler, R. Blatt, Quantum state engineering on an optical transition and decoherence in a Paul trap. Phys. Rev. Lett. **83**, 4713 (1999)
18. A.H. Myerson, D.J. Szwer, S.C. Webster, D.T.C. Allcock, M.J. Curtis, G. Imreh, J.A. Sherman, D.N. Stacey, A.M. Steane, D.M. Lucas, High-fidelity readout of trapped-ion qubits. Phys. Rev. Lett. **100**, 200502 (2008)
19. D. Leibfried, R. Blatt, C. Monroe, D. Wineland, Quantum dynamics of single trapped ions. Rev. Mod. Phys. **75**, 281 (2003)
20. D.J. Wineland, C. Monroe, W.M. Itano, D. Leibfried, B.E. King, D.M. Meekhof, Resolved-sideband Raman cooling of a bound atom to the 3D zero-point energy. J. Res. Natl. Inst. Stand. Technol. **103**(3), 259 (1998)
21. T.W. Hänsch, A.L. Schawlow, Cooling of gases by laser radiation. Opt. Comm. **13**, 68 (1975)
22. D.J. Wineland, H.G. Dehmelt, Proposed $10^{14}\nu < \Delta\nu$ laser fluoresence spectroscopy of the Tl^+ monoion oscillator III (side band cooling). Bull. Am. Phys. Soc. **20**, 637 (1975)
23. S. Stenholm, Semiclassical theory of laser cooling. Rev. Mod. Phys. **58**, 699 (1986)
24. J. Eschner, G. Morigi, F. Schmidt-Kaler, R. Blatt, Laser cooling of trapped ions. J. Opt. Soc. Am. B **20**, 1003 (2003)
25. F. Diedrich, J.C. Bergquist, W.M. Itano, D.J. Wineland, Laser cooling to the zero-point energy of motion. Phys. Rev. Lett. **62**, 403 (1989)
26. D.M. Meekhof, C. Monroe, B.E. King, W.M. Itano, D.J. Wineland, Generation of nonclassical motional states of a trapped atom. Phys. Rev. Lett. **76**, 1796 (1996)
27. F. Mintert, C. Wunderlich, Ion-trap quantum logic using long-wavelength radiation. Phys. Rev. Lett. **87**, 257904 (2001)
28. D.L. Moehring, P. Maunz, S. Olmschenk, K.C. Younge, D.N. Matsukevich, L.M. Duan, C. Monroe, Entanglement of single-atom quantum bits at a distance. Nature **449**, 68 (2007)
29. A. Sørensen, K. Mølmer, Entanglement and quantum computation with ions in thermal motion. Phys. Rev. A **62**, 022311 (2000)
30. G.J. Milburn, S. Schneider, D.F. James, Ion trap quantum computing with warm ions. Fortschr. Phys. **48**, 801 (2000)
31. C. Monroe, D.M. Meekhof, B.E. King, W.M. Itano, D.J. Wineland, Demonstration of a fundamental quantum logic gate. Phys. Rev. Lett. **75**, 4714 (1995)
32. M. Riebe, K. Kim, P. Schindler, T. Monz, P.O. Schmidt, T.K. Körber, W. Hänsel, H. Häffner, C.F. Roos, R. Blatt, Process tomography of ion trap quantum gates. Phys. Rev. Lett. **97**, 220407 (2006)
33. P.J. Lee, K.A. Brickman, L. Deslauriers, P.C. Haljan, L. Duan, C. Monroe, Phase control of trapped ion quantum gates. J. Opt. B **7**, 371 (2005)
34. T. Monz, P. Schindler, J.T. Barreiro, M. Chwalla, D. Nigg, W.A. Coish, M. Harlander, W. Hänsel, M. Hennrich, R. Blatt, 14-qubit entanglement: creation and coherence. Phys. Rev. Lett. **106**, 130506 (2011)
35. C.F. Roos, M. Riebe, H. Häffner, W. Hänsel, J. Benhelm, G.P.T. Lancaster, C. Becher, F. Schmidt-Kaler, R. Blatt, Control and measurement of three-qubit entangled states. Science **304**, 1478 (2004)
36. M. Riebe, H. Häffner, C.F. Roos, W. Hänsel, J. Benhelm, G.P.T. Lancaster, T.W. Körber, C. Becher, F. Schmidt-Kaler, D.F.V. James, R. Blatt, Deterministic quantum teleportation with atoms. Nature **429**, 734 (2004)
37. M.D. Barrett, J. Chiaverini, T. Schaetz, J. Britton, W.M. Itano, J.D. Jost, E. Knill, C. Langer, D. Leibfried, R. Ozeri, D.J. Wineland, Deterministic quantum teleportation of atomic qubits. Nature **429**, 737 (2004)
38. J. Chiaverini, J. Britton, D. Leibfried, E. Knill, M.D. Barrett, R.B. Blakestad, W.M. Itano, J.D. Jost, C. Langer, R. Ozeri, T. Schaetz, D.J. Wineland, Implementation of the semiclassical quantum Fourier transform in a scalable system. Science **308**, 997 (2005)

39. R. Reichle, D. Leibfried, E. Knill, J. Britton, R.B. Blakestad, J.D. Jost, C. Langer, R. Ozeri, S. Seidelin, D.J. Wineland, Experimental purification of two-atom entanglement. Nature **443**, 838 (2006)
40. M. Riebe, T. Monz, A.S. Villar, P. Schindler, M. Chwalla, M. Hennrich, R. Blatt, Deterministic entanglement swapping with an ion trap quantum computer. Nat. Phys. **4**, 839 (2008)
41. H. Häffner, C. Roos, R. Blatt, Quantum computing with trapped ions. Phys. Rep. **469**, 155 (2008)
42. R. Blatt, D. Wineland, Entangled states of trapped atomic ions. Nature **453**, 1008 (2008)
43. M. Riebe, M. Chwalla, J. Benhelm, H. Häffner, W. Hänsel, C.F. Roos, R. Blatt, Teleportation with atoms: quantum process tomography. New J. Phys. **9**, 211 (2007)
44. R. Feynman, Simulating physics with computers. Int. J. Theoret. Phys. **21**, 467 (1982)
45. S. Lloyd, Universal quantum simulators. Science **273**, 1073 (1996)
46. M. Johanning, A.F. Varón, C. Wunderlich, Quantum simulations with cold trapped ions. J. Phys. B **42**, 154009 (2009)
47. R. Blatt, C.F. Roos, Quantum simulations with trapped ions. Nat. Phys. **8**, 277 (2012)
48. H. Friedenauer, H. Schmitz, J. Glueckert, D. Porras, T. Schaetz, Simulating a quantum magnet with trapped ions. Nat. Phys. **4**, 757 (2008)
49. K. Kim, M. Chang, R. Islam, S. Korenblit, L. Duan, C. Monroe, Entanglement and tunable spin-spin couplings between trapped ions using multiple transverse modes. Phys. Rev. Lett. **103**, 120502 (2009)
50. R. Islam, E.E. Edwards, K. Kim, S. Korenblit, C. Noh, H. Carmichael, G.-D. Lin, L.M. Duan, C.C.J. Wang, J.K. Freericks, C. Monroe, Onset of a quantum phase transition with a trapped ion quantum simulator. Nat. Commun. **2**, 377 (2011)
51. J.J. Bollinger, W.M. Itano, D.J. Wineland, D.J. Heinzen, Optimal frequency measurements with maximally correlated states. Phys. Rev. A **54**(6), R4649 (1996)
52. D. Leibfried, M.D. Barrett, T. Schaetz, J. Britton, J. Chiaverini, W.M. Itano, J.D. Jost, C. Langer, D.J. Wineland, Toward Heisenberg-limited spectroscopy with multiparticle entangled states. Science **304**, 1476 (2004)
53. S.F. Huelga, C. Macchiavello, T. Pellizzari, A.K. Ekert, M.B. Plenio, J.I. Cirac, Improvement of frequency standards with quantum entanglement. Phys. Rev. Lett. **79**, 3865 (1997)
54. C.F. Roos, M. Chwalla, K. Kim, M. Riebe, R. Blatt, 'Designer atoms' for quantum metrology. Nature **443**, 316 (2006)
55. P.O. Schmidt, T. Rosenband, C. Langer, W.M. Itano, J.C. Bergquist, D.J. Wineland, Spectroscopy using quantum logic. Science **309**, 749 (2005)
56. T. Rosenband, D.B. Hume, P.O. Schmidt, C.W. Chou, A. Brusch, L. Lorini, W.H. Oskay, R.E. Drullinger, T.M. Fortier, J.E. Stalnaker, S.A. Diddams, W.C. Swann, N.R. Newbury, W.M. Itano, D.J. Wineland, J.C. Bergquist, Frequency ratio of Al^+ and Hg^+ single-ion optical clocks; Metrology at the 17th decimal place. Science **319**, 1808 (2008)
57. C.W. Chou, D.B. Hume, J.C.J. Koelemeij, D.J. Wineland, T. Rosenband, Frequency comparison of two high-accuracy Al^+ optical clocks. Phys. Rev. Lett. **104**, 070802 (2010)
58. C. Langer, R. Ozeri, J.D. Jost, J. Chiaverini, B. DeMarco, A. Ben-Kish, R.B. Blakestad, J. Britton, D.B. Hume, W.M. Itano, D. Leibfried, R. Reichle, T. Rosenband, T. Schaetz, P.O. Schmidt, D.J. Wineland, Long-lived qubit memory using atomic ions. Phys. Rev. Lett. **95**, 060502 (2005)
59. Q.A. Turchette, Kielpinski, B.E. King, D. Leibfried, D.M. Meekhof, C.J. Myatt, M.A. Rowe, C.A. Sackett, C.S. Wood, W.M. Itano, C. Monroe, D.J. Wineland, Heating of trapped ions from the quantum ground state. Phys. Rev. **A 61**, 063418 (2000)
60. J. Labaziewicz, Y. Ge, P. Antohi, D. Leibrandt, K. Brown, I. Chuang, Suppression of heating rates in cryogenic surface-electrode ion traps. Phys. Rev. Lett. **100**, 013001 (2008)
61. D.A. Hite, Y. Colombe, A.C. Wilson, K.R. Brown, U. Warring, R. Jördens, J.D. Jost, K.S. McKay, D.P. Pappas, D. Leibfried, D.J. Wineland, 100-fold reduction of electric-field noise in an ion trap cleaned with in situ argon-ion-beam bombardment. Phys. Rev. Lett. **109**, 103001 (2012)

62. K.R. Brown, C. Ospelkaus, Y. Colombe, A.C. Wilson, D. Leibfried, D.J. Wineland, Coupled quantized mechanical oscillators. Nature **471**, 196 (2011)
63. M. Harlander, R. Lechner, M. Brownnutt, R. Blatt, W. Hänsel, Trapped-ion antennae for the transmission of quantum information. Nature **471**, 200 (2011)
64. R. Bowler, J. Gaebler, Y. Lin, T.R. Tan, D. Hanneke, J.D. Jost, J.P. Home, D. Leibfried, D.J. Wineland, Coherent diabatic ion transport and separation in a multizone trap array. Phys. Rev. Lett. **109**, 080502 (2012)
65. A. Walther, F. Ziesel, T. Ruster, S.T. Dawkins, K. Ott, M. Hettrich, K. Singer, F. Schmidt-Kaler, U. Poschinger, Controlling fast transport of cold trapped ions. Phys. Rev. Lett. **109**, 080501 (2012)
66. C. Ospelkaus, U. Warring, Y. Colombe, K.R. Brown, J.M. Amini, D. Leibfried, D.J. Wineland, Microwave quantum logic gates for trapped ions. Nature **476**, 181 (2011)
67. A. Khromova, C. Piltz, B. Scharfenberger, T.F. Gloger, M. Johanning, A.F. Varón, C. Wunderlich, Designer spin pseudomolecule implemented with trapped ions in a magnetic gradient. Phys. Rev. Lett. **108**, 220502 (2012)
68. M. Müller, L. Liang, I. Lesanovsky, P. Zoller, Trapped Rydberg ions: from spin chains to fast quantum gates. New J. Phys. **10**, 093009 (2008)
69. J. Mizrahi, C. Senko, B. Neyenhuis, K.G. Johnson, W.C. Campbell, C.W.S. Conover, C. Monroe, Ultrafast spin-motion entanglement and interferometry with a single atom. Phys. Rev. Lett. **110**, 203001 (2013)

Chapter 9
Optical Transitions in Highly Charged Ions for Detection of Variations in the Fine-Structure Constant

A. Ong, J. C. Berengut and V. V. Flambaum

Abstract In this review, we explore a class of optical transitions in highly charged ions that have very high sensitivity to variation of the fine-structure constant, α. An atomic clock based on such a transition could place strong limits on α-variation, and could be sensitive enough to corroborate astronomical studies that suggest cosmological spatial variations in α. We discuss how to find the ions which have these optical transitions, the source of the high sensitivity to α-variation, and some scaling laws that suggest that a highly charged ion clock could have better systematics than existing singly-ionized trapped ion clocks. Finally we give an overview of atomic spectra calculations as applied in highly charged ions.

9.1 Introduction

Atomic clocks are the most accurate devices in physics. The cesium fountain clock—the primary frequency standard used to define the second—is accurate to parts in 10^{-16} (see, e.g., [1–4]). More recently, two different clock technologies based on optical transitions have enabled stable frequency determination at the 10^{-17} level: optical lattice clocks, such as the Sr clock [5]; and ion clocks based on forbidden optical transitions in a single Al^+ ion interrogated with a logic ion [6]. In this chapter we consider the use of highly charged ions (HCIs) as the basis of an optical-frequency atomic clock. Electrons in highly charged ions (HCIs) are far more tightly bound than their near-neutral equivalents, and so systematic shifts in atomic clocks based on HCIs could be much smaller. We will show that optical transitions from the ground state may be found in particular HCIs, and estimate the systematic shifts. Furthermore we will show that such clocks can have very high sensitivity to variation of fundamental constants, which enables tests of fundamental physics. We start by

A. Ong (✉) · J. C. Berengut · V. V. Flambaum
School of Physics, University of New South Wales, Sydney, NSW 2052, Australia
e-mail: ong.andrew@gmail.com

W. Quint and M. Vogel (eds.), *Fundamental Physics in Particle Traps*,
Springer Tracts in Modern Physics 256, DOI: 10.1007/978-3-642-45201-7_9,

discussing variations in the fundamental constants of nature, and the astronomical results that motivated the discovery of these HCI transitions.

The Standard Model of Particle Physics contains around 25 dimensionless fundamental constants, such as coupling constants and mass ratios, that cannot be predicted by the theory and must be determined from experiment. It is widely believed that the Standard Model is a low energy manifestation of a more complete theory that unifies the forces of the Standard Model (electromagnetism and the strong and weak nuclear forces) with gravity. Several such theories predict variations in the fundamental constants in an evolving Universe [7–11]. The example that concerns us in this chapter is the fine-structure constant, $\alpha = e^2/\hbar c \approx 1/137$: a dimensionless fundamental constant of nature that determines the strength of the electromagnetic interaction.[1]

The value of α at different times and places in the Universe can be probed using the fine-structure splitting of observed atomic spectra, such as alkali-doublets, which depend on the value of the fine-structure constant. This "alkali-doublet" method was used to determine the value of α first in the emission spectra of distant galaxies [12] and later in quasar absorption spectra [13–15]. The many-multiplet (MM) method was developed in 1999 as a generalization of the alkali-doublet method [16]. The method has some advantages including: use of lines with higher sensitivity to α-variation; use of all lines seen in a quasar absorption spectrum in the analysis (higher statistical significance); and use of both positive and negative shifters that shift in opposite directions if α changes, which helps control systematics. Taken together, these allow for more than an order of magnitude improvement in the determination of α from quasar spectra. The downside is that it requires a more sophisticated approach to determine the α-sensitivity of lines used in the analysis, which is similar to that presented in Sect. 9.2.

An astronomical study of around 300 different quasar absorption systems observed using the Keck Telescope in Hawaii [17–19] and the Very Large Telescope in Chile [20, 21] indicate a spatial gradient in the value of α across the Universe. In one direction on the sky α appears to have been larger than the current laboratory value, while in the opposite direction it appears to have been smaller. The data from the two telescopes independently agree on the direction and magnitude of this variation. This result is known as the spatial α-dipole (or "Australian dipole") and it has the value of

$$\frac{\delta\alpha}{\alpha_0} = (1.10 \pm 0.25) \times 10^{-6} r \cos\psi \text{ Gly}^{-1}, \tag{9.1}$$

where $\delta\alpha/\alpha_0 = (\alpha(\mathbf{r}) - \alpha_0)/\alpha_0$ is the relative variation of α at a particular place $\mathbf{r}$ in the Universe (relative to Earth at $\mathbf{r} = 0$). The geometrical term $r \cos\psi$, where ψ is the angle between the direction of the measurement and the axis of the spatial dipole, gives the position of the quasar absorption system along the dipole axis. The axis direction corresponds to (17.4 (0.9) h, −58 (9)°) in equatorial coordinates.

[1] The variation of α is unrelated to the well-known "running" of α at high energies, rather, it is a variation of α at zero-momentum transfer across space and/or time.

The distance function is the equivalent light-travel distance $r = ct$, measured in giga-lightyears.

Our goal now is to translate the observed spatial gradient in values of α into terrestrially observable measurements in order to corroborate the result in the laboratory. If we note that the Earth moves with respect to the rest frame of the cosmic microwave background (CMB), then we can assume that this motion will have a component along the spatial α dipole [22]. Thus, we would expect α to vary in the terrestrial frame. Of course, we are making the assumption that the CMB should be taken as the rest frame of the α-variation, as it would be if the source of the variation is a co-moving scalar field. To compute the expected terrestrial variation, we start by neglecting the annual motion of the Earth around the Sun, which is averaged out over the course of a year. The velocity of the Sun in the CMB rest frame is known to high accuracy from the CMB itself, and is 369 $\mathrm{kms^{-1}}$ in the direction $(168°, -7°)$ [23]. This is almost perpendicular to the observed direction of the spatial dipole, and is therefore sensitive to the exact angle the motion makes with the dipole. This astronomical result thus translates into an α variation at the level of [22]

$$\frac{\dot{\alpha}}{\alpha} = 1.35 \times 10^{-18} \cos\theta \ \mathrm{yr}^{-1}, \tag{9.2}$$

where θ is the angle the motion of our solar system makes with the α-dipole axis. The best fit value for the spatial dipole results in a value of $\cos\psi \sim 0.07$, but this is uncertain to the level of $\sim$0.1. Naturally, the full result must include the modulation of the Earth's annual motion around the Sun, with the angle between the ecliptic plane and the spatial dipole approximately 35°. This modulation results in an α-variation of

$$\frac{\delta\alpha}{\alpha} = 1.4 \times 10^{-20} \cos\omega t, \tag{9.3}$$

where ω refers to the angular frequency of the yearly orbital and $\delta\alpha/\alpha$ takes its maximal ($\cos\omega t = 1$) value on around 15 June each year. This modulation can be used to show that a drift in α such as Eq. (9.2) is due to a spatial, rather than temporal, variation.

The implications of α variation would manifest in a number of different systems [22], but of interest here are the shifts in atomic transition frequencies observable by terrestrial atomic clocks. The current best limit on the α dipole is $(-1.6 \pm 2.3) \times 10^{-17}\ \mathrm{yr}^{-1}$ and comes from comparing the transition frequencies in the Al^{+} and Hg^{+} ion clocks [24]. Furthermore, a new generation of atomic clocks based on highly charged ions has been proposed due to their increased level of accuracy. These clocks would have the advantage of having a measurable drift in transition frequencies in time frames ranging from one year to a decade, depending on the choice of ion and the transitions used. Comparing this best limit with the astronomical result in Eq. (9.2), we see that atomic clocks would need to be 2 to 3 orders of magnitude more sensitive in order to detect α-variation within a year of data collection. Fortunately, HCI-based atomic clocks have the potential to provide this level of increase in accuracy.

HCIs have been produced in electron beam ion trap (EBIT) for around 30 years, however trapping and cooling of HCIs is difficult, partly because of the lack of strong E1 transitions. Furthermore, transitions in these ions are typically in the x-ray region, which is not suitable for performing accurate spectroscopy. However, with the refinement of EBIT technology and higher beam energies, the production of a wide range of HCIs has become feasible [25]. Cooling and spectroscopy techniques for HCIs in EBITs are improving [26–29], and in this chapter we will show how level crossings can be used to overcome the problem of x-ray transition energies.

In the following sections, we first begin by introducing the method used to determine the α-variation sensitivity of an energy level in an atom or ion (Sect. 9.2). Then, we will discuss in further detail the concept of level crossings (Sect. 9.3), which introduces useful optical $E1$, $M1$ and $E2$ transitions that can be used to probe α-variation. In doing so, we show how HCIs can provide the additional two orders of magnitude required to bridge this result with the astronomical limit previously mentioned. After that, we will introduce scaling laws to help extend our intuition from near-neutral systems to highly charged ions. Finally, we provide a brief overview of the computational methods that are used to accurately determine the theoretical properties of HCIs.

9.2 Sensitivity of Atomic Transitions to α-Variation

In this section, we quantify the sensitivity to α-variation of various atomic transitions so that we may directly compare the sensitivity of different systems. Our aim is to assign a value to each energy level or transition that measures the magnitude and direction of the change in atomic transition frequency for a fixed change in the value of α. This can be done by parametrizing the frequency shift using the quantity q, defined as

$$q = \left.\frac{d\omega}{dx}\right|_{x=0}, \tag{9.4}$$

where $x = \alpha^2/\alpha_0^2 - 1 \approx 2(\alpha - \alpha_0)/\alpha_0$ is the fractional change of the square of the fine-structure constant, α, from its current value α_0. q is the therefore a measure of the rate of change of a transition frequency with respect to the fractional change in the fine-structure constant, with all quantities measured in atomic units ($e = m_e = \hbar = 1$).

Let us consider the leading relativistic corrections (order v^2/c^2), which are most important near the origin, to the central-field Schrödinger equation for a valence electron, which can be written as (see, e.g. [30])

$$q = -\frac{\alpha^2}{2}\int R^2(r)(E_0+V(r))^2 r^2 dr + \frac{\alpha^2}{4}\int R(r)\left(\frac{dR}{dr} - X\frac{R}{r}\right)\frac{dV}{dr} r^2 dr, \tag{9.5}$$

where $R(r)$ and E_0 are the non-relativistic radial wavefunction and energy respectively, $V(r)$ is the potential and $X = j(j+1) - l(l+1) - s(s+1)$. Near the origin, the potential $V(r) \approx Z/r$ as it is unscreened by the electrons. Furthermore, the wavefunction squared for a many-electron ion with atomic number Z and ion charge Z_{ion} is [31–33]

$$\psi(0)^2 \approx \frac{Z(Z_{ion}+1)^2}{\nu^3} = \frac{ZZ_a^2}{\nu n^2}, \tag{9.6}$$

where ν is the effective principal quantum number and we have introduced an effective charge Z_a, defined by the relation $Z_a/n = (Z_{ion}+1)/\nu$. With the appropriate substitution, we are then able to write

$$q_{nj} \approx -I_n \frac{(Z\alpha)^2}{\nu(j+1/2)}. \tag{9.7}$$

Here I_n is the (positive) ionization energy of an electron in that atomic orbital, Z is the atomic charge, ν is the non-integer effective principal quantum number and j the angular momentum of the orbital.

Alternatively, we may arrive at the same result if we consider the fine-structure energy correction (the non-relativistic energy does not have an explicit dependence on α) to the Dirac equation for a hydrogen-like atom or ion in a pure, central Coulomb field, given by

$$E_{f.s.} = -\frac{\alpha^2 Z^4}{2n^3}\left(\frac{1}{j+1/2} - \frac{3}{4n}\right), \tag{9.8}$$

we see that as a consequence of the proportionality of the wavefunction near the origin above, we must have [34]

$$E_{f.s.} = -\frac{Z_a^2}{2n^2}\frac{\alpha^2 Z_a^2}{\nu}\left(\frac{1}{j+1/2} - \frac{Z_a}{nZ}\left(1 - \frac{Z_a\nu}{4nZ}\right)\right). \tag{9.9}$$

Noting that $I_n = -E_n = Z_a^2/2n^2$ and that in HCIs $1/(j+1/2) \gg Z_a/(nZ)$ we arrive again at Eq. (9.7).

If we label the lower state in the clock transition i and the upper state f, then the sensitivity of the transition to α-variation is $q = q_f - q_i$. Measuring a variation in α requires the comparison of the frequencies of two atomic clocks. As an example we consider the ions used to obtain the current best limit on α-variation [24] mentioned in the introduction. The transition used as the optical frequency standard in the Al^+ clock is from the $3s^{2\,1}S_0$ state to the $3s3p^3P_0$ state, this transition is relatively insensitive to α variation (small q). On the other hand, the Hg^+ clock uses a transition which is strongly dependent on α variation (large q). In fact, the Hg^+ transition is a strong negative shifter, $q \approx -52{,}000\,\mathrm{cm}^{-1}$. The ratio of these transition frequencies would therefore change if α were changing. By comparing how the ratio of these frequencies changes over a year, a terrestrial limit was placed on α-variation of

$\dot{\alpha}/\alpha = (-1.6 \pm 2.3) \times 10^{-17}$ yr^{-1}. This limit could in principle be improved by using a strongly positive shifting clock transition, rather than the Al^+ clock transition.

Equation (9.7) is an analytical approximation useful for estimating the size of the q value of atomic levels. However, we calculate the value of q for a atomic levels by directly varying the value of α in *ab initio* atomic codes that calculate the ion spectrum (our energy calculations are outlined in Sect. 9.6). The value of ω in Eq. (9.4) is calculated for the values of $x = -0.01, 0.0, 0.01$; q is then calculated from the numerical derivative using Eq. (9.4). The value of q calculated using any two of the three computed values should remain within a few percentage points of the others—any further deviation from this amount indicates that it is possible that repulsion between levels is occurring, at least in the numerical computation. Special care must be taken to compute q in these cases, such as computing ω for two extra values of x. It is possible to resolve the q values in cases where level crossings occur using experimental g factors, if they are available [35].

9.3 Level Crossings in Highly Charged Ions

Using Eq. (9.7), we see that the sensitivity to α-variation can be maximized by using highly charged ions because of their large ionization energy I_n. This can be further enhanced by using heavier elements to provide a large Z. Finally, significant differences in the configuration of the states involved allows for larger ranges of q. To maximize $|q| = |q_f - q_i|$, Eq. (9.7) should be used to look for maximal q values of both levels with opposite signs. However at the same time we also wish to constrain the transition energy to be within the range of optical lasers so that accurate optical spectroscopy techniques can be used. This is problematic for highly charged ions, since just like I_n the transition frequencies also grow with ion charge as $\sim Z_a^2$. Fortunately we can solve this problem using the phenomenon of level crossings [36].

A level crossing in an ion occurs when the energy ordering of two or more orbitals changes. In hydrogen-like ions in the non-relativistic limit, the energy levels are sorted only by principal quantum number n. That is, the energy of levels with the same principal quantum number n are degenerate—we will refer to this as the Coulomb degenerate limit.

In many-electron atoms or ions, the order of filling of the orbitals is different from the ordering implied by the Coulomb degenerate limit. This is because the presence of the first electron modifies the electric potential experienced by the second electron, which modifies the energy of the second electron. Electrons near the nucleus of the ion tend to reduce the "visible" charge experienced by the external electrons—an effect known as screening. The result of this effect is that orbitals with a smaller effective radius become more energetically favoured than those with a larger effective radius.

K^+ has completely filled $1s$, $2s$, $2p$, $3s$, and $3p$ orbitals. In the Coulomb degenerate limit, the $3d$ orbital would be filled next as it is expected to have the lowest energy amongst unfilled orbitals. However, in neutral potassium we see that the $4s$ level is filled first, because as mentioned above, an electron in the $3d$ orbital would tend to spend a large amount of time further away from the nucleus, hence experiencing a smaller screened charge. As a result, the electron is more tightly bound to the nucleus in the $4s$ orbital than the $3d$ orbital, and correspondingly the $4s$ orbital is more energetically favourable than the $3d$ orbital. Thus we see that in the periodic table the energy levels fill in a different order from the Coulomb degenerate limit; for most atoms the ordering follows the Madelung rule (also known as the Klechkowski rule).

Consider an isoelectronic sequence, which is created by adding additional protons to a nucleus while keeping the number of electrons constant. Successively adding protons brings the ion closer to the Coulomb degenerate limit, where screening effects no longer dominate the energy ordering. Returning to neutral potassium, this implies that there is some point along its isoelectronic sequence where having the valence electron in the $3d$ orbital becomes more energetically favorable than the $4s$ orbital. We call the point at which this occurs a level crossing. For the isoelectronic sequence of neutral potassium, the ground state of Sc^{2+} is $3d$—here the level crossing occurs at a relatively low ion charge.

Near such level crossings, the ionisation energies of the two levels can be very similar, so that the energy of transitions between the crossing orbitals may be within the range of optical lasers even in highly charged ions. This is of significant importance because the energy intervals in highly charged ions are usually in the x-ray regime. On the other hand, fine-structure transitions may also be within optical range, but such transitions would most likely have relatively small q-values, as predicted by Eq. (9.7). The most promising level crossings in HCIs possessing a high sensitivity to α-variation have been identified as the $4f - 5s$, $4f - 5p$ and $5f - 6p$ crossings [37], which take place at ion charges of between 8+ and 20+. See [37] for an exhaustive listing of all available level crossings in the periodic table.

As an example, we consider the $5s - 4f$ (see Fig. 9.1) level crossing in the Ag isoelectronic sequence. In neutral Ag ($Z = N = 47$) the valence electron is in the $5s$ orbital, with the $4f$ lying above energetically. As the nuclear charge is increased along the isoelectronic sequence, the $4f$ orbital eventually goes below the $5s$ orbital; this occurs at $Z \approx 61$. In Ref. [36] the energy levels of the single-valence-electron HCIs Nd^{13+}, Pm^{14+}, and Sm^{15+} were calculated. The $4f - 5s$ transitions were indeed found to be within laser range, even though the ionisation energies are ~250 eV. The spectrum of the two-valence-electron ion Sm^{14+} was also calculated in [36]. This ion has $E1$ transitions that could potentially be used for trapping and cooling, as well as weak clock transitions with very high sensitivity to α-variation. Indeed, comparison of two different transitions in Sm^{14+} would have sensitivity to α of $\Delta q \approx 260{,}000\,\text{cm}^{-1}$, around five times larger than the sensitivity of the Hg^{+} ion clock.

The potential of high-precision HCI clocks using Nd^{13+} and Sm^{15+} was explored in [38]; there a potential fractional clock accuracy of $\sim 10^{-19}$ was estimated. This level of accuracy would allow for determination of α-variation to within parts in 10^{-20} per year, competitive with astronomical bounds on spatial variation. A summary of potential candidate HCIs for detecting α-variation is presented in Table 9.1.

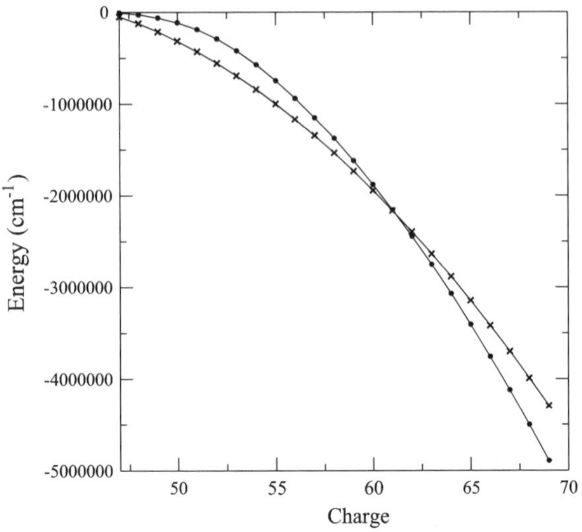

Fig. 9.1 Plot of the binding energy of the $4f$ and $5s$ orbitals in the Ag isoelectronic sequence. The $5s$ orbital is more energetically favorable compared to the $4f$ orbital until the level crossing at $Z \approx 61$. In the vicinity of this level crossing, the difference in the binding energies of the two orbitals is within the range of optical lasers

Table 9.1 List of one-, two-, and three-valence-electron HCIs near selected level crossings that may have optical transitions from the ground state

Crossing	One electron	Two electrons	Three electrons
$4f$–$5p$	Ce^{9+}, Pr^{10+}, Nd^{11+}	Ce^{8+}, Pr^{9+}	Pr^{8+}, Nd^{9+}, Pm^{10+}
$4f$–$5s$	Pm^{14+}	Pm^{13+}, Sm^{14+}, Eu^{15+}	Sm^{13+}, Eu^{14+}, Gd^{15+}, Tb^{16+}
$5f$–$6p$	Bk^{16+}, Cf^{17+}, Es^{18+}	Bk^{15+}, Cf^{16+}, Es^{17+}, Fm^{18+}	Cm^{13+}, Bk^{14+}, Cf^{15+}, Es^{16+}, Fm^{17+}

9.4 Hole Crossings

Within the isoelectronic sequences of highly charged ions near a given level crossing, which ions would be the most sensitive to α-variation? We will address this question in this section. In our search for further enhancement to the sensitivity coefficient q, it is helpful to notice that the number of allowed angular momentum combinations for one or two holes in otherwise closed shells is equivalent to that of the electron case. However, consider how Eq. (9.7) should be modified for system of two holes with the same quantum numbers as a system of two electrons near the same level crossing. In particular, the ionization energy would be larger, due to the hole state spending a larger amount of time closer to the nucleus, resulting in a larger relativistic shift. In Ref. [39] it was shown that $q \sim I_n^{3/2}$ for holes, rather than the $\sim I_n$ of Eq. (9.7). This makes it possible to achieve higher sensitivity to α variation without complicating the electronic level structure of a particular ion by using holes.

From an experimental perspective, a two hole ion is not particularly more complex than the two-valence-electron ion. Some of the most promising candidates for building an atomic clock for the detection of α-variation studied so far are W^{7+}, W^{8+}, Ir^{16+} and Ir^{17+}, and they are representative of one and two hole systems with $4f - 5p$ and $4f - 5s$ crossings, respectively [39]. Comparison of two transitions in Ir^{17+} could yield a total sensitivity to α-variation up to $\Delta q \approx 730{,}000\,\mathrm{cm}^{-1}$. The $5f - 6p$ level crossing for holes is not considered here because it occurs well beyond the range of stable atomic nuclei, and would be unsuitable for a long-lived α-variation experiment.

One complication that using holes presents is that it greatly increases the complexity of the configuration interaction calculations performed that are normally used to calculate the theoretical energy levels. In the case of two electrons above closed shells, the closed shells can be neglected for a reasonably accurate physical representation of the system. With two holes, however, one needs to treat all the electrons in the two orbitals that are near the level crossing as valence electrons. This is because the energy difference between these two orbitals are similar, by construction of taking ions near level crossings. The result of this is that in the $p - f$ crossing, it effectively becomes an 18-valence-electron problem. Nevertheless, the benefits of increased q coefficients conferred by hole crossings make them worth studying.

9.5 Scaling Laws for Atomic Clocks Based on Highly Charged Ions

In the previous sections, we saw that HCI-based atomic clocks are highly sensitive to α-variation. The next step would then be to consider the feasibility of constructing such a clock by examining the size of systematic errors experimentalists would expect to encounter in HCIs compared to near-neutral systems. In this section we generalise the size of systematic effects from existing ion clock systems towards highly charged ions, and examine if there is a general trend these systematics follow in highly charged systems. To do so, we first parametrize the evolution of an ion from a neutral system to a highly charged one and write approximate scaling laws which can be used in conjunction with the already-known properties of near-neutral ions to compute properties of highly charged ions. A previous topical review on highly charged ions [40] provides estimates for the scaling of some of these effects in hydrogen-like ions. In this review, we will instead derive scaling laws applicable to multi-electron ions.

It is first necessary to sensibly parametrize the evolution from neutral atom to a highly charged ion. There are two natural ways of doing this: starting with a neutral atom and then removing electrons one at a time (isonuclear sequence); or considering starting with a neutral atom and then adding protons one at a time (isoelectronic sequence). While the former way is how one would actually produce highly charged ions (such as the production of U^{92+} in an electron beam ion trap by removing electrons from neutral uranium one at a time), it turns out that the use of isoelectronic sequences is more convenient for calculations because the electronic structure of the ion is kept roughly constant. For example, it is meaningful to ask how a particular $6s - 6p$ transition changes along the isoelectronic sequence, but the same transition may not be present along the corresponding isonuclear sequence once the electronic structure changes.

It is tempting to use the ion charge, Z_{ion}, as a parameter for traversing along an isoelectronic sequence. However, the ion charge does not take into consideration the changes to the screening that the valence electron experiences. A different approach would be to use the effective charge experienced by the valence electron, Z_a. In deriving this parameter, we assume that the electron is moving in a screened Coulomb potential

$$V(r) \sim -\frac{Z_a}{r} \tag{9.10}$$

If we recall the formula for the nonrelativistic energy of an electron in such a potential,

$$E_n = -\frac{Z_a}{2n^2} = -\frac{(Z_{ion}+1)^2}{2\nu^2}, \tag{9.11}$$

where n is the integer principal quantum number. The second equality relates this equation to the formalism where an effective (non-integer) principal quantum number ν is used to parametrize the screening. We see that multi-electron ions require two parameters to characterise the effective interaction—the bare nuclear charge Z (important for short-range effects where the nucleus is unscreened), and the effective charge Z_a that the electron sees at larger distances. For example, the fine-structure scaling cannot be described by Z_a alone since it is a relativistic effect and depends on the wavefunction at the nucleus and hence depends on Z as well.

The value of Z_a for a particular orbital in an ion can be extracted from the Dirac-Fock energy of the level via $E_n = -Z_a^2/2n^2$. This approximation yields reasonable results whenever the dominant contribution of the potential experienced by the electron is well described by the mean field. Alternatively, one may turn to tabulated tables for Z_a, such as those presented in [37]. The evolution of Z_a from the neutral atom to highly charged ions can be fit to a linear function of the ion charge Z_{ion} over restricted intervals ($Z_a = AZ_{ion} + B$). We find that near-neutral ions ($0 < Z_{ion} < 5$) of the isoelectronic sequence can be described by a different set of coefficients A and B than ions in the same sequence with $5 < Z_{ion} < 30$.

9.5.1 Scaling of the Sensitivity to α-Variaition

In a previous section, we presented a formula that can be used to estimate the value of the q coefficient that measures the sensitivity of an energy level to α variation. The relation $q_{nj} \approx -I_n \frac{(Z\alpha)^2}{\nu(j+1/2)}$ was derived using the Schrödinger equation with some

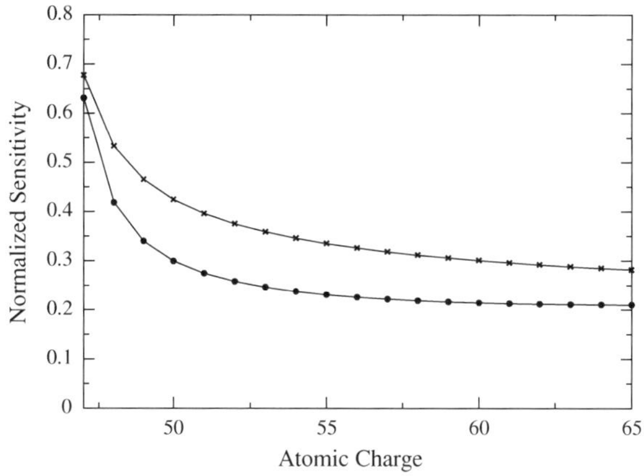

Fig. 9.2 Comparison of the normalized sensitivity, q_{norm}, of the 5s orbital for the Ag isoelectronic sequence. Circles: q_{norm} calculated from Eq. (9.12) using Dirac-Fock values of q and I_n. Crosses: q_{norm} approximated using the scaling formula, $Z_a/(Z_{ion}+1)n$, where Z_a is extracted from the calculated Dirac-Fock value of I_n. The method of scaling reproduces the trend of the diagram correctly

modifications, and it would be interesting to examine how closely the our convention for using Z_i is able to predict the results of fully relativistic calculations. We can define a 'normalized' version of the q coefficient, q_{norm} [39], defined by

$$q_{norm} = \left| \frac{q}{Z^2 \alpha^2 I_n} \right| . \tag{9.12}$$

Using this equation, we find that the normalized q coefficient scales as $\frac{Z_a}{(Z_{ion}+1)n}$. Figure 9.2 compares the result of using this scaling and the values derived from a Dirac-Fock calculation of the 5s orbital in the Ag isoelectronic sequence (ions that have the same number of electrons as neutral Ag but with any charge $Z \geq 47$). In the figure, the points marked with crosses represent the value of q_{norm} calculated using our scaling formulation, whereas the points marked with filled circles are the values obtained from direct Dirac-Fock calculation. We see that the simple scaling model is able to correctly reproduce the trend, but the value of the approximation decreases when trying to accurately predict the q value due to compounding errors. This is because the initial approximation overestimates the ratio of $q_{norm,Z+1}/q_{norm,Z}$, thus further points include the overestimation of the previous points (although the point-to-point error in $q_{norm,Z+1}/q_{norm,Z}$ decreases as we approach higher Z).

9.5.2 Scaling of *EJ* and *MJ* Transition Matrix Elements

In this section we present the approximate analytical scaling of the *EJ* and *MJ* transition matrix elements. As a starting point, we first examine the scaling of the non-relativistic $E1$ matrix element,

$$\langle nl|r|n'l'\rangle = \int P_{nl} r P_{n'l'} dr. \tag{9.13}$$

We can approximate the radial wavefunction P_{nl} by the non-relativistic single-electron wavefunction in a central Coulomb potential with effective charge Z_a by

$$P_{nl} = N_{nl}\left(\frac{2Z_a r}{n}\right)^{2l+1} e^{-\frac{Z_a r}{n}} F\left(-n+l+1, 2l+2, \frac{2Z_a r}{n}\right) \tag{9.14}$$

where

$$N_{nl} = \frac{1}{n(2l+1)!}\sqrt{\frac{Z_a(n+l)!}{(n-l-1)!}} \tag{9.15}$$

is a normalization coefficient and $F(a, b, x) = \sum_{m=0}^{\infty} a^{(m)} x^m / b^{(m)} m!$ are confluent hypergeometric functions. This radial wavefunction which is a reasonable approximation at distances far from the nucleus where the electron wavefunction has its peak.

The Z_a dependence of the integral in Eq. (9.13) can be calculated as

$$\int r P_i P_j dr \sim \frac{1}{Z_a}. \tag{9.16}$$

This is simply a formal way of stating that the ion gets smaller in size as its charge increases. As a result, the strength of $E1$ matrix elements is reduced in highly charged ions. On the other hand, $M1$ matrix elements are proportional to the magnetic moment (and hence angular momentum) of the external electron, which does not reduce as the ion gets smaller. Therefore the $M1$ matrix element does not scale with the charge. In neutral atoms, $E1$ transitions are usually the strongest transitions available. Our scaling law, however, tells us that in highly charged ions, $E1$ transitions grow weaker with increasing ion charge, yet $M1$ transitions remain relatively constant in strength.

For higher multipoles, one obtains higher powers of the atomic radius,

$$\langle r^n\rangle \sim \left(\frac{a_B}{Z_a}\right)^n, \tag{9.17}$$

where a_B is the Bohr radius, a characteristic length in atomic systems. The general scaling law for *EJ* matrix elements is

$$\langle \kappa_i || q_J^{(E)} || \kappa_j \rangle \sim (Z_a)^{-J} \tag{9.18}$$

and for MJ matrix elements is

$$\langle \kappa_i || q_J^{(M)} || \kappa_j \rangle \sim (Z_a)^{1-J}. \tag{9.19}$$

9.5.3 Scaling of Polarizability and Blackbody Radiation Shift

The blackbody radiation shift, in the lowest order, is proportional to the difference in the dipole polarizabilities of the two energy levels involved in the atomic clock transition. For an adiabatic system, this can be evaluated using the formula

$$\delta E = -\frac{1}{2}(831.9\ V/m)^2 \left[\frac{T(K)}{300}\right] \alpha_0(1+\eta) \tag{9.20}$$

where $T(K)$ is the temperature in Kelvins, α_0 is the static dipole polarizability and η is a small dynamic correction due to the frequency distribution, which will be disregarded for our purposes. The valence scalar polarizability of an atom in a state v can be expressed as a sum over all excited intermediate states n allowed by $E1$ selection rules,

$$\alpha_0 = \frac{2}{3(2j_v+1)} \sum_n \frac{\langle v||r||n\rangle\langle n||r||v\rangle}{E_n - E_v}. \tag{9.21}$$

We are interested in the approximate scaling of the blackbody radiation shift in highly charged ions. The reduced matrix elements in Eq. (9.21) have been shown to scale as $1/Z_a$ in Eq. (9.16). The non-relativistic energy scales as Z_a^2 using Eq. (9.11). This means that all terms in the summation have the same $1/Z_a^4$ scaling, so that the total polarizability must also scale as

$$\delta E \sim \alpha_0 \sim \left(\frac{1}{Z_a}\right)^4. \tag{9.22}$$

Equation (9.22) implies that the blackbody shifts in highly charged ions (large Z_a) would be largely suppressed compared to near-neutral ions.

9.5.4 Scaling of the Hyperfine Structure

Operators with large negative powers of radius will not follow the Coulomb radius scaling, Eq. (9.17), since the wavefunction at small distances cannot be described by P_{nl} that we introduced in Sect. 9.5.2. The hyperfine interaction is an example of such an interaction, with operators proportional to large negative powers of the radius r.

Table 9.2 Magnetic dipole hyperfine coefficients A and their scaling with increasing Z along the lithium isoelectronic sequence

Isotope	I	g_I	A (MHz)	Z_a	$\frac{(A/g_I)}{(A/g_I)\vert_p}$	$\frac{\frac{ZZ_a^3}{Z_{ion}+1}}{\frac{ZZ_a^3}{Z_{ion}+1}\vert_p}$
${}^{7}_{3}$Li	3/2	2.1709	399.34	1.25		
${}^{9}_{4}Be^{+}$	3/2	−0.7850	−625.55	2.31	4.35	4.20
${}^{11}_{5}B^{2+}$	3/2	1.7924	3603.77	3.33	2.53	2.49
${}^{13}_{6}C^{3+}$	1/2	1.4048	5642.40	4.35	2.00	2.00
${}^{15}_{7}N^{4+}$	1/2	−0.5664	−3973.68	5.36	1.74	1.74
${}^{17}_{8}O^{5+}$	5/2	−0.7575	−8474.13	6.36	1.59	1.59
${}^{19}_{9}F^{6+}$	1/2	5.2578	88106.93	7.37	1.50	1.50

Values of Z_a were obtained from Dirac-Fock calculations using the relation $Z_a = \sqrt{|2n^2 E|}$ where n is the principal quantum number and E is the orbital energy in atomic units. The notation $|_p$ means to use the values in the previous row of the table

Instead we must use the approach of Fermi-Segré [32, 33] where the normalised squared wavefunction at the origin is shown to be proportional to $Z(Z_{ion}+1)^2/\nu^3$. Note that since $\nu = n(Z_{ion}+1)/Z_a$, we can substitute these results into the formulae for the non-relativistic magnetic dipole hyperfine constant [41] A of a one-valence-electron system:

$$A = \begin{cases} \frac{2}{3} g_I \lim_{r\to 0}\left(\frac{P_{nl}(r)}{r}\right), & \text{for } l = 0 \\ \frac{l(l+1)}{j(j+1)} g_I \int_0^\infty P_{nl}(r) \frac{1}{r^3} P_{nl}(r), & \text{for } l \neq 0 \end{cases} \tag{9.23}$$

to obtain the following scaling law

$$\frac{A}{g_I} \sim \frac{ZZ_a^3}{(Z_{ion}+1)}. \tag{9.24}$$

Here we have factored out the nuclear g-factor g_I which varies greatly between nuclei to facilitate comparison with experimental results. We compare this scaling law with experimental data in Table 9.2. The table shows that the scaling law in this case very accurately predicts the experimental result. A similar result may be derived for the electric quadrupole hyperfine constant B. We should also point out that the widths of hyperfine transitions will scale as $\omega^3 \sim A^3$, therefore relaxation of hyperfine structure will occur much faster in HCIs.

9.5.5 Summary of Scaling Laws

We present a summary of the scaling laws in Table 9.3. One may extrapolate from known values of systematic shifts in near-neutral ions to estimate their values in HCIs; to do this one only needs to know the relevant Z_a values which are tabulated in [37].

Table 9.3 Scaling dependences for HCIs for various sources of systematic shifts in optical clocks

2nd order Stark shift	$\sim 1/Z_a^4$
Blackbody shift	$\sim 1/Z_a^4$
2nd order Zeeman shift	suppressed[1]
Electric quadrupole shift	$\sim 1/Z_a^2$
Fine-structure	$\sim Z^2 Z_a^3/(Z_{ion}+1)$
Hyperfine *A* coefficient	$\sim Z Z_a^3/(Z_{ion}+1)$

The second-order Zeeman effect is slightly more complicated, since it may depend on the particular fine or hyperfine structure of the ion. Nevertheless it will generally be suppressed in HCIs due to a larger energy denominator [42]. As a result of the reduced systematic effects, atomic clocks using HCIs and fine-structure transitions could reach 10^{-19} relative accuracy [43].

9.6 Atomic Calculations for Highly Charged Ions

In the following sections, we give a quick overview of the techniques we have used in computing the energy levels and other properties in HCIs. Further details can be found in Ref. [44]. We will pay particular attention to the difficulties that may arise when using these techniques in HCIs, especially when near a level crossing.

9.6.1 Hartree-Fock and Relativistic Hartree-Fock

The Hartree-Fock (HF) method is used to determine the ground state wave function and energies of a system that can be approximated by a single Slater determinant of orbitals (labelled $1s$, $2s$, $2p$, etc. in atomic systems). Relativistic Hartree-Fock, also known as Dirac-Fock (DF), is an extension of the regular Hartree-Fock method that makes use of spin-orbitals instead ($1s_{1/2}$, $2s_{1/2}$, $2p_{3/2}$, $2p_{1/2}$, etc.). Here, the equation being solved is the Dirac equation as opposed to the Schrödinger equation in the non-relativistic variant.

For highly charged ions, the nuclear charge is generally large enough such that the spin-orbit (fine-structure) energy splitting incurred by the relativistic terms is significant. For example, the $6p_{1/2}$ and $6p_{3/2}$ orbitals in $_{98}Cf^{17+}$ may be considered separately, as their energy separation exceeds 200,000 cm^{-1} (approximately 25 eV), far greater than the separation of $6p_{1/2}$ and $5f_{5/2}$ which is less than 10,000 cm^{-1}. Thus, while non-relativistic methods may suffice for light neutral atoms and near-neutral ions, relativistic methods should be employed in order to ensure meaningful results when working with heavy, highly charged ions.

9.6.2 Configuration Interaction

In the Hartree-Fock approximation, we assume that the electrons move independently in the self-consistent average field of the other electrons. The configuration interaction (CI) method is introduced to account for the correlation between electrons in different orbitals. The lowest order of the wave function is fixed by the Hartree-Fock determinant, further corrections to the wave function are generated by allowing progressively more excitations from the spin-orbitals present in the Hartree-Fock determinant. In contrast to the Hartree-Fock method where the basic building blocks are single electron spin-orbitals, the configuration interaction uses linear combinations the Hartree-Fock spin-orbitals to form configuration state functions. These configuration state functions are then taken in further linear combinations to form the final configuration interaction wave function.

The technical requirements of a full configuration interaction calculation generally limits it to being used in simple systems with a small number of electrons. To apply the technique to ions with many occupied spin-orbitals, configuration state functions may instead be generated by characterised by the number of spin-orbitals they contain which are swapped with virtual orbitals from one or more Hartree-Fock determinants (excitations). For example, if only a single spin-orbital differs, this is referred to as a configuration state function with a single excitation from the original configuration. In our calculations, the configuration state function is usually truncated at the level of double excitations. This is a particularly limiting approximation for the accuracy of the configuration interaction method in highly charged ions, as many interesting highly charged ions are heavy and thus tend to have many core electrons. When excitations are limited to those generated from only the valence orbitals, the core-core electron correlations and the core-valence electron correlations are poorly accounted for. In the next section, we will discuss a method that is particularly well-suited for use in such systems.

In general, one may expect the CI method to produce results for ionization energies that are roughly within the region of 1–10 % of the experimental value. However, in highly charged ions with level crossings, we will see in our sample Cf^{16+} calculation that this level of accuracy is insufficient for predicting the ordering of levels. This is because the ordering of energy levels is derived from the cancellation of the ionization energies at the fourth or fifth significant digit (because of the large ionization energies in HCIs). This means that our margin of error is comparable to the actual separation between levels, and thus is unable to predict the ordering of levels decisively.

9.6.3 Combining Many-Body Perturbation Theory with Configuration Interaction

In principle, a full configuration interaction calculation solves the Schrödinger (or Dirac) equation exactly, however it is not possible to perform a full calculation for any but the simplest atoms and ions. As previously mentioned, in the configuration interaction a small subset of electrons in an ion is often chosen (usually only the electrons in the valence orbitals are used), neglecting core-core and core-valence electron correlation effects. Many-body perturbation theory, on the other hand, is not accurate in describing valence-valence interactions, but accurately accounts for the core-valence interactions. A practical method has been developed [44] to combine the two techniques in order to leverage both their strengths, aptly called the combined configuration interaction and many-body perturbation theory (CI + MBPT) method.

The idea is to first use many-body perturbation theory to form an effective Hamiltonian, which includes accurate core-valence corrections as computed by many-body perturbation theory. The next step is to use this corrected Hamiltonian in the configuration interaction calculation instead of the original, unperturbed Hamiltonian. The method of the calculation remains otherwise the same as that was described in 9.6.2. The method produces highly accurate results that can be compared to other techniques such as the coupled-cluster method. As we will see, the corrections introduced by the inclusion of many-body perturbation theory in highly charged ions is usually significant, more so for such ions near level crossings, as the corrections may ultimately result in a reordering of energy levels. This usually translates to an ionization energy accurate at the 0.1–1 % level, compared to the 1–10 % accuracy of the CI-only method in Sect. 9.6.2.

9.6.4 *Cf^{16+}: A Sample Calculation*

In this section, we include a sample calculation for the highly charged Cf^{16+} ion, following [45]. The implementation of the CI + MBPT method used here, named AMBiT, is described in Refs. [46–48]. Californium is one of the last elements in the period table with a long-lived isotope (~900 years). This is advantageous in that it provides one of the highest charges available together with a level crossing. The Cf^{16+} ion itself contains two valence electrons above closed shells, near a $5f - 6p_{1/2}$ level crossing. The level structure of this ion is relatively simple, because the relatively large energy interval of the $6p_{1/2} - 6p_{3/2}$ causes mixed fine-structure states to be pushed well out of optical range. Therefore, the leading configurations of the lowest-lying states only involve $6p_{1/2}$. In order to highlight the significance of the inclusion of MBPT effects, we present two calculations side-by-side in Table 9.4—a CI-only calculation that includes all single and double excitations (from the valence configuration $6p^2$) up to a $n = 16, l = 3$, and a second CI + MBPT calculation that includes the CI-only calculation in addition to an MBPT calculation that includes states up to $n = 40, l = 6$.

From the table, we see that the effect of adding MBPT into the calculation causes the ordering of energy levels to change. We have already mentioned the importance of correctly accounting for the core-valence correlations, which the CI method fails to take into account particularly in Cf^{16+}. The 80 core electrons that have been excluded from the CI calculation must summarily be accounted for using the MBPT

Table 9.4 Table of Cf^{16+} Levels, all levels of interest have even parity. g-factors have been included to assist in the identification of levels

Configuration	J	g	Energy $(\text{cm})^{-1}$ CI	CI + Σ	q $(\text{cm})^{-1}$
$5f^16p^1$	3	0.82989	0	0	0
$6p^2$	0		−7429	5267	−370928
$5f^16p^1$	2	0.84818	7313	6104	106124
$5f^2$	4	0.85345	28746[1]	9711	414876
$5f^16p^1$	4	1.0481	21415[2]	24481	162126
$5f^2$	2	0.75319	38674	24483	354444
$5f^16p^1$	3	1.1776	23979	25025	59395
$5f^2$	5	1.0333	43097	29588	451455
$5f^2$	3	1.0954	50953	37467	393755
$5f^2$	4	1.1197	53229	42122	319216
$5f^2$	6	1.1371	57220	44107	459347
$5f^2$	0		68192	51425	380986
$5f^2$	2	1.1672	67267	51471	446045
$5f^2$	4	1.1198	69475	58035	461543
$5f^2$	1	1.5	75018	58132	449977
$5f^2$	6	1.0296	78739	63175	460416
$5f^2$	2	1.2672	89580	75041	465293
$5f^2$	0		127521	114986	446376
$5f^16p^1$	3	0.97651	211414	212632	323435
$6p^2$	1	1.4963	198879	213864	−113277

Levels marked[1,2] are heavily mixed in the CI calculation, resulting in a dominant contribution from $5f^2$ (63 and 51 % respectively) in both, while in the CI + Σ calculation the first state is 96 % $5f^2$ and the second state is 63 % $5f^16p^1$. In order to account for the total number of physical states, we have labelled the state with the largest $5f^16p^1$ contribution as the $5f^16p^1$ state

method. These corrections constitute about 1 % of the ionization energy of these levels. However, because the level crossing results in the intervals between the crossing states becoming anomalously small—the transition energies between crossing levels involve the calculation of the differences between large ionization energies in order to arrive at a difference in the fourth to sixth digit. In that regard, a physically accurate picture implies the necessity for calculations that are far more accurate than in atoms or ions not near level crossing. The exclusion of quantum electrodynamical corrections and the Breit interaction is noted here, as they may make a difference to the ionization energy at this level of accuracy.

The transition with the highest sensitivity to α variation will have the $6p^2 J = 0$ state as the lower (metastable) ground state. This is due to the level being a large negative shifter. Suitable upper levels include any of the $5f^2$ levels that have $J \leq 2$ to ensure the transition is at least of $E2$ type. The accuracy of an atomic clock derived from this ion will be dictated in part by the transition rates in Table 9.5. The transition rate controls the number of discrete excitation (or relaxation) events that can be counted, and the shorter the time between the discrete events, the greater the resolution of the device. Excitation events may be recorded through the use of a

Table 9.5 Selected transitions between states, identified by energy in relation to Table 9.4

Config.	J	E $(\mathrm{cm})^{-1}$	Config.	J	E $(\mathrm{cm})^{-1}$	S	$g.A$ $(\mathrm{s})^{-1}$
$5f^16p^1$	3	0	$6p^2$	0	5267	0.92401_{M3}	4.3519E-18
			$5f^16p^1$	2	6104	0.14553_{M1}	0.89281
			$5f^2$	4	9711	0.16895_{M1}	4.1731
			$5f^16p^1$	4	24481	2.5836_{M1}	1022.5
			$5f^16p^1$	4	24481	0.20237_{E2}	0.1993
			$5f^16p^1$	3	25025	0.071521_{M1}	30.235
			$5f^2$	1	58132	0.0023199_{E2}	0.17249
$6p^2$	0	5267	$5f^16p^1$	2	6104	0.54123_{E2}	2.4845E-8
			$5f^2$	2	24483	0.019556_{E2}	0.0057377
			$5f^2$	1	58132	0.1195_{M1}	476.2
$5f^16p^1$	2	6104	$5f^2$	4	9711	0.1839_{E2}	1.2569E-5
			$5f^16p^1$	4	24481	0.028394_{E2}	0.0066653
			$5f^16p^1$	3	25025	3.1428_{M1}	574.24
			$5f^16p^1$	3	25025	0.033026_{E2}	0.0089701
			$5f^2$	1	58132	0.0082881_{M1}	31.485
			$5f^2$	1	58132	0.043495_{E2}	1.8571
$5f^16p^1$	4	24481	$5f^16p^1$	3	25025	0.99502_{M1}	0.0043219
$5f^2$	2	24483	$5f^2$	1	58132	0.024099_{M1}	24.766
			$5f^2$	1	58132	0.18074_{E2}	0.87324
$5f^16p^1$	3	25025	$5f^16p^1$	4	58132	0.084098_{E2}	0.37462
$5f^2$	3	37467	$5f^2$	1	58132	0.084098_{E2}	0.37462
$5f^2$	0	51425	$5f^2$	1	58132	1.6793_{M1}	13.671
$5f^2$	2	51471	$5f^2$	1	58132	1.0698_{M1}	8.5301
			$5f^2$	1	58132	0.58299_{E2}	0.00085652

The included $6p^2$ state is metastable with respect to spontaneous decay into the ground state

separate logic ion, such as in [24]. In such a scheme, the clock ion and the logic ion are trapped together and a laser is used to pump the clock transition. The logic ion is then probed at a fixed rate, and it will be in a different state depending on whether the excitation of the clock ion occurred or otherwise.

A further advantage of using the $6p_{1/2}^2$ $J = 0$ state is that the quadrupole moment shift for such states, which is proportional to the stretched $E2$ diagonal matrix element $\langle 6p_{1/2}^2 M_J = 0|E2|6p_{1/2}^2 M_J = 0\rangle$ is zero due to angular momentum considerations. This means, however, that the quadrupole moment shift of the other transition state should also be kept small to minimize the total observable shift, which is just the difference between the two individual shifts of the two states. In theory, it is possible to work around a large quadrupole moment shift by achieving cancellation between the individual shifts of the upper and lower transition states.

References

1. R. Li, K. Gibble, K. Szymaniec, Improved accuracy of the NPL-CsF$_2$ primary frequency standard: evaluation of distributed cavity phase and microwave lensing frequency shifts. Metrologia **48**, 283 (2011)
2. V. Gerginov, N. Nemitz, S. Weyers, R. Schröder, D. Griebsch, R. Wynands, Uncertainty evaluation of the caesium fountain clock PTB-CSF2. Metrologia **47**, 65 (2010)
3. T.P. Heavner, T.E. Parker, J.H. Shirley, P.D. Kunz, S.R. Jefferts, in *PTTI Conference proceedings* (VA, Rastin, 2010), p. 457
4. J. Guéna, M. Abgrall, D. Rovera, P. Rosenbusch, M.E. Tobar, Ph Laurent, A. Clarion, S. Bize, Improved tests of local position invariance using ^{87}Rb and ^{133}Cs fountains. Phys. Rev. Lett. **109**, 080801 (2012)
5. T.L. Nicholson, M.J. Martin, J.R. Williams, B.J. Bloom, M. Bishof, M.D. Swallows, S.L. Campbell, J. Ye, Comparison of two independent Sr optical clocks with 1×10^{-17} stability at 10^3s. Phys. Rev. Lett. **109**, 230801 (2012). doi:10.1103/PhysRevLett.109.230801
6. C.W. Chou, D.B. Hume, J.C.J. Koelemeij, D.J. Wineland, T. Rosenband, Frequency comparison of two high-accuracy Al^+ optical clocks. Phys. Rev. Lett. **104**, 070802 (2010). doi:10.1103/PhysRevLett.104.070802
7. W.J. Marciano, Time variation of the fundamental 'constants' and kaluza-klein theories. Phys. Rev. Lett. **52**, 489 (1984)
8. P. Langacker, G. Segré, M.J. Strassler, Implications of gauge unification for time variation of the fine structure constant. Phys. Lett. B **528**, 121 (2002)
9. X. Calmet, H. Fritzsch, The cosmological evolution of the nucleon mass and the electroweak coupling constants. Eur. Phys. J. C **24**, 639 (2002)
10. C. Wetterich, Probing quintessence with time variation of couplings. J. Cosmol. Astropart. Phys. **2003** (2003). JCAP10 002
11. T. Dent, M. Fairbairn, Time-varying coupling strengths, nuclear forces and unification. Nucl. Phys. B **653**, 256 (2003)
12. M.P. Savedoff, Physical constants in extra-galactic nebulae. Nature **178**, 688 (1956)
13. J.N. Bahcall, W.L.W. Sargent, M. Schmidt, Astrophys. J. **149**, L11 (1967)
14. Y. Fenner, M.T. Murphy, B.K. Gibson, On variations in the fine-structure constant and stellar pollution of quasar absorption systems. Mon. Not. R. Astron. Soc. **358**, 468 (2005)
15. H. Chand, P. Petitjean, R. Srianand, B. Aracil, Probing the time-variation of the fine-structure constant: Results based on Si IV doublets from a UVES sample. Astron. Astrophys. **430**, 47 (2005)
16. V.A. Dzuba, V.V. Flambaum, J.K. Webb, Space-time variation of physical constants and relativistic corrections in atoms. Phys. Rev. Lett. **82**, 888 (1999)
17. M.T. Murphy, J.K. Webb, V.V. Flambaum, Further evidence for a variable fine-structure constant from Keck/HIRES QSO absorption spectra. Mon. Not. R. Astron. Soc. **345**, 609(2003). doi:10.1046/j.1365-8711.2003.06970.x
18. M.T. Murphy, V.V. Flambaum, J.K. Webb, V.A. Dzuba, J.X. Prochaska, A. M. Wolfe, Constraining variations in the fine-structure constant, quark masses and the strong interaction. Lect. Notes Phys. **648**, 131 (2004). doi:10.1007/978-3-540-40991-5_9
19. J.K. Webb, V.V. Flambaum, C.W. Churchill, M.J. Drinkwater, J.D. Barrow, Search for time variation of the fine structure constant. Phys. Rev. Lett. **82**, 884 (1999). doi:10.1103/PhysRevLett.82.884
20. J.K. Webb, J.A. King, M.T. Murphy, V.V. Flambaum, R.F. Carswell, M.B. Bainbridge, Indications of a spatial variation of the fine structure constant. Phys. Rev. Lett. **107**, 191101 (2011). doi:10.1103/PhysRevLett.107.191101
21. J.A. King, J.K. Webb, M.T. Murphy, V.V. Flambaum, R.F. Carswell, M.B. Bainbridge, M.R. Wilczynska, F.E. Koch, Spatial variation in the fine-structure constant new results from VLT/UVES. Mon. Not. R. Astron. Soc. **422**, 3370 (2012). doi:10.1111/j.1365-2966.2012.20852.x

22. J.C. Berengut, V.V. Flambaum, Manifestations of a spatial variation of fundamental constants on atomic clocks, Oklo, meteorites, and cosmological phenomena. Europhys. Lett. **97**, 20006 (2012). doi:10.1209/0295-5075/97/20006
23. G. Hinshaw, J.L. Weiland, R.S. Hill, N. Odegard, D. Larson, C.L. Bennett, J. Dunkley, B. Gold, M.R. Greason, N. Jarosik, E. Komatsu, M.R. Nolta, L. Page, D.N. Spergel, E. Wollack, M. Halpern, A. Kogut, M. Limon, S.S. Meyer, G.S. Tucker, E.L. Wright, Five-year wilkinson microwave anisotropy probe observations: data processing, sky maps, and basic results. Astrophys. J. Suppl. Ser. **180**, 225 (2009)
24. T. Rosenband, D.B. Hume, P.O. Schmidt, C.W. Chou, A. Brusch, L. Lorini, W.H. Oskay, R.E. Drullinger, T.M. Fortier, J.E. Stalnaker, S.A. Diddams, W.C. Swann, N.R. Newbury, W.M. Itano, D.J. Wineland, J.C. Bergquist, Frequency ratio of Al^+ and Hg^+ single-ion optical clocks; metrology at the 17th decimal place. Science **319**, 1808 (2008). doi:10.1126/science.1154622
25. R.E. Marrs, S.R. Elliott, D.A. Knapp, Production and trapping of hydrogenlike and bare uranium ions in an electron beam ion trap. Phys. Rev. Lett. **72**, 4082 (1994). doi:10.1103/PhysRevLett.72.4082
26. V. Mäckel, R. Klawitter, G. Brenner, J.R. Crespo López-Urrutia, J. Ullrich, Laser spectroscopy on forbidden transitions in trapped highly charged Ar^{13+} ions. Phys. Rev. Lett. **107**, 143002 (2011)
27. M. Schwarz, O.O. Versolato, A. Windberger, F.R. Brunner, T. Ballance, S.N. Eberle, J. Ullrich, P.O. Schmidt, A.K. Hansen, A.D. Gingell, M. Drewsen, J.R. Crespo López-Urrutia, Cryogenic linear Paul trap for cold highly charged ion experiments. Rev. Sci. Instrum. **83**, 083115 (2012)
28. O.O. Versolato, M. Schwarz, A. Windberger, J. Ullrich, P.O. Schmidt, M. Drewsen, J.R. Crespo López-Urrutia, Cold highly charged ions in a cryogenic Paul trap. Hyp. Int. **214**, 189 (2013)
29. A. Windberger, M. Schwarz, O.O. Versolato, T. Baumann, H. Bekker, L. Schmöger, A.K. Hansen, A.D. Gingell, L. Klosowski, S. Kristensen, P.O. Schmidt, J. Ullrich, M. Drewsen, J.R. Crespo López-Urrutia, Coulomb crystals in a cryogenic paul trap for sympathetic cooling of molecular ions and highly charged ions. AIP Conf. Proc. **1521**, 250 (2013)
30. I.I. Sobelman, *Introduction to the Theory of Atomic Spectra* (Pergamon Press, NEw York, 1972)
31. I.B. Khriplovich, *Parity Nonconservation in Atomic Phenomena* (Taylor & Francis Group, London, 1991)
32. L.L. Foldy, Fermi-segre formula. Phys. Rev. **111**, 1093 (1958). doi:10.1103/PhysRev.111.1093
33. Z.R. Iwinski, Generalized fermi-segre formula. Phys. Rev. A **22**, 1358 (1980). doi:10.1103/PhysRevA.22.1358
34. V.A. Dzuba, V.V. Flambaum, J.K. Webb, Calculations of the relativistic effects in many-electron atoms and space-time variation of fundamental constants. Phys. Rev. A **59**, 23 (1999). doi:10.1103/PhysRevA.59.230
35. V.A. Dzuba, V.V. Flambaum, M.G. Kozlov, M. Marchenko, α Dependence of transition frequencies for ions Si II, Cr II, Fe II, Ni II, and Zn II. Phys. Rev. A **66**, 022501 (2002)
36. J.C. Berengut, V.A. Dzuba, V.V. Flambaum, Enhanced laboratory sensitivity to variation of the fine-structure constant using highly charged ions. Phys. Rev. Lett. **105**, 120801 (2010). doi:10.1103/PhysRevLett.105.120801
37. J.C. Berengut, V.A. Dzuba, V.V. Flambaum, A. Ong, Highly charged ions with E1, M1, and E2 transitions within laser range. Phys. Rev. A **86**, 022517 (2012). doi:10.1103/PhysRevA.86.022517
38. V.A. Dzuba, A. Derevianko, V.V. Flambaum, Ion clock and search for the variation of the fine structure constant using optical transitions in Nd^{13+} and Sm^{15+}, arXiv: 1208.4157 (2012).
39. J.C. Berengut, V.A. Dzuba, V.V. Flambaum, A. Ong, Electron-hole transitions in multiply charged ions for precision laser spectroscopy and searching for variations in α. Phys. Rev. Lett. **106**, 210802 (2011). doi:10.1103/PhysRevLett.106.210802
40. J.D. Gillaspy, Highly charged ions. J. Phys. B: At. Mol. Opt. Phys. **34**, R93–R130 (2001)
41. W.R. Johnson, *Atomic Structure Theory* (Springer, Berlin, 2007), pp. 133

42. H.F. Beyer, V.P. Shevelko, *Introduction to the Physics of Highly Charged Ions* (CRC Press, London, 2002), pp. 245–246
43. A. Derevianko, V.A. Dzuba, V.V. Flambaum, Highly charged ions as a basis of optical Aaomic clockwork of exceptional accuracy. Phys. Rev. Lett. **109**, 180801 (2012)
44. V.A. Dzuba, V.V. Flambaum, M.G. Kozlov, Combination of the many-body perturbation theory with the configuration-interaction method. Phys. Rev. A **54**, 3948 (1996). doi:10.1103/PhysRevA.54.3948
45. J.C. Berengut, V.A. Dzuba, V.V. Flambaum, A. Ong, Optical transitions in highly charged californium ions with high sensitivity to variation of the fine-structure constant. Phys. Rev. Lett. **109**, 070802 (2012). doi:10.1103/PhysRevLett.109.070802
46. J.C. Berengut, V.V. Flambaum, M.G. Kozlov, Calculation of isotope shifts and relativistic shifts in C I, C II, C III, and C IV. Phys. Rev. A **73**, 012504 (2006)
47. J.C. Berengut, V.V. Flambaum, M.G. Kozlov, Calculation of isotope shifts and relativistic shifts in C I, C II, C III, and C IV. Phys. Rev. A **72**, 04450 (2005)
48. J.C. Berengut, V.V. Flambaum, M.G. Kozlov, Isotope shift calculations in Ti II. J. Phys. B **41**, 235702 (2008)

Chapter 10
Emission and Laser Spectroscopy of Trapped Highly Charged Ions in Electron Beam Ion Traps

José R. Crespo López-Urrutia and Zoltán Harman

Abstract Research with highly charged ions addresses bound state quantum electrodynamics and relativistic atomic theory at their frontiers. Electron beam ion traps have provided practical means to study not only these fundamental fields, but also the physics of extremely hot plasmas in stars, active galactic nuclei, and fusion research plasma devices. Starting from the X-ray region, where the first experiments took place, this chapter will visit various regions of the electromagnetic spectrum and give a review of essential contributions of different groups to this field.

10.1 Introduction

Wherever high concentrations of energy are present, be it inside stars, in active galactic nuclei containing black holes, or fusion devices, highly charged ions (HCI) play a dominating role in radiation transport. They also tenuously fill vast regions of space as the warm–hot intergalactic medium (WHIM). Knowledge of the spectroscopy of HCI is essential for diagnosing astrophysical plasmas. In HCI, the tightly bound active electrons show enormously magnified quantum electrodynamics, relativistic and nuclear size effects compared to their outer-shell counterparts in neutrals. Today, their accurate description still challenges theory, and their study offers new avenues for an improved understanding of fundamental interactions also governing the physics of inner-shell electrons in both neutral atoms and ions in low charge states. Isoelectronic sequences of all elements multiply the number of objects of

J. R. Crespo López-Urrutia (✉) · Z. Harman
Max-Planck-Institut für Kernphysik, Saupfercheckweg 1, 69117 Heidelberg, Germany
e-mail: crespojr@mpi-hd.mpg.de

Z. Harman
e-mail: z.harman@mpi-hd.mpg.de

W. Quint and M. Vogel (eds.), *Fundamental Physics in Particle Traps*,
Springer Tracts in Modern Physics 256, DOI: 10.1007/978-3-642-45201-7_10,

study available for experiments. Because of the different scaling laws of the various interactions, their relative strengths can be tuned along such series, providing a tool to disentangle their effects.

10.2 Quantum Electrodynamics Studies with Trapped HCI

Both quantum electrodynamics (QED) effects and relativistic contributions to the electron binding energies show a steep $\propto Z^4$ dependence on the atomic number Z. Another central scaling law of QED energy corrections is their $\propto 1/n^3$ dependence, with n being the principal quantum number. Therefore, electrons in deeper shells of heavy elements acquire the largest binding energy contributions. As a consequence, the perturbative treatment of QED, which is well suited for small QED corrections breaks down, as the series expansions in terms of α^n has to be extended to include terms in the form of $(Z\alpha)^n$ that do not rapidly converge at high charges and require a completely different mathematical treatment.

In principle, these points also apply to neutral atoms, as well as to ions in any charge state. But for spectroscopy it is crucial both to reduce the complexity of the system and to extend the lifetime of the states which have to be interrogated. Therefore, heavy atoms are less suitable for these studies, since their filled, neutral electronic shells introduce correlations effects that are difficult to disentangle from the QED contributions. Furthermore, their inner-shell excitations usually decay by fast Auger processes that broaden the energy spectrum of both the competing radiative decay and Auger electron energy. It thus seems logical to use ions of heavy elements with the lowest possible number of bound electrons, found in the hydrogen-like, the helium-like, and the lithium-like isoelectronic sequences. The simplicity of the first makes it ideal for the study of a bound two-body system. The second one constitutes the simplest example of a three-body system including two strongly correlated fermions. In the third one, a quasi one-electron system again, the calculable, reduced overlap of the electronic wave function with the nucleus allows for complementary investigations of QED and nuclear size (NS) effects. Important theoretical questions which have been experimentally investigated are the one-loop, two-loop, and screened (many-electron) QED contributions, the treatment of the negative energy spectrum of the bound electrons, of the Breit interaction, as well as the calculation of electron correlations using multi-configuration Dirac-Fock (MCDF), relativistic many-body perturbation theory (RMBPT), and relativistic configuration interaction (RCI) schemes. Effects arising from nuclear structure such as finite nuclear size (NS) effects, nuclear recoil, hyperfine interactions and their anomalies represent another very interesting field of research with HCI, which show extremely magnified contributions of these effects. An interesting summary of these issues and references to the relevant literature can be found in Refs. [1, 256].

10.3 Historical Development of Experiments with Electron Beam Ion Traps

Laboratory work with HCI requires practical sources of these species. From a spectroscopical point of view, a line-shaped radiation source of the highest possible density and at a very low temperature would seem ideal. For HCI sources, however, compromises are necessary, since HCI production encompasses tremendous concentration of energy in various forms upon the educt atoms in any given target ensemble. Thus, the product HCI will show high kinetic energies and temperatures. Stationary conditions can be difficult to achieve, as the presence of steep density as well as temperature gradients leads to a rapid expansion of the targets. The different mobilities of electrons and ions in a plasma also induces strong electromagnetic fields with a fast temporal evolution. Transient sources using electrical discharges or laser-produced plasmas suffer from these limitations.

Fortunately, these obstacles can effectively be circumvented since the introduction of the electron beam ion trap (EBIT). Its principle, ion production and trapping by a focused electron beam, quickly leads to a steady-state spatial distribution of HCI occupying a small cylindrical volume with a large aspect ratio [2, 3]. This arrangement is very well suited for coupling to spectroscopic equipment, as shown in Fig. 10.1. Mutual ion repulsion is compensated by the negative space charge potential of the electron beam, which is focused by a strong magnetic field overcoming its own tendency to expand. By virtue of the large mass difference between projectile electrons and atomic targets, momentum transfer is rather slow, and the heating proceeds by orders of magnitude less abruptly than in other HCI sources [4]. At the same time, ions of different charge states evaporate from the trapping potential at enormously disparate rates, and so some species can act as coolants for others. These advantages have been essential for the many applications of EBITs to spectroscopy of HCI. Starting from the X-ray region, where the first experiments took place, this chapter will visit all regions of the electromagnetic spectrum and present essential contributions of different groups to this field of research.

After the introduction of the electron beam ion source (EBIS) [5–7], it became obvious that production of HCI by means of focused electron beams was rather effective. However, EBIS could not generate the highest possible charge states. It was presumed that this was due to plasma instabilities heating up the trapped ions and reducing their interaction time with the beam. Moreover, the EBIS design did not offer adequate diagnostic ports for spectroscopic work. Both problems were overcome with the invention of the EBIT (see Fig. 10.1) by Levine and Marrs [2, 3, 8–11] at Lawrence Livermore National Laboratory (LLNL). By reducing the total length of the device and changing the electron beam design, a stable and more compressed beam was achieved, while the Helmholtz configuration of the superconducting coils introduced in the design granted optical access to the center of the region where the ions interact with the beam. At the same time, the beam was sufficiently stable to allow for long trapping times without ion losses by heating. It was soon realized how important evaporative cooling was to keep species of interest from leaving the trap,

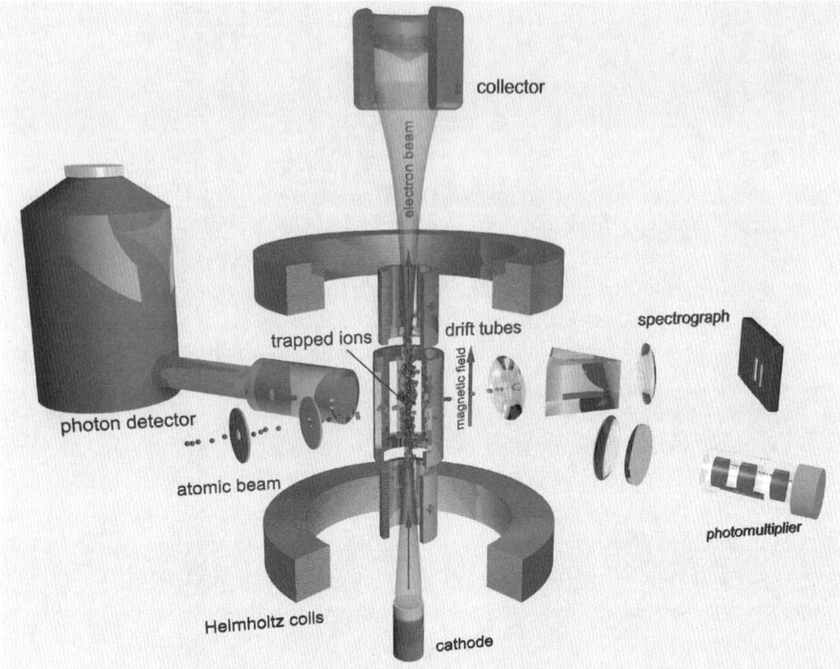

Fig. 10.1 Principle of an electron beam ion trap. Electrons emitted by a thermoionic cathode at the *bottom* are accelerated and launched into the magnetic field generated by the superconducting Helmholtz coils depicted. The Lorentz force acting on electrons counteracts their mutual repulsion and compresses the beam. As it passes through the central drift tube, the beam ionizes neutrals injected as an atomic beam (*left*), and traps the ions thereby produced by virtue of its negative space charge. Ion escape in the axial direction is suppressed by positive potentials applied to the drift tubes below and above of the central one. X-ray detectors or other spectrometers gain optical access to the trapped ion cloud through slotted apertures in the central drift tube

and careful modelling by Penetrante [4] pointed to the underlying principles of the time evolution of charge states and ion temperature, which were soon experimentally confirmed [12, 13].

From its introduction in the year 1986 on, EBITs have become the most useful instruments for the investigation of the properties of HCI. Their larger competitors, accelerators and ion storage rings, can indeed also perform very well, and even outperform EBITs in certain tasks, but the scope and variety of experimental techniques associated to EBITs is remarkable.

Initially, ion production was monitored in EBITs by X-ray photon detectors, soon complemented by crystal spectrometers. Later, the soft X-ray range was also covered by similar instruments and also grating-based devices. Optical spectrometers followed soon thereafter. With the inception of lifetime-measuring techniques [14] and the magnetic trapping mode [15], EBITs became an established tool for investigation of quantum processes in time scales spanning from the femtosecond range of the fast X-ray transitions to the millisecond domain in which forbidden optical transitions take place, an amazing twelve orders of magnitude.

Particularly fruitful have been experiments utilizing the variable electron beam energy as a tool to resonantly excite inner-shell electrons by photo-recombination. As the time reversal of the Auger process, dielectronic recombination plays a crucial role in the temperature and ionization balance of hot astrophysical and laboratory plasmas, and in the opacity of stellar interiors.

Finally, a compact scientific tool was available for investigating the, in the Universe ubiquitous, HCI in the laboratory. Application of the novel device to fundamental studies started immediately at LLNL. Soon, the importance of the development became clear in the wider community, and other groups followed rapidly along the same path. Two EBITs based on the LLNL blueprint were built at Oxford University [16, 17] for that group and for the one based at the National Institute for Standards and Technology (NIST) [18, 19]. A device of the same design was purchased for the Max-Planck Institute for Plasma Physics subsidiary in Berlin from a company based in Livermore. In Japan, strong efforts led to the construction of another superconducting EBIT inspired by the Livermore model [20–22]. A French group around J. P. Briand developed a very compact electron beam ion trap using permanent magnet technology which showed very promising capabilities [23]. Meanwhile, strong scientific activity went on at the LLNL team and the younger then existing groups. Since the commissioning of the high energy SuperEBIT [2], even bare uranium ions, which hitherto had only been accessible at relativistic velocities in high-end accelerators, became available in a trap. Beiersdorfer has recompiled the history of those pioneering LLNL experiments until the year 2008 in a review paper [24].

The cross sections for the production of heavy hydrogen-like ions were determined by analyzing the beam geometry and the ion yield [25]. Lifetimes of excited levels in the fs range could be determined based on the natural line width of X-ray transitions [26]. With the magnetic trapping mode [15], it became possible to reduce the background caused by electron impact excitation (EIE) as well as bremsstrahlung in certain types of measurements, and also to suppress the quenching of metastable states allowing for precise lifetime measurements [14, 27–35]. A wide variety of experiments with HCI [36] had become possible, addressing important questions of QED in strong fields (for an introduction to these theoretical questions in the perspective of EBIT experiments see e.g., the short reviews by Cheng [37] and Sapirstein [1]).

By the end of the 1990s, in addition to the original vertical design, a horizontal type was introduced at the Freiburg University EBIT (FreEBIT) [38] to allow for more convenient ion extraction and for the use of commercially available superconducting magnet technology with much lower liquid helium consumption than the first devices. Plans were also drafted there for an EBIT geared towards investigations at free-electron lasers and synchrotrons, which later became FLASH-EBIT [39], and for another twin device with applications to charge breeding of short-lived isotopes, the TRIUMF-EBIT. Elsewhere, vertical EBITs were planned and came into operation in Shanghai [40] and Stockholm [41]. Commercial models achieving some of the performance of the cryogenic devices appeared on the market [42]. Quite recently, small, compact EBITs used for visible and VUV emission spectroscopy have been introduced by the Tokyo and Shanghai EBIT groups [43, 44].

In recent years, laser spectroscopy with EBIT trapped ions has been applied to the study of the Lamb shift in hydrogen-like and lithium-like ions, of accurate wavelengths of coronal lines, to the QED contributions to the binding energy. Combination of EBITs with free-electron lasers [39] and synchrotron radiation sources [45–47] has resulted in the first examples of laser spectroscopy in the X-ray region [48, 49], with applications to both fundamental and applied physics. And the inherent stability of quantum states in HCI against external perturbations has lead to the first attempts for using trapped HCI to investigating the time variation of fundamental constants.

Given the large literature body which has originated from research with EBITs, we will have to restrict ourselves in this chapter to a partial view focused on a few experiments that aimed at simple electronic systems with few bound electrons, and exclude experiments with straightforward applications to the study of the electronic structure of complex ions. This latter field has also central position in atomic physics, since HCI are truly essential and abundant constituents of stars and astrophysical plasmas (for a review see Ref. [50]), such as those found in the powerhouses of active galactic nuclei, in accretion disks around black holes, and spectroscopic diagnostics are key for understanding the physics of such objects.

10.4 Production and Trapping of Highly Charged Ions by Electron Beams

HCI require energetic collisions for their production. The most efficient mechanism is electron impact ionization (EII), for which the cross sections at threshold are zero, and grow with the energy, peaking at approximately two and a half times the ionization potential. However, in presence of a trapping mechanism, the small cross section at threshold is compensated by the long interaction times.

As its name already states, an EBIT starts with an intense beam of electrons. Inside the trapping region (see Fig. 10.2), the electron beam is compressed by a strong magnetic field which produces a magnetic pressure keeping the beam from expanding due to the mutual repulsion of its particles. A very high density beam (reaching values of thousands of amperes per square centimeter) is the result, which induces a negative space charge potential arising from the 'line charge' (see Fig. 10.3). At typical operation parameters, the potential difference between the central axis of the electron beam and its boundary is of the order of tens to hundreds of volts. Across the distance to the inner wall of the central drift tube, the potential difference may reach kilovolt values [51].

Small EII cross sections call for the highest possible current density. Indeed, the electron beams realized at EBITs belong to the most intense examples available. Their ancestors were the ones used in travelling wave tubes and other microwave-generating devices. By perfecting electron beam injection into the EBIT and optimizing its

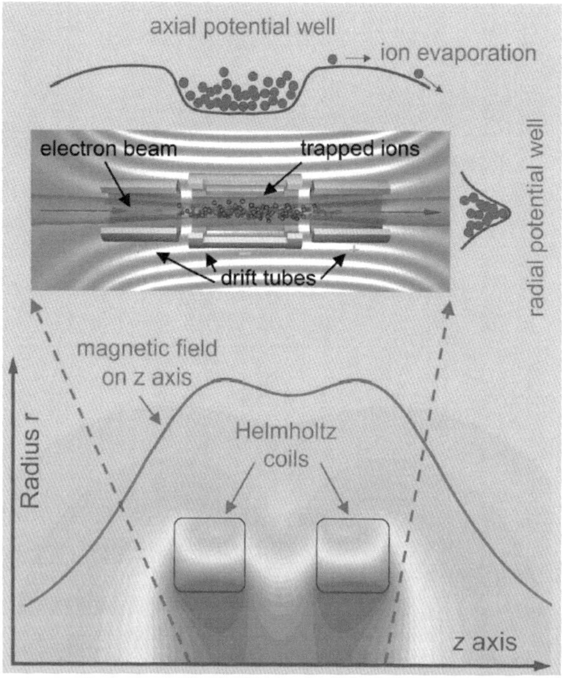

Fig. 10.2 Principle of operation of an electron beam ion trap. A strong magnetic field (value at axis given by *red curve*) generated by Helmholtz coils (*bottom*, false colour map) is used to compress an electron beam launched into its axis. The negative space charge potential (*top insert*) induced by the beam confines the motion of positive ions radially. Its value is sketched as a function of the separation from the axis (*top, right*). An axial electrostatic potential well (*top center*) produced by voltages applied to the drift tubes traps the ions axially. At a typical current of 200 mA, a potential difference of ≈200 V between the center and the edge of the electron beam (typical radius ≈70 μm) arises from the negative charge

transport with superconducting magnets, the currently achieved current area densities of more than 10^4 A/cm^2 imply that atoms are hit millions of times every second. With electron concentrations of 10^{13}/cm^3 at its center, the beam represents a 0.3-mbar pressure-equivalent concentration of negative charges, and also causes a nearly as dense positive ionic charge accumulation. All this occurs within an ultrahigh vacuum environment which is roughly ten orders of magnitude thinner. It is here, at conditions rather similar to those prevailing in astrophysical plasmas widely found in stellar coronae, where the physics of ionization and recombination between these two classes of partners gives rise to a wide variety of processes that have to be understood in the first place. When a neutral atom or molecule moving at typical thermal velocities crosses the electron beam at the center drift tube (see Figs. 10.2 and 10.4), it is ionized with a high probability in a time scale of microseconds or less. The negative space charge potential then traps the resulting ion within the electron beam, and the repeated interactions with it strip more and more electrons from the

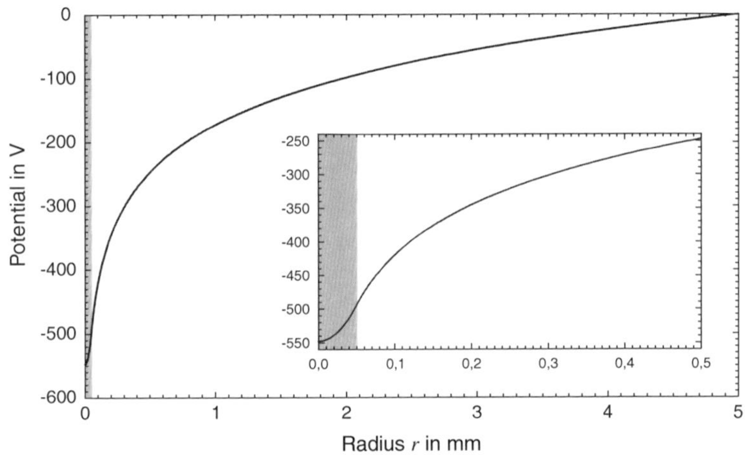

Fig. 10.3 Space charge potential generated by an electron beam as a function of the radial separation from the beam axis. The green shaded region indicates the beam radius. The insert shows a magnified view of the potential at a close distance from the electron beam. Even inside the beam itself, a potential difference of 50 V arises from the presence of the negative charges

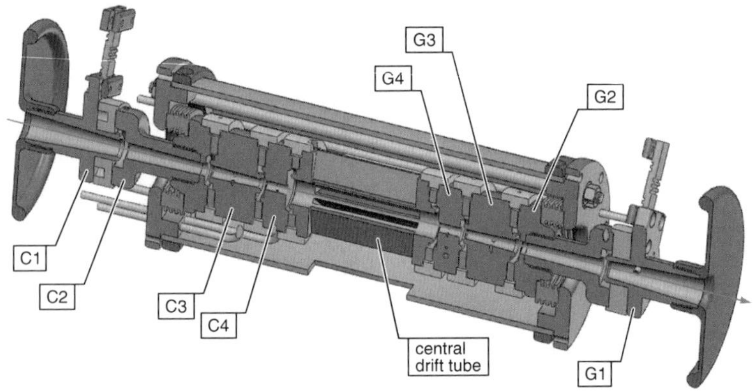

Fig. 10.4 Section through the trap electrodes, or drift tubes, of a cryogenic electron beam ion trap (FLASH EBIT [39]). Sapphire insulators are used to separate the different elements. Slotted apertures in the central electrode allow for viewing the ions trapped within it. The whole assembly is approximately 350 mm long,and the inner diameter of the drift tubes (labelled G1–G4, C1–G4, and the central one) varies from 5 to 15 mm

trapped ion. Although the cross sections for EII rapidly decrease with the removal of the external electronic shells, the trapping time amply compensates for their reduction. The average charge state of the ions rises in time, keeping a rather narrow distribution. This can be used to measure the cross sections for ion production, as, e.g., in the case of hydrogen-like ions [25, 52, 53]. Eventually, no more electrons can be stripped, when the binding energy of the remaining ones is higher than the kinetic energy of the beam electrons. The choice of this energy is essential for the charge states being generated. However, a pure ensemble of ions in a single charge state is never fully achieved. Quite frequently, a continuous injection of neutrals as an atomic beam is used. In this case, a wide distribution of charge states results, since the interaction time of the trapped species with the beam varies [13, 54]. This effect is not present when pulsed injection of atoms or ions is applied. In cases where very heavy elements are injected and very high charge states are aimed at, the recombination processes (radiative and dielectronic recombination as well as charge exchange with residual gas) efficiently compete with EII. Then, due to ion-electron recombination, the charge state distribution encompasses also several charge states.

10.5 Physical Processes in the Trap Region

Whenever electrons and ions encounter, their collisions induce dynamic processes that are paramount for the energetics and composition of plasmas in general. First of all, EII drives the ion production. While the direct process can bring electrons into the continuum always at any projectile energy surpassing the binding forces, its cross section is quite often dwarfed by the one corresponding to indirect pathways following electron impact excitation of an inner-shell electron to an outer shell, and the subsequent collapse of the electronic shell with emission of Auger electrons.

At the same time, the reversibility of quantum processes at the atomic level leads to the appearance of radiative recombination (RR), the time reversal of photo-ionization. Here, photons are emitted instead of being absorbed, and their energy is larger than that of the projectile electron in the ionic reference frame. Once some of the electronic shells offers vacancies, electrons from deeper levels are continuously cycled in EIE and photonic as well as auto-ionization decay (AI). Again, the strong resonant time-reversed process, called dielectronic recombination (DR), competes with the up-charging mechanisms and generates strong emission lines. In both directions, electronic correlations have proven to be a key element for the dynamic evolution of the charge states.

Independently of the presence of free electrons, residual gas offers for HCI a potential reservoir for replenishment of the missing negative charges. In most EBITs, cryogenic pumping due to the cold surfaces at near 4 K lowers the pressure into the $\approx 10^{-13}$ mbar range, reducing the neutral particle density accordingly, and therefore strongly suppressing charge exchange. However, it is also possible to intentionally enhance charge exchange by injecting a dense atomic beam in a pulsed mode into the trap region [55]. Analogously to the situation in the atmospheres of planets and comets [56] under the effects of the solar wind, recombination of HCI by charge exchange produces not only ions in lower charge states but also recognizable X-ray spectral lines [15, 56] which can be used for plasma diagnostics or for a background-free study of resonance transitions.

The individual mechanisms driving the trap physics, but also governing the radiation balance of hot plasmas in astrophysics and tokamaks, are described in the following subsections.

10.5.1 Electron Impact Ionization

Under typical EBIT operating conditions, the current density in the beam reaches values of $10^{22}\, e/(\mathrm{s}\cdot\mathrm{cm}^2)$, the equivalent of several thousand amperes per square centimeter. For an atom with an assumed typical EII cross section of $10^{-16}\,\mathrm{cm}^2$, the initial ionization rates lie in the MHz range, much faster than all other processes. Thus, this process dominates the charge evolution within the trap region. When the kinetic energy of the impacting electron E_e is higher than the ionization potential I_p of the least-bound electron, its ejection can result with a given probability. Since energy is conserved,

$$A^{q+} + e^-(E_e) \rightarrow A^{(q+1)+} + e^-(E_1) + e^-(E_2) \quad \text{and} \quad E_e - I_p = E_1 + E_2. \quad (10.1)$$

Here, the notation A^{q+} stands for an ion of the element A with a net positive charge q. A rough estimate for the EII cross section can be obtained utilizing the semi-empirical formula given by Lotz [57]:

$$\sigma_i^L[cm^2] = 4.49 \times 10^{-14} \frac{N}{I_p^2} \frac{\ln(u+1)}{u+1}\,, \quad (10.2)$$

with $u = E_e/I_p - 1$ and N being the number of equivalent electrons in the same shell.

10.5.2 Electron Impact Excitation

After the initial ionization steps, the availability of open shells closer to the ground state gives rise to much stronger couplings between initial and the now possible, unoccupied final states by collisional interactions. Excitation at a given projectile electron kinetic energy E_e of the incident electron follows when this is higher than the energy difference from the level n to n'. The complete process, involving subsequent radiative decay of the excited state $\left[A^{q+}\right]^*$, can be represented as follows:

$$A^{q+} + e^- \rightarrow \left[A^{q+}\right]^* + e^- \rightarrow A^{q+} + \hbar\omega\,. \quad (10.3)$$

Here, $\hbar$ is the reduced Planck constant, and ω is the angular frequency of the emitted photon. For the case of resonant excitation resulting in a hole state:

$$A^{q+} + e^{-} \rightarrow \left[A^{(q-1)+}\right]^{**}, \tag{10.4}$$

and relaxation can proceed either through auto-ionization following the scheme $\rightarrow \left[A^{q+}\right]^{*} + e^{-}$ or by photon emission $\rightarrow A^{(q-1)+} + \hbar\omega$. Already for ions in moderately charged states, the rates for photon emission compete and overcome the typical Auger rates, which are rather independent of the charge state.

An useful approximation for the excitation cross sections is represented for an electric dipole transition with an energy $E_{n,n'}$ by the Bethe formula [58]:

$$\sigma_{ex} \approx \frac{A}{E_e} + B\frac{\ln E_e}{E_e}, \quad E_e \gg E_{n,n'}, \tag{10.5}$$

where A is a constant, and the parameter B is linked to the corresponding oscillator strength. Basically, this results from the fact that electronic collisions can be seen as a pulse of virtual photons coupling with the bound electrons through dipole–dipole interaction. This connection is applied in the empirical Van Regemorter formula [59] for calculation of electric dipole-allowed cross sections:

$$\sigma_{ex}^{\mathrm{VR}}(x)[cm^2] = 2.36 \times 10^{-13} \frac{1}{E_{n,n'}^{2}} \frac{G(x)}{x} f_{n,n'}, \tag{10.6}$$

where the dimensionless parameter x is defined for a given transition $n \rightarrow n'$ as the ratio between the kinetic energy of the incoming electron and the threshold energy $E_{n,n'}$: $x = E_e/E_{n,n'}$. The value $f_{n,n'}$ is the photon absorption oscillator strength. The term $G(x)$ is called the effective Gaunt factor, which scales the classical oscillator strength to that of a quantum mechanical system. Van Regemorter [59] uses for this factor another empirical approximation as

$$G(x) = 0.349\ \ln(x) + 0.0988 + 0.455\, x^{-1}. \tag{10.7}$$

10.5.3 Radiative Recombination

When an ion captures a free electron under emission of a photon, and lowering its charge state by one,

$$A^{q+} + e^{-} \rightarrow A^{(q-1)} + \hbar\omega_{RR}, \tag{10.8}$$

energy conservation requires that the photon energy be equal to the sum of the kinetic energy of the projectile and the binding energy of the ion in the final state. The condition is stated as:

$$\hbar\omega_{\mathrm{RR}} = E_e + I_p. \tag{10.9}$$

In order to obtain the energy-differential RR cross section, one can apply the principle of detailed balance to the photo-ionization cross sections (σ_{ph}):

$$g_q \sigma_{\mathrm{ph}}(\omega) = \frac{2 m_e c^2 E_e}{\hbar^2 \omega^2} g_{q+1} \sigma_{\mathrm{RR}}(E_e), \tag{10.10}$$

where the statistical weights for each principal quantum number n are given by $g_q = 2n^2$ for the final state and $g_{q+1} = 1$ for the initial one. σ_{ph} can be estimated by means of the Bethe and Salpeter formula [60]. Kramers obtained for the RR cross section the following expression:

$$\sigma_{\mathrm{RR}}^{\mathrm{Kramers}} = \frac{32\pi a_0^2 \alpha^3}{3\sqrt{3}} \frac{Z|E_0|^{3/2}}{(E_0 + E_e)\, E_e}, \qquad E_0 = Z^2 Ry/n^2. \tag{10.11}$$

In order to obtain more accurate values, the use of advanced atomic structure methods and computer codes is required. Useful modules for calculating RR and photo-ionization cross sections are part of the Flexible Atomic Code by Gu [61] and the RATIP package by Fritzsche [62].

10.5.4 Dielectronic Recombination

In dielectronic recombination, the free electron is captured into a vacant excited state of the ion A^{q+} transferring non-radiatively the energy difference to a core electron with energy E_1, which is simultaneously promoted to a higher-laying excited state with energy E_2 of the ion, thus forming an intermediate (singly, doubly or multiply excited) state, as shown on Fig. 10.5. This step is then followed by the emission of a photon (or several photons):

$$A^{q+} + e^- \rightarrow \left[A^{(q-1)+}\right]^* \rightarrow \left[A^{(q-1)+}\right] + \hbar\omega_{DR}. \tag{10.12}$$

This resonant process can only happen when the energy difference between the core electron state and the state in which the second electron is excited is equal to the kinetic energy of the free electron E_e plus the binding energy I_p of the recombined state. This resonance condition is written as

$$\Delta E = E_2 - E_1 = E_e + I_p\,. \tag{10.13}$$

In the second step, the formed excited state $\left[A^{(q-1)+}\right]^*$ may as well decay to the ground state via an Auger auto-ionization process, where again there is a change in the ion charge as the ion returns to its initial state. However, in HCI, radiative de-excitation via emission of one or more photons strongly dominates. While the Auger rates are proportional to the electron–electron interaction and show only a

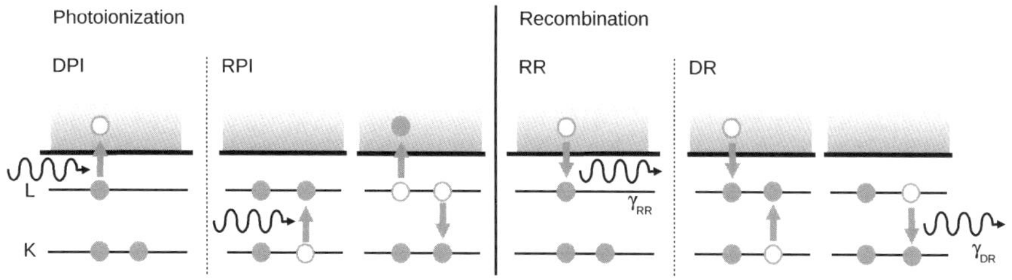

Fig. 10.5 Photo-ionization (*PI*) can proceed both directly (*DPI*) and resonantly (*RPI*). On the *right side*, the time-reversed processes, namely, radiative recombination (*RR*) and dielectronic recombination (*DR*) are depicted [63]

weak dependence on Z, the radiative decay rates increase with Z^4 for dipole-allowed transitions, and with up to very high powers of Z for magnetic, two-photon or higher-order transitions. In this way, radiative stabilization of the excited intermediate states is usually the dominant channel in HCI, in contrast to neutrals, where Auger processes dominate.

Dielectronic recombination is the time-reversed Auger process and, thus, the corresponding pathways are labelled accordingly. For instance, in a KLL DR resonance for a helium-like initial state, as shown in Fig. 10.5, the free electron is captured into the L shell of an ion, while a bound electron is promoted from the K shell to the L shell to form an excited $1s2l2l'$ state (with l, l' denoting single-electron angular momentum states). This intermediate excited state decays radiatively:

$$\begin{aligned} e^- + A^{(Z-2)+}\ (1s^2) &\rightarrow \left[A^{(Z-3)+}\right]^*\ (1s2l2l') \\ &\rightarrow A^{(Z-3)+}\ (1s^2 2l \text{ or } 1s^2 2l') + \hbar\omega(K_\alpha). \end{aligned} \tag{10.14}$$

To a good approximation, the energy dependence of the cross section of resonant recombination is described by a Lorentzian peak. The magnitude of the process is best characterized by the area of the resonance peak, which is customarily named resonance strength. Cross sections and resonance strengths for DR will be discussed later in Sect. 10.7.1.

10.6 Photon Spectroscopy Under Electron Beam Excitation

The first observations of ions trapped in an EBIT were carried out using solid-state photon detectors to analyze their X-ray emission upon EIE, generating characteristic bound–bound radiation. Furthermore, radiative recombination of mono-energetic electrons into open shells produces narrow transitions which immediately reveal the principal quantum number of the vacancies into which the electrons are captured. An example of a recorded photon emission spectrum is given in Fig. 10.6. Far higher

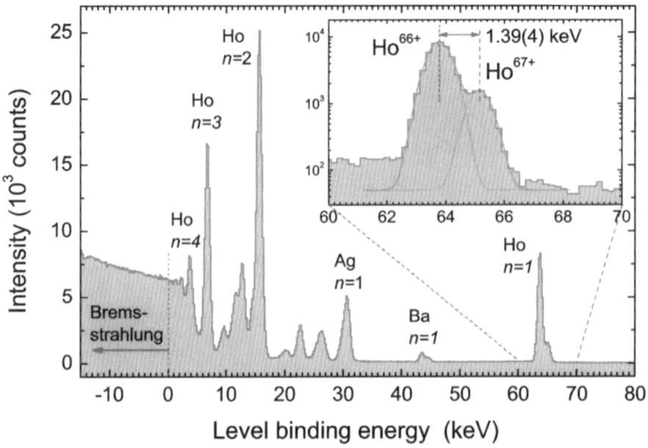

Fig. 10.6 X-ray photon spectrum of trapped holmium ions under excitation by an electron beam with an energy of 132.7 keV. This value has been subtracted from the photon energy to display the energy additionally liberated during radiative recombination (*RR*) into the different open shells (see labels) of the trapped holmium ions, which corresponds to the binding energy of the corresponding level. *Insert* RR peak of the $n = 1$ shell. The only vacancies in this shell occur in the cases of the bare Ho^{67+} ion and of hydrogen-like Ho^{66+} ions. The emitted photons have an energy difference (*indicated*) equal to the two-electron contribution to the binding energy

spectral resolution was soon reached at LLNL [64] by using crystal spectrometers of various kinds [65, 66]. A distinct advantage of the EBIT as a radiation source is the narrow confinement of the emitting particles in a narrow cylindrical volume with a diameter on the order of 100 μm [51], which constitutes the geometric equivalent of an entrance slit for the spectrometer. With the introduction of appropriate grating instruments, spectroscopy in the soft X-ray and vacuum ultraviolet region also became possible [67–73]. The more recent introduction of X-ray micro-calorimeters at EBITs (see e.g. [56, 74]) offers simultaneous spectral coverage over much wider regions than it is possible with crystal spectrometers, while achieving a much higher resolving power than typical solid-state photon detectors.

10.6.1 The X-ray Region: Lyman-α Transitions of Hydrogen-Like Ions

The Lyman-α lines of hydrogenic ions are the touchstone of atomic structure theory. Due to the comparatively simple structure of such single-electron systems, bound-state QED allows an accurate prediction of transition energies. Within relativistic quantum mechanics, the energy levels of hydrogenic ions are given by the Sommerfeld formula. Radiative corrections beyond the Dirac theory are described by self-energy diagrams, representing the energy shift of the bound electron due to its interaction with the vacuum fluctuations of the electromagnetic field, and by vacuum polarization diagrams, which depict the virtual creation and annihilation of electron–positron pairs in the field of the nucleus. These diagrams on the one-loop level

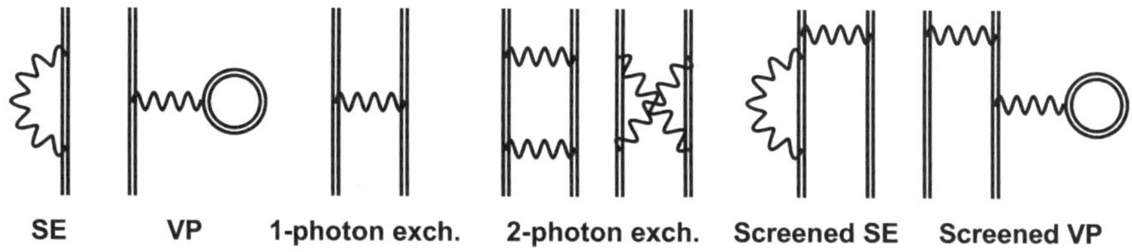

Fig. 10.7 Furry-picture [84] Feynman diagrams representing: (*left*) one-loop self-energy (*SE*) and vacuum polarization (*VP*) corrections to the energy levels in hydrogenic ions; (*center*) exchange diagrams with one and two photons in many-electron ions; and (*right*) two typical QED screening diagrams in many-electron systems. The *double lines* indicate wave functions and propagators for the electrons moving in the Coulomb field of the nucleus, and *wavy lines* depict virtual photons

(i.e. in first order in α (fine-structure constant) expansion) are shown on the left side of Fig. 10.7. In light hydrogenic ions, these corrections can be evaluated to a large extent analytically by employing truncated expansions in powers of the interaction strength $Z\alpha$ (see e.g. [75, 76]). In highly charged ions, such an approach fails as $Z\alpha$ cannot be regarded as a small parameter, and a theoretical ansatz is called for which treats the interaction with the nucleus in an all-order fashion. The calculation of one-loop QED corrections in all order in $Z\alpha$ has started decades ago and has been refined over the course of time [77–84]. Due to their comparatively high accuracy, theoretical transition energies of hydrogen-like ions, as for instance given in the tables by Johnson and Soff [79], are often simply taken as calibration standards [85, 86].

The improvement of experimental techniques with regard to accuracy and the accessible range of nuclear charges necessitated the calculation of two-loop QED corrections (see e.g. the work of Yerokhin, Indelicato and Shabaev [87]). Besides QED terms, calculations also have to account for effects arising from finite nuclear size [82, 88] and mass [84, 89, 90]. In inner-shell transitions of heavy ions, the polarization of the structured nucleus by the field of the orbiting electron gives rise to further small level shifts [91]. Figure 10.8 shows the dependence of the different contributions on the atomic number Z and compares one-loop and two-loop corrections and nuclear-size effects. Reviews by Mohr, Plunien and Soff [82], Shabaev [84] and Cheng et al. [256] summarize the progress of high-precision QED calculations in highly charged ions.

Early experimental work with HCI by Briand et al. at the Bevalac accelerator facility was dedicated to hydrogen-like Ar^{17+} and Fe^{25+} [92, 93]. Källne et al. and Richard et al. measured Cl^{16+} [94, 95], Tavernier at al. [96] hydrogenic Kr^{35+}, and Marmar determined the Lamb shift of the Ar^{17+} ion [97]. At GSI, Deslattes and Beyer studied the Lamb shift in hydrogen-like Ar^{17+} [98, 99] and Ni^{27+} [100] by observing the radiative decay of recoil ions in the charge state that had been produced by energetic collisions with relativistic heavy ions [101].

Independently of wavelength measurements calibrated with emission lines from solid targets excited by electron impact, for which various systematic effects are present, there has been a small number of absolute energy determinations based on crystallographic standards. Beiersdorfer carried out EBIT investigations of the

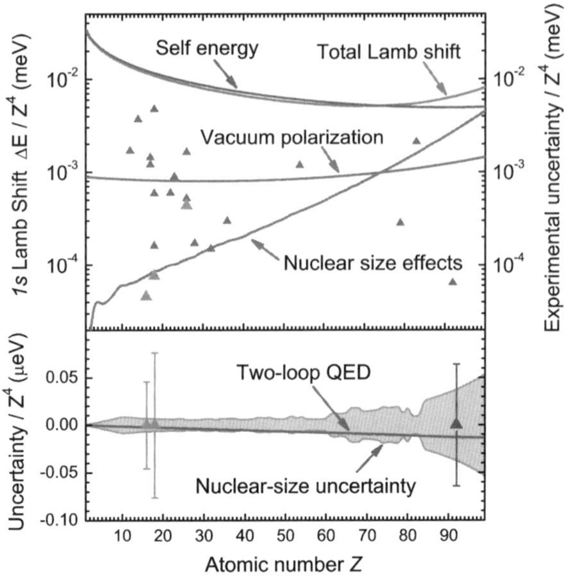

Fig. 10.8 *Top* scaling of one-loop QED contributions (self-energy, vacuum polarization) and nuclear-size effects as a function of the atomic number Z for hydrogen-like ions according to Johnson and Soff [79]. *Bottom* Comparison of the current uncertainty level due to nuclear-size effects (*grey shaded area*) with reported experimental uncertainties (*triangles*) for the measurement of the total Lamb shift in hydrogen-like ions and with the predicted magnitude of two-loop QED contributions to the binding energy (*blue curve*). For heavy elements, nuclear-size effects currently preclude a direct determination of two-loop QED corrections based solely on measurements of the hydrogen-like Lyman-α transitions. All energy values have been normalized by a factor $1/Z^4$ accounting for the first-order scaling of one-electron QED

wavelength of the two Lyman-α transitions in Mg^{11+} [102] using a quasi-monolith [103] with two crystal mounted in parallel which had been metrologically characterized. The two reflections from each line appear separated by a distance which allows determining the Bragg angle solely based on the crystal separation. An absolute value for the transition energy resulted with an error bar of 24 ppm. In order to infer the Lamb shift, the Dirac energy is simply calculated and subtracted from the experimental result. In the case of Mg^{11+}, this yields 0.246(44) and 0.351(70) eV for the Lamb shift correction to Lyman-α 1 and Lyman-α 2, respectively. For Si^{13+}, the value obtained in similar measurement had a larger error bar of 70 ppm [104].

Gillaspy et al. [105] measured at the NIST EBIT the energies for Lyman-α_1 ($1s-2p_{3/2}$) and Lyman-α_2 ($1s-2p_{1/2}$) in hydrogen-like V^{22+} to be 5443.95(25) and 5431.10(25) eV, respectively, in fair agreement with the theoretical values 5443.63 and 5430.70 eV. Nakamura and et al. [106] observed the intensity ratio between the two Lyman$_\alpha$ transitions in hydrogenic Ti^{21+} at the Tokyo EBIT. The Oxford group utilized a Johann crystal spectrometer [66] calibrated using the Kα lines of neutral vanadium [107] for investigating the $1s$ Lamb shift of hydrogen-like Ti^{21+}. Their determination yielded a value of 2.29(14) eV, in good agreement with the predictions (Fig. 10.9).

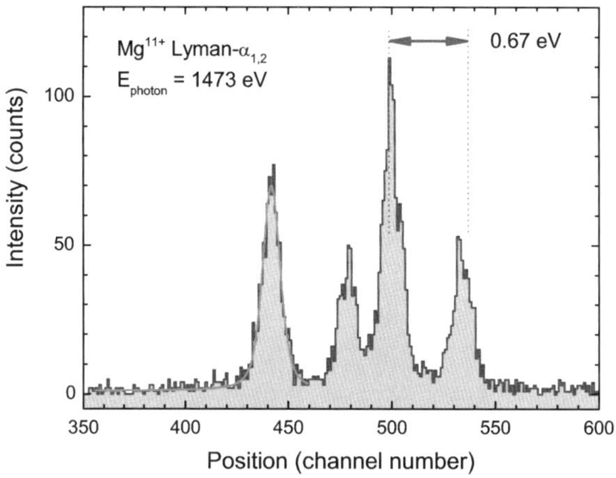

Fig. 10.9 High-resolution spectrum of the Lyman-α transitions of hydrogen-like Mg^{11+} [102] using a quasi-monolith consisting of two crystals mounted in parallel. Due to this arrangement, every line appears twice, showing a geometric splitting which is used for an absolute energy determination. The experimental resolution $E/\Delta E$ of nearly 9,000 already allows to separate the Lorentz natural line width ($\tau \approx 77$ fs) from the thermal broadening contribution in a Voigt fit

A recent approach to absolute transition energy measurements has been the application of the Bond method [108] to EBIT spectroscopy. In this technique, a flat crystal is turned from one Bragg reflection of the line of interest to the mirror symmetric reflection. By using two detectors and measuring the angle difference, the absolute Bragg angle can be obtained without offset correction. Since the lattice constant of silicon [109] is a secondary representation of the meter in the International System of Units (SI), the Bragg angle yields the wavelength, and thus the transition energy without the need for calibration lines. Bruhns et al. [110–114] implemented this method, which was later improved by Kubiček et al. [115–117] by including two laser beams as angular fiducials into the geometric scheme. By locating the source of the laser beams virtually at the position of the emitting ions within the trap region, the angular fiducials allowed to determine the X-ray position on the X-ray CCD detectors without needing to know the relative positions of source, crystal, or detectors. All the rays emanating from the X-ray source are reflected by the crystal which becomes a symmetry plane, and then they intersect the detector camera. The ratios of the respective separations between each fiducial and the X-ray line on the detector plane are in this configuration independent from the total distance between origin and detector. This setup eliminates the major sources of uncertainty typically found in Bond method determinations, while allowing to dispense with using slits to fix the incidence direction of the X rays, an approach which reduces the available intensity. With this technique, systematic measurements of the Lyman-α transitions in S^{15+}, Ar^{17+}, and Fe^{25+} were performed (see Fig. 10.10 and Table 10.1). The final uncertainty for S^{15+} reached a level of 1.4 ppm, not yet sufficiently small to allow for benchmarking predictions of two-loop QED contributions at this value of Z.

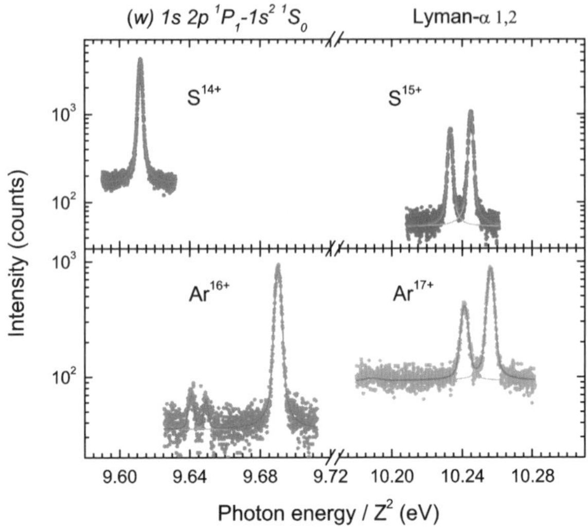

Fig. 10.10 Resonance transitions $2p \to 1s$ in helium-like (*left*) and hydrogen-like (*right*) sulphur and argon ions as observed in high-precision crystal spectrometer measurements [115–117]. The photon energies have been scaled by $1/Z^2$ in order to represent them in a common plot. This scaling removes the first-order Coulomb term from the transition energy and shows shifts due to correlation, relativistic and QED terms. The $1s2p\,^1P_1 \to 1s^2\,^1S_0$ shows a larger shift than the Lyman-α transitions due to interelectronic correlations, which are not present in the hydrogen-like system

Table 10.1 Experimental energies ($E_{exp.}$) for the Ly-α_1 and w ($1s^2\,^1S_0 \to 1s\,2p\,^1P_1$) transitions and inferred values for the $1s$ ground-state Lamb shift (LS) obtained at the Heidelberg EBIT [115, 117] compared with earlier measurements [99, 116, 135, 138, 139] ($E_{prev.\ exp.}$), unpublished data for Fe^{24+} from Rudolph [140] obtained with FLASH EBIT, and predictions ($E_{theo.}$) from references [141] (H-like),[124] (UM), [118] (all order in $1/Z$), [131] (BSQED)

Transition	$E_{exp.}$ (eV)	$E_{prev.\ exp.}$ (eV)	References	$E_{theo.}$ (eV)	References
Ly-α_1 S^{15+}	2622.693(3)			2622.6995(1)	[141]
LS S^{15+}	0.828(3)			0.8212(1)	[141]
Ly-α_1 Ar^{17+}	3322.993(8)	3322.989(17)	[99]	3322.9929(1)	[141]
LS Ar^{17+}	1.201(8)			1.2009(1)	[141]
Ly-α_1 Fe^{25+}	6973.18(66)	6972.732(609)	[139]	6973.1804(8)	[141]
LS Fe^{25+}	4.06(66)			4.0574(8)	[141]
w S^{14+}	2460.627(3)	2460.641(32)	[116]	2460.626(1)	[124]
				2460.629	[118]
				2460.6292(3)	[131]
w Ar^{16+}	3139.581(5)	3139.583(6)	[138]	3139.576(2)	[124]
				3139.581	[118]
				3139.5821(4)	[131]
w Fe^{24+}	6700.442(23)	6700.725(201)	[135]	6700.401(7)	[124]
		6700.513(80)	[140]	6700.424	[118]
				6700.4347(12)	[131]

10.6.2 The Heliumlike Sequence: A Test for Interelectronic Correlation

In many-electron ions, correlation effects increase the complexity of the theoretical description. Helium-like ions with their two electrons represent an archetype many-body system, therefore, a large literature list of theoretical work exists for the electronic structure of such ions.

Most atomic structure methods use the Hartree-Fock (HF) approximation, or, its relativistic generalization, the Dirac-Hartree-Fock (DHF) approximation as their starting ground. Here, the two-electron wave function is represented by an antisymmetrized product of single-electron orbital functions, i.e. the particles are assumed to be independent. Correlation contributions, defined as electron–electron interaction effects beyond the (D)HF approximation, can be either treated perturbatively or variationally.

The former approach is generally referred to as relativistic many-body perturbation theory (RMBPT). Plante, Johnson and Sapirstein carried out RMBPT calculations for helium-like ions, with corrections taken into account to all orders (AO method, Ref. [118]). In variational calculations, typically, a multiconfiguration approach is employed: the many-electron wave function $\Psi(JP)$ is represented as linear combination of Slater determinants $\Phi(JP)$ coupled to a given total angular momentum quantum number J and possessing a given parity P:

$$\Psi(JP) = \sum_{i=1}^{n_c} c_i \Phi_i(JP) . \tag{10.15}$$

Here, n_c is the number of configurations taken into account. In case of the multiconfiguration Dirac-Fock (MCDF) method, all single-electron orbitals used to construct the configuration state functions $\Phi(JP)$ are variationally optimized [119]. MCDF calculations for heliumlike ions were performed by Indelicato, with the inclusion of additional radiative corrections [120, 121]. In case of the relativistic configuration interaction method (RCI), the total Hamiltonian is diagonalized by using a fixed set of single-electron orbitals. Such calculations for two-electron ions were performed by Cheng et al. [122, 123, 256]. Both MCDF and RCI essentially include electronic correlations in all orders.

In a completely different approach, non-relativistic explicitly correlated wave functions are used, and relativistic and QED corrections are calculated perturbatively with these. This unified method (UM) of Drake [124–126] is rather successful since it is able to predict the dominant electronic correlation contributions better than earlier works.

Especially in heavy two-electron ions, a quantum electrodynamic treatment of electron interaction effects is mandatory. Here, the interelectronic interaction is described to lowest order in $1/Z$ by the one-photon exchange diagram (see Fig. 10.7), which, unlike the instantaneous Coulomb interaction, takes into account magnetic and retardation corrections. The QED expansion is further improved by taking into

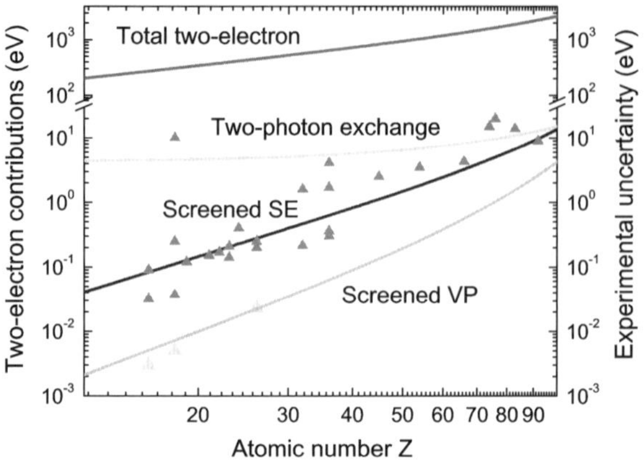

Fig. 10.11 Break-up of the absolute values of the different contributions to the binding energy of the $1s^2\,{}^1S_0$ ground state along the helium-like isoelectronic sequence resulting from electron–electron correlations and QED. The curves follow calculations by Artemyev [131]. The experimental uncertainties of different measurements are indicated as triangles for comparison. Recent work [115–117] (*green triangles*) allows to test screened vacuum polarization predictions

account two-photon exchange contributions (Fig. 10.7). Lindgren performed detailed QED calculations for the helium-like isoelectronic sequence [127, 128]. Essential aspects of the QED treatment, like the use of the no-pair Hamiltonian, the role of negative energy states are covered in the work of Johnson et al. [129] and by Sapirstein [130].

The exchange of virtual photons between the bound electrons also modifies the QED effects. Such corrections are described by the QED screening diagrams, containing both a SE or VP loop as well as an exchange photon connecting different electrons. Examples of such Feynman diagrams can be seen on Fig. 10.7. The most advanced QED predictions, containing results for QED screening, two-photon exchange as well as higher-order correlation effects were carried out by Artemyev and colleagues [131, 132]. Figure 10.11 shows the magnitude of the different physical contributions along the isoelectronic sequence.

Experiments in the range from $Z = 16$ to 39 were reported by Schleinkofer [133], Aglitsky [134] and Beiersdorfer [135] based on tokamak observations. They were based on plasmas with very high excitation rates, and contamination by satellite lines could sometimes have compromised the accuracy of those results. Alternatively, and for the same reasons, the helium-like lines and their satellites can be used for plasma diagnostic purposes, and Biedermann carried out experiments at the Berlin EBIT to that end [136]. The helium-like spectrum of argon [107] and vanadium [137] were also obtained.

Kubiček [115–117] determined with very small uncertainty absolute Bragg angles also for several helium-like electronic transitions. These latest measurements with improved resolution confirmed bound-state QED predictions by Artemyev et al. [131]. These results for He-like and H-like ions, together with earlier measurements and theoretical predictions, are summarized in Table 10.1. Here, for completeness, we also introduce a result for helium-like iron from synchrotron radiation measurements (see Sect. 10.8.2).

10.6.3 The QED-Sensitive $2s_{1/2} \rightarrow 2p_{3/2,1/2}$ Transitions in Lithiumlike Ions

In this section, we focus on results obtained using trapped ions. For measurements of the $2s_{1/2} \rightarrow 2p_{3/2,1/2}$ transitions in lithium-like heavy ions using storage rings, we refer to the the work at GSI [142–146] and references therein.

Studies on the Lamb shift of U ions form U^{82+} to U^{89+} were carried out at LLNL by Beiersdorfer et al. already in 1993 [147]. A value of -47.39 ± 0.35 eV has been determined for the $2s$ Lamb shift, in agreement with the theoretical result of -47.58 eV by Blundell [148], based on the evaluation of radiative corrections with an approximate potential describing the screening of the $2s$ and $2p$ electrons by the $1s^2$ core. A value of 4025.23 ± 0.14 eV was found for the $2s_{1/2} - 2p_{3/2}$ transition in lithium-like Th^{87+} [149]. This value was sensitive to the effect of nuclear polarization and confirmed predictions of a 36-eV QED contribution to within 0.4 %. A similar test of QED and atomic structure calculations was also performed by measuring twelve $2s_{1/2} - 2p_{3/2}$ transitions in the neighbouring charge states. The measured transition energies in beryllium-like Th^{86+} through neonlike Th^{80+} were compared with theoretical predictions from MCDF calculations of the electron-correlation terms.

By means of a high resolution crystal spectrometer, Beiersdorfer et al. measured the $2s_{1/2} - 2p_{3/2}$ line in $^{209}Bi^{80+}$ [150]. Due to the large nuclear magnetic moment, a sizable hyperfine splitting of 0.820 ± 0.026 eV appears between the states $F = 4, 5$ in the $1s^2\,2s_{1/2}$ ground configuration, as shown in Fig. 10.12. The measurement had a very high accuracy due to its careful calibration with hydrogen-like transitions. Its averaged result of 2788.139 ± 0.039 eV for the transition constituted at that time an essential test of QED correction terms and motivated the evaluation of two-loop QED corrections for heavy elements [150].

For the lithium-like ion W^{71+}, Podpaly et al. determined a value of 1697.34 ± 1.03 eV for the $1s^2\,2s_{1/2}(J = 1/2) \rightarrow 1s^2\,2p_{3/2}(J = 3/2)$ transition using a flat crystal spectrometer [151]. This result is in agreement with the multiconfiguration Dirac-Fock prediction of 1696.0 eV by Kim et al. [152] and the configuration interaction value of 1697.6 eV [151]. However, the error bars were still rather large for rigorously testing QED corrections. Besides the lithium-like one, a range of charge states of tungsten was investigated, with applications to the diagnostics of fusion plasmas [151].

Beiersdorfer determined the two-loop contributions to the binding energy of the $1s_{1/2}$ electron [153] by scaling the nuclear-size contributions from lithium-like U^{89+} [154] to hydrogen-like U^{91+}. This inferred value is in perfect agreement with the theoretical prediction of Yerokhin et al. [87]. As for the transition energy of the

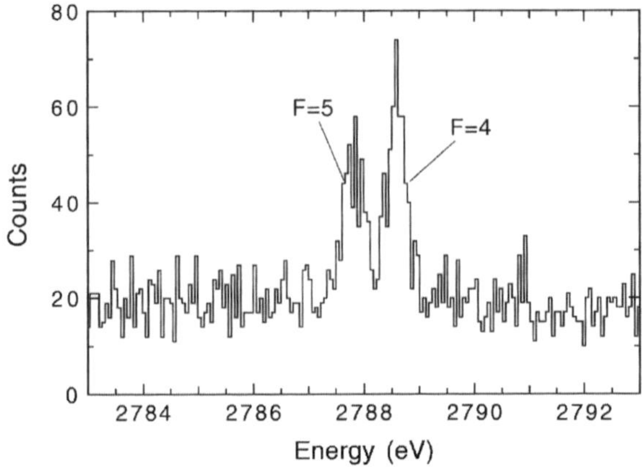

Fig. 10.12 X-ray spectrum of the $2s_{1/2} \rightarrow 2p_{3/2}$ transition in lithium-like $^{209}Bi^{80+}$ showing the two components due to the $F = 4$ and $F = 5$ hyperfine levels present in the ground state [150]. A very high resolution crystal spectrometer and efficient evaporative cooling were essential for this result obtained using SuperEBIT operating at 100 keV beam energy

lithium-like charge state, also a good agreement with bound-state QED predictions has been found [155]. Analogous experimental studies on the $3s$ electron QED shifts were performed for the $3s_{1/2} - 3p_{3/2}$ resonance line in sodium-like U^{81+} [156].

More recently, the energy of the $1s^2\,2p_{3/2} \rightarrow 1s^2\,2s_{1/2}$ transition in Pb^{79+} was measured with different isotopes at the Tokyo electron beam ion trap by Zhang et al. [157]. Due to its nuclear spin (I $= 1/2$), the transition in ^{207}Pb shows three blended hyperfine components. The transition energy result of 2642.260 ± 0.100 eV yields a QED contribution of 24.990 ± 0.100 eV which was compared with different calculations. The predicted isotopic shifts relative to ^{208}Pb are 0.1034, 0.0516, and 0.0258 eV for $^{204,206,207}Pb$, respectively.

10.6.4 Spectroscopy of Forbidden Transitions in the Visible Range

Spectroscopic experiments determining optical transition wavelengths and lifetimes of few-electron ions represent a continuing challenge for the theoretical understanding, due to the complex interplay of electron correlation, relativistic, and QED effects. In the region of intermediate atomic numbers, these effects have comparable magnitudes (as measured, i.e., by their contribution to binding energies) and are intertwined. Measurements of lifetimes are especially informative in benchmarking structure theories since the radiative transition matrix elements are particularly sensitive to the long-range behaviour of the electronic probability density. Furthermore, experimental investigations of optical isotope shifts benefit from an increased sensitivity of HCI to effects arising from nuclear properties. Therefore, experimental mass shifts in these systems provide on an ideal testing ground of relativistic nuclear recoil effects.

The first forbidden transition in an EBIT was observed by Morgan at the NIST EBIT. He detected the emission from the line connecting two low-lying states of the $3d^4$ configuration in titanium-like Ba^{34+} [158]. Later, Porto et al. continued these observations [159] along the isoelectronic sequence in ytterbium, tungsten and bismuth. Studies on the forbidden lines of Ba and Kr ions followed [27, 160]. Watanabe performed a systematic study of the M1 transitions in the titanium-like sequence [161] at the Tokyo EBIT. Other studies in this isoelectronic sequence are found in Ref. [162]. Bieber studied coronal lines in argon and barium ions at the Oxford EBIT [17]. Several measurements were performed at LLNL by Crespo et al. [163, 164]. Draganič [165] carried out very precise wavelength measurements in $Ar^{13+,14+}$ and Ba^{34+} and was able to resolve the Zeeman splitting of the transitions due to the magnetic field of the EBITs in Freiburg and Heidelberg. Soria Orts continued these investigations. An interesting outcome was her experimental confirmation [166] of relativistic mass shifts [89, 167], and also the determination of the g factor on the same ion [168]. An overview on the development of the field of visible transitions in HCI is given in [169].

Lifetime measurements of the $1s^2 2s^2 2p\ ^2P_{3/2}$ level of boron-like Ar^{13+} were performed [31, 170] at the Heidelberg EBIT [171]. The electron beam was switched on periodically to excite the trapped ions to the $^2P_{3/2}$ state and, then, switched off to detect the M1 decay to the ground state. Lifetimes were determined by recording the decay curve, i.e. the decrease of fluorescence intensity versus time. The experimental result of $\tau = 9.573(4)(5)$ ms (stat)(syst) agrees well with measured data of 9.70(15) ms [32], while being by more than an order of magnitude more accurate than previous experiments [32, 172–174]. Figure 10.13 displays a comparison with various theoretical and experimental results. The most recent calculation using the configuration interaction Dirac–Fock–Sturmian method (CI–DFS, Ref. [175]) yields a lifetime value $\tau = 9.538$ ms. The transition rate (Einstein coefficient) is proportional to the third power of the transition frequency. The corresponding wavelength has been accurately measured in Ref. [165], resulting in $\lambda = 441.2559(1)$ nm. In Fig. 10.13, theoretical results are displayed in different ways: (i) as published, (ii) corrected for the more accurate experimental transition frequency, and (iii) excluding and (iv) including the effect of the leading radiative correction in the electron anomalous magnetic moment (EAMM) approximation. The EAMM correction leads to a decrease of the lifetime by a factor $1 - 2\,\alpha/\pi$. Calculated lifetimes appear to cluster around a value of 9.53 ms, which is within a 3-σ disagreement with the experimental value. The reason of this discrepancy is not yet understood, triggering further theoretical calculations [176] and measurements with different ions and transitions to get to the root of this issue.

By means of a similar emission spectroscopy technique, the isotope shift of the above transition in Ar^{13+} and that of the $1s^2\, 2s\, 2p\, ^3P_1$–$^3P_2$ line in Ar^{14+} was studied by measuring and comparing the wavelengths with the isotopes ^{36}Ar and ^{40}Ar. The isotope shifts were determined based on the wavelength differences, which had a sub-ppm accuracy. The details of the measurement can be found in Refs. [166, 183].

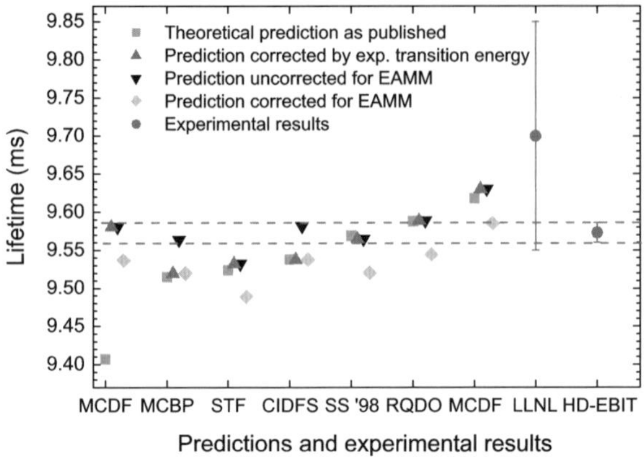

Fig. 10.13 A selection of predicted lifetimes of the ${}^2P_{3/2}$ first excited state in Ar^{13+} (see in corresponding order [175, 177–182]) and the experimental results of Ref. [31, 32] (*red dots* with error bars). *Dashed lines* uncertainty level. In all calculations, the electron anomalous magnetic moment (*EAMM*) contribution had been ignored, except in Refs. [175, 178]. See text for further details

Such isotope shifts are due to the isotopic change of the nuclear mass and the nuclear size, modifying the electronic wave function at short distances and thus shifting the binding energy. The mass shift, or recoil, contribution, is customarily separated into the normal and the specific mass shift terms. The relativistic recoil operator for a many-electron state is given by [89, 167, 184]

$$R_{ij} = \frac{\mathbf{p}_i \cdot \mathbf{p}_j}{2M} - \frac{Z\alpha}{2Mr_i}\left(\boldsymbol{\alpha}_i + \frac{(\boldsymbol{\alpha}_i \cdot \mathbf{r}_i)\mathbf{r}_i}{r_i^2}\right) \cdot \mathbf{p}_j\,, \tag{10.16}$$

where relativistic units are used, and $\mathbf{r}_i$ and $\mathbf{p}_i$ are the position and the momentum, respectively, of a given electron indexed by i, and α_i stands for the Dirac matrices acting on the wave function of that electron. M denotes the nuclear mass and α is the fine-structure constant. The normal mass shift contribution is defined as the expectation value of $\sum_i R_{ii}$, and the specific mass shift term is given as the sum of the non-diagonal elements $\langle \sum_{i \neq j} R_{ij} \rangle$. Restricting Eq. (10.16) to the first term only yields the familiar non-relativistic mass shift operator. Atomic structure calculations of recoil effects based on the above operator are presented in Ref. [185]. A listing of the normal and specific mass shift terms NMS and SMS, together with the respective relativistic corrections RNMS and RSMS in Ar^{13+} is given in Table 10.2. These terms define the nuclear recoil effect up to the order $(\alpha Z)^4 m_e^2/M$, with m_e being the electron mass. The calculation of the QED contributions not included in the recoil operator is based on the article [90]. For an evaluation of these radiative terms, the use of QED beyond the Breit approximation is mandatory (see e. g., [184, 186]). The calculated shifts including relativistic recoil terms are in an excellent agreement with the experimental values. A calculation involving the non-relativistic recoil operator

Table 10.2 Numerical contributions of relativistic recoil and QED recoil terms to the total mass shift, and the field shift effect ($^{36}Ar - {}^{40}Ar$) in the ground-state M1 transitions of Ar^{13+} and Ar^{14+} ions, in units of cm^{-1}. Results from [166]

Contribution	Ar^{13+}	Ar^{14+}
NMS	0.1052	0.0796
RNMS	−0.0822	−0.0627
One-electron QED	0.0002	0.0002
SMS	−0.0741	−0.0697
RSMS	0.1151	0.0887
Two-electron QED	−0.0008	−0.0015
FS	−0.0005	−0.0001
Sum	0.0629	0.0345

only would disagree with the experiment by a factor of approx. 2 (for Ar^{13+}) or 3 (for Ar^{14+}), as Table 10.2 shows.

10.6.5 The Hyperfine Structure of Hydrogenlike Ions

The most tightly bound electron, the $1s_{1/2}$, can give rise to an optical transition due to the hyperfine splitting (HFS) in heavy hydrogen-like ions in the presence of a nuclear spin. For heavy elements, the transition energy scales from the microwave region up to the visible spectrum, and even into the ultraviolet region in the case of bismuth. The magnetic hyperfine splitting of a hydrogenic ion with the electron being in a state described by the usual quantum numbers n, l and j may be generally represented in the form [187]

$$\Delta E_\mu = \frac{\alpha(Z\alpha)^3}{n^3} g_I \frac{m}{m_p} mc^2 \frac{F(F+1) - I(I+1) - j(j+1)}{2j(j+1)(2l+1)} \times \left[A_\mu(\alpha Z)(1-\delta)(1-\varepsilon) + x_{rad}\right]. \tag{10.17}$$

Here, Z is the atomic number, α is the fine-structure constant, m and m_p denote the mass of the electron and the proton, respectively, and c is the speed of light. I is the nuclear spin, F with $|I - j| \leq F \leq I + j$ is the total angular momentum quantum number of the ion, and $g_I = \mu/(\mu_N I)$ is the g factor of the nucleus, expressed in terms of the nuclear magnetic moment μ and the nuclear magneton μ_N. $A_\mu(\alpha Z)$ stands for the relativistic factor

$$A_\mu(\alpha Z) = \frac{n^3(2l+1)\kappa(2\kappa(\gamma + n_r) - N)}{N^4\gamma(4\gamma^2 - 1)}, \tag{10.18}$$

with the relativistic angular momentum quantum number $\kappa = 2(l - j)(j + 1/2)$, and the further quantities $\gamma = \sqrt{\kappa^2 - (\alpha Z)^2}$, $N = \sqrt{n_r^2 + 2n_r\gamma + \kappa^2}$, $n_r = n - |\kappa|$. A_μ converges to unity in the non-relativistic limit. Equation (10.17) with keeping only the analytical relativistic factor $A_\mu(\alpha Z)$ in the square brackets describes the

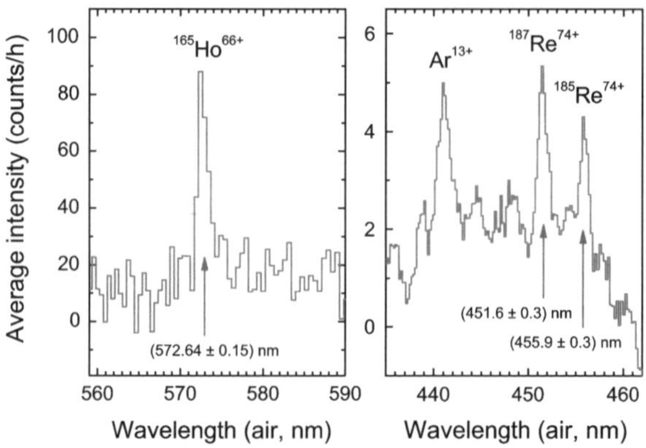

Fig. 10.14 Spectra from the hyperfine transition between the $F = 4 \rightarrow F = 3$ states in hydrogenlike Ho^{66+} (*left*) and $F = 3 \rightarrow F = 2$ states in hydrogenlike $^{185,187}Re^{74+}$ (*right*) [29, 189]

hyperfine splitting in the point-like nucleus approximation. Further corrections are introduced by the other parameters in the square brackets: δ is the nuclear charge distribution correction, ε is the nuclear magnetization (Bohr-Weisskopf) correction, and x_{rad} stands for the radiative correction, represented by self-energy and vacuum polarization diagrams. Further corrections arise due to the finite mass of the nucleus. The parameters δ and ε can be calculated assuming some nuclear model for the charge and magnetization distribution of the nucleus [187].

The first observation of HFS in HCI was achieved using the ESR storage ring at GSI [188]. Accelerated $^{209}Bi^{82+}$ ions produced by stripping were stored in the ring and excited by a laser which was tuned across the expected range. A fluorescence signal was obtained when the Doppler-shifted photon energy hit the $F = 4 \rightarrow F = 5$ transition between the ground state hyperfine levels. Shortly after, experiments at Super EBIT measured the corresponding transitions in Ho^{66+} (see Fig. 10.14), $^{173,175}Re^{74+}$, and $^{183,185}Tl^{80+}$ ions [189–191]. Later, the $^{207}Pb^{81+}$ ion was investigated, again at GSI [192]. Resonant laser excitation of the ground-state hyperfine transition in lithium-like bismuth performed at the storage ring ESR has been reported very recently [193]. Further studies will be required to reduce the uncertainties resulting from the Doppler shift.

Initially, the goal of the experiments had been to test QED contributions by means of the HFS of the hydrogenic systems: in contrary to measuring transition energies, which are mainly determined by the properties of the bound electron wave function at atomic distances (typically, a_0/Z, with a_0 being the Bohr radius), the value of the hyperfine splitting is dominantly determined by the behaviour of the electron wave function in the vicinity of the nucleus. Since the nuclear Coulomb field is strongest in this regime, measurements of the HFS can provide tests of QED in electric fields much stronger than those typical for binding energies. Experimental errors in hyperfine splitting energies are also well below the magnitude of QED corrections.

However, very soon it turned out that uncertainties in the nuclear magnetization distribution, due to nuclear structural effects, were already sufficient to compromise the QED tests, since knowledge about the so-called hyperfine anomaly was not sufficiently accurate [194, 195]. The Bohr-Weisskopf effect resulting from this magnetization distribution was larger than the total QED correction, thus an investigation of QED terms by directly comparing experimental and theoretical hyperfine splitting is not possible. Additional problems might arise from the fact that nuclear magnetic moments were usually determined under conditions in which the diamagnetic correction, the shielding of external magnetic fields by the electronic shell, played a large role. While this correction plays less of a role in cases where atomic beam measurements are the source of the nuclear magnetic moment data than in the cases where nuclear magnetic resonance measurement with condensed probes were used, the calculation of this large correction factor also affects the final result. For these reasons, the measured wavelengths were used to determine the nuclear magnetization radii under different assumptions for the model magnetization distribution. In this way it became possible to quantify this elusive parameter, which otherwise can only be studied using muonic atoms. As an example, in the case of thallium the RMS value for the nuclear magnetic radius was determined with a fivefold increase in accuracy in comparison with previous work, to be 5.83 ± 0.14 fm and 5.89 ± 0.14 fm for the ^{203}Tl and ^{203}Tl isotopes, respectively [29].

As shown by Shabaev [196], the unwanted effect of not sufficiently well understood nuclear structural properties can be circumvented by employing a specific combination of hyperfine splitting values in hydrogen-like and lithium-like ions. In the following difference of the hyperfine splitting values of the different charge states,

$$\Delta' E = \Delta E^{(2s)} - \xi \Delta E^{(1s)}, \tag{10.19}$$

the parameter ξ can be chosen such that the Bohr-Weisskopf correction is largely cancelled. ξ can be calculated with sufficient accuracy, as it depends dominantly on the bound electron wave function and hardly on nuclear properties. For example, for bismuth ($Z = 83$), with $\xi = 0.16885$, a test of strong-field quantum electrodynamics on the percent level can be reached via $\Delta' E$, provided that the HFS values in both charge states can be measured with a relative accuracy of approx. 10^{-6} [196]. A similar procedure may be applied using boron-like and hydrogen-like ions of the same heavy element [197].

10.7 Resonant Photorecombination Processes

By choosing exactly the energy needed to excite dielectronic recombination resonances, far more selective excitation spectra can be obtained. Early EBIT measurements using this method were carried out by Beiersdorfer [198] on helium-like

Fe^{24+} and by Knapp et al. [199]. For the Fe ions, the energy resolution allowed to test dielectronic resonance strengths for individual resonances and compare them with predictions. Normalization was performed by comparison with the radiative recombination rate, which as a non-resonant process could be predicted theoretically with much smaller uncertainty.

10.7.1 Dielectronic Recombination

In the DR process, the cross section $\sigma^{\mathrm{DR}}_{i\to d\to f}(E)$ indicates the probability for electronic capture from an initial state i to a final state f under excitation of an intermediate, and often doubly-excited, state d:

$$\sigma^{\mathrm{DR}}_{i\to d\to f}(E) = \frac{2\pi^2\hbar^4}{p^2}\frac{A_r^{d\to f}}{\Gamma_d}L_d(E)V_{\mathrm{DC}}^{i\to d} . \tag{10.20}$$

Here, $p = \sqrt{(E/c)^2 - m^2c^2}$ represents the momentum of the projectile electron with a kinetic energy E. The energy dependence is conveniently represented with a Lorentzian profile

$$L_d(E) = \frac{\Gamma_d/(2\pi)}{(E_i + E - E_d)^2 + \frac{\Gamma_d^2}{4}} . \tag{10.21}$$

In this equation, Γ_d is the natural width of the intermediate state resulting from both radiative and auto-ionization decay channels: $\Gamma_d = \hbar A_r^d + \hbar A_a^d$. As mentioned above, the dielectronic capture rate V_{DC} and the Auger process rate are connected by the principle of detailed balance:

$$V_{\mathrm{DC}}^{i\to d} = \frac{2J_d + 1}{2(2J_i + 1)}A_a^{d\to i} . \tag{10.22}$$

The different density of initial and final states is accounted for with the multiplicities which follow from the total angular momenta of the intermediate J_d and the initial ionic state J_i, and the spin degeneracy of the initial continuum electron. Knowing that the Lorentzian function (10.21) is normalized to unity on the energy scale, i.e. $\int dEL_d(E) = 1$, and assuming for the electron momentum its resonance value $p_{res} = \sqrt{(E_d - E_i)^2/c^2 - m^2c^2}$, the strength of a DR resonance, i.e. the integrated area of a peak can be written as

$$S^{\mathrm{DR}}_{i\to d\to f} = \frac{2\pi^2\hbar^4}{p_{res}^2}\frac{2J_d + 1}{2(2J_i + 1)}\frac{A_r^{d\to f}A_a^{d\to i}}{\Gamma_d} . \tag{10.23}$$

If the final states are not resolved in an experiment, a summation over all possible such states, i.e. radiative decay pathways of the auto-ionizing states, has to be performed.

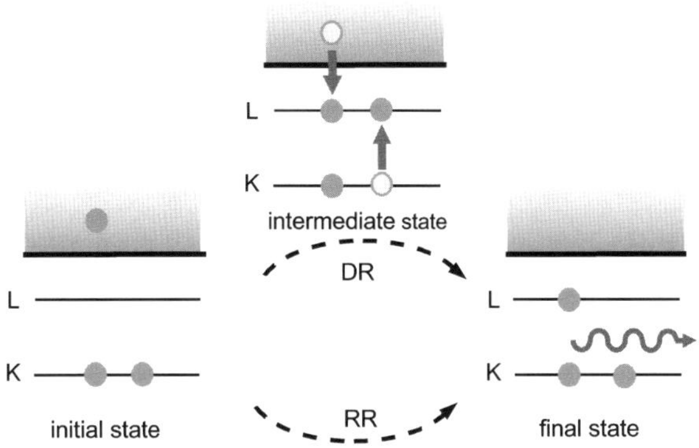

Fig. 10.15 Schematic representation of dielectronic (*top*) and radiative recombination (*bottom*), exemplified for KLL transitions in helium-like ions. The initial state of the total system consists of the ground-state ion and a continuum electron. After dielectronic capture, the ion is in an autoionizing intermediate state, which then relaxes by photon emission to a final state (which is not necessarily a ground state). The figure also illustrates that the initial and final states of RR and DR are indistinguishable, thus an interference of the two quantum pathways may occur

Also, usually in an EBIT experiment, it is not the above total cross section which is measured, but the photon intensity at a given angle with respect to the direction of the incoming electron beam. Therefore, the above formulas have to be corrected for the angular distribution of the X-ray emission [200–205] (Fig. 10.15).

Equation (10.23) allows one to derive a simple scaling law for the DR resonance strength with respect to the atomic number Z. As electron kinetic energies are much lower than the rest energy mc^2, the non-relativistic approximation may be used for the electron momentum, thus $1/p^2 \propto 1/E \propto 1/Z^2$. As noted before, for E1 transitions radiative rates are approximately proportional to Z^4, and the Auger rates are independent on Z, i.e. $A_a \propto V_{\mathrm{DC}} \propto Z^0$. Combining these, a non-relativistic scaling law of the following form can be derived for the DR resonance strength:

$$S^{\mathrm{DR}}_{i\to d\to f} \propto \frac{1}{aZ^2 + bZ^{-2}}, \tag{10.24}$$

with a and b being parameters depending on the actual rates and transition energy of the recombination channel. This function increases for low Z values, peaks usually in the intermediate Z region, and then decreases as $\propto Z^{-2}$.

The notion of dielectronic recombination (DR) was introduced by Massey and Bates in 1942 [206]. In 1964, Burgress used this process for the explanation of the temperature distribution of the solar corona. For a detailed historical treatment of DR until 1976, see the review article of Seaton and Storey [207].

The helium-like ions trapped in the LLNL EBIT were investigated with an electron beam of variable energy and an event-by-event data collection of the emitted X-rays and their energy as a function of the electron acceleration voltage [199, 208].

These experiments were reproduced and extended to other species in different laboratories. Investigations at the Berlin EBIT addressed channel-specific paths of dielectronic recombination in krypton ions including the heliumlike resonances [209].

Dielectronic resonances of helium-like Fe^{24+} [210], Xe^{52+} [37], I^{53+} [211] and heliumlike Ti^{20+}[212] were investigated by the Tokyo group, together with other isoelectronic sequences involving recombination into the $n = 2$ shell. The resonance strengths in the DR process with Ar^{16+} ions were measured by Smith et al. [213] at LLNL. He also studied the high n satellites of the Kα transition [214]. Tarbutt measured the dielectronic satellite transitions of the same ion under resonant excitation conditions [107], and Biedermann carried out similar measurements with the goal of improving plasma diagnostics in fusion research [136]. Relaxation of the doubly-excited intermediate state $1s^1\,2s^2\,J = 1/2$ through a two-electron one-photon transition in the KLL resonance array of Ar^{16+} were studied by Zou [215]. Zhang performed an investigation on Ge^{32+} [216].

Paralleling the experimental developments, extensive non-relativistic theories for the description of DR have been developed in the late 1970s and early 1980s, with the distorted wave formalism of Hahn et al. being an example [217, 218] (see also the review [207]). An extension to the relativistic case was carried out in several works, e.g. by Zimmerer et al. [219–221], by Chen and Scofield [37, 200], Pindzola [222]. The theoretical formalisms of Shabaev [84, 187] and Labzowsky et al. [223, 224] are based on relativistic field theory and allow a systematic treatment of radiative corrections and quantum interference effects.

10.7.2 Higher-Order Interactions in Photorecombination: Trielectronic and Quadruelectronic Terms

For the KLL resonances, the dielectronic recombination process had been generally understood to be the strongest photo-recombination channel. Beyond DR, however, higher-order processes may occur, in which the recombining electron simultaneously excites two or more bound electrons. These channels may be termed as trielectronic and quadruelectronic recombination (TR and QR, respectively), as three or four electrons are actively participating in the transition. These processes may be presumed to be weaker, since additional electron–electron interactions are needed to populate the intermediate states which are more highly excited than in the case of DR. However, experiments by Beilmann [225] showed that for some isoelectronic sequences, those processes were actually rather strong and in some cases dominant (see Fig. 10.16). The carbon-like ions in the range of atomic number $12 < Z < 26$ show trielectronic resonances as strong or stronger than the dielectronic ones.

For higher-order recombination processes, also the scaling of the resonance strength with Z is modified. In contrast to the dielectronic capture rates (10.22) which are largely independent on Z in the non-relativistic regime, trielectronic capture rates scale roughly as $1/Z^2$, since the additional Coulomb interaction among the

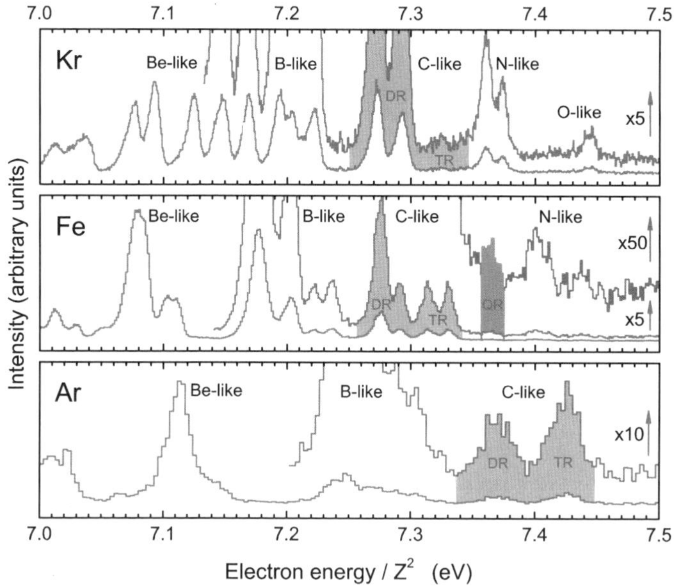

Fig. 10.16 Comparison of the X-ray fluorescence spectra following resonance recombination with carbon-like Ar, Fe, and Kr ions ($Z = 18, 26$, and 36). Trielectronic recombination (*TR*), barely resolvable in the case of Kr, gathers a larger total resonance strength than the dielectronic channel (*DR*) in the case of Ar (and other lighter elements). DR, TR and partially QR lines are marked by *orange*, *green* and *red* colours, respectively. The electron energies are normalized by Z^2 for a better comparison

electrons introduces a perturbative factor of $1/Z$ to the transition amplitude of the recombination step [225]. Thus, the scaling of the TR strength can be given as

$$S^{\mathrm{TR}} \propto \frac{1}{aZ^{-2} + b + cZ^4}\,. \tag{10.25}$$

This behavior has been confirmed experimentally by measuring the ratio of certain TR and DR strengths along the carbon-like isoelectronic series at different Z values ($Z = 18, 26, 36$) [225].

Theoretical models have to include the more complex intermediate states populated by the higher-order processes, and configuration mixing between the resonances corresponding to DR, TR, and QR in order not to neglect a fraction of the photo-recombination resonance strength. This has implications for the energy transfer within the radiative zone of stellar interiors, and for the ionization equilibrium of those and other astrophysical plasmas.

10.7.3 The Effect of the Breit Interaction in the Dielectronic Process

In atomic structure calculations and in the modelling of dynamical processes, the interaction between the electrons is usually approximated by the instantaneous Coulomb interaction. As known from the Bohr model, the normalized velocity v/c of atomic electrons can be approximated as $v/c = Z\alpha/n$, with n being the principal quantum number. This shows that in HCI with higher nuclear charge numbers, and, especially, in transitions involving inner-shell electrons, relativistic effects have to be taken into account. This does not only concern the bound-electron wave function, but also the inter-electronic interaction: as it is known even from classical considerations, the Lorentz force coupling a moving charged particle to the magnetic field induced by an other is proportional to v/c. Besides such current–current coupling effects, retardation also plays a role. A rigorous derivation of the electron–electron interaction can be made in the framework of QED. To the lowest order, the interaction of electrons is described by the one-photon exchange Feynman diagrams. In the Coulomb gauge, this yields to an effective interaction operator, which can be represented as the sum of the Coulomb and generalized Breit interactions:

$$V_{ij}(\omega) = V_{ij}^C + V_{ij}^B(\omega)\,, \tag{10.26}$$

with $V_{ij}^C = 1/r_{ij}$ and r_{ij} being the distance between two electrons. The generalized (or frequency-dependent) Breit interaction $V_{ij}^B(\omega)$ can be decomposed into the following parts:

$$V_{ij}^B(\omega) = V_{ij}^{\text{magn}} + V_{ij}^{\text{ret}} + V_{ij}^{\text{ret}}(\omega)\,. \tag{10.27}$$

Here, the magnetic term is

$$V_{ij}^{\text{magn}} = -\frac{\boldsymbol{\alpha}_i \cdot \boldsymbol{\alpha}_j}{r_{ij}}\,, \tag{10.28}$$

and the retardation of the scalar and the transverse interaction in the Breit approximation is described by the operator

$$V_{ij}^{\text{ret}} = \frac{\boldsymbol{\alpha}_i \cdot \boldsymbol{\alpha}_j}{2r_{ij}} - \frac{(\boldsymbol{\alpha}_i \cdot \mathbf{r}_{ij})(\boldsymbol{\alpha}_j \cdot \mathbf{r}_{ij})}{2r_{ij}^3}\,. \tag{10.29}$$

The sum of V_{ij}^{magn} and V_{ij}^{ret} is the usual (frequency-independent) Breit interaction V_{ij}^B. Additional retardation effects are described by a term which depends explicitly on the frequency of the exchanged virtual photon: ω:

$$V_{ij}^{\text{ret}}(\omega) = -\boldsymbol{\alpha}_i \cdot \boldsymbol{\alpha}_j \frac{\cos(\omega r_{ij}) - 1}{r_{ij}} + (\boldsymbol{\alpha}_i \cdot \boldsymbol{\nabla}_i)(\boldsymbol{\alpha}_j \cdot \boldsymbol{\nabla}_j) \frac{\cos(\omega r_{ij}) - 1 + \omega^2 r_{ij}^2/2}{\omega^2 r_{ij}} . \tag{10.30}$$

Here, $\boldsymbol{\nabla}_i$ denotes differentiation with respect to the position vector of the ith electron. It has been recognized soon that such effects largely influence the energy levels in heavy atoms and ions [226]. Cheng, Johnson et al. [123, 129, 227] found that even higher-order Breit interactions contribute at the level of 0.1 eV and that inclusion of frequency-dependent and other high-order many-body corrections was necessary in order to extract QED contributions from the accurate experimental data.

A strong influence of the Breit interaction to the dielectronic capture rates have been predicted by different groups [200, 219]. This effect was experimentally confirmed in EBIT measurements by Nakamura and his team [43] through comparison of the resonance strengths to calculations with and without the inclusion of the Breit interaction in the capture matrix elements. In a recent experiment, also at the Tokyo-EBIT [228], a measurement of the intensity of an X-ray decay following dielectronic capture into lithium-like Au ions at 90° has confirmed the dominance of the Breit interaction in the angular distribution of certain transitions, predicted by Fritzsche et al. [205].

10.7.4 Quantum Interference Terms Between Dielectronic and Radiative Recombination

For very high charge states, the RR cross Eq. 10.11 become very large. Intuitively, this results from the strong nuclear field. At the same time, electron–electron correlations do not change much with the ionic charge, and, as also Eq. (10.24) shows, dielectronic resonances decrease and partly lose their dominating role in photo-recombination. Thus, the quantum mechanical transition amplitudes for the direct and resonant recombination channels may become comparable. Since the initial and final states of both channels may be the same, as shown in Fig. 10.15, quantum interference of the two processes can occur. This situation is analogous to the case of photo-ionization, for which a modification of the line shapes due to interference was described by Fano [229]. The total cross section of photo-recombination including interference effects is given by [221, 230–232]

$$\sigma_{i\to d\to f}^{\text{PR}}(E) = S_{i\to d\to f}^{\text{DR}} \frac{2}{\Gamma_d \pi} \frac{1 - 1/q^2 + 2\varepsilon/q}{1+\varepsilon^2} , \tag{10.31}$$

with q being the Fano interference parameter [221, 229–231], and the unitless energy is $\varepsilon = 2(E + E_i - E_d)/\Gamma_d$. For large q values, the asymmetric Fano line shape converges to the symmetric Lorentzian one, i.e. $\lim_{q\to\infty} \sigma^{\text{PR}} = \sigma^{\text{DR}}$. Calculations for

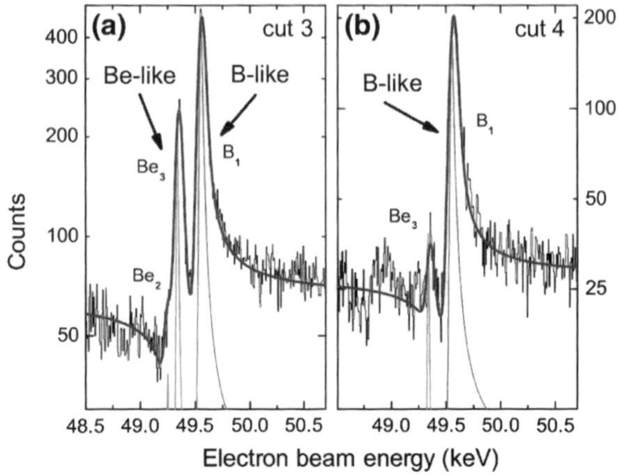

Fig. 10.17 Fano *line shapes* observed in the X-ray fluorescence spectra following KLL photo-recombination with beryllium- and boron-like Hg ions. The experimental data is shown in *black*, and a fit curve in *red* (Fano profile convoluted with a Gaussian function to account for experimental line broadening). Normalized non-convoluted theoretical cross sections are shown in *blue* and *green* for the beryllium- and boron-like charge state, respectively. Figure taken from Ref. [63]

the photo-recombination of highly charged heavy ions taking into account quantum interference effects between DR and RR have been performed by several groups [221–223, 233, 234].

Since the increase of the radiative widths renders the resonances also rather broad in the high-Z range, studies of resonance profiles by tuning the electron beam energy became possible. Observations were made for helium- to boronlike uranium ions in the Super-EBIT [235], and with highly charged mercury ions using the Heidelberg EBIT [63], as shown in Fig. 10.17. Further experimental observation of the interference effect has been performed with beryllium-like bismuth ions using the Tokyo-EBIT [236]. The q parameters of the Fano profiles was extracted and compared with theory in Ref. [237].

10.7.5 Resonant Photo-Ionization Processes

The first demonstration of photo-ionization of HCI using an EBIT quickly gave way to detailed studies in the soft X-ray region [45–47]. Much higher charge states were accessed than in earlier work using synchrotron radiation, where merged ion-photon beams were used to study resonant K-shell PI of C^{4+} at photon energies near 300 eV, and also charge states as high as 7+ in Xe. However, ionization potentials were limited to around 150 eV due to insufficient ion beam area densities. With the EBIT method, Fe^{14+} and Ar^{12+} were addressed, with four times higher ionization potentials and with high spectral resolution. The results for iron (see Fig. 10.18 showed remarkable

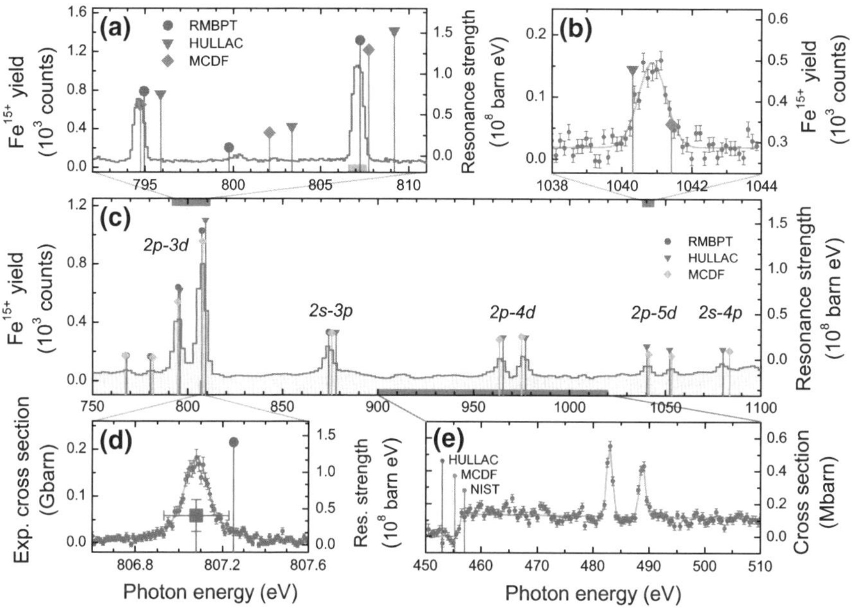

Fig. 10.18 Photoion yield after irradiation of Fe^{13+} with synchrotron radiation as a function of the photon energy [47]. Panel **c** overview scan; **a**, **b**, **d** detailed scans of various resonances; **e** ionization edge and second-order photon lines. Theoretical values: RMBPT: *blue circles* (Ref. [238]); HULLAC: *brown triangles*; MCDF: *orange circles* (both from Ref. [47])

disagreement with several predictions, but also a very consistent agreement with the relativistic many-body perturbation theory (RMBPT) calculations of Gu [238] at the level of meV.

Fano quantum interference of direct and resonant photoionization in the X-ray regime, analogously to interferences in recombination processes discussed above, has been observed with EBIT methods by Simon et al. [46].

10.8 Photonic Excitation: Laser Spectroscopy from the Visible to the X-ray Range

One of the great advantages of EBITs is the linear structure of the ensemble of trapped ions. The ion density can be estimated simply by assuming that the negative space charge of the electron beam of a density n_e (which can be as high as $n_e = 10^{13}$ cm^{-3}) is compensated in part by the positive ions of charge state q_{ion} it traps. Thus, a typical ion density is $n_{ion} < 0.3 \times n_e/q_{ion}$. If the interacting beam overlaps with the ion cloud axially, the ion target area density results from integrating over a trap length of a few centimeters, giving thus values of 10^{10} cm^{-2} ions. By comparison, the density of the residual gas background at pressures down to 10^{-13} mbar found in a cryogenic EBIT can be as low as 3,000 neutrals/cm^3. Therefore, the photon beam interacts mainly with the ion target and very little with the residual gas.

10.8.1 Laser Spectroscopy in the Soft X-ray Region at Free-Electron Lasers

While laser spectroscopy had become by the end of the twentieth century an established method in atomic physics, and its applications to various fields of science had unleashed very rapid progress in the understanding of the electronic structure of atoms and molecules, metrology, remote sensing and many other branches, the vacuum ultraviolet region had only been touched by two-photon laser spectroscopy of the Lyman-α lines in the hydrogen atom. The cause for this status was primarily the lack of practical lasers in this spectral range. The few demonstrations of X-ray lasers based on collisional excitation within plasmas generated by pulsed optical lasers in the kilojoule range [240, 241] or table-top devices using electrical discharges [242] did not provide tunable beams, and required a very complex operation. The situation changed with the introduction of free-electron lasers (FEL) capable of generating the sought-after short wavelengths [243]. By changing the acceleration and undulator parameters, it was now possible to scan the photon energy across a wide range, and repetition rates on the order of 100 Hz were within reach. The Tesla Test Facility, later renamed FLASH (Free-Electron LASer in Hamburg) opened its doors as an experimental user facility in 2005 and enabled a large variety of novel experiments [244]. The Freiburg EBIT group had already started with the design of an experiment involving an EBIT as a target of HCI for FEL laser spectroscopic studies. The so-called Tesla (later FLASH) EBIT, a transportable device, came into operation in the year 2006 [39] and was installed at FLASH. There, the Fe^{23+} ions produced and stored within the device were exposed to the FLASH pulses (see Fig. 10.19), and the transition $2s_{1/2} \rightarrow 2p_{1/2}$ was resonantly excited by them at a photon energy of 48.6 eV. A coincidence detection method enormously reduced the background. Since the bandwidth of the radiation from FLASH was broad, an in-line grating monochromator had to be used to reduce it to an acceptable level [39, 239]. The experiment with Fe^{23+} achieved an excellent level of statistical accuracy, but calibration and stability of the in-line monochromator were not commensurate, and a systematic calibration error of 0.15 eV remained. However, due to the high ion production rate of that experiment, an accompanying measurement using a flat-field spectrometer could improve earlier results and brought the total uncertainty to a level of 4 ppm.

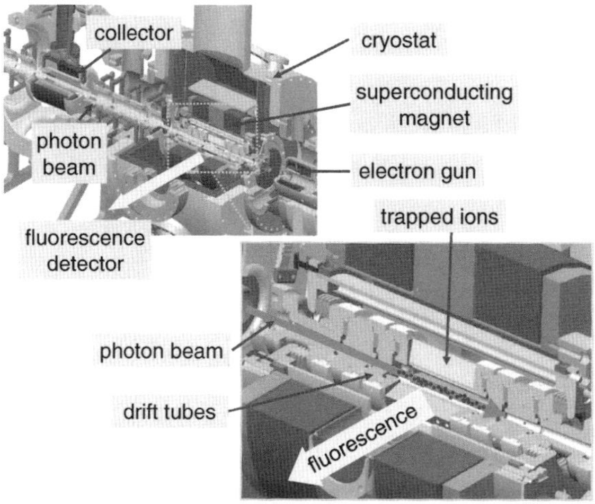

Fig. 10.19 X-ray laser spectroscopy setup with FLASH EBIT [39, 48, 239]. The monochromatic photon beam from a free-electron laser or synchrotron radiation source enters through the hollow collector and interacts with the trapped ions in the central drift tube region (zoomed insert). X-ray resonance fluorescence is detected through the viewports arranged around the cryostat

10.8.2 Extending Laser Spectroscopy into the X-ray Domain

The next step followed with the commissioning of the Linac Coherent Light Source (LCLS) at SLAC in Menlo Park, California. The FLASH EBIT was attached to the SXR monochromator beamline in order to analyze the strongest transitions in Fe^{16+}, an ion of utmost astrophysical importance. The question being addressed here was the existing discrepancy between advanced theory and astrophysical as well as plasma observations with regard to the intensity ratio of the strongest transitions in its spectrum [48, 49] (Figs. 10.20, 10.21).

Very recently, this method has been demonstrated for excitation of the strongest transitions in helium-like Fe^{24+} and in lithium-like Fe^{23+} by Rudolph and colleagues [140], and for Kr^{34+} by Steinbrügge et al. Fig. 10.22 shows an overview spectrum and individual line scans obtained with FLASH EBIT using a hard X-ray beamline equipped with a double undulator [245] at PETRA-III. The stability of the available monochromator permitted high resolution scans, yielding accurate energies (see Table 10.3 and also line widths (see plot on the bottom of Fig. 10.22). The measured energies confirm bound-state QED predictions by Artemyev et al. for the helium-like [131] and lithiumlike [141] charge states, and point out the need of extending such calculations to ions with more electrons. By using two detectors, one of them parallel to the polarization plane of the incoming radiation, and one perpendicular, the fluorescence emission characteristics of the different transitions can be recorded. Since for helium-like Fe^{24+}, the intercombination line y has a narrower natural line width than the resonance transition w, the line profiles are distinct and can

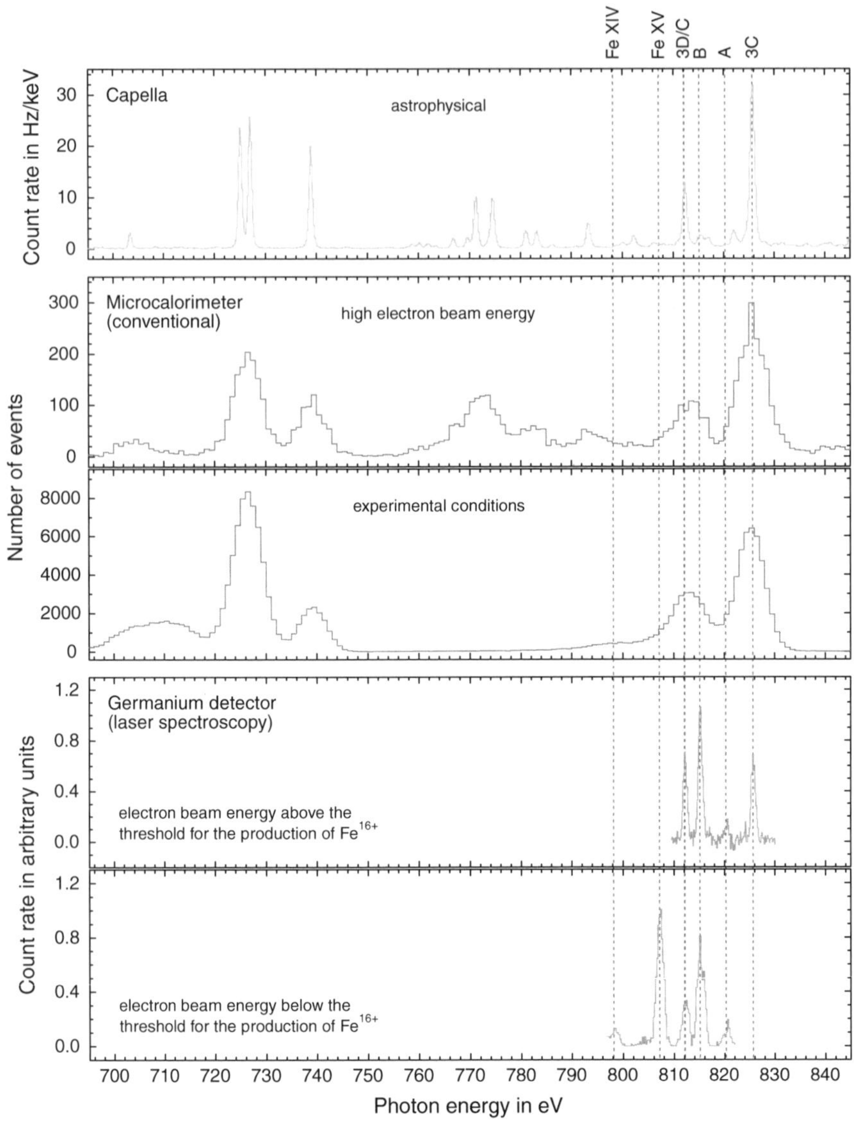

Fig. 10.20 Comparison of astrophysical emission spectra of highly charged iron ions (from Capella) with data from FLASH EBIT. *Blue* Grating spectrometer data from Capella. *Red* Microcalorimeter spectra obtained under different EBIT conditions. *Green* Soft X-ray laser spectroscopy results [48, 49] obtained at the LCLS free-electron laser facility

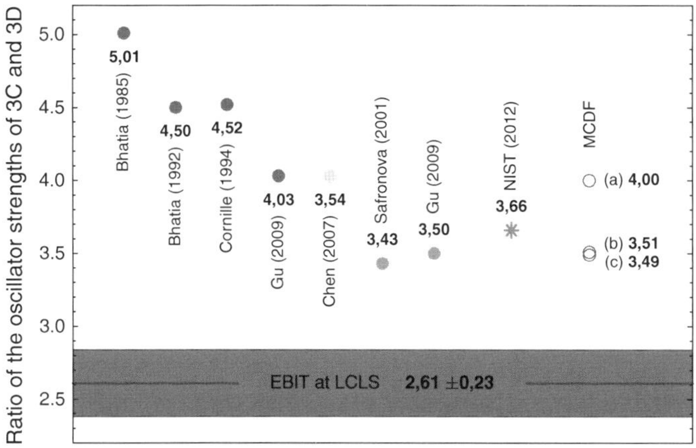

Fig. 10.21 Ratio 3C/3D of the oscillator strengths of the strongest soft X-ray transitions in neonlike Fe^{16+} obtained with various theoretical methods, in comparison with the X-ray laser spectroscopy result obtained at LCLS [48]. The significant and consistent discrepancy points to a probably insufficient inclusion of electron–electron correlation contributions in the theoretical approaches

be fitted with a Voigt profile, yielding a Gaussian component resulting from instrumental effects and ion temperature, and a Lorentzian width corresponding to the transition rate. The example shown indicates a trapped ion temperature of less than 30 eV, resulting from strong evaporative cooling. These initial results clearly that in the near future, combining synchrotron radiation filtered by high-resolution monochromators with trapped HCI will help overcoming current limitations of emission spectroscopy, which has always to compromise spectral resolution for signal strength.

10.8.3 Laser Excitation of Forbidden Transitions in the Visible Range

Many years there has been a strong interest in laser spectroscopic studies with highly charged ions. Preparations for an experiment using beryllium-like Ar^{14+} were carried out already at the Oxford EBIT [248]. The RETRAP program at LLNL [249] had the aim of sympathetically cooling HCI for their use in laser spectroscopic studies. Laser spectroscopy on trapped HCI was first achieved in Heidelberg by Mäckel et al. [250] using Ar^{13+} ions. Here, a pulsed tunable laser was used in a collinear geometry for excitation of the E1-forbidden transition $2p\,^2P_{3/2} \rightarrow 2p\,^2P_{1/2}$ within the ground-state configuration of the ion. Due to its M1 character with a long decay time of 9.7 ms [31], detection of fluorescence radiation had to rely on the magnetic-trapping mode operation of the EBIT. The setup is sketched in Fig. 10.23. For the measurement the following cyclic scheme was applied: After ion production, the electron beam is turned off, and after a suitable decay time for the electron-impact excited ions, the

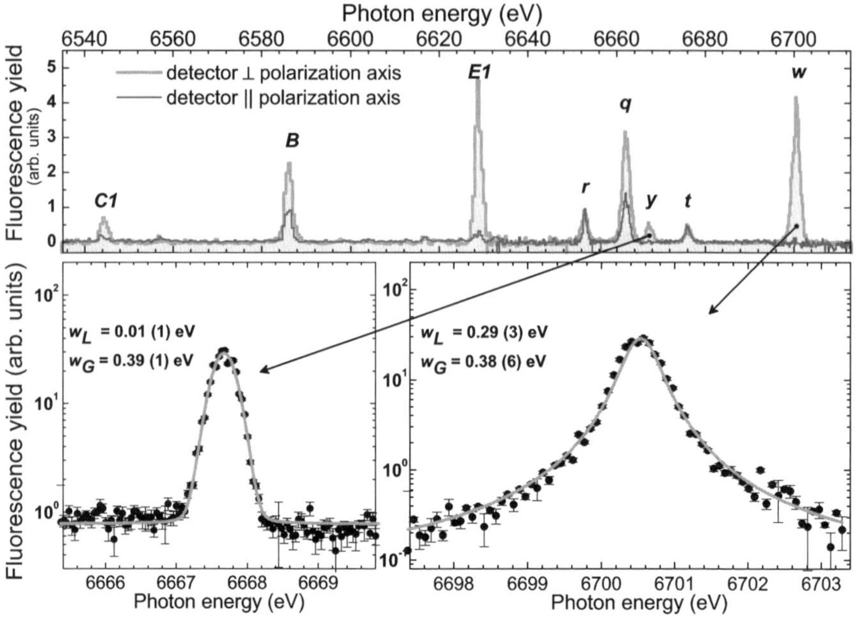

Fig. 10.22 *Top* Overview fluorescence spectrum of helium-like, lithium-like and beryllium-like Fe ions using two detectors aligned differently to the polarization plane of the incoming photon beam: (*red*) perpendicular to polarization axis, and (*blue*) parallel to polarization axis. *Bottom left* Scan over the intercombination line (y) $1s^2\,{}^1S_0 \rightarrow 1s\,2p\,{}^3P_1$. *Bottom right* Scan over the resonance transition (w) $1s^2\,{}^1S_0 \rightarrow 1s\,2p\,{}^1P_1$. The fitted Gaussian and the Lorentzian widths w_G and w_L of the transitions are shown [140]

fluorescence intensity after the laser excitation pulses is recorded as a function of their wavelength. Simultaneously, evaporative cooling [12] was applied in order to reduce the thermal width of the lines, which is very high due to the deep ion trapping potential in the EBIT. In this way, the translational ion temperature was lowered from values above 10^6 K down to the range of 10^5 K. Figure 10.24 shows the time evolution of the fluorescence signal and the ion temperature during the wavelength scans. One can observe how the width of four Zeeman components becomes smaller during the cooling cycle. The resulting wavelength of 441.25568(26) nm agrees excellently with earlier measurements carried out with grating spectrometers (441.2559(1) nm [165], and 441.2556(1) nm [168]). The currently most elaborate theoretical prediction of 441.261(70) nm [176] has a much larger uncertainty. Higher resolution would be possible if the trapped ions could be cooled even more. However, temperatures low enough (in the mK and lower range) for very high resolution laser spectroscopy are extremely difficult to achieve in this EBIT environment. A direct application of laser cooling schemes is currently not possible due to the long lifetimes of the forbidden transitions, and also to the lack of allowed transitions in the optical range. Most likely, future experiments will have to rely on the use of sympathetic cooling using external ion traps coupled to an EBIT.

Table 10.3 X-ray transitions of helium-like, lithium-like, beryllium-like, boron-like and carbon-like transitions resonantly excited from the ground state with synchrotron radiation. X-ray fluorescence was detected as a function of photon energy. Energies are given in units of eV; the calibration is based on the absorption edge technique [140]

	Transition	Experiment	Theory	Reference
w	$1s^2\,{}^1S_0 \rightarrow 1s\,2p\,{}^1P_1$	6700.549(5)(70)	6700.4347(11)	[131]
			6700.490	[118]
y	$1s^2\,{}^1S_0 \rightarrow 1s\,2p\,{}^3P_1$	6667.671(3)(69)	6667.5786(12)	[131]
			6667.629	[118]
t	$1s^2\,2s\,{}^2S_{1/2} \rightarrow 1s\,2s\,2p\,{}^2P_{1/2}$	6676.202(3)(69)	6676.129(47)	[141]
			6675.87	[246]
q	$1s^2\,2s\,{}^2S_{1/2} \rightarrow 1s\,2s\,2p\,{}^2P_{3/2}$	6662.240(6)(69)	6662.188(11)	[141]
			6661.88	[246]
r	$1s^2\,2s\,{}^2S_{1/2} \rightarrow 1s\,2s\,2p\,{}^2P_{1/2}$	6652.826(3)(69)	6652.776(25)	[141]
			6652.58	[246]
E	$1s^2\,2s^2\,{}^1S_0 \rightarrow 1s\,2s^2\,2p\,{}^1P_1$	6628.804(5)(68)	6631.057	[247]
B	$1s^2\,2s^2\,2p\,{}^2P_{1/2} \rightarrow 1s\,2s^2\,2p^2\,{}^2P_{1/2},\,{}^2D_{3/2}$	6586.085(7)(67)	6586.7	[140]
C	$1s^2\,2s^2\,2p^2\,{}^3P_0 \rightarrow 1s\,2s^2\,2p^3\,{}^3D_1$	6544.225(4)(66)	6544.8	[140]

A similar scheme was recently applied by Schnorr et al. [251] to the study of the so-called Fe XIV*green coronal line*, yielding a result of 530.2801(4) nm. This astrophysically essential line was discovered in the solar corona during the solar eclipse of 1869 [252], and originally even a new element, *coronium*, was postulated as its source. Eventually, it was explained in the 1940s by Grotrian [253] and Edlén [254] as arising from the Fe^{13+} ion, thus giving the first hint on the high temperature of the solar corona. However, in spite of its widespread use in astrophysical plasma diagnostics, no laboratory measurement of its wavelength had hitherto been reported.

10.9 Ion Cooling Schemes for Spectroscopy with HCI

In recent years, singly charged trapped ions have shown their ideal suitability as excellent frequency standards [255, 257–260], in particular if combined with frequency combs [261–263]. Sympathetic cooling (see e.g., [264] allows for preparing extremely cold samples of ions for this purpose. Among the number of open scientific questions which are aimed at, the constancy of Nature's laws stands out. A possible time variation of fundamental constants, and very prominently of the fine-structure constant α appears in various theories [265]. HCI offer new possibilities [266] due to their intrinsic, advantageously low polarizability and strong α binding energy dependence.Several schemes that make use of particular isoelectronic sequences have been proposed [266–273].

A central problem for laser spectroscopy with EBITs arises from the high translational temperature of the trapped ion ensemble, which is the result of the ionization by

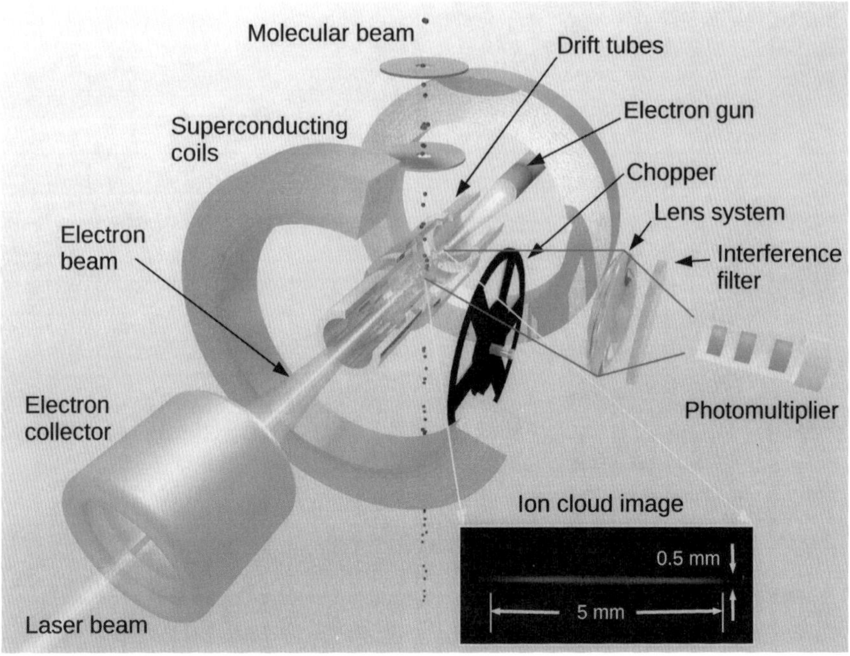

Fig. 10.23 Experimental setup for laser spectroscopy [251] of the forbidden transition $1s^2 2s^2 2p^2 P_{3/2} \rightarrow ^2 P_{1/2}$ in Ar^{13+}. *Side* photograph of the ion cloud. A pulsed tunable laser irradiates the trapped ions along the trap axis. Their fluorescence upon resonant excitation of forbidden transitions decays slowly with a lifetime of several ms, and is collected by a hollow light guide and detected with a photomultiplier. A chopper and an interference filter are used to reduce the background during and after the pulses

electron impact. Evaporative cooling [250] does not solve this problem, even though the the ion temperature is largely reduced. Since the cooling starts at $T > 10^6$ K, the decrease to $T < 10^5$ K does not represent a substantial advantage for metrology purposes. Here, a different approach has to be implemented. The RETRAP experiment [249] implemented resistive ion cooling and later succeeded in sympathetically cooling HCI within a laser-cooled crystal of Be^+ ions [274] in a Penning trap. However, that method does not provide the advantages of Paul traps, which offer a field-free region at the trap center and no shifts due to the Zeeman effect. For this purpose, the Heidelberg group has initiated a program using a novel cryogenic linear Paul trap, CryPTEx [275] built for that purpose [276]. Both the storage of HCI over time scales amenable for laser spectroscopy, and reduction of interactions with blackbody radiation set very exacting requirements for this experiments (see Fig. 10.25). Quantum-logic methods [278] and similar schemes based on interrogating the coherent motional states of co-trapped ions ([279]) would offer a tool for detection of extremely long-lived states [280] in the optical domain. Furthermore, applications to the X-ray region have been proposed [281].

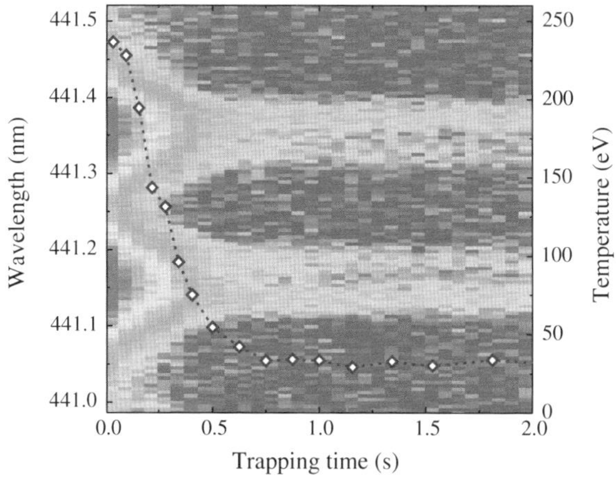

Fig. 10.24 Intensity of the resonant fluorescence of the forbidden M1 transition $1s^2 2s^2 2p^2 P_{3/2} \rightarrow {}^2P_{1/2}$ in Ar^{13+} while scanning the laser wavelength (*vertical axis*) versus trapping time under conditions of evaporative cooling [250]. The Zeeman π components are observed in perpendicular direction to the magnetic field axis. Their linewidth decreases with increasing trapping time as the ion temperature (*right scale*) is reduced by evaporative cooling. At the same time, the related ion losses degrade the signal-to-noise ratio

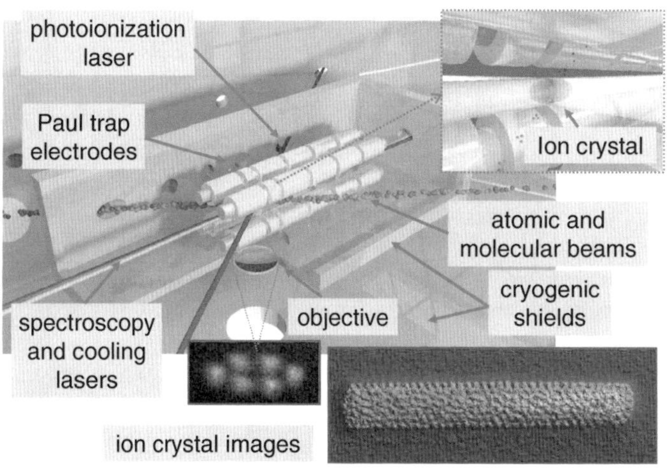

Fig. 10.25 Principle of operation of the cryogenic linear Paul trap CryPTEx [275]. Be^+ ions are produced by photo-ionization of Be atoms from an atomic beam. Laser-cooled ion crystals of Be^+ (images show a Mg^+ and a mixed $Mg^+ + MgH^+$ ion crystal) are used to sympathetically cool highly charged ions produced in an EBIT. The 4 K cryogenic environment enables a vacuum on the order of 10^{-14} mbar and decouples the trapped ions from thermal radiation causing undesired shifts in the laser transitions of interest. One long term goal of this experiment is studying the possible time variation of the fine-structure constant α using quantum-logic readout of forbidden transitions in the trapped highly charged ions [277]. Predictions indicate that the frequencies of some of those transitions have an extremely high sensitivity to the value of α [266]

10.10 Summary and Future Directions

Nearly three decades of spectroscopy with EBITs have boosted our knowledge of QED and relativistic effects in the electronic structure of the ubiquitous HCI. New directions have surged with the advent of free-electron lasers and extremely intense X-ray sources. An interesting new field which has not yet been exploited for HCI research is the use of laser-generated high harmonics in the femtosecond and attosecond range. These compact sources of VUV radiation will soon become a workhorse of atomic physics, and their combination with trapped HCI seems a logical consequence. Cold crystals offer interesting possibilities not only in the visible and UV regions, but also for the detection of X-ray photons, since the heating of the crystal by the recoil due to photon emission and absorption in the X-ray domain becomes appreciable [282]. In combination with synchrotron radiation sources and X-ray free-electron lasers, this might prove to be a very powerful technique for X-ray spectroscopy with sympathetically cooled ions.

Nuclear excitation effects resulting from interactions with the electronic shell are an important and nearly unexplored field [283]. High resolution laser spectroscopy of fine and hyperfine structure will achieve a far higher sensitivity to those effects as the overlap of electronic wave functions with the nucleus is strongly enhanced in HCI. An interesting alternative approach are proposed investigations of circular states in Rydberg states of HCI [284], in which nuclear size effects can be made weaker in order to address other physical parameters of e.g., QED with lower uncertainties.

X-ray metrology is currently bound to the use of crystals as standards of length. However, their suitability beyond the parts-per-billion accuracy is seriously compromised by various solid state effects. The most serious problem is thermal expansion, implying that a temperature stability on the order of mK is needed for achieving such an accuracy. In contrast, electronic transitions in HCI are per se insensitive to external effects to a much higher degree. Even the allowed X-ray transitions in HCI, in spite of their rather large natural width, provide truly stable standards in this region. Forbidden transitions of different types would allow for an stability far beyond any other standards currently in sight, including Mössbauer transitions in nuclei, since they are sensitive to the strong fields within their host crystals. Trapped HCI can provide a very large number of transitions to be used as standards at much higher frequencies (10^{20} Hz) than the current optical clocks (10^{16} Hz). The possible applications that such fast-running clocks would afford for both technology and fundamental scientific studies can only be dreamed of.

Acknowledgments The authors are indebted to a large number of colleagues which whom they had the opportunity and pleasure to collaborate with over the years. Explicitly, I (JRCLU) would like to thank to Peter Beiersdorfer, who has been a steady source of ideas for me and for this field in general. I also would like to thank to other EBIT colleagues at LLNL, like Greg V. Brown and Klaus Widmann, who worked closely with him and me. In both Freiburg and Heidelberg, EBIT science was always strongly backed by Joachim Ullrich and also by my colleages Robert Moshammer, Alexander Dorn and Claus-Dieter Schröter. Ideas, contributions and constant work of former and current EBIT students and postdocs have been the basis of the experiments in Freiburg and Heidelberg: Here I would like to mention Bhas Bapat, Andreas Werdich, Ilija Draganič, Christina Dimopolou, Panlin

Guo, Xuemei Zhang, Alexandra Rohr, Antonio González Martínez, Rosario Soria Orts, Johannes Braun, Hjalmar Bruhns, Vladimir Mironov, Sascha Epp, Günther Sikler, Alain Lapierre, Günter Brenner, Volkhard Mäckel,Katharina Kubiček, Lodeweijk Arntzen, Martin Simon, Guiyun Liang, Renee Klawitter, Rainer Ginzel, Benjamin Schmitt, Maria Schwarz, Thomas Baumann, Kirsten Schnorr, Franziska Brunner, Stuart Higgins, Christian Beilmann, Sven Bernitt, Oscar Versolato, Sebastian Georgi, Jan Rudolph, René Steinbrügge, Alexander Windberger, James Harries, and Hendrik Bekker. Profs. Yaming Zou, Hiroyuki Tawara and Paul Mokler have been our long-time guests, and both their help and stimulating discussions are greatly appreciated. I would also like to mention my collaborations in this field with the groups of Klaus Blaum at MPIK, Jens Dilling at TRIUMF, Fritz Aumayr at TU Vienna, Michael Drewsen in Aarhus, and Piet Schmidt at PTB. Crucially, a strong theory support by the division led by Christoph Keitel was essential for these works. Last but not least, nothing could have been achieved without the craftsmanship of the technicians of our mechanical workshops and work group. Of those companions, Karl Bechberger, Norbert Müller and Christian Kaiser deserve our particular gratitude for their year-long commitment. ZH would like to thank in addition for fruitful collaboration and insightful conversations with Natalia Oreshkina, Jacek Zatorski, Adriana Pálffy, Roxana Schiopu, Stefano Cavaletto, Octavian Postavaru, Ioannis Stasinopoulos, Matthias Gail, Stephan Balliel-Zakowicz, Anton Artemyev, Ilya Tupitsyn, Vladimir Yerokhin, Vladimir Shabaev, Ulrich Jentschura, Werner Scheid and Christoph Keitel.

References

1. J. Sapirstein, K.T. Cheng, Tests of quantum electrodynamics with EBIT. Can. J. Phys. **86**, 25 (2008)
2. R.E. Marrs, S.R. Elliott, D.A. Knapp, Production and trapping of hydrogenlike and bare uranium ions in an electron beam ion trap. Phys. Rev. Lett. **72**, 4082 (1994)
3. R.E. Marrs, P. Beiersdorfer, D. Schneider, The electron-beam ion trap. Phys. Today **47**, 27 (1994)
4. B.M. Penetrante, J.N. Bardsley, D. Dewitt, M. Clark, D. Schneider, Evolution of ion-charge-state distributions in an electron-beam ion trap. Phys. Rev. A: At. Mol. Opt. Phys. **43**, 4861 (1991)
5. E.D. Donets, The electron beam method of production of highly charged ions and its applications. Phys. Scr. T **3**, 11 (1983)
6. E.D. Donets, Electron beam ion sources and associated physics at JINR. Nucl. Instrum. Methods Phys. Res. Sect. B **9**(4), 522 (1985)
7. J. Arianer, C. Collart, C. Goldstein, H. Laurent, M. Malard, Status report of the Orsay EBIS program. Phys. Scr. T **3**, 35 (1983)
8. M.A. Levine, R.E. Marrs, J.R. Henderson, D.A. Knapp, M.B. Schneider, The electron beam ion trap: A new instrument for atomic physics measurements. Phys. Scr. **22**, 157 (1988)
9. M.A. Levine, R.E. Marrs, J.N. Bardsley, P. Beiersdorfer, C.L. Bennett, M.H. Chen, T. Cowan, D. Dietrich, J.R. Henderson, D.A. Knapp, A. Osterheld, B.M. Penetrante, M.B. Schneider, J.H. Scofield, The use of an electron beam ion trap in the study of highly charged ions. Nucl. Instrum. Methods Phys. Res. Sect. B **43**, 431 (1989)
10. D. Schneider, M.W. Clark, B.M. Penetrante, J. McDonald, D. Dewitt, J.N. Bardsley, Production of high-charge-state thorium and uranium ions in an electron-beam ion trap. Phys. Rev. A: At. Mol. Opt. Phys. **44**, 3119 (1991)
11. D.A. Knapp, R.E. Marrs, S.R. Elliott, E.W. Magee, R. Zasadzinski, A high-energy electron beam ion trap for production of high-charge high-*Z* ions. Nucl. Instrum. Methods Phys. Res. Sect. A **334**, 305 (1993).
12. B.M. Penetrante, J.N. Bardsley, M.A. Levine, D.A. Knapp, R.E. Marrs, Evaporative cooling of highly charged dysprosium ions in an enhanced electron-beam ion trap. Phys. Rev. A: At. Mol. Opt. Phys. **43**, 4873 (1991)

13. P. Beiersdorfer, V. Decaux, S.R. Elliott, K. Widmann, K. Wong, Temperature of the ions produced and trapped in an electron-beam ion trap. Rev. Sci. Instr. **66**, 303 (1995)
14. B.J. Wargelin, P. Beiersdorfer, S.M. Kahn, Radiative lifetime of the long-lived $1s2s\,^3S_1$ state in heliumlike neon by electron-beam excitation of trapped ions. Phys. Rev. Lett. **71**, 2196 (1993)
15. P. Beiersdorfer, L. Schweikhard, J.R. Crespo López-Urrutia, K. Widmann, The magnetic trapping mode of an electron beam ion trap: New opportunities for highly charged ion research. Rev. Sci. Instr. **67**, 3818 (1996)
16. J.D. Silver, A.J. Varney, H.S. Margolis, P.E.G. Baird, I.P. Grant, P.D. Groves, W.A. Hallett, A.T. Handford, P.J. Hirst, A.R. Holmes, D.J.H. Howie, R.A. Hunt, K.A. Nobbs, M. Roberts, W. Studholme, J.S. Wark, M.T. Williams, M.A. Levine, D.D. Dietrich, W.G. Graham, I.D. Williams, R. O'Neil, S.J. Rose, The Oxford electron-beam ion trap: A device for spectroscopy of highly charged ions. Rev. Sci. Instr. **65**, 1072 (1994)
17. D.J. Bieber, H.S. Margolis, P.K. Oxley, J.D. Silver, Studies of magnetic dipole transitions in highly charged argon and barium using an electron beam ion trap. Phys. Scr. T **73**, 64–66 (1997)
18. J.D. Gillaspy, EBIT spectra of highly stripped ions from the visible to the X-ray. Phys. Scr. T **65**, 168 (1996)
19. J.D. Gillaspy, Y. Aglitskiy, E. Bell, C.M. Brown, C. Chantler, R.D. Deslattes, U. Feldman, L. Hudson, J.M. Laming, E.S. Meyer, C.A. Morgan, J.R. Roberts, F.G. Serpa, J. Sugar, E. Takacs, First results from the EBIT at NIST. Phys. Scr. T **71**, 99 (1997)
20. F.J. Currell, J. Asada, K. Ishii, A. Minoh, K. Motohashi, N. Nakamura, K. Nishizawa, S. Ohtani, K. Okazaki, M. Sakurai, H. Shiraishi, S. Tsurubuchi, H. Watanabe, A new versatile electron-beam ion trap. J. Phys. Soc. Jpn. **65**, 3186 (1996)
21. H. Watanabe, J. Asada, F.J. Currell, T. Fukami, T. Hirayama, K. Motohashi, N. Nakamura, E. Nojikawa, S. Ohtani, K. Okazaki, M. Sakurai, H. Shimizu, N. Tada, S. Tsurubuchi, Characteristics of the Tokyo electron-beam ion trap. J. Phys. Soc. Jpn. **66**, 3795 (1997)
22. N. Nakamura, J. Asada, F.J. Currell, T. Fukami, T. Hirayama, K. Motohashi, T. Nagata, E. Nojikawa, S. Ohtani, K. Okazaki, M. Sakurai, H. Shiraishi, S. Tsurubuchi, H. Watanabe, An overview of the Tokyo electron beam ion trap. Phys. Scr. T **73**, 362 (1997)
23. H. Khodja, J.P. Briand, A warm electron beam ion trap: the micro-EBIT. Phys. Scr. T **71**, 113 (1997)
24. P. Beiersdorfer, A "brief" history of spectroscopy on EBIT. Can. J. Phys. **86**, 1 (2008)
25. R.E. Marrs, S.R. Elliott, J.H. Scofield, Measurement of electron-impact ionization cross sections for hydrogenlike high-Z ions. Phys. Rev. A: At. Mol. Opt. Phys. **56**, 1338 (1997)
26. P. Beiersdorfer, A.L. Osterheld, V. Decaux, K. Widmann, Observation of lifetime-limited X-ray linewidths in cold highly charged ions. Phys. Rev. Lett. **77**, 5353 (1996)
27. F.G. Serpa, C.A. Morgan, E.S. Meyer, J.D. Gillaspy, E. Träbert, D.A. Church, E. Takács, Measurement of a magnetic-dipole transition probability in Xe^{32+} using an electron-beam ion trap. Phys. Rev. A: At. Mol. Opt. Phys. **55**, 4196 (1997)
28. G.S. Stefanelli, P. Beiersdorfer, V. Decaux, K. Widmann, Measurement of the radiative lifetime of the $1s2s\,^3S_1$ level in heliumlike magnesium. Phys. Rev. A: At. Mol. Opt. Phys. **52**, 3651 (1995)
29. J.R. Crespo López-Urrutia, P. Beiersdorfer, D.W. Savin, K. Widmann, Precision measurement of the lifetime of the $1s2s\,^3S_1$ metastable level in heliumlike O^{6+}. Phys. Rev. A: At. Mol. Opt. Phys. **58**, 238 (1998)
30. P. Beiersdorfer, E. Träbert, E.H. Pinnington, Experimental transition rate of the green coronal line of Fe XIV. Astrophys. J. Lett. **587**, 836 (2003)
31. A. Lapierre, U.D. Jentschura, J.R. Crespo, López-Urrutia, J. Braun, G. Brenner, H. Bruhns, D. Fischer, A.J. González Martínez, Z. Harman, W.R. Johnson, C.H. Keitel, V. Mironov, C.J. Osborne, G. Sikler, R. Soria Ort, V. Shabaev, H. Tawara, I.I. Tupitsy, J. Ullrich, A. Volotka, Relativistic electron correlation, quantum electrodynamics, and the lifetime of the $1s^22s^22p$ $^2P^o_{3/2}$ level in boronlike argon. Phys. Rev. Lett. **95**, 183001 (2005)

32. E. Träbert, P. Beiersdorfer, S.B. Utter, G.V. Brown, H. Chen, C.L. Harris, P.A. Neill, D.W. Savin, A.J. Smith, Experimental M1 transition rates of coronal lines from Ar X, Ar XIV, and Ar XV. Astrophys. J. Lett. **541**, 506 (2000)
33. E. Träbert, S.B. Utter, P. Beiersdorfer, Visible-range spectroscopy and a lifetime measurement on Kr^{22+} in an electron beam ion trap. Phys. Lett. A **272**, 86–92 (2000)
34. E. Träbert, P. Beiersdorfer, G.V. Brown, K. Boyce, R.L. Kelley, C.A. Kilbourne, F.S. Porter, A. Szymkowiak, Time-resolved soft-X-ray spectroscopy of a magnetic octupole transition in nickel-like xenon, cesium, and barium ions. Phys. Rev. A: At. Mol. Opt. Phys. **73**, 022508 (2006)
35. G. Brenner, J.R. Crespo, López-Urrutia, Z. Harman, P.H. Mokler, J. Ullrich, Lifetime determination of the Fe XIV $3s^2 3p\ {}^2P_{3/2}$ metastable level. Phys. Rev. A: At. Mol. Opt. Phys. **75**, 032504 (2007)
36. J.D. Gillaspy, Highly charged ions. J. Phys. B: At. Mol. Opt. Phys. **34**, R93 (2001)
37. W.D. Chen, J. Xiao, Y. Shen, Y.Q. Fu, F.C. Meng, C.Y. Chen, B.H. Zhang, Y.J. Tang, R. Hutton, Y. Zou, Precise studies on resonant energies of the first intershell (KLL) dielectronic recombination processes for He- up to O-like xenon. Phys. Plasmas **15**, 083301 (2008)
38. J.R. Crespo López-Urrutia, A. Dorn, R. Moshammer, J. Ullrich, The Freiburg electron beam ion trap/source project FREEBIT. Phys. Scr. T **80**, 502 (1999)
39. S.W. Epp, J.R. Crespo, López-Urrutia, G. Brenner, V. MäckelcKel, P.H. Mokler, R. Treusch, M. Kuhlmann, M.V. Yurkov, J. Feldhaus, J.R. Schneider, M. Wellhöfer, M. Martins, W. Wurth, J. Ullrich, Soft X-ray laser spectroscopy on trapped highly charged ions at FLASH. Phys. Rev. Lett. **98**, 183001 (2007)
40. X. Zhu, Y. Liu, X. Wang, Y. Liu, P. Guo, D. Lu, X. Zhang, W. Hu, M. He, L. Liljeby, Y. Zou, Status of the Shanghai EBIT. Nucl. Instrum. Methods Phys. Res. Sect. B **235**, 509 (2005)
41. S. Böhm, A. Enulescu, T. Fritio, I. Orban, S. Tashenov, R. Schuch, First results from the Stockholm electron beam ion trap. J. Phys: Conf. Ser. **58**, 303 (2007)
42. U. Kentsch, G. Zschornack, F. Gromann, V.P. Ovsyannikov, F. Ullmann, S. Fritzsche, A. Surzhykov, Production of bare argon, manganese, iron and nickel nuclei in the Dresden EBIT. Nucl. Instrum. Methods Phys. Res. Sect. B **187**, 238 (2002)
43. N. Nakamura, A.P. Kavanagh, H. Watanabe, H.A. Sakaue, Y. Li, D. Kato, F.J. Currell, S. Ohtani, Evidence for strong Breit interaction in dielectronic recombination of highly charged heavy ions. Phys. Rev. Lett. **100**, 073203 (2008)
44. N. Nakamura, H. Kikuchi, H.A. Sakaue, T. Watanabe, Compact electron beam ion trap for spectroscopy of moderate charge state ions. Rev. Sci. Instr. **79**, 063104 (2008)
45. M.C. Simon, M. Schwarz, B.L. Schmitt, C. Beilmann, S.W. Epp, T.M. Baumann, K. Kubiček, R. Ginzel, S.G. Higgins, R. Klawitter, V. Mäckel, S. Bernitt, P.H. Mokler, J. Ullrich, J.R. Crespo, López-Urrutia, Photoionization of ions in arbitrary charge states by synchrotron radiation in an electron beam ion trap. J. Phys: Conf. Ser. **194**, 012009 (2009)
46. M.C. Simon, M. Schwarz, S.W. Epp, C. Beilmann, B.L. Schmitt, Z. Harman, T.M. Baumann, P.H. Mokler, S. Bernitt, R. Ginzel, S.G. Higgins, C.H. Keitel, R. Klawitter, K. Kubiček, V. Mäckel, J. Ullrich, J.R. Crespo, López-Urrutia, Photoionization of N^{3+} and Ar^{8+} in an electron beam ion trap by synchrotron radiation. J. Phys. B: At. Mol. Opt. Phys. **43**, 065003 (2010)
47. M.C. Simon, J.R. Crespo López-Urrutia, C. Beilman, M. Schwarz, Z. Harman, S.W. Epp, B.L. Schmitt, T.M. Baumann, E. Behar, S. Bernitt, R. Follath, R. Ginzel, C.H. Keitel, R. Klawitte, K. Kubiček, V. Mäckel, P.H. Mokle, G. Reichardt, O. Schwarzkopf, J. Ullric, Resonant and near-threshold photoionization cross sections of Fe^{14+}. Phys. Rev. Lett. **105**, 183001 (2010)
48. S. Bernitt, G.V. Brown, J.K. Rudolph, R. Steinbrügge, A. Graf, M. Leutenegger, S.W. Epp, S. Eberle, K. Kubiček, V. Mäckel, M.C. Simon, E. Träbert, E.W. Magee, C. Beilmann, N. Hell, S. Schippers, A. Müller, S.M. Kahn, A. Surzhykov, Z. Harman, C.H. Keitel, J. Clementson, F.S. Porter, W. Schlotter, J.J. Turner, J. Ullrich, P. Beiersdorfer, J.R. Crespo, López-Urrutia, An unexpectedly low oscillator strength as the origin of the FeXVII emission problem. Nature **492**, 225 (2012)

49. S. Bernitt, G.V. Brown, J.R. Crespo, López-Urrutia, J. Rudolph, R. Steinbrügge, A. Graf, M. Leutenegger, C. Beilmann, S. Eberle, S.W. Epp, K. Kubiček, V. Mäckel, S. Schippers, W. Schlotter, M.C. Simon, E. Träbert, J. Turner, S.M. Kahn, E.W. Magee, A. Müller, F.S. Porter, A. Rasmussen, P. Beiersdorfer, J. Ullrich, X-ray laser spectroscopy with an electron beam ion trap at the free electron laser LCLS. J. Phys: Conf. Ser. **388**, 032037 (2012)
50. P. Beiersdorfer, Laboratory X-ray astrophysics. Ann. Rev. Astron. Astrophys. **41**, 343 (2003)
51. S.B. Utter, P. Beiersdorfer, J.R. Crespo López-Urrutia, K. Widmann, Position and size of the electron beam in the high-energy electron beam ion trap. Nucl. Instrum. Methods Phys. Res. Sect. A **428**, 276 (1999)
52. B. O'Rourke, F.J. Currell, H. Kuramoto, Y.M. Li, S. Ohtani, X.M. Tong, H. Watanabe, Electron-impact ionization of hydrogen-like iron ions. J. Phys. B: At. Mol. Opt. Phys. **34**, 4003 (2001)
53. H. Watanabe, F.J. Currell, H. Kuramoto, S. Ohtani, B.E. O'Rourke, X.M. Tong, Electron impact ionization of hydrogen-like molybdenum ions. J. Phys. B: At. Mol. Opt. Phys. **35**, 5095 (2002)
54. P. Beiersdorfer, V. Decaux, K. Widmann, Measurement of the temperature of cold highly charged ions produced in an electron beam ion trap. Nucl. Instrum. Methods Phys. Res. Sect. A **98**, 566 (1995)
55. L. Schweikhard, P. Beiersdorfer, G.V. Brown, J.R. Crespo López-Urrutia, S.B. Utter, K. Widmann, Pulsed gas injection for X-ray spectroscopy of highly charged ions stored in the magnetic trapping mode of an electron beam ion trap. Nucl. Instrum. Methods Phys. Res. Sect. A **142**, 245 (1998)
56. P. Beiersdorfer, K.R. Boyce, G.V. Brown, H. Chen, S.M. Kahn, R.L. Kelley, M. May, R.E. Olson, F.S. Porter, C.K. Stahle, W.A. Tillotson, Laboratory simulation of charge exchange–produced X-ray emission from comets. Science **300**, 1558 (2003)
57. W. Lotz, An empirical formula for the electron-impact ionization cross-section. Zeitschrift für Physik **206**, 205 (1967)
58. H.A. Bethe, E.E. Salpeter, *Quantum Mechanics of One- and Two-Electron Atoms* (Springer, Berlin, 1957)
59. H. van Regemorter, Rate of collisional excitation in stellar atmospheres. Astrophys. J. **136**, 906 (1962)
60. E.E. Salpeter, H.A. Bethe, A relativistic equation for bound-state problems. Phys. Rev. **84**, 1232 (1951)
61. M.F. Gu, The flexible atomic code. Can. J. Phys. **86**, 675 (2008)
62. S. Fritzsche, The RATIP program for relativistic calculations of atomic transition, ionization and recombination properties. Comput. Phys. Commun. **183**(7), 1525 (2012)
63. A.J. González Martínez, J.R. Crespo López-Urrutia, J. Braun, G.Brenner, H. Bruhns, A. Lapierre, V. Mironov, R. Soria Orts, H. Tawara, M. Trinczek, J. Ullrich, J.H. Scofield, State-selective quantum interference observed in the recombination of highly charged $Hg^{75+...78+}$ mercury ions in an electron beam ion trap. Phys. Rev. Lett. **94**, 203201 (2005)
64. P. Beiersdorfer, R.E. Marrs, J.R. Henderson, D.A. Knapp, M.A. Levine, D.B. Platt, M.B. Schneider, D.A. Vogel, K.L. Wong, High-resolution x-ray spectrometer for an electron beam ion trap. Rev. Sci. Instrum. **61**, 2338 (1990)
65. J.W.M. DuMond, A high resolving power, curved-crystal focusing spectrometer for the short wave-length x-rays and gamma-rays. Rev. Sci. Instrum. **18**, 626 (1947)
66. H.H. Johann, Die Erzeugung lichtstarker Röntgenspektren mit Hilfe von Konkavkristallen. Zeitschrift für Physik **69**, 185 (1931)
67. P. Beiersdorfer, B.J. Wargelin, Low energy x-ray spectrometer for an electron beam ion trap. Rev. Sci. Instrum. **65**, 13 (1994)
68. P. Beiersdorfer, J.R. Crespo, López-Urrutia, E. Förster, J. Mahiri, K. Widmann, Very high resolution soft x-ray spectrometer for an electron beam ion trap. Revi. Sci. Instrum. **68**, 1077 (1997)
69. P. Beiersdorfer, J.R. Crespo, López-Urrutia, P. Springer, S.B. Utter, K.L. Wong, Spectroscopy in the extreme ultraviolet on an electron beam ion trap. Rev. Sci. Instrum. **70**, 276 (1999)

70. G.V. Brown, P. Beiersdorfer, K. Widmann, Wide-band, high-resolution soft x-ray spectrometer for the electron beam ion trap. Rev. Sci. Instrum. **70**, 280 (1999)
71. S.B. Utter, G.V. Brown, P. Beiersdorfer, E.J. Clothiaux, N.K. Podder, Grazing-incidence measurements of L-shell line emission from highly charged Fe in the soft x-ray region. Rev. Sci. Instrum. **70**, 284 (1999)
72. C. Biedermann, R. Radtke, J.-L. Schwob, P. Mandelbaum, R. Doron, T. Fuchs, G. Fußmann, EUV spectroscopy of highly charged tungsten ions relevant to hot plasmas. Phys. Scr. T **92**, 85 (2001)
73. P. Beiersdorfer, E.W. Magee, E. Träbert, H. Chen, J.K. Lepson, M.-F. Gu, M. Schmidt, Flat-field grating spectrometer for high-resolution soft x-ray and extreme ultraviolet measurements on an electron beam ion trap. Rev. Sci. Instrum. **75**, 3723 (2004)
74. F.S. Porter, G.V. Brown, K.R. Boyce, R.L. Kelley, C.A. Kilbourne, P. Beiersdorfer, H. Chen, S. Terracol, S.M. Kahn, A.E. Szymkowiak, The astro-e2 x-ray spectrometer/EBIT microcalorimeter x-ray spectrometer. Rev. Sci. Instrum. **75**, 3772 (2004)
75. M.I. Eides, H. Grotch, V.A. Shelyuto, Theory of light hydrogenlike atoms. Phys. Rep. **342**, 63 (2001)
76. K. Pachucki, Higher-order binding corrections to the Lamb shift. Ann. Phys. **226**(1), 1 (1993)
77. P.J. Mohr, Self-energy radiative corrections in hydrogen-like systems. Ann. Phys. **88**, 26 (1974)
78. P.J. Mohr, Quantum electrodynamics of high-Z few-electron atoms. Phys. Rev. A: At. Mol. Opt. Phys. **32**, 1949 (1985)
79. W.R. Johnson, G. Soff, The Lamb shift in hydrogen-like atoms, $1 < Z < 110$. At. Data Nucl. Data Tables **33**, 405 (1985)
80. P.J. Mohr, Self-energy correction to one-electron energy levels in a strong Coulomb field. Phys. Rev. A: At. Mol. Opt. Phys. **46**, 4421 (1992)
81. P.J. Mohr, G. Soff, Nuclear size correction to the electron self-energy. Phys. Rev. Lett. **70**, 158 (1993)
82. P.J. Mohr, G. Plunien, G. Soff, QED corrections in heavy atoms. Phys. Rep. **293**, 227 (1998)
83. V.M. Shabaev, A.N. Artemyev, V.A. Yerokhin, QED and nuclear effects in high-Z few-electron atoms. Phys. Scr. T **86**, 7 (2000)
84. V.M. Shabaev, Two-time Green's function method in quantum electrodynamics of high-Z few-electron atoms. Phys. Rep. **356**, 119 (2002)
85. J.L. Flowers, H.A. Klein, D.J. E. Knight, H.S. Margolis, Hydrogenic systems for calculable frequency standards: Status and options. NPL Report CBTLM 11, March 2001
86. R. Deslattes, E.G. Kessler, X-ray transition energies: new approach to a comprehensive evaluation. Rev. Mod. Phys. **75**, 35 (2003)
87. V.A. Yerokhin, P. Indelicato, V.M. Shabaev, Two-loop self-energy correction in high-Z hydrogenlike ions. Phys. Rev. Lett. **91**, 073001 (2003)
88. V.M. Shabaev, Finite nuclear size corrections to the energy levels of the multichargedions. J. Phys. B: At. Mol. Opt. Phys. **26**, 1103 (1993)
89. V.M. Shabaev, Mass corrections in a strong nuclear field. Theor. Math. Phys. **63**, 588 (1985)
90. A.N. Artemyev, V.M. Shabaev, V.A. Yerokhin, Relativistic nuclear recoil corrections to the energy levels of hydrogenlike and high-Z lithiumlike atoms in all orders in αZ. Phys. Rev. A: At. Mol. Opt. Phys. **52**, 1884 (1995)
91. G. Plunien, B. Müller, W. Greiner, G. Soff, Nuclear polarization contribution to the Lamb shift in heavy atoms. Phys. Rev. A: At. Mol. Opt. Phys. **39**, 5429 (1989)
92. J.P. Briand, M. Tavernier, P. Indelicato, R. Marrus, H. Gould, High-precision spectroscopic studies of Lyman-α lines of hydrogenlike iron: A measurement of the 1s Lamb shift. Phys. Rev. Lett. **50**, 832 (1983)
93. J.P. Briand, M. Tavernier, R. Marrus, J.P. Desclaux, High-precision spectroscopic study of heliumlike iron. Phys. Rev. A: At. Mol. Opt. Phys. **29**, 3143 (1984)
94. E. Källne, J. Källne, P. Richard, M. Stöckli, Precision measurement of the H-like x-ray spectrum of Cl and the 1s Lamb shift. J. Phys. B: At. Mol. Opt. Phys. **17**, L115 (1984)

95. P. Richard, M. Stöckli, R.D. Deslattes, P. Cowan, R.E. LaVilla, B. Johnson, K. Jones, M. Meron, R. Mann, K. Schartner, Measurement of the 1*s* Lamb shift in hydrogenlike chlorine. Phys. Rev. A: At. Mol. Opt. Phys. **29**, 2939 (1984)
96. M. Tavernier, J.P. Briand, P. Indelicato, D. Liesen, P. Richard, Measurement of the (1s) Lamb shift of hydrogen-like krypton. J. Phys. B: At. Mol. Opt. Phys. **18**, L327–L330 (1985)
97. E.S. Marmar, J.E. Rice, E. Källne, J. Källne, R.E. LaVilla, Precision measurement of the 1*s* Lamb shift in hydrogenlike argon. Phys. Rev. A: At. Mol. Opt. Phys. **33**, 774 (1986)
98. R.D. Deslattes, H.F. Beyer, F. Folkmann, Precision x-ray wavelength measurements in helium-like argon recoil ions. J. Phys. B: At. Mol. Opt. Phys. **17**, L689 (1984)
99. H.F. Beyer, R.D. Deslattes, F. Folkmann, R.E. LaVilla, Determination of the 1s Lamb shift in one-electron argon recoil ions. J. Phys. B: At. Mol. Opt. Phys. **18**, 207–215 (1985)
100. H.F. Beyer, P. Indelicato, K.D. Finlayson, D. Liesen, R.D. Deslattes, Measurement of the 1s Lamb shift in hydrogenlike nickel. Phys. Rev. A: At. Mol. Opt. Phys. **43**, 223 (1991)
101. R.D. Deslattes, R. Schuch, E. Justiniano, Application of decelerated bare nuclei to precision spectroscopy of one-electron ions. Phys. Rev. A: At. Mol. Opt. Phys. **32**, 1911 (1985)
102. G. Hölzer, E. Förster, D. Klöpfel, P. Beiersdorfer, G.V. Brown, J.R. Crespo, López-Urrutia, K. Widmann, Absolute wavelength measurement of the Lyman-α transitions of hydrogenic Mg^{11+}. Phys. Rev. A: At. Mol. Opt. Phys. **57**, 945 (1998)
103. D. Klöpfel, G. Hölzer, E. Förster, P. Beiersdörfer, A quartz quasimonolith for absolute x-ray wavelength measurements. Rev. Sci. Instrum. **68**, 3669 (1997)
104. J. Tschischgale, D. Klöpfel, P. Beiersdorfer, G.V. Brown, E. Förster, H. Schulte-Schrepping, S.B. Utter, Absolute wavelength measurement of the Lyman-α transition of hydrogen-like silicon. Can. J. Phys. **80**, 867 (2002)
105. J.D. Gillaspy, C.T. Chantler, D. Paterson, L.T. Hudson, F.G. Serpa, E. Takcs, First measurement of Lyman-α x-ray lines in hydrogen-like vanadium: Results and implications for precision wavelength metrology and tests of QED. J. Phys. B: At. Mol. Opt. Phys. **43**, 074021 (2010)
106. N. Nakamura, D. Kato, N. Miura, T. Nakahara, S. Ohtani, Intensity ratio between Lyman-α1 and Lyman-α2 lines of hydrogenlike titanium observed in an electron-beam ion trap. Phys. Rev. A: At. Mol. Opt. Phys. **63**, 024501 (2001)
107. M.R. Tarbutt, R. Barnsley, N.J. Peacock, J.D. Silver, Wavelength measurements of the satellite transitions to the n = 2 resonance lines of helium-like argon. J. Phys. B: At. Mol. Opt. Phys. **34**, 3979 (2001)
108. W.L. Bond, Precision lattice constant determination. Acta Crystallogr. **13**, 814 (1960)
109. P. Becker, K. Dorenwendt, G. Ebeling, R. Lauer, W. Lucas, R. Probst, H.J. Rademacheer, G. Reim, P. Seyfried, H. Siegert, Absolute measurement of the (220) lattice plane spacing in a silicon crystal. Phys. Rev. Lett. **46**, 1540–1543 (1981)
110. Johannes Braun, *"Entwicklung eines Kristallspektrometers für röntgenspektroskopische Untersuchungen an Hochgeladenen Ionen* (Universität Heidelberg, Master's thesis, 2003)
111. J. Braun, H. Bruhns, M. Trinczek, J.R. Crespo, López-Urrutia, J. Ullrich, Novel technique for high-precision Bragg-angle determination in crystal x-ray spectroscopy. Rev. Sci. Instrum. **76**, 073105 (2005)
112. J. Braun, *Präzisionsröntgenspektroskopie zur absoluten Wellenlängenbestimmung an hochgeladenen Ionen* (Diss, Max-Planck-Institut für Kernphysik, 2006)
113. H. Bruhns, High-precision x-ray spectroscopy on highly charged argon ions. PhD thesis, Universität Heidelberg, Germany, Dezember 2005
114. H. Bruhns, J. Braun, K. Kubiček, J.R. Crespo, López-Urrutia, J. Ullrich, Testing QED screening and two-loop conributions with He-like ions. Phys. Rev. Lett. **99**, 113001 (2007)
115. K. Kubiček, *Absolute determination of the* $n = 1$ *to* $n = 2$ *transition energies in hydrogen- and helium-like* S^{14+}, S^{15+}, Ar^{16+} *and* Ar^{17+} *ions* (Universität Göttingen, Master's thesis, 2006)
116. K. Kubiček, H. Bruhns, J. Braun, J.R. Crespo, López-Urrutia, J. Ullrich, Two-loop QED contributions tests with mid-*Z* He-like ions. J. Phys: Conf. Ser. **163**, 012007 (2009)
117. K. Kubiček, J. Braun, H. Bruhns, J.R. Crespo, López-Urrutia, P. H. Mokler, J. Ullrich, High-precision laser-assisted absolute determination of x-ray diffraction angles. Rev. Sci. Instrum. **83**, 013102 (2012)

118. D.R. Plante, W.R. Johnson, J. Sapirstein, Relativistic all-order many-body calculations of the n=1 and n=2 states of heliumlike ions. Phys. Rev. A: At. Mol. Opt. Phys. **49**, 3519 (1994)
119. K.G. Dyall, I.P. Grant, C.T. Johnson, F.A. Parpia, E.P. Plummer, Grasp: A general-purpose relativistic atomic structure program. Comput. Phys. Commun. **55**, 425 (1989)
120. P. Indelicato, O. Gorceix, M. Tavernier, J.P. Briand, J.P. Desclaux, R. Marrus, M. Prior, Experimental and theoretical study of QED corrections in the ground state of heliumlike iron. Zeitschrift für Physik D: At. Mol. Clusters **2**, 149 (1986)
121. P. Indelicato, O. Gorveix, J.P. Desclaux, Multiconfiguration Dirac-Fock studies of two-electron ions: II. Radiative corrections and comparison with experiment. J. Phys. B: At. Mol. Opt. Phys. **20**, 651 (1987)
122. K.T. Cheng, M.H. Chen, W.R. Johnson, J. Sapirstein, Relativistic configuration-interaction calculations for the ground state and $n = 2$ singlet states in heliumlike ions. Phys. Rev. A: At. Mol. Opt. Phys. **50**, 247 (1994)
123. K.T. Cheng, M.H. Chen, Energy levels of the low-lying states of mid-Z heliumlike ions. Phys. Rev. A: At. Mol. Opt. Phys. **61**, 044503 (2000)
124. G.W.F. Drake, Theoretical energies for the n=1 and 2 states of the helium isoelectronic sequence up to Z=100. Can. J. Phys. **66**, 586 (1988)
125. Z.-C. Yan, G.W.F. Drake, High precision calculation of fine structure splittings in helium and He-like ions. Phys. Rev. Lett. **74**, 4791 (1995)
126. G.W.F. Drake, Progress in helium fine-structure calculations and the fine-structure constant. Can. J. Phys. **80**, 1195–1212 (2002)
127. I. Lindgren, H. Persson, S. Salomonson, Full QED calculations of two-photon exchange for heliumlike-systems: Analysisin the Coulomb and Feynman gauges. Phys. Rev. A: At. Mol. Opt. Phys. **51**, 1167 (1995)
128. I. Lindgren, B. Asen, S. Salomonson, A.M. Mårtensson-Pendrill, QED procedure applied to the quasidegenerate fine-structure levels of He-like ions. Phys. Rev A: At. Mol. Opt. Phys. **64**, 062505 (2001)
129. W.R. Johnson, K.T. Cheng, M.H. Chen, Chapter 3: Accurate relativistic calculations including QED contributions for few-electron systems. Theor. Comput. Chem. **14**, 120 (2004)
130. J. Sapirstein, K.T. Cheng, M.H. Chen, Potential independence of the solution to the relativistic many-body problem and the role of negative-energy states in heliumlike ions. Phys. Rev. A: At. Mol. Opt. Phys. **59**, 259 (1999)
131. A.N. Artemyev, V.M. Shabaev, V.A. Yerokhin, G. Plunien, G. Stoff, QED calculation of the n = 1 and n = 2 energy levels in He-like ions. Phys. Rev. A: At. Mol. Opt. Phys. **71**, 062104 (2005)
132. O.Y. Andreev, L.N. Labozowsky, G. Plunien, G. Stoff, Calculation of quasidegenerate energy levels of two-electron ions. Phys. Rev. A: At. Mol. Opt. Phys. 69 (2004)
133. L. Schleinkofer, F. Bell, H.-D. Betz, G. Trollmann, J. Rothermel, Precision wavelength determination of $2^1P_1 - 1^1S_0$ and $2^3P_1 - 1^1S_0$ transitions in helium-like sulfur ions. Phys. Scr. **25**, 917 (1982)
134. E.V. Aglitsky, P.S. Antisiferov, S.L. Mandelstam, A.M. Panin, U.I. Safronova, S.A. Ulitin, Comparison of calculated and measured wavelengths of resonance transitions in He-like ions for Z=16-39. Phys. Scr. **38**, 136 (1988)
135. P. Beiersdorfer, M. Bitter, S. von Goeler, K.W. Hill, Experimental study of the x-ray transitions in the heliumlike isoelectronic sequence. Phys. Rev. A: At. Mol. Opt. Phys. **40**, 150 (1989)
136. C. Biedermann, R. Radtke, K.B. Fournier, Spectroscopy of heliumlike argon resonance and satellite lines for plasma temperature diagnostics. Phys. Rev. E: Stat. Nonlin. Soft Matter Phys. **66**, 066404 (2002)
137. C.T. Chantler, D. Paterson, L.T. Hudson, F.G. Serpa, J.D. Gillaspy, E. Takács, Absolute measurement of the resonance lines in heliumlike vanadium on an electron-beam ion trap. Phys. Rev. A: At. Mol. Opt. Phys. **62**, 042501 (2000)
138. H. Bruhns, J. Braun, K. Kubiček, J.R. Crespo, López-Urrutia, J. Ullrich, Testing QED screening and two-loop contributions with He-like ions. Phys. Rev. Lett. **99**, 113001 (2007)

139. C.T. Chantler, J.M. Laming, D.D. Dietrich, W.A. Hallett, R. McDonald, J.D. Silver, Hydrogenic Lamb shift in iron Fe^{25+} and fine-structure Lamb shift. Phys. Rev. A: At. Mol. Opt. Phys. **76**, 042116 (2007)
140. J.K. Rudolph, J.R. Crespo López-Urrutia, et al., X-Ray Resonant Photoexcitation: Linewidths and Energies of $K\alpha$ Transitions in Highly Charged Fe Ions. Phys. Rev. Lett. **111**(10), 103002 (2013)
141. V.A. Yerokhin, A. Surzhykov, Relativistic configuration-interaction calculation of energy levels of core-excited states in lithiumlike ions: Argon through krypton. Phys. Rev. A: At. Mol. Opt. Phys. **86**, 042507 (2012)
142. U. Staude, Ph. Bosselmann, R. Büttner, D. Horn, K.-H. Schartner, F. Folkmann, A.E. Livingston, T. Ludziejewski, P.H. Mokler, Measurements of $2s\ ^2S_{1/2}$–$2p\ ^2P_{3/2,1/2}$ transition energies in lithiumlike heavy ions: Experiments and results for Ni^{25+} and Zn^{27+}. Phys. Rev. A: At. Mol. Opt. Phys. **58**, 3516 (1998)
143. Ph Bosselmann, U. Staude, D. Horn, K.-H. Schartner, F. Folkmann, A.E. Livingston, P.H. Mokler, Measurements of $2s\ ^2S_{1/2}$–$2p\ ^2P_{3/2,1/2}$ transition energies in lithiumlike heavy ions. II. experimental results for Ag and discussion along the isoelectronic series. Phys. Rev. A: At. Mol. Opt. Phys. **59**, 1874 (1999)
144. D. Feili, Ph Bosselmann, K.-H. Schartner, F. Folkmann, A.E. Livingston, E. Träbert, X. Ma, P.H. Mokler, Measurements of $2s\ ^2S_{1/2}$–$2p\ ^2P_{1/2}$ transition energies in lithiumlike heavy ions. III. experimental results for Sn^{47+} and Xe^{51+}. Phys. Rev. A: At. Mol. Opt. Phys. **62**, 022501 (2000)
145. C. Brandau, T. Bartsch, A. Hoffknecht, H. Knopp, S. Schippers, W. Shi, A. Müller, N. Grün, W. Scheid, T. Steih, F. Bosch, B. Franzke, C. Kozhuharov, P.H. Mokler, F. Nolden, M. Steck, T. Stöhlker, Z. Stachura, High Rydberg resonances in dielectronic recombination of Pb^{79+}. Phys. Rev. Lett. **89**, 053201 (2002)
146. C. Brandau, C. Kozhuharov, A. Müller, W. Shi, S. Schippers, T. Bartsch, S. Böhm, C. Böhme, A. Hoffknecht, K. Knopp, N. Grün, W. Scheid, T. Steih, F. Bosch, B. Franzke, P.H. Mokler, F. Nolden, T. Stöhlker, Z. Stachura, Precise determination of the $2s_{1/2} - 2p_{1/2}$ splitting in very heavy lithiumlike ions utilizing dielectronic recombination. Phys. Rev. Lett. **91**, 073202 (2003)
147. P. Beiersdorfer, D. Knapp, R.E. Marrs, S.R. Elliott, M.H. Chen, Structure and Lamb shift of $2s_{1/2}$–$2p_{1/2}$ levels in lithiumlike U^{89+} through neonlike U^{82+}. Phys. Rev. Lett. **71**, 3939 (1993)
148. S.A. Blundell, Accurate screened QED calculations in high-Z many-electron ions. Phys. Rev. A: At. Mol. Opt. Phys. **46**, 3762 (1992)
149. P. Beiersdorfer, A. Osterheld, S.R. Elliott, M.H. Chen, D. Knapp, K. Reed, Structure and lamb shift of $2s_{1/2}$–$2p_{3/2}$ levels in lithiumlike Th^{87+} through neonlike Th^{80+}. Phys. Rev. A: At. Mol. Opt. Phys. **52**, 2693 (1995)
150. P. Beiersdorfer, A.L. Osterheld, J.H. Scofield, J.R. Crespo, López-Urrutia, K. Widmann, Measurement of QED and hyperfine splitting in the $2s_{1/2}$–$2p_{3/2}$ x-ray transition in Li-like $^{209}Bi^{80+}$. Phys. Rev. Lett. **80**, 3022 (1998)
151. Y. Podpaly, J. Clementson, P. Beiersdorfer, J. Williamson, G.V. Brown, M.F. Gu, Spectroscopy of $2s_{1/2}$–$2p_{3/2}$ transitions in W^{65+} through W^{71+}. Phys. Rev. A: At. Mol. Opt. Phys. **80**, 052504 (2009)
152. Y.-K. Kim, D.H. Baik, P. Indelicato, J.P. Desclaux, Resonance transition energies of Li-, Na-, and Cu-like ions. Phys. Rev. A: At. Mol. Opt. Phys. **44**, 148 (1991)
153. P. Beiersdorfer, Testing QED and atomic-nuclear interactions with high-Z ions. J. Phys. B: At. Mol. Opt. Phys. **43**, 074032 (2010)
154. P. Beiersdorfer, H. Chen, D.B. Thorn, E. Träbert, Measurement of the two-loop Lamb shift in lithiumlike U^{89+}. Phys. Rev. Lett. **95**, 233003 (2005)
155. A.N. Artemyev, V.M. Shabaev, M.M. Sysak, V.A. Yerokhin, T. Beierand, G. Plunien, G. Soff, Evaluation of the two-photon exchange diagrams for the $(1s)^2 2p_{3/2}$ electron configuration in Li-like ions. Phys. Rev. A: At. Mol. Opt. Phys. **67**, 062506 (2003)

156. P. Beiersdorfer, E. Träbert, H. Chen, M.-H. Chen, M.J. May, A.L. Osterheld, Measurement of the $3s_{1/2}$–$3p_{3/2}$ resonance line in Na-like U^{81+}. Phys. Rev. A: At. Mol. Opt. Phys. **67**, 052103 (2003)
157. X. Zhang, N. Nakamura, Ch. Chen, M. Andersson, Y. Liu, Sh Ohtani, Measurement of the QED energy shift in the $1s^2 2p_{3/2} 1s^2 2s_{1/2}$ x-ray transition in Li-like $^{208}Pb^{79+}$. Phys. Rev. A: At. Mol. Opt. Phys. **78**, 032504 (2008)
158. C.A. Morgan, F.G. Serpa, E. Takcs, E.S. Meyer, J.D. Gillaspy, J. Sugar, J.R. Roberts, C.M. Brown, U. Feldman, Observation of visible and UV magnetic dipole transitions in highly charged xenon and barium. Phys. Rev. Lett. **74**, 1716 (1995)
159. J.V. Porto, I. Kink, J.D. Gillaspy, UV light from the ground term of Ti-like ytterbium, tungsten, and bismuth. Phys. Rev. A: At. Mol. Opt. Phys. **61**, 545011 (2000)
160. F.G. Serpa, E.W. Bell, E.S. Meyer, J.D. Gillaspy, J.R. Roberts, Kr spectra from an electron-beam ion trap: 300 nm to 460 nm. Phys. Rev. A: At. Mol. Opt. Phys. **55**, 1832 (1997)
161. H. Watanabe, D. Crosby, F.J. Currell, T. Fukami, D. Kato, S. Ohtani, J.D. Silver, C. Yamada, Magnetic dipole transitions in titaniumlike ions. Phys. Rev. A: At. Mol. Opt. Phys. **63**, 425131–425136 (2001)
162. S.B. Utter, P. Beiersdorfer, G.V. Brown, Measurement of an unusual M1 transition in the ground state of Ti-like W^{52+}. Phys. Rev. A: At. Mol. Opt. Phys. **61**, 305031 (2000)
163. J.R. Crespo López-Urrutia, P. Beiersdorfer, K. Widmann, V. Decaux, Visible spectrum of highly charged ions: The forbidden optical lines of Kr, Xe, Ba ions in the Ar I to Ni I isoelectronic sequence. Phys. Scr. T **80**, 448 (1999)
164. J. R. Crespo López-Urrutia, P. Beiersdorfer, K. Widmann, V. Decaux, Visible spectrum of highly charged ions: The forbidden optical lines of Kr, Xe, and Ba ions in the Ar I to Kr I isoelectronic sequence. Can. J. Phys. **80**, 1687 (2002)
165. I. Draganič, J.R. Crespo, López-Urrutia, R. DuBois, S. Fritzsche, V.M. Shabaev, R. Soria Orts, I.I. Tupitsyn, Y. Zou, J. Ullrich, High precision wavelength measurements of QED-sensitive forbidden transitions in highly charged argon ions. Phys. Rev. Lett. **91**, 1830011 (2003)
166. R.S. Soria Orts, Z. Harman, J.R. Crespo López-Urrutia, A.N. Artemyev, H. Bruhns, A.J. González Martínez, U.D. Jentschura, C.H. Keitel, A. Lapierre, V. Mironov, V.M. Shabaev, H. Tawara, I.I. Tupitsyn, J. Ullrich, A.V. Volotka, Exploring relativistic many-body recoil effects in highly charged ions. Phys. Rev. Lett. **97**, 103002 (2006)
167. C.W.P. Palmer, Reformulation of the theory of the mass shift. J. Phys. B: At. Mol. Phys. **20**, 5987 (1987)
168. R. Soria Orts, J.R. Crespo López-Urrutia, H. Bruhns, A.J. González Martínez, Z. Harman, U.D. Jentschura, C.H. Keitel, A. Lapierre, H. Tawara, I.I. Tupitsyn, J. Ullrich, A.V. Volotka, Zeeman splitting and g factor of the $1s^2 2s^2 2p\,^2P_{3/2}$ and $^2P_{1/2}$ levels in Ar^{13+}. Phys. Rev. A: At. Mol. Opt. Phys. **76**, 052501 (2007)
169. J.R. Crespo, López-Urrutia, The visible spectrum of highly charged ions: A window to fundamental physics. Can. J. Phys. **86**, 111 (2008)
170. A. Lapierre, J.R. Crespo, López-Urrutia, J. Braun, G. Brenner, H. Bruhns, D. Fischer, A. J. González Martínez, V. Mironov, C. Osborne, G. Sikler, R. Soria Orts, H. Tawara, J. Ullrich, V. M. Shabaev, I. I. Tupitsyn, A. Volotka, Lifetime measurement of the Ar XIV $1s^2 2s^2 2p\,^2P^o_{3/2}$ metastable level at the Heidelberg electron-beam ion trap. Phys. Rev. A: At. Mol. Opt. Phys. **73**, 052507 (2006)
171. J.R. Crespo López-Urrutia, B. Bapat, I. Draganič, B. Feuerstein, D. Fischer, H. Lörch, R. Moshammer, J. Ullrich, R.D. Dubois, Y. Zou, Physics with highly-charged ions in an EBIT. Hyperfine Interact. **146**, 109 (2003)
172. F.G. Serpa, J.D. Gillaspy, E. Träbert, Lifetime measurements in the ground configuration of Ar^{13+} and Kr^{22+} using an electron beam ion trap. J. Phys. B: At. Mol. Opt. Phys. **31**, 3345 (1998)
173. D.P. Moehs, D.A. Church, Magnetic dipole transition rates from measured lifetimes of levels of Be-like and B-like argon ions. Phys. Rev. A: At. Mol. Opt. Phys. **58**, 1111 (1998)
174. E. Träbert, Precise atomic lifetime measurements with stored ion beams and ion traps. Can. J. Phys. **80**, 1481 (2002)

175. I.I. Tupitsyn, A.V. Volotka, D.A. Glazov, V.M. Shabaev, G. Plunien, J.R. Crespo, López-Urrutia, A. Lapierre, J. Ullrich, Magnetic-dipole transition probabilities in B-like and Be-like ions. Phys. Rev. A: At. Mol. Opt. Phys. **72**, 062503 (2005)
176. A.N. Artemyev, V.M. Shabaev, I.I. Tupitsyn, G. Plunien, V.A. Yerokhin, QED calculation of the $2p_{3/2}$-$2p_{1/2}$ transition energy in boronlike argon. Phys. Rev. Lett. **98**, 173004 (2007)
177. K.T. Cheng, Y.K. Kim, J.P. Desclaux, Electric dipole, quadrupole, and magnetic dipole transition probabilities of ions isoelectronic to the first-row atoms, Li through F. At. Data Nucl. Data Tables **24**, 111 (1979)
178. C. Froese-Fischer, Multiconfiguration Hartree-Fock Breit-Pauli results for $2p_{1/2}$–$2p_{3/2}$ transitions in the boron sequence. J. Phys. B: At. Mol. Opt. Phys. **16**, 157 (1983)
179. B. Warner, Transition probabilities in np and np^5configurations. Zeitschrift für Astrophysik **69**, 399 (1968)
180. M.E. Galavís, C. Mendoza, C.J. Zeippen, Atomic data from the iron project. Astron. Astrophys. Suppl. **133**, 245 (1998)
181. E. Charro, S. López-Ferrero, I. Martín, Forbidden emission coefficients for intraconfiguration transitions $2p_{3/2}{\rightarrow}2p_{1/2}$ along the boron sequence. J. Phys. B: At. Mol. Opt. Phys. **34**, 4243 (2001)
182. T.R. Verhey, B.P. Das, W.F. Perger, Multiconfiguration Dirac-Fock calculation of the forbidden $2p_{1/2}$–$2p_{3/2}$ ground-state M1 transition in the boron isoelectronic sequence. J. Phys. B: At. Mol. Opt. Phys. **20**, 3639 (1987)
183. Z. Harman, U.D. Jentschura, C.H. Keitel, A. Lapierre, R.S. Orts, J.R. Crespo, López-Urrutia, H. Tawara, J. Ullrich, A.N. Artemyev, I.I. Tupitsyn, A.V. Volotka, V.M. Shabaev, Correlation and quantum electrodynamic effects on the radiative lifetime and relativistic nuclear recoil in Ar^{13+} and Ar^{14+} ions. J. Phys: Conf. Ser. **58**, 133 (2007)
184. V.M. Shabaev, QED theory of the nuclear recoil effect in atoms. Phys. Rev. A: At. Mol. Opt. Phys. **57**, 59 (1998)
185. I.I. Tupitsyn, V.M. Shabaev, J.R. Crespo, López-Urrutia, I. Draganič, R. Soria Orts, J. Ullrich, Relativistic calculations of isotope shifts in highly charged ions. Phys. Rev. A: At. Mol. Opt. Phys. **68**, 225111 (2003)
186. A.N. Artemyev, V.M. Shabaev, V.A. Yerokhin, Nuclear recoil corrections to the $2p_{3/2}$ state energy of hydrogen-like and high-Z lithium-like atoms in all orders in αZ. J. Phys. B: At. Mol. Opt. Phys. **28**, 5201 (1995)
187. V.M. Shabaev, Hyperfine structure of hydrogen-like ions. J. Phys. B: At. Mol. Opt. Phys. **27**, 5825 (1994)
188. I. Klaft, S. Borneis, T. Engel, B. Fricke, R. Grieser, G. Huber, T. Kühl, D. Marx, R. Neumann, S. Schröder, P. Seelig, L. Völker, Precision laser spectroscopy of the ground state hyperfine splitting of hydrogenlike ^{209} Bi^{82+}. Phys. Rev. Lett. **73**, 2425 (1994)
189. J.R. Crespo López-Urrutia, P. Beiersdorfer, D.W. Savin, K. Widmann, Direct observation of the spontaneous emission of the hyperfine transition $F=4$ to $F=3$ in ground state hydrogenlike $^{165}Ho^{66+}$ in an electron beam ion trap. Phys. Rev. Lett. **77**, 826 (1996)
190. J.R. Crespo López-Urrutia, P. Beiersdorfer, K. Widmann, B.B. Birkett, A.-M. Mårtensson-Pendrill, M.G.H. Gustavsson, Nuclear magnetization distribution radii determined by hyperfine transitions in the 1s level of H-like ions $^{185}Re^{74+}$ and $^{187}Re^{74+}$. Phys. Rev. A: At. Mol. Opt. Phys. **57**, 879 (1998)
191. P. Beiersdorfer, S.B. Utter, K.L. Wong, J.R. Crespo, López-Urrutia, J.A. Britten, H. Chen, C.L. Harris, R.S. Thoe, D.B. Thorn, E. Träbert, M.G.H. Gustavsson, C. Forssén, A.-M. Mårtensson-Pendrill, Hyperfine structure of hydrogenlike thallium isotopes. Phys. Rev. A: At. Mol. Opt. Phys. **64**, 032506 (2001)
192. P. Seelig, S. Borneis, A. Dax, T. Engel, S. Faber, M. Gerlach, C. Holbrow, G. Huber, T. Kühl, D. Marx, K. Meier, P. Merz, W. Quint, F. Schmitt, M. Tomaselli, L. Völker, H. Winter, M. Würtz, K. Beckert, B. Franzke, F. Nolden, H. Reich, M. Steck, T. Winkler, Ground state hyperfine splitting of hydrogenlike $^{207}Pb^{81+}$ by laser excitation of a bunched ion beam in the GSI experimental storage ring. Phys. Rev. Lett. **81**, 4824 (1998)

193. W. Nörtershäuser, M. Lochmann, R. Jöhren, C. Geppert, Z. Andjelkovic, D. Anielski, B. Botermann, M. Bussmann, A. Dax, N. Frömmgen, M. Hammen, V. Hannen, T. Kühl, Y. A. Litvinov, J. Mader, T. Stöhlker, R. Thompson, C. Weinheimer, W. Wen, E. Will, D. Winters, R.M. Sánchez, Phys. Scr. 2013. 16th International Conference on the Physics of Highly Charged Ions (in print)
194. M.G.H. Gustavsson, A.-M. Mårtensson-Pendrill, Four decades of hyperfine anomalies. Adv. Quantum Chem. **30**(C), 343 (1998)
195. M.G.H. Gustavsson, A.-M. Mårtensson-Pendrill, Need for remeasurements of nuclear magnetic dipole moments. Phys. Rev. A: At. Mol. Opt. Phys. **58**, 3611 (1998)
196. V.M. Shabaev, A.N. Artemyev, V.A. Yerokhin, O.M. Zherebtsov, G. Soff, Towards a test of QED in investigations of the hyperfine splitting in heavy ions. Phys. Rev. Lett. **86**, 3959 (2001)
197. N.S. Oreshkina, D.A. Glazov, A. Volotka, V.M. Shabaev, I.I. Tupitsyn, Radiative and interelectronic-interaction corrections to the hyperfine splitting in highly charged B-like ions. Phys. Lett. A **372**(5), 675 (2008)
198. P. Beiersdorfer, T.W. Phillips, K.L. Wong, R.E. Marrs, D.A. Vogel, Measurement of level-specific dielectronic-recombination cross sections of heliumlike Fe XXV. Phys. Rev. A: At. Mol. Opt. Phys. **46**, 3812 (1992)
199. D.A. Knapp, R.E. Marrs, M.B. Schneider, M.H. Chen, M.A. Levine, P. Lee, Dielectronic recombination of heliumlike ions. Phys. Rev. A: At. Mol. Opt. Phys. **47**, 2039 (1993)
200. M.H. Chen, J.H. Scofield, Relativistic effects on angular distribution and polarization of dielectronic satellite lines of hydrogenlike ions. Phys. Rev. A: At. Mol. Opt. Phys. **52**, 2057 (1995)
201. M. Gail, N. Grün, W. Scheid, Angular distribution of radiation emitted after resonant transfer and excitation. J. Phys. B: At. Mol. Opt. Phys. **31**, 4645 (1998)
202. S. Zakowicz, Z. Harman, N. Grün, W. Scheid, Angular distribution of hypersatellite and satellite radiation emitted after resonant transfer and excitation into U^{91+} ions. Phys. Rev. A: At. Mol. Opt. Phys. **68**, 042711 (2003)
203. S. Zakowicz, W. Scheid, N. Grün, Dielectronic recombination into hydrogen-like heavy ions with emission of two photons. J. Phys. B: At. Mol. Opt. Phys. **37**, 131 (2004)
204. S. Fritzsche, N.M. Kabachnik, A. Surzhykov, Angular distribution of the dielectronic satellite lines from relativistic high-Z ions: Multipole-mixing effects. Phys. Rev. A: At. Mol. Opt. Phys. **78**, 032703 (2008)
205. S. Fritzsche, A. Surzhykov, Th Stöhlker, Dominance of the Breit interaction in the x-ray emission of highly charged ions following dielectronic recombination. Phys. Rev. Lett. **103**, 113001 (2009)
206. H.S.W. Massey, D.R. Bates, The properties of neutral and ionized atomic oxygen and their influence on the upper atmosphere. Rep. Progr. Phys. **9**, 62 (1942)
207. M.J. Seaton, P.J. Storey, *Di-electronic recombination* (North-Holland Publ. Co., Amsterdam, Netherlands, 1976)
208. D.A. Knapp, R.E. Marrs, M.A. Levine, C.L. Bennett, M.H. Chen, J.R. Henderson, M.B. Schneider, J.H. Scofield, Dielectronic recombination of heliumlike nickel. Phys. Rev. Lett. **62**, 2104 (1989)
209. T. Fuchs, C. Biedermann, R. Radtke, E. Behar, R. Doron, Channel-specific dielectronic recombination of highly charged krypton. Phys. Rev. A: At. Mol. Opt. Phys. **58**, 4518 (1998)
210. H. Watanabe, F.J. Currell, H. Kuramoto, Y.M. Li, S. Ohtani, B. O'Rourke, X.M. Tong, The measurement of the dielectronic recombination in He-like Fe ions. J. Phys. B: At. Mol. Opt. Phys. **34**, 5095 (2001)
211. H. Watanabe, H. Tobiyama, A.P. Kavanagh, Y.M. Li, N. Nakamura, H.A. Sakaue, F.J. Currell, S. Ohtani, Dielectronic recombination of He-like to C-like iodine ions. Phys. Rev. A: At. Mol. Opt. Phys. **75**, 012702 (2007)
212. B.E. O'Rourke, H. Kuramoto, Y.M. Li, S. Ohtani, X.M. Tong, H. Watanabe, F.J. Currell, Dielectronic recombination in He-like titanium ions. J. Phys. B: At. Mol. Opt. Phys. **37**, 2343 (2004)

213. A.J. Smith, P. Beiersdorfer, K. Widmann, M.H. Chen, J.H. Scofield, Measurement of resonant strengths for dielectronic recombination in heliumlike Ar^{16+}. Phys. Rev. A: At. Mol. Opt. Phys. **62**, 052717 (2000)
214. A.J. Smith, P. Beiersdorfer, V. Decaux, K. Widmann, K.J. Reed, M.H. Chen, Measurement of the contributions of high-n satellite lines to the $K\alpha$ lines of He-like Ar^{16+}. Phys. Rev. A: At. Mol. Opt. Phys. **54**, 462 (1996)
215. Y. Zou, J.R. Crespo, López-Urrutia, J. Ullrich, Observation of dielectronic recombination through two-electron-one-photon correlative stabilization in an electron-beam ion trap. Phys. Rev. A: At. Mol. Opt. Phys. **67**, 427031 (2003)
216. X. Zhang, J.R. Crespo, López-Urrutia, P. Guo, V. Mironov, X. Shi, A.J.G. Martinez, H. Tawara, J. Ullrich, Experimental study of the deep-lying dielectronic recombination resonances of He-like germanium ions. J. Phys. B: At. Mol. Opt. Phys. **37**, 2277 (2004)
217. Y. Hahn, Theory of dielectronic recombination. Adv. At. Mol. Opt. Phys. **21**(C), 123 (1985)
218. Y. Hahn, K.J. Lagattuta, Dielectronic recombination and related resonance processes. Phys. Rep. **166**, 195 (1988)
219. P. Zimmerer, N. Grün, W. Scheid, Auger rates for dielectronic recombination cross sections with highly charged relativistic heavy ions. Phys. Lett. Sect. A: Gen. At. Solid State Phys. **148**, 457 (1990)
220. P. Zimmerer, N. Grün, W. Scheid, Scaling of relativistic Auger rates with Z for ions with two electrons. J. Phys. B: At. Mol. Opt. Phys. **24**, 2633 (1991)
221. M. Zimmermann, N. Grün, W. Scheid, Photo recombination on highly charged few-electron uranium ions. J. Phys. B: At. Mol. Opt. Phys. **30**, 5259 (1997)
222. M.S. Pindzola, F.J. Robicheaux, N.R. Badnell, M.H. Chen, M. Zimmermann, Photorecombination of highly charged uranium ions. Phys. Rev. A: At. Mol. Opt. Phys. **52**, 420 (1995)
223. L.N. Labzowsky, A.V. Nefiodov, Radiative interference effects in the dielectronic-recombination process of an electron with hydrogenlike uranium. Phys. Rev. A: At. Mol. Opt. Phys. **49**, 236 (1994)
224. O.Y. Andreev, L.N. Labzowsky, G. Plunien, D.A. Solovyev, QED theory of the spectral line profile and its applications to atoms and ions. Phys. Rep. **455**, 135 (2008)
225. C. Beilmann, P.H. Mokler, S. Bernitt, C.H. Keitel, J. Ullrich, J.R. Crespo, López-Urrutia, Z. Harman, Prominent higher-order contributions to electronic recombination. Phys. Rev. Lett. **107**, 143201 (2011)
226. J.B. Mann, W.R. Johnson, Breit interaction in multielectron atoms. Phys. Rev. A: At. Mol. Opt. Phys. **4**, 41 (1971)
227. K.T. Cheng, M.H. Chen, J. Sapirstein, Quantum electrodynamic corrections in high-Z Li-like and Be-like ions. Phys. Rev. A: At. Mol. Opt. Phys. **62**, 054501 (2000)
228. H. Zhimin, X. Han, Y. Li, D. Kato, X. Tong, N. Nakamura, Experimental demonstration of the Breit interaction which dominates the angular distribution of x-ray emission in dielectronic recombination. Phys. Rev. Lett. **108**, 073002 (2012)
229. U. Fano, Effects of configuration interaction on intensities and phase shifts. Phys. Rev. **124**, 1866 (1961)
230. V.L. Jacobs, Theory of radiative corrections to Auger and fluorescence yields and dielectronic satellite line intensities. Phys. Rev. A: At. Mol. Opt. Phys. **31**, 383 (1985)
231. S.L. Haan, V.L. Jacobs, Projection-operator approach to the unified treatment of radiative and dielectronic recombination. Phys. Rev. A: At. Mol. Opt. Phys. **40**, 80 (1989)
232. S. Schippers, S. Kieslich, A. Müller, G. Gwinner, M. Schnell, A. Wolf, A. Covington, M.E. Bannister, L.-B. Zhao, Interference effects in the photorecombination of argonlike Sc^{3+} ions: Storage-ring experiment and theory. Phys. Rev A: At. Mol. Opt. Phys. **65**, 042723 (2002)
233. V.V. Karasiov, L.N. Labzowsky, A.V. Nefiodov, V.M. Shabaev, Overlapping resonances in the process of recombination of an electron with hydrogenlike uranium. Phys. Lett. Sect. A: Gen. At. Solid State Phys. **161**, 453 (1992)
234. A.V. Nefiodov, V.V. Karasiev, V.A. Yerokhin, Interference effects in the recombination process of hydrogenlike lead. Phys. Rev. A: At. Mol. Opt. Phys. **50**, 4975 (1994)

235. D.A. Knapp, P. Beiersdorfer, M.H. Chen, J.H. Scofield, D. Schneider, Observation of interference between dielectronic recombination and radiative recombination in highly charged uranium ions. Phys. Rev. Lett. **74**, 54 (1995)
236. N. Nakamura, A.P. Kavanagh, H. Watanabe, H.A. Sakaue, Y. Li, D. Kato, F.J. Currell, X.-M. Tong, T. Watanabe, S. Ohtani, Asymmetric profiles observed in the recombination of Be^{79+}: A benchmark for relativistic theories involving interference. Phys. Rev. A: At. Mol. Opt. Phys. **80**, 014503 (2009)
237. A. J. González Martínez, J.R. Crespo López-Urrutia, J. Braun, G. Brenner, H. Bruhns, A. Lapierre, V. Mironov, R. Soria Orts, H. Tawara, M. Trinczek, J. Ullrich, A.N. Artemyev, Z. Harman, U.D. Jentschura, C.H. Keitel, J.H. Scofield, I.I. Tupitsyn, Benchmarking high-field few-electron correlation and QED contributions in Hg^{75+} to Hg^{78+} ions. I. Experiment. Phys. Rev. A: At. Mol. Opt. Phys. **73**, 052710 (2006)
238. M.F. Gu, T. Holczer, E. Behar, S.M. Kahn, Inner-shell absorption lines of Fe VI–Fe XVI: A many-body perturbation theory approach. Astrophys. J. **641**, 1227 (2006)
239. S.W. Epp, J.R. Crespo, López-Urrutia, M. C. Simon, T. Baumann, G. Brenner, R. Ginzel, N. Guerassimova, V. Mäckel, P.H. Mokler, B.L. Schmitt, H. Tawara, J. Ullrich, X-ray laser spectroscopy of highly charged ions at FLASH. J. Phys. B: At. Mol. Opt. Phys. **43**, 194008 (2010)
240. D.L. Matthews, P.L. Hagelstein, M.D. Rosen, M.J. Eckart, N.M. Ceglio, A.U. Hazi, H. Medecki, B.J. MacGowan, J.E. Trebes, B.L. Whitten, E.M. Campbell, C.W. Hatcher, A.M. Hawryluk, R.L. Kauffman, L.D. Pleasance, G. Rambach, J.H. Scofield, G. Stone, T.A. Weaver, Demonstration of a soft x-ray amplifier. Phys. Rev. Lett. **54**, 110 (1985)
241. M.D. Rosen, P.L. Hagelstein, D.L. Matthews, E.M. Campbell, A.U. Hazi, B.L. Whitten, B. MacGowan, R.E. Turner, R.W. Lee, G. Charatis, Gar. E. Busch, C. L. Shepard, P. D. Rockett, Exploding-foil technique for achieving a soft x-ray laser. Phys. Rev. Lett. **54**, 106 (1985)
242. J.J. Rocca, V. Shlyaptsev, F.G. Tomasel, O.D. Cortzar, D. Hartshorn, J.L.A. Chilla, Demonstration of a discharge pumped table-top soft-x-ray laser. Phys. Rev. Lett. **73**, 2192 (1994)
243. J. Andruszkow, B. Aune, V. Ayvazyan, N. Baboi, R. Bakker, V. Balakin, D. Barni, A. Bazhan, M. Bernard, A. Bosotti, J.C. Bourdon, W. Brefeld, R. Brinkmann, S. Buhler, J.-P. Carneiro, M. Castellano, P. Castro, L. Catani, S. Chel, Y. Cho, S. Choroba, E.R. Colby, W. Decking, P. Den Hartog, M. Desmons, M. Dohlus, D. Edwards, H.T. Edwards, B. Faatz, J. Feldhaus, M. Ferrario, M.J. Fitch, K. Flöttmann, M. Fouaidy, A. Gamp, T. Garvey, C. Gerth, M. Geitz, E. Gluskin, V. Gretchko, U. Hahn, W.H. Hartung, D. Hubert, M. Hüning, R. Ischebek, M. Jablonka, J.M. Joly, M. Juillard, T. Junquera, P. Jurkiewicz, A. Kabel, J. Kahl, H. Kaiser, T. Kamps, V.V. Katelev, J.L. Kirchgessner, M. Körfer, L. Kravchuk, G. Kreps, J. Krzywinski, T. Lokajczyk, R. Lange, B. Leblond, M. Leenen, J. Lesrel, M. Liepe, A. Liero, T. Limberg, R. Lorenz, L.H. Hua, L.F. Hai, C. Magne, M. Maslov, G. Materlik, A. Matheisen, J. Menzel, P. Michelato, W.-D. Möller, A. Mosnier, U.-C. Müller, O. Napoly, A. Novokhatski, M. Omeich, H.S. Padamsee, C. Pagani, F. Peters, B. Petersen, P. Pierini, J. Pflüger, P. Piot, B. Phung, Ngoc, L. Plucinski, D. Proch, K. Rehlich, S. Reiche, D. Reschke, I. Reyzl, J. Rosenzweig, J. Rossbach, S. Roth, E. L. Saldin, W. Sandner, Z. Sanok, H. Schlarb, G. Schmidt, P. Schmüser, J. R. Schneider, E. A. Schneidmiller, H.-J. Schreiber, S. Schreiber, P. Schütt, J. Sekutowicz, L. Serafini, D. Sertore, S. Setzer, S. Simrock, B. Sonntag, B. Sparr, F. Stephan, V.A. Sytchev, S. Tazzari, F. Tazzioli, M. Tigner, M. Timm, M. Tonutti, E. Trakhtenberg, R. Treusch, D. Trines, V. Verzilov, T. Vielitz, V. Vogel, G.V. Walter, R. Wanzenberg, T. Weiland, H. Weise, J. Weisend, M. Wendt, M. Werner, M.M. White, I. Will, S. Wolff, M.V. Yurkov, K. Zapfe, P. Zhogolev, F. Zhou, First observation of self-amplified spontaneous emission in a free-electron laser at 109 nm wavelength. Phys. Rev. Lett. **85**, 3825 (2000)
244. C. Bostedt, H.N. Chapman, J.T. Costello, J.R. Crespo, López-Urrutia, S. Düsterer, S.W. Epp, J. Feldhaus, A. Föhlisch, M. Meyer, T. Möller, R. Moshammer, M. Richter, K. Sokolowski-Tinten, A. Sorokin, K. Tiedtke, J. Ullrich, W. Wurth, Experiments at FLASH. Nucl. Instrum. Methods Phys. Res. Sect. A **601**, 108 (2009)

245. H.-C. Wille, H. Franz, R. Röhlsberger, W. Caliebe, F.-U. Dill, Nuclear resonant scattering at PETRA III : Brillant opportunities for nano– and extreme condition science. J. Phys: Conf. Ser. **217**, 012008 (2009)
246. U.I. Safronova, A.S. Safronova, W.R. Johnson, Relativistic many-body calculations of dielectronic satellite spectra created by autoionizing $1s2l2l'$ states in Li-like ions. J. Phys. B: At. Mol. Opt. Phys. **43**(14), 144001 (2010)
247. A.S. Shlyaptseva, R.C. Mancini, P. Neill, P. Beiersdorfer, J.R. Crespo, López-Urrutia, K. Widmann, Polarization-dependent spectra of x-ray dielectronic satellite lines of Be-like Fe. Phys. Rev. A: At. Mol. Opt. Phys. **57**, 888 (1998)
248. T.V. Back, H.S. Margolis, P.K. Oxley, J.D. Silver, E.G. Myers, Laser spectroscopy of the $1s^2 2s2p\,^3P_2$–$^3P_1$ transition in beryllium-like argon using the Oxford EBIT. Hyperfine Interact. **114**, 203 (1998)
249. D. Schneider, D.A. Church, G. Weinberg, J. Steiger, B. Beck, J. McDonald, E. Magee, D. Knapp, Confinement in a cryogenic penning trap of highest charge state ions from EBIT. Rev. Sci. Instrum. **65**, 3472 (1994)
250. V. Mäckel, R. Klawitter, G. Brenner, J.R. Crespo, López-Urrutia, J. Ullrich, Laser spectroscopy on forbidden transitions in trapped highly charged Ar^{13+} ions. Phys. Rev. Lett. **107**, 143002 (2011)
251. K. Schnorr, V. Mäckel, N.S. Oreshkina, S. Augustin, F. Brunner, Z. Harman, C.H. Keitel, J. Ullrich, J.R. Crespo López-Urrutia, Coronium in the laboratory: measuring the Fe XIV green coronal line by laser spectroscopy. Astrophys. J. **776**(2), 121 (2013)
252. C.A. Young, The Sun and the phenomena of its atmosphere (C.C. Chatfield & Co., New Haven, Connecticut, 1872)
253. W. Grotrian, Zur Frage der Deutung der Linien im Spektrum der Sonnenkorona. Naturwissenschaften **27**(13), 214 (1939)
254. B. Edlén, Die Deutung der Emissionslinien im Spektrum der Sonnenkorona. Mit 6 Abbildungen. Zeitschrift für Astrophysik **22**, 30 (1943)
255. S.A. Diddams, Th Udem, J.C. Bergquist, E.A. Curtis, R.E. Drullinger, L. Hollberg, W.M. Itano, W.D. Lee, C.W. Oates, K.R. Vogel, D.J. Wineland, An optical clock based on a single trapped $^{199}Hg^+$ ion. Science **293**, 825 (2001)
256. K.T. Cheng, M.H. Chen, W.R. Johnson, J. Sapirstein, High-precision relativistic atomic structure calculations and the EBIT: Tests of quantum electrodynamics in highly charged ions. Can. J. Phys. **86**, 33 (2008)
257. T. Rosenband, P.O. Schmidt, D.B. Hume, W.M. Itano, T.M. Fortier, J.E. Stalnaker, K. Kim, S.A. Diddams, J.C.J. Koelemeij, J.C. Bergquist, D.J. Wineland, Observation of the $^1S_0$–$^3P_0$ clock transition in Al^+. Phys. Rev. Lett. **98**, 220801 (2007)
258. C.W. Chou, D.B. Hume, J.C.J. Koelemeij, D.J. Wineland, T. Rosenband, Frequency comparison of two high-accuracy Al^+ optical clocks. Phys. Rev. Lett. **104**, 070802 (2010)
259. C.J. Campbell, A.G. Radnaev, A. Kuzmich, V.A. Dzuba, V.V. Flambaum, A. Derevianko, Single-ion nuclear clock for metrology at the 19th decimal place. Phys. Rev. Lett. **108**, 120802 (2012)
260. K. Hosaka, S.A. Webster, A. Stannard, B.R. Walton, H.S. Margolis, P. Gill, Frequency measurement of the $^2S_{1/2}$–$^2F_{7/2}$ electric octupole transition in a single $^{171}Yb^+$ ion. Phys. Rev. A: At. Mol. Opt. Phys. **79**, 033403 (2009)
261. A.L. Wolf, S.A. Van Den Berg, W. Ubachs, K.S.E. Eikema, Direct frequency comb spectroscopy of trapped ions. Phys. Rev. Lett. **102**, 223901 (2009)
262. D.Z. Kandula, C. Gohle, T.J. Pinkert, W. Ubachs, K.S.E. Eikema, Extreme ultraviolet frequency comb metrology. Phys. Rev. Lett. **105**, 063001 (2010)
263. A. Cingaz, D.C. Yost, T.K. Allison, A. Ruehl, M.E. Fermann, I. Hartl, J. Ye, Direct frequency comb spectroscopy in the extreme ultraviolet. Nature **482**, 68 (2012)
264. J.B. Wübbena, S. Amairi, O. Mandel, P.O. Schmidt, Sympathetic cooling of mixed-species two-ion crystals for precision spectroscopy. Phys. Rev. A: At. Mol. Opt. Phys. **85**, 043412 (2012)

265. V.V. Flambaum, J.C. Berengut, Space-time variation of coupling constants and fundamental masses. Int. J. Modern Phys. A **24**, 3342 (2009)
266. J.C. Berengut, V.A. Dzuba, V.V. Flambaum, Enhanced laboratory sensitivity to variation of the fine-structure constant using highly charged ions. Phys. Rev. Lett. **105**, 120801 (2010)
267. S. Schiller, Hydrogenlike highly charged ions for tests of the time independence of fundamental constants. Phys. Rev. Lett. **98**, 180801 (2007)
268. J.C. Berengut, V.A. Dzuba, V.V. Flambaum, A. Ong, Electron-hole transitions in multiply charged ions for precision laser spectroscopy and searching an α: for variations in α. Phys. Rev. Lett. **106**, 210802 (2011)
269. J.C. Berengut, V.A. Dzuba, V.V. Flambaum, A. Ong, Optical transitions in highly charged californium ions with high sensitivity to variation of the fine-structure constant. Phys. Rev. Lett. **109**, 070802 (2012)
270. J.C. Berengut, V.A. Dzuba, V.V. Flambaum, A. Ong, Highly charged ions with E1, M1, and E2 transitions within laser range. Phys. Rev. A: At. Mol. Opt. Phys. **86**, 022517 (2012)
271. V.A. Dzuba, A. Derevianko, V.V. Flambaum, Ion clock and search for the variation of the fine-structure constant using optical transitions in Nd^{13+} and Sm^{15+}. Phys. Rev. A: At. Mol. Opt. Phys. **86**, 054502 (2012)
272. V.A. Dzuba, A. Derevianko, V.V. Flambaum, High-precision atomic clocks with highly charged ions: Nuclear-spin-zero f^{12}-shell ions. Phys. Rev. A: At. Mol. Opt. Phys. **86**, 054501 (2012)
273. A. Derevianko, V.A. Dzuba, V.V. Flambaum, Highly charged ions as a basis of optical atomic clockwork of exceptional accuracy. Phys. Rev. Lett. **109**, 180801 (2012)
274. L. Gruber, J.P. Holder, J. Steiger, B.R. Beck, H.E. DeWitt, J. Glassman, J.W. McDonald, D.A. Church, D. Schneider, Evidence for highly charged ion coulomb crystallization in multicomponent strongly coupled plasmas. Phys. Rev. Lett. **86**, 636 (2001)
275. M. Schwarz, O.O. Versolato, A. Windberger, F.R. Brunner, T. Ballance, S.N. Eberle, J. Ullrich, P.O. Schmidt, A.K. Hansen, A.D. Gingell, M. Drewsen, J.R. Crespo, López-Urrutia. Cryogenic linear Paul trap for cold highly charged ion experiments. Rev. Sci. Instrum. **83**, 083115 (2012)
276. O. O. Versolato, M. Schwarz, A.K. Hansen, A,D. Gingell, A. Windberger, L. Klosowski, J. Ullrich, F. Jensen, J.R. Crespo, M. López-Urrutia Drewsen, Decay rate measurement of the first vibrationally excited state of MgH^+ in a cryogenic Paul trap. Phys. Rev. Lett. **111**(5), 053002 (2013)
277. O.O. Versolato, M. Schwarz, A. Windberger, J. Ullrich, P.O. Schmidt, M. Drewsen, J.R. Crespo, López-Urrutia, Cold highly charged ions in a cryogenic paul trap. Hyperfine Interact. **214**, 189 (2013)
278. P.O. Schmidt, T. Rosenband, C. Langer, W.M. Itano, J.C. Bergquist, D.J. Wineland, Physics: Spectroscopy using quantum logic. Science **309**(5735), 749 (2005)
279. D.B. Hume, C.W. Chou, D.R. Leibrandt, M.J. Thorpe, D.J. Wineland, T. Rosenband, Trapped-ion state detection through coherent motion. Phys. Rev. Lett. **107**, 243902 (2011)
280. D.J. Wineland, D. Leibfried, Quantum information processing and metrology with trapped ions. Laser Phys. Lett. **8**, 175 (2011)
281. B.W. Adams, C. Buth, S.M. Cavaletto, J. Evers, Z. Harman, C.H. Keitel, A. Pálffy, A. Picón, R. Röhlsberger, Y. Rostovtsev, K. Tamasaku, X-ray quantum optics. J. Modern Opt. **60**, 2–21 (2013)
282. C.R. Clark, J.E. Goeders, Y.K. Dodia, C.R. Viteri, K.R. Brown, Detection of single-ion spectra by Coulomb-crystal heating. Phys. Rev. A: At. Mol. Opt. Phys. **81**, 043428 (2010)
283. A. Pálffy, Nuclear effects in atomic transitions. Contemp. Phys. **51**, 471 (2010)
284. J.N. Tan, S.M. Brewer, N.D. Guise, Experimental efforts at NIST towards one-electron ions in circular Rydberg states. Phys. Scr. T **144**, 014009 (2011)

Chapter 11
Tests of Theory in Rydberg States of One-Electron Ions

Joseph N. Tan and Peter J. Mohr

Abstract Comparison of optical frequency measurements to predictions of quantum electrodynamics (QED) for Rydberg states of one-electron ions can test theory and allow new determinations of constants of nature to be made. Simplifications in the QED theory of high-angular-momentum states reduces the uncertainty in the prediction of transition frequencies to a level where a new value of the Rydberg constant which is independent of the proton radius can be determined. Since the energy-level spacing between neighbouring Rydberg states grows as the square of the nuclear charge number, it is possible to study transitions with optical frequencies that are accessible to femtosecond laser frequency combs. Recently at the US National Institute of Standards and Technology (NIST), highly charged ions (including bare nuclei) created in an Electron Beam Ion Trap (EBIT) were extracted and captured in a novel compact Penning trap. An ongoing experiment aims to produce one-electron ions isolated in an ion trap designed for laser spectroscopy. Tests of theory in a regime free of nuclear effects would be valuable in shedding light on the puzzle surrounding the large discrepancy in the value of the proton radius inferred from the observed Lamb shift in muonic hydrogen as compared to the value deduced from hydrogen and deuterium spectroscopy and electron scattering measurements.

11.1 Introduction

Quantum electrodynamics (QED) is the first satisfactory quantum description of the interaction of charged particles (and antiparticles) via the exchange of photons and of the creation and annihilation of elementary particles. QED makes precise predictions

J. N. Tan (✉) · P. J. Mohr
National Institute of Standards and Technology, 100 Bureau Drive,
Gaithersburg, MD 20899, USA
e-mail: joseph.tan@nist.gov

P. J. Mohr
e-mail: mohr@nist.gov

W. Quint and M. Vogel (eds.), *Fundamental Physics in Particle Traps*, Springer Tracts in Modern Physics 256, DOI: 10.1007/978-3-642-45201-7_11, for various physical quantities, and these predictions have been tested across a vast array of phenomena. Spectroscopic measurements in atoms have played a crucial role in spurring the development of QED to be the most accurate physical theory yet invented.

This success seems all the more remarkable because, as the quantum field theory of electrons and photons, QED does not remove the divergences well known in Maxwell's classical theory of electromagnetism; on the contrary, new infinities are found in QED, associated with virtual processes in the vacuum. Sensible, finite results are obtainable only after renormalisation; this is an art inextricably tied to the introduction of fundamental constants, such as the electron mass and charge, into the theory.

One-electron atomic systems are among the simplest quantum-mechanical objects that have enabled very stringent tests of QED to be made, yielding some of the most precise measurements of fundamental constants. For example, the value of the Rydberg constant recommended by the Committee on Data for Science and Technology (CODATA), an interdisciplinary scientific committee of the International Council for Science (ICSU), is obtained from precise measurements of various transitions in hydrogen and its stable isotope deuterium taken together with electron scattering measurements [1].

For over a hundred years, the study of hydrogen-like atoms has contributed to our understanding of the quantum world. Figure 11.1 illustrates the eight orders-of-magnitude improvement in the accuracy of the Rydberg constant in a 130-year history that includes many theoretical and experimental discoveries and developments [2, 3]. As another example, assuming the validity of quantum electrodynamics (QED) calculations at the tenth order, the fine-structure constant α (the universal constant of electromagnetic interactions) is most precisely obtained from the electron $|g|-2$ experiment, which measures the anomalous magnetic moment of a single electron stored in a Penning trap (dubbed *geonium*, an engineered atom consisting of one electron bound to the earth).

The first tests of the emerging formalism of QED came shortly after World War II. Reference [11] reported the first measurement of the anomalous magnetic moment of the electron; in the same year, Ref. [12] presented the first measurement of the Lamb shift of the $2s$ level of hydrogen [13], another departure from the Dirac theory of the hydrogen atom. These discoveries have led to more stringent tests of QED, with remarkable progress over six decades. On the one hand, with control and minimization of cavity effects that limited early geonium $|g|-2$ experiments [14, 15], the magnetic moment of the electron $|g| = 2(1 + a_e)$ has been measured recently at Harvard University with a relative uncertainty of 2.8×10^{-13} using a single electron isolated in a cylindrical Penning trap [16]; by comparison, the calculated magnetic moment of the electron has a relative uncertainty of 5.2×10^{-12} coming mainly from the uncertainty of the best independent determination of the fine-structure constant. On the other hand, the hydrogen $1s - 2s$ transition has been measured with a relative uncertainty of 1.4×10^{-14} [17]; here a test of theory is hampered by uncertainties in the nuclear-size corrections.

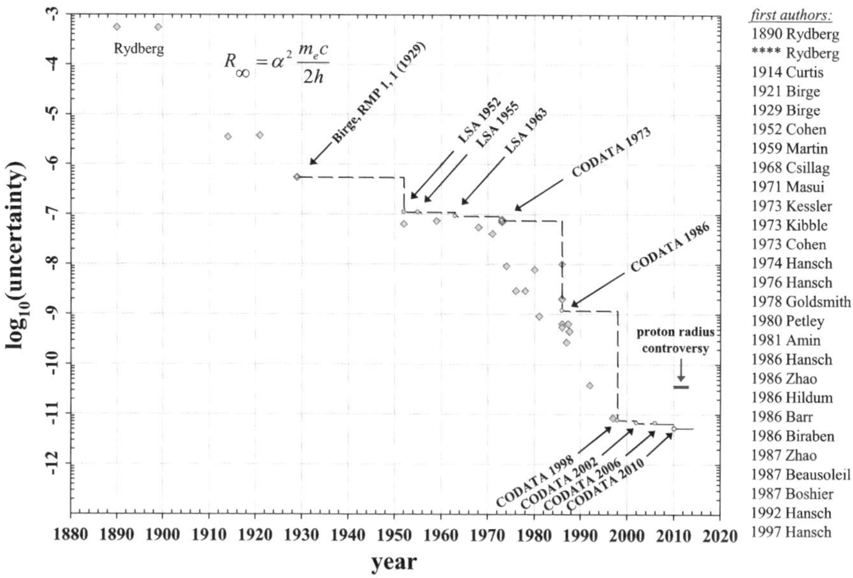

Fig. 11.1 Uncertainty in the determinations of the Rydberg constant over its history. Its current uncertainty of 5×10^{-12} is beset by a puzzling discrepancy in the proton radius that is about 7 times as large as the uncertainty of the disagreement. LSA refers to a least-squares adjustment of fundamental constants

Currently the accuracy in the determination of the Rydberg constant has a relative uncertainty of 5.0×10^{-12} [1]. It seems astonishing that QED should have attained such high accuracy based on abstractions employed to represent physical objects and measurements, particularly when one of the pioneers in its development has noted the 'mathematical inconsistencies and renormalized infinities swept under the rug' [18].

Looking to the future, in Sect. 11.5, we consider possible determinations of fundamental constants and tests of theory in Rydberg states of one-electron ions. This has been discussed in [19], where it was pointed out that the problems that limit the theoretical predictions in low-angular-momentum states are strongly suppressed in Rydberg states with high angular momentum (high l), because the electron has a very small probability of being near the nucleus in such states. Remarkably, the simplification of higher-order QED terms in Rydberg states yields an overall theoretical uncertainty that is smaller than the uncertainties propagated from the fundamental constants. In particular, even with its currently-assigned uncertainty of 5.0×10^{-12}, the Rydberg constant is the leading source of uncertainty in this regime, larger than the uncertainties due to other constants by a factor of 100. Hence, if precise measurements can be made for comparison with theory, one-electron ions in Rydberg states can be useful in testing QED in a regime with negligible nuclear effects. Assuming that QED remains valid, this could enable a Rydberg constant determination that is independent of the proton radius. As discussed in Sect. 11.3, interest in such a determination is heightened by the large discrepancy in the proton radius determined from the recent measurement of the muonic hydrogen Lamb shift compared to the determinations of the radius by other methods, which potentially would be impacted by an independent measurement of the Rydberg constant. In Sect. 11.6, we describe an experimental effort at NIST to make one-electron ions in Rydberg states via charge transfer from highly-excited atom to bare nuclei isolated in a trap.

11.2 The Rydberg Constant

The Rydberg constant is a combination of fundamental constants:

$$R_\infty = \alpha^2 m_e c/2h, \tag{11.1}$$

where h is the Planck constant, c is the speed of light, m_e is the electron mass, and α is the fine-structure constant. It plays a pivotal role in the CODATA least-square adjustments. Historically, it first appeared as the overall coefficient in the mathematical formula connecting the light frequency (or wavelength) to the numbers characterizing the atomic states involved in the light emission. This relationship was first discovered phenomenologically in the spectral line emissions of hydrogen by Johann Balmer, who showed that the visible lines in the series named after him are related by a simple formula involving integers. Inspired by this discovery, Johannes Rydberg showed that a more general formula which included small non-integral offsets (now called 'quantum defects') has much broader application by accounting for perturbations due to core electrons when studying atoms with one highly-excited electron (Rydberg atoms) [20]. Figure 11.1 shows the improvement in the determinations of the Rydberg constant going back over one hundred years; a compilation of the experimental values can be found in Ref. [3].

The CODATA recommended value of the Rydberg constant has been obtained primarily by comparing theory and experiment for twenty-four transition frequencies or pairs of frequencies in hydrogen and deuterium [1]. The theoretical value for each transition is the product of the Rydberg constant and a calculated factor based on QED that also depends on other constants. While the most accurately measured transition frequency in hydrogen (1S–2S) has a relative uncertainty of 1.4×10^{-14} [17], the recommended value of the Rydberg constant has a larger relative uncertainty of 5.0×10^{-12} which is essentially the uncertainty in the theoretical factor which includes the proton radius. The main source of uncertainty in the theoretical factor is from the charge radius of the proton with additional uncertainty due to uncalculated or partially calculated higher-order terms in the QED corrections.

11.3 The Proton Radius Puzzle: Is QED in Trouble?

The uncertainty introduced into the theory of hydrogen and deuterium transition frequencies due to uncertainty in the proton and deuteron charge radii can be reduced by employing accurate independent determinations of the radii. Electron-proton and electron-deuteron scattering experiments are important sources of information about the radii. Potentially more accurate information is provided by comparison of theory and experiment for muonic hydrogen. In this section, the various sources of information on the charge radii are briefly reviewed, and the discrepancy that leads to the proton radius puzzle is described.

11.3.1 Spectroscopic Data and the Rydberg Constant

Values of the Rydberg constant and the proton and deuteron charge radii follow from precise spectroscopic measurements on hydrogen and deuterium. For the 2010 CODATA determination of the constants, 24 frequencies or differences of frequencies were taken into account. These data are listed in Table 11.1. Comparison of theory and experiment, which does not include electron scattering values for the radii, leads to a value of the Rydberg constant given by [1]

$$R_\infty = 10\,973\,731.568\,521(82)\ \mathrm{m}^{-1}\,, \tag{11.2}$$

which has a relative uncertainty of 7.4×10^{-12}, and a value for the proton radius given by [1]

$$r_\mathrm{p} = 0.8758(77)\ \mathrm{fm}\,. \tag{11.3}$$

The theoretical input used in the analysis is described in [1]. The main source of uncertainty, as has been mentioned, is the uncertainty in the proton charge radius. A smaller, but not negligible, uncertainty arises from uncalculated theoretical contributions.

11.3.2 Scattering Determinations of the Radii

The accuracy of the Rydberg constant extracted from hydrogen and deuterium spectroscopic data can be improved by including information on the nuclear charge radii from electron scattering experiments in the analysis. Electron-proton scattering data values of the proton radius r_p and an electron-deuteron scattering data value of the deuteron radius r_d used as input data in the 2010 CODATA adjustment are

Table 11.1 Measured frequencies ν considered in the LSA determination of the Rydberg constant R_∞ (H is hydrogen and D is deuterium)

	Frequency interval(s)	Reported value ν/kHz	Rel. stand. uncert. u_r
	MPQ 2004 (Fischer et al. [23])		
1	$\nu_H(1S_{1/2} - 2S_{1/2})$	2 466 061 413 187.080(34)	1.4×10^{-14}
	MPQ 1995 (Weitz et al. [24])		
2	$\nu_H(2S_{1/2} - 4S_{1/2}) - \frac{1}{4}\nu_H(1S_{1/2} - 2S_{1/2})$	4 797 338(10)	2.1×10^{-6}
3	$\nu_H(2S_{1/2} - 4D_{5/2}) - \frac{1}{4}\nu_H(1S_{1/2} - 2S_{1/2})$	6 490 144(24)	3.7×10^{-6}
4	$\nu_D(2S_{1/2} - 4S_{1/2}) - \frac{1}{4}\nu_D(1S_{1/2} - 2S_{1/2})$	4 801 693(20)	4.2×10^{-6}
5	$\nu_D(2S_{1/2} - 4D_{5/2}) - \frac{1}{4}\nu_D(1S_{1/2} - 2S_{1/2})$	6 494 841(41)	6.3×10^{-6}
	MPQ 2010 (Parthey et al. [25])		
6	$\nu_D(1S_{1/2} - 2S_{1/2}) - \nu_H(1S_{1/2} - 2S_{1/2})$	670 994 334.606(15)	2.2×10^{-11}
	LKB/SYRTE 1997 (de Beauvoir et al. [25])		
7	$\nu_H(2S_{1/2} - 8S_{1/2})$	770 649 350 012.0(8.6)	1.1×10^{-11}
8	$\nu_H(2S_{1/2} - 8D_{3/2})$	770 649 504 450.0(8.3)	1.1×10^{-11}
9	$\nu_H(2S_{1/2} - 8D_{5/2})$	770 649 561 584.2(6.4)	8.3×10^{-12}
10	$\nu_D(2S_{1/2} - 8S_{1/2})$	770 859 041 245.7(6.9)	8.9×10^{-12}
11	$\nu_D(2S_{1/2} - 8D_{3/2})$	770 859 195 701.8(6.3)	8.2×10^{-12}
12	$\nu_D(2S_{1/2} - 8D_{5/2})$	770 859 252 849.5(5.9)	7.7×10^{-12}
	LKB/SYRTE 1999 (Schwob et al. [26])		
13	$\nu_H(2S_{1/2} - 12D_{3/2})$	799 191 710 472.7(9.4)	1.2×10^{-11}
14	$\nu_H(2S_{1/2} - 12D_{5/2})$	799 191 727 403.7(7.0)	8.7×10^{-12}
15	$\nu_D(2S_{1/2} - 12D_{3/2})$	799 409 168 038.0(8.6)	1.1×10^{-11}
16	$\nu_D(2S_{1/2} - 12D_{5/2})$	799 409 184 966.8(6.8)	8.5×10^{-12}
	LKB 2010 (Arnoult et al. [27])		
17	$\nu_H(1S_{1/2} - 3S_{1/2})$	2 922 743 278 678(13)	4.4×10^{-12}
	LKB 1996 (Bourzeix et al. [28])		
18	$\nu_H(2S_{1/2} - 6S_{1/2}) - \frac{1}{4}\nu_H(1S_{1/2} - 3S_{1/2})$	4 197 604(21)	4.9×10^{-6}
19	$\nu_H(2S_{1/2} - 6D_{5/2}) - \frac{1}{4}\nu_H(1S_{1/2} - 3S_{1/2})$	4 699 099(10)	2.2×10^{-6}
	Yale 1995 (Berkeland et al. [29])		
20	$\nu_H(2S_{1/2} - 4P_{1/2}) - \frac{1}{4}\nu_H(1S_{1/2} - 2S_{1/2})$	4 664 269(15)	3.2×10^{-6}
21	$\nu_H(2S_{1/2} - 4P_{3/2}) - \frac{1}{4}\nu_H(1S_{1/2} - 2S_{1/2})$	6 035 373(10)	1.7×10^{-6}
	Harvard 1994 (Hagley and Pipkin [30])		
22	$\nu_H(2S_{1/2} - 2P_{3/2})$	9 911 200(12)	1.2×10^{-6}
	Harvard 1986 (Lundeen and Pipkin [31])		
23	$\nu_H(2P_{1/2} - 2S_{1/2})$	1 057 845.0(9.0)	8.5×10^{-6}
	U. Sussex 1979 (Newton et al. [32])		
24	$\nu_H(2P_{1/2} - 2S_{1/2})$	1 057 862(20)	1.9×10^{-5}

[1] MPQ: Max-Planck-Institut für Quantenoptik, Garching. LKB: Laboratoire Kastler-Brossel, Paris. SYRTE: Systèmes de référence Temps Espace, Paris, formerly Laboratoire Primaire du Temps et des Fréquences (LPTF)

$$r_p = 0.895(18) \text{ fm}, \tag{11.4}$$

$$r_p = 0.8791(79) \text{ fm}, \tag{11.5}$$

$$r_d = 2.130(10) \text{ fm}. \tag{11.6}$$

The value of r_p in Eq. (11.4) is based on Sick's analysis of the world data available at the time [21]. Equation (11.5) gives a value based on measurements made at the Mainz Microtron MAMI and analysed using a variety of form-factor models [4]. The result for r_d in Eq. (11.6) is the result of an analysis of the world data by Sick [22].

More recent values for the proton radius from scattering data have been given since the 2010 CODATA cutoff date for input data. Precise measurement of the ratio of the electric to magnetic form factors over a range of Q^2 was made at the Thomas Jefferson National Accelerator facility in Virginia (JLab). These data, combined with selected earlier data, yield the value $r_\mathrm{p} = 0.875(10)$ fm [5, 33]. Sick has obtained $r_\mathrm{p} = 0.894(8)$ fm [6] and $r_\mathrm{p} = 0.886(8)$ fm [7] by supplementing the scattering data with a calculation of the shape of the large radius proton charge distribution. The analysis leading to the latter value for r_p includes data from Ref. [4]. An analysis of existing data with an analytic form-factor model has yielded a smaller radius $r_\mathrm{p} = 0.8489(69)$ fm [8].

The consensus of these values is consistent with the 2010 CODATA recommended value for the proton radius. Exceptions are the result of [6] which is slightly larger and the result of [8] which is significantly smaller. The various treatments of the electron scattering data and the difficulties in arriving at a unique result for the proton radius are discussed in Ref. [34].

11.3.3 Proton Radius from Muonic Hydrogen

It is expected that a better value for the proton radius, which would lead to a better value for the Rydberg constant, can be obtained from muonic hydrogen, an atom consisting of a negative muon and a proton. Due to its larger mass, the Bohr radius of the muon is about 207 times smaller than the electron Bohr radius. As a result, the splitting of the 2S and 2P states in muonic hydrogen is quite sensitive to the size of the proton, which contributes about 2 % to the total Lamb shift. (Because of the large electron-vacuum-polarization effect, the $2\mathrm{S}_{1/2}$ level is below both the $2\mathrm{P}_{3/2}$ and $2\mathrm{P}_{1/2}$ levels.)

In measurements carried out at the Paul Scherrer Institute (PSI), Villigen, Switzerland, the $2\mathrm{S}_{1/2}(F = 1) - 2\mathrm{P}_{3/2}(F = 2)$ and $2\mathrm{S}_{1/2}(F = 0) - 2\mathrm{P}_{3/2}(F = 1)$ transitions in muonic hydrogen have been accurately measured [9, 35, 36]. These results, when combined with the theory for the transition, lead to [36]

$$r_\mathrm{p} = 0.840\,87(39)\ \mathrm{fm}\,. \tag{11.7}$$

Because of the discrepancy between this value of the proton radius and the values obtained from spectroscopic data or electron scattering data, the theory for the transition frequency has been reviewed many times. Recent reviews, which also list earlier reviews, are given in references [34, 37–43].

11.3.4 Comparison of the Determinations of the Proton Radius

Values of the proton radius discussed in Sects. 11.3.1–11.3.3 are shown in Fig. 11.2. Based on both spectroscopic data and the electron scattering data, the CODATA recommended value of the Rydberg constant is [1]

$$R_\infty = 10\,973\,731.568\,539(55)\ \mathrm{m}^{-1} \tag{11.8}$$

which has a relative uncertainty of 5.0×10^{-12}, and the recommended value of the proton radius is [1]

$$r_\mathrm{p} = 0.8775(51)\ \mathrm{fm}\,. \tag{11.9}$$

If the proton radius from muonic hydrogen in Eq. (11.7) is compared to the CODATA 2010 recommended value in Eq. (11.9), there is a $7\,\sigma$ disagreement. If it is compared to the value in Eq. (11.3), based on only H and D spectroscopic data, the disagreement is $4.5\,\sigma$.

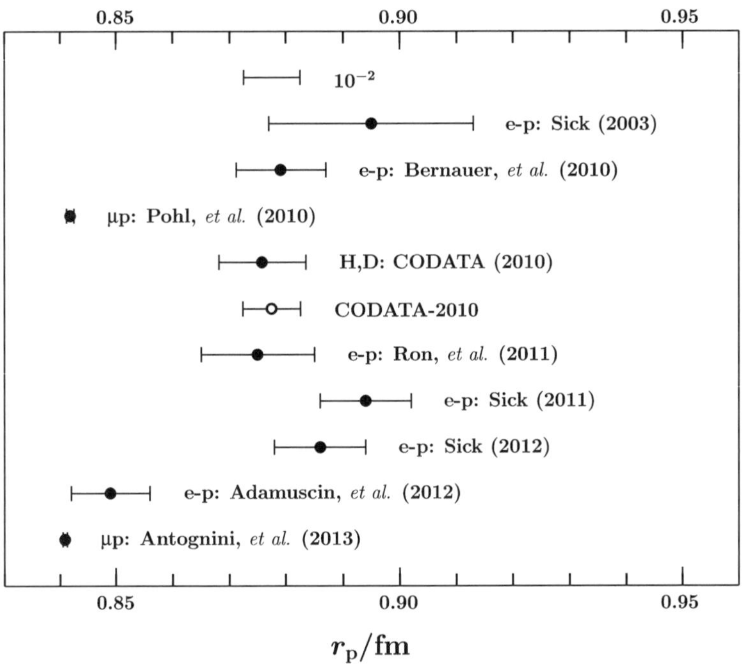

Fig. 11.2 Values of the proton radius r_p given in Sects. 11.3.1–11.3.3. The entry at the top sets the scale for the graph. The values, from top to bottom, are from Refs. [1, 4–10, 36]. *Error bars* represent 1 standard uncertainty

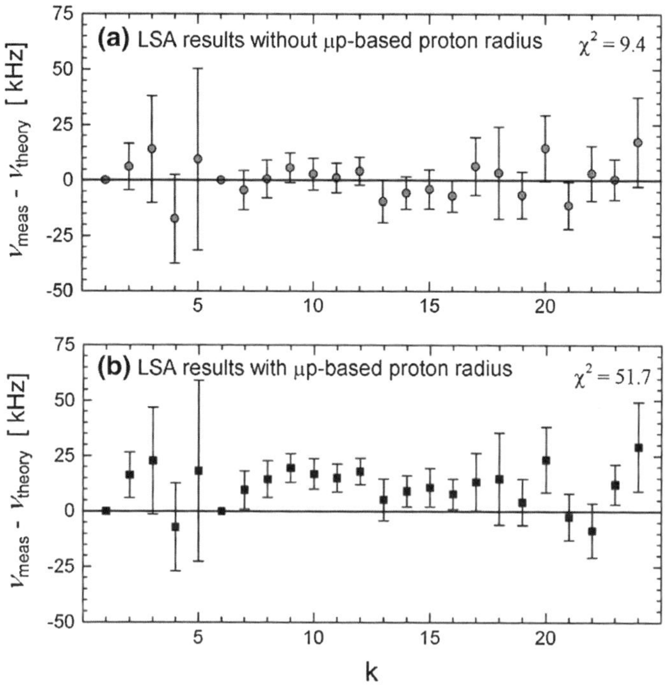

Fig. 11.3 Transition frequencies in hydrogen and deuterium. Deviations from theory are large when the proton radius deduced from the muonic hydrogen Lamb-shift is included among the input data. *Error bars* represent 1 standard uncertainty

Alternatively, if the proton radius derived from muonic hydrogen is included among the input data for a least-squares adjustment of the spectroscopic data, the result is an apparent inconsistency between the fitted values of the transition frequencies and the experimental values used as the input data, as shown in Fig. 11.3 [1]. An additional problem with including the muonic hydrogen proton radius in a least-squares analysis of the spectroscopic data is that it leads to a value of the fine-structure constant given by $\alpha^{-1} = 137.035\,881(35)$ $[2.6 \times 10^{-7}]$, which differs from the 2010 recommended value by $3.4\,\sigma$ [1]. The value of R_∞ from such an adjustment is [1]

$$R_\infty = 10\,973\,731.568\,175(12)\ \text{m}^{-1}\,. \tag{11.10}$$

This value of the Rydberg constant differs from the 2010 recommended value by $6.5\,\sigma$.

These discrepancies constitute the proton radius puzzle. At this time, there is no explanation for the disagreements between the muonic hydrogen Lamb shift data and both the H and D spectroscopic data and the electron-proton scattering data. It may be necessary to modify QED to properly account for the muon data, or there still may be a contribution from conventional QED that has not been taken into account properly.

11.3.5 Comparison of the Determinations of the Rydberg Constant

As shown in the previous sections, there is disagreement between the values of the Rydberg constant determined from the H and D spectroscopic data with and without the inclusion of the proton radius data from muonic hydrogen. The reason for this is that S states play an important role in the determination of the Rydberg constant in the least-squares adjustments, and these states are relatively strongly affected by the size of the nucleus.

On the other hand, there is the possibility of using Rydberg states to determine the Rydberg constant. Such a determination would be virtually independent of the proton or nuclear radius, because high-ℓ states have a negligible overlap with the nucleus. Thus such an independent value for the Rydberg constant might favour one of the values associated with either including or not including the muonic hydrogen proton radius in the analysis, thereby shedding light on the proton radius puzzle.

In this regard, it is of interest to note that there is an existing unpublished measurement of the Rydberg constant made in Rydberg states [44]

$$R_\infty = 10\,973\,731.56834(23)\ \mathrm{m}^{-1} \tag{11.11}$$

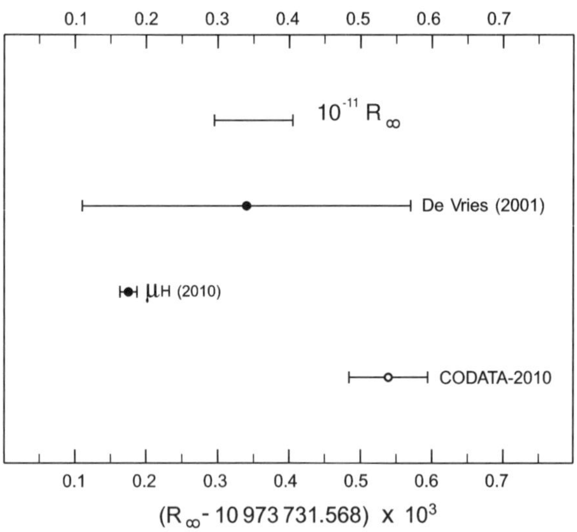

Fig. 11.4 Values of the Rydberg constant based on microwave transitions in circular states of hydrogen, the value obtained by including the muonic hydrogen proton radius in the least-squares adjustment of H and D data, and the 2010 recommended value. First entry sets the scale of the graph. *Error bars* represent 1 standard uncertainty

Values for the Rydberg constant in Eqs. (11.8), (11.10), and (11.11) are shown in Fig. 11.4. It is apparent from that figure that an independent determination of the Rydberg constant from Rydberg states with even a modest accuracy might distinguish between the values involving S states with or without the muonic hydrogen data included.

11.4 Optical Transitions Between Rydberg States

Optical frequency combs [45] can be used to make precise measurements of optical transitions between Rydberg states. For example, laser spectroscopy of antiprotonic helium together with theoretical calculations of the atomic structure have been used to weigh the antiproton [46]. On the other hand, if *CPT* symmetry is assumed, the experiments and theory can be interpreted as a determination of the electron mass.

For one-electron ions in Rydberg states, the optical frequency combs have the potential of making precise absolute frequency measurements which can lead to a determination of the Rydberg constant. Although transitions between Rydberg states in neutral atoms are typically in the microwave frequency range, in highly-ionized ions, the frequencies can be in the optical range, as shown in Fig. 11.5. In that figure, the frequencies corresponding to transitions between adjacent Bohr energy levels (n to $n-1$) in hydrogen-like ions are shown as a function of the principle quantum number of the upper level n and the charge number of the nucleus Z. The coloured bands indicate the (optical) colours corresponding to the frequencies in that figure. The pink band is approximately the region of infra-red frequencies.

Much of this parameter space is accessible to optical frequency synthesizers based on mode-locked femtosecond lasers, which readily provide ultra-precise reference rulers (optical frequency combs) spanning the near-infrared and visible region of the optical spectrum (530–2100 nm). Even when the absolute accuracy is limited by the primary frequency standard (a few parts in 10^{16}), optical frequency combs can enable relative frequency measurements with uncertainties approaching 1 part in 10^{19} over

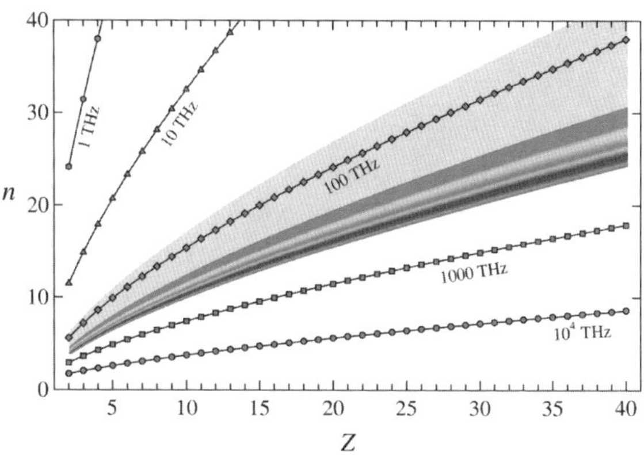

Fig. 11.5 Iso-frequency plots showing values of Z and approximate n that give a specified value of the frequency for transitions changing the principal quantum number n by 1 in a one-electron ion with nuclear charge Z. *Colour bands* represent frequencies in the near infrared and visible region

100 THz of bandwidth [47]—making comb-based measurements generally more precise than other existing technologies. The precisely controlled pulse train from a femtosecond laser can also be used directly to probe the global atomic structure, thus integrating the optical, Tera-Hertz, and radio-frequency domains [48].

This is an advantage, because heavier hydrogen-like ions with a higher nuclear charge Z have larger fine-structure separations, which scale as Z^4. As a consequence, the Rydberg states are far more robust against perturbations such as Stark mixing than Rydberg states in the microwave regime. In addition, the charged ions repel each other, so the effect of short range collisions can be expected to be less important.

11.5 Theory of Rydberg States

In this section, we consider the theory of one-electron ions in Rydberg states, focusing on aspects unique to the high-ℓ case. In particular, simplifications in the theory reduce the uncertainties and allow accurate predictions of the energy levels to be made. Radiative transitions are also discussed.

11.5.1 Simplification of Nuclear Size and Higher-Order QED Effects

The largest source of uncertainty in determining the Rydberg constant from spectroscopy of H and D is the uncertainty in the nuclear radius, which translates into uncertainty in the theoretical values of the transition frequencies. For states with high orbital angular momenta ℓ, this uncertainty is relatively smaller than in lower-ℓ states, because in the high-ℓ case, the probability $P(r)$ for the electron to be found within a short distance r of the nucleus is very low. As a consequence, the nuclear-size correction to the energy level is small, and knowledge of the nuclear size is unimportant. For a Rydberg state of a hydrogen-like ion with charge number Z, high principal quantum number n, and angular momentum $\ell = n - 1$, this probability is

$$P(r) = \int_{|\boldsymbol{x}|<r} d\boldsymbol{x} |\psi(\boldsymbol{x})|^2 \approx \frac{1}{(2n+1)!} \left(\frac{2Zr}{na_0} \right)^{2n+1}, \tag{11.12}$$

where a_0 is the Bohr radius. If r is taken to be the nuclear radius, then the high-power factor, together with the factorial in the denominator, leads to an almost complete suppression of nuclear effects for circular or near-circular Rydberg states. For example, if $n = 5$ and $Z = 5$, then $P(1\ \text{fm}) \approx 10^{-56}$.

Another advantage of high-ℓ states for theory is the fact that the higher-order corrections in the QED theory of the levels are relatively smaller for these states compared to low-ℓ states. Thus, the perturbation expansion of the theoretical expressions for the energy levels, as a function of $Z\alpha$, provides relatively more accurate results with a given number of terms.

Since these two portions of the theory have the largest associated uncertainties, their elimination significantly reduces the relative uncertainty for the theory of the transition frequencies to the extent that in certain cases the Rydberg constant is the largest source of uncertainty. Then, a comparison of theory and experiment for the transitions can provide information on the Rydberg constant and potentially improve its accuracy.

11.5.2 High-ℓ-State Energy Levels in Hydrogen-Like Atoms

In this section, we list the known theoretical expressions for the energy levels of hydrogen-like ions and give numerical results with estimates of the uncertainties for special cases. Reviews of the theory and references to original work are given in [1, 49, 50]. Here the theory is given only for $\ell \geq 2$.

The energy levels can be written as

$$E_n = E_{\mathrm{DM}} + E_{\mathrm{RR}} + E_{\mathrm{QED}}\,, \tag{11.13}$$

a sum of the Dirac energy with nuclear motion corrections E_{DM}, relativistic recoil corrections E_{RR}, and radiative corrections E_{QED}.

For the first term, the difference between the Dirac eigenvalue E_{D} and the electron rest energy, is proportional to

$$\alpha^2 D\, m_{\mathrm{e}}c^2 = E_{\mathrm{D}} - m_{\mathrm{e}}c^2 = \left(\left[1 + \frac{(Z\alpha)^2}{(n-\delta)^2}\right]^{-1/2} - 1\right) m_{\mathrm{e}}c^2\,, \tag{11.14}$$

$$\alpha^2 D = -\frac{(Z\alpha)^2}{2n^2} + \left(\frac{3}{8n} - \frac{1}{2j+1}\right)\frac{(Z\alpha)^4}{n^3} + \dots\,, \tag{11.15}$$

where j is the total angular momentum quantum number, $\delta = |\kappa| - \sqrt{\kappa^2 - (Z\alpha)^2}$, and $\kappa = (-1)^{\ell+j+1/2}(j+1/2)$ is the Dirac spin-angular quantum number. The expansion in Eq. (11.15) shows the leading Schrödinger and fine-structure terms. The energy level, taking into account the leading nuclear motion effects, is given by [50]

$$E_{\mathrm{DM}} = 2hcR_\infty\left[\mu_{\mathrm{r}} D - \frac{r_{\mathrm{N}}\mu_{\mathrm{r}}^3\alpha^2}{2}D^2 + \frac{r_{\mathrm{N}}^2\mu_{\mathrm{r}}^3 Z^4\alpha^2}{2n^3\kappa(2\ell+1)}\right], \tag{11.16}$$

where $r_{\mathrm{N}} = m_{\mathrm{e}}/m_{\mathrm{N}}$ is the ratio of the electron mass to the nucleus mass, and $\mu_{\mathrm{r}} = 1/(1+r_{\mathrm{N}})$ is the ratio of the reduced mass to the electron mass.

Relativistic corrections to Eq. (11.16), associated with motion of the nucleus are

$$E_{\mathrm{RR}} = 2hcR_\infty \frac{r_{\mathrm{N}} Z^5 \alpha^3}{\pi n^3} \left\{ \mu_{\mathrm{r}}^3 \left[-\frac{8}{3} \ln k_0(n,\ell) - \frac{7}{3l(\ell+1)(2\ell+1)} \right] + \pi Z\alpha \left[3 - \frac{\ell(\ell+1)}{n^2} \right] \frac{2}{(4\ell^2-1)(2\ell+3)} + \dots \right\}, \tag{11.17}$$

where $\ln k_0(n,\ell)$ is the Bethe logarithm. We assume that the uncertainty due to uncalculated higher-order terms is $Z\alpha \ln (Z\alpha)^{-2}$ times the last term in Eq. (11.17).

Quantum electrodynamics (QED) corrections for high-ℓ states are

$$E_{\mathrm{QED}} = 2hcR_\infty \frac{Z^4 \alpha^2}{n^3} \left\{ -\mu_{\mathrm{r}}^2 \frac{a_{\mathrm{e}}}{\kappa(2\ell+1)} + \mu_{\mathrm{r}}^3 \frac{\alpha}{\pi} \left[-\frac{4}{3} \ln k_0(n,\ell) + \frac{32}{3} \frac{3n^2 - \ell(\ell+1)}{n^2} \frac{(2\ell-2)!}{(2\ell+3)!} (Z\alpha)^2 \ln \left[\frac{1}{\mu_{\mathrm{r}}(Z\alpha)^2} \right] + (Z\alpha)^2 G(Z\alpha) \right] \right\}, \tag{11.18}$$

where a_{e} is the electron magnetic moment anomaly and $G(Z\alpha)$ is a function that contains higher-order QED corrections. Equation (11.18) contains no explicit vacuum polarization contribution because of the damping of the wave function near the origin where the polarization effect is largest. The quantity a_{e} replaces the order-by-order terms in the theory that are equivalent to the free-electron magnetic moment anomaly. Instead, the experimental value is used for the sum of all such terms in order to eliminate uncertainty associated with the calculated values and higher-order omitted terms. We employ the value $a_{\mathrm{e}} = 1.159\,652\,180\,73(28) \times 10^{-3}$ obtained in a recent experiment [51].

The dominant terms in $G(Z\alpha)$ are expected to be of the form

$$G(Z\alpha) = A_{60} + A_{81}(Z\alpha)^2 \ln (Z\alpha)^{-2} + A_{80}(Z\alpha)^2 + \dots + \frac{\alpha}{\pi} B_{60} + \dots + \left(\frac{\alpha}{\pi}\right)^2 C_{60} + \dots, \tag{11.19}$$

where the letters A, B, and C indicate contributions arising from one-, two-, and three-photon Feynman diagrams, respectively. The function $G(Z\alpha) = G_{\mathrm{SE}}(Z\alpha) + G_{\mathrm{VP}}(Z\alpha)$ is the sum of contributions from self-energy (SE) diagrams and vacuum-polarization (VP) diagrams. The coefficients A_{60} and A_{81} arise from the self energy, and A_{80} arises from both the self energy and the long-range component of the vacuum polarization.

The first two terms in Eq. (11.19) come from the one-loop self energy

$$G_{\mathrm{SE}}(Z\alpha) = A_{60} + A_{81}(Z\alpha)^2 \ln (Z\alpha)^{-2} + \dots, \tag{11.20}$$

and the leading contribution is given by the term A_{60}. Calculated values for this coefficient are given in Table 11.2 [19, 52]. Results of a complete calculation of the one-loop self energy, given in Table 11.3, are consistent with the calculation of

Table 11.2 Calculated values for the coefficient A_{60} from [19, 52]. The numbers in parentheses are standard uncertainties in the last figure

n	$\ell = n-2,\ j = \ell - \frac{1}{2}$	$\ell = n-1,\ j = \ell - \frac{1}{2}$
9	$7.018\,373(5) \times 10^{-5}$	$3.860\,349(5) \times 10^{-5}$
10	$3.655\,111(5) \times 10^{-5}$	$2.158\,923(5) \times 10^{-5}$
11	$2.008\,438(5) \times 10^{-5}$	$1.259\,580(5) \times 10^{-5}$
12	$1.019\,187(5) \times 10^{-5}$	$0.759\,620(5) \times 10^{-5}$
13	$0.679\,575(5) \times 10^{-5}$	$0.469\,973(5) \times 10^{-5}$
14	$0.410\,825(5) \times 10^{-5}$	$0.296\,641(5) \times 10^{-5}$
15	$0.252\,108(5) \times 10^{-5}$	$0.189\,309(5) \times 10^{-5}$
16	$0.155\,786(5) \times 10^{-5}$	$0.121\,749(5) \times 10^{-5}$
n	$\ell = n-2,\ j = \ell + \frac{1}{2}$	$\ell = n-1,\ j = \ell + \frac{1}{2}$
9	$28.939\,225(5) \times 10^{-5}$	$14.918\,400(5) \times 10^{-5}$
10	$16.589\,245(5) \times 10^{-5}$	$9.141\,150(5) \times 10^{-5}$
11	$10.111\,871(5) \times 10^{-5}$	$5.882\,197(5) \times 10^{-5}$
12	$6.331\,080(5) \times 10^{-5}$	$3.940\,256(5) \times 10^{-5}$
13	$4.318\,998(5) \times 10^{-5}$	$2.729\,475(5) \times 10^{-5}$
14	$2.979\,937(5) \times 10^{-5}$	$1.945\,279(5) \times 10^{-5}$
15	$2.116\,050(5) \times 10^{-5}$	$1.420\,631(5) \times 10^{-5}$
16	$1.540\,181(5) \times 10^{-5}$	$1.059\,674(5) \times 10^{-5}$

Table 11.3 Calculated values for the coefficient $G_{\mathrm{SE}}(Z\alpha)$ with $Z = 14,\ 16$ from [53]. The numbers in parentheses are standard uncertainties in the last figure

n	Z	$\ell = n-2,\ j = \ell - \frac{1}{2}$	$\ell = n-1,\ j = \ell - \frac{1}{2}$
13	14	$0.676(9) \times 10^{-5}$	$0.469(9) \times 10^{-5}$
13	16	$0.682(4) \times 10^{-5}$	$0.468(7) \times 10^{-5}$
14	14	$0.403(8) \times 10^{-5}$	$0.296(9) \times 10^{-5}$
14	16	$0.408(6) \times 10^{-5}$	$0.296(9) \times 10^{-5}$
15	14	$0.243(9) \times 10^{-5}$	$0.184(7) \times 10^{-5}$
15	16	$0.249(3) \times 10^{-5}$	$0.191(9) \times 10^{-5}$
n	Z	$\ell = n-2,\ j = \ell + \frac{1}{2}$	$\ell = n-1,\ j = \ell + \frac{1}{2}$
13	14	$4.317(5) \times 10^{-5}$	$2.728(9) \times 10^{-5}$
13	16	$4.321(2) \times 10^{-5}$	$2.728(5) \times 10^{-5}$
14	14	$2.974(5) \times 10^{-5}$	$1.944(9) \times 10^{-5}$
14	16	$2.978(3) \times 10^{-5}$	$1.945(9) \times 10^{-5}$
15	14	$2.107(9) \times 10^{-5}$	$1.415(7) \times 10^{-5}$
15	16	$2.114(2) \times 10^{-5}$	$1.423(9) \times 10^{-5}$

the leading term, which provides an independent check against substantial errors in either calculation [53]. The good agreement also indicates that the higher-order terms are not significantly larger than expected. The uncertainty in $G_{\mathrm{SE}}(Z\alpha)$ is assumed to be given by $A_{60}(Z\alpha)^2 \ln{(Z\alpha)^{-2}}$, where the coefficient A_{81} in the first omitted term is replaced by A_{60} for the estimate.

The vacuum-polarization contribution to A_{80} is extremely small [54]. The two-photon coefficient B_{60} has not been calculated for high-ℓ states. However, a comparison of calculated values of B_{60} [55] and A_{60} [56] for $\ell \leq 5$, suggests it is of the order of $4|A_{60}|$, which is taken to be its uncertainty. The three-photon term C_{60} is expected to be the next term, and is assumed to be negligible.

Table 11.4 Transition frequencies between the highest-j states with $n = 14$ and $n = 15$ in hydrogen-like helium and hydrogen-like neon

Term	$^4He^+ \nu$(THz)	$^{20}Ne^{9+} \nu$(THz)
E_{DM}	8.652 370 766 016(43)	216.335 625 5749(11)
E_{RR}	0.000 000 000 000	0.000 000 000 1
E_{QED}	−0.000 000 001 894	− 0.000 001 184 1
Total	8.652 370 764 122(43)	216.335 624 3909(11)

Table 11.5 Sources and estimated relative standard uncertainties in the theoretical value of the transition frequency between the highest-j states with $n = 14$ and $n = 15$ in hydrogen-like helium and hydrogen-like neon

Source	He^+	Ne^{9+}
Rydberg constant	5.0×10^{-12}	5.0×10^{-12}
Fine-structure constant	3.3×10^{-16}	8.2×10^{-15}
Electron-nucleus mass ratio	5.5×10^{-14}	1.1×10^{-14}
a_e	5.1×10^{-20}	1.3×10^{-18}
Theory: E_{RR} higher order	6.2×10^{-17}	2.4×10^{-14}
Theory: E_{QED} A_{81}	1.7×10^{-18}	1.6×10^{-14}
Theory: E_{QED} B_{60}	8.6×10^{-18}	5.4×10^{-15}

11.5.3 Transition Frequencies and Uncertainties

To put the theory presented above into perspective, we list the numerical predictions for transition frequencies for two different ions, and indicate the various sources of uncertainties. We consider the frequency of the transition between the state with $n = 14$, $\ell = 13$, $j = \frac{27}{2}$ and the state with $n = 15$, $\ell = 14$, $j = \frac{29}{2}$ in the hydrogen-like ions He^+ and Ne^{9+}. The constants used in the evaluation are the 2010 CODATA recommended values [1], with the exception of the neon nucleus mass $m(^{20}Ne^{10+})$ which is taken from the neon atomic mass [57], corrected for the mass of the electrons and their binding energies. Values of the various contributions and the total are given as frequencies in Table 11.4. Standard uncertainties are listed with the numbers where they are non-negligible. The largest uncertainty arises from the Rydberg frequency cR_∞, which is a common factor in all of the contributions. There is no uncertainty from the Planck constant since $\nu = (E_{15} - E_{14})/h$, and h drops out when calculating the frequency.

Table 11.5 gives sources and estimates of the various known uncertainties in the theory. By way of comparison, in hydrogen, the relative uncertainty from the two-photon term B_{60} for the 1S–2S transition is of the order of 10^{-12} whereas in the $n = 14$ to $n = 15$ Rydberg transitions the QED relative uncertainties are orders of magnitude smaller, as indicated in that table. The convergence of the expansion of the QED corrections in powers of $Z\alpha$ is significantly faster for Rydberg transitions, because of the smallness of the higher-order terms for the high-ℓ states compared to S states.

11.5.4 Natural Line Widths

The advantages of using optical transitions between Rydberg states come with experimental trade-offs associated with a large spontaneous emission rate. Natural decay linewidths tend to be small for states from which electric dipole (E1) decay is forbidden (as in the case of the 2S level). In contrast, the spontaneous decay rate for a circular Rydberg state is dominated by an electric dipole E1 transition from the highest-ℓ value of the state n to the highest-ℓ value of the state $n-1$.

Formally, the QED level shift given by Eq. (11.18) is the real part of the radiative correction. The complete radiative correction to the level can be written as $\mathcal{E}_{\mathrm{QED}} = E_{\mathrm{QED}} - \mathrm{i}\,\Gamma/2$ which is complex with an imaginary part proportional to the decay rate $A = \Gamma/\hbar$ of the level. In the non-relativistic framework, for a state with principal quantum number n and angular momentum $\ell = n-1$, the dominant decay mode is an E1 transition to state with $n' = n-1$ and $\ell' = n'-1$ [58]. Reference [58] gives the corresponding expression for the decay rate, which is the non-relativistic limit of the imaginary part of the level shift:

$$\Gamma_n \to hcR_\infty Z^4 \alpha^3 \frac{4n^{2n-4}(n-1)^{2n-2}}{3\left(n-\frac{1}{2}\right)^{4n-1}}. \tag{11.21}$$

The imaginary part of the level shift gives the natural width of the level, and when this is taken into account, the resonant frequency of the transition between states with quantum numbers n and $n-1$, as a function of the frequency of the applied laser field, has a full width at half maximum given by $\Gamma_n + \Gamma_{n-1}$. A measure of the quality of the line for making precision measurements is the ratio of the transition energy to the resonance width, given by

$$Q = \frac{E_n - E_{n-1}}{\Gamma_n + \Gamma_{n-1}} \to \frac{3n^2}{4\alpha(Z\alpha)^2} + \dots, \tag{11.22}$$

where the expression on the right is the asymptotic form as $n \to \infty$ of the non-relativistic value. Figure 11.6 gives a contour plot of the values of n and Z that give a specified value of Q based on Eq. (11.21).

In addition to a width of the resonance, there can be an asymmetry in the line shape that affects the relationship between the measured frequency distribution and the level splitting. Such effects have been shown to be small by Low [59], of order $\alpha(Z\alpha)^2\,E_{\mathrm{QED}}$. For example, for the 1S–2S transition in hydrogen, they are completely negligible at the current level of experimental accuracy [60]. However, for Rydberg states of hydrogen-like ions, particularly at higher-Z, asymmetries in the line shape could be more significant. Such effects, some of which might depend on the detailed configuration of the experiment, can be calculated if necessary [19].

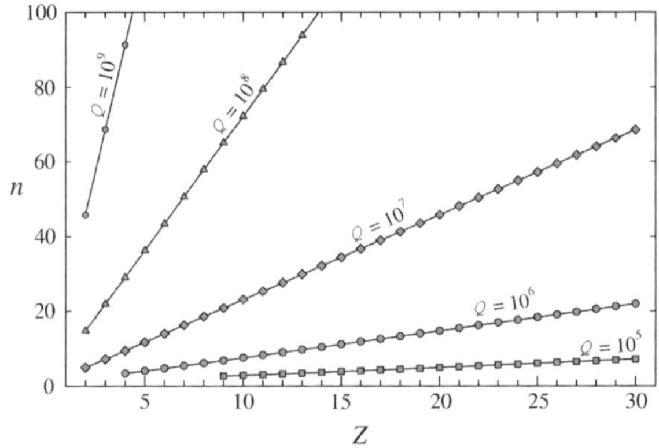

Fig. 11.6 Graph showing combinations of nuclear charge Z and approximate n values that give the same ratio of the transition frequency to the natural linewidth of the transition resonance between circular states of one-electron ions with principal quantum number n and $n - 1$

11.6 Experiment

As discussed in Sect. 11.5, Rydberg states of hydrogen-like ions with $\ell > 2$ essentially avoid a number of problems associated with either higher-order binding corrections to QED interactions or the nuclear size effect in Lamb shift predictions. In the cases considered for planned experiments, the higher-order QED corrections for Rydberg states are smaller by a factor of about 10^7 compared to S states while the nuclear size effect is completely negligible, providing significant advantages in making accurate predictions.

Of particular interest are Rydberg states with maximum angular momentum or *circular states*, so-called because the probability density is significant only in an annular region centred on the nucleus. In addition to aforementioned advantages, other useful features of circular states include: (1) the longest lifetime in a given shell n, and (2) suppressed Stark effect. In fact, circular Rydberg states of hydrogen in the microwave regime have been used in very precise measurements of transition frequencies [44]; as a result, a determination of the Rydberg constant has been made with a 2.1×10^{-11} relative uncertainty [44]. In Fig. 11.4 we compare this result based upon circular-state transitions with more recent determinations of the Rydberg constant. Unfortunately, the precision of the hydrogen circular-state measurements is not sufficient to help resolve the large discrepancy between the μp measurement and the hydrogen measurements (CODATA). In retrospect, the large dipole moments in circular Rydberg states of hydrogen would significantly increase the measurement uncertainty due to the possible perturbations from the dipole-dipole interaction between Rydberg atoms [61]. In contrast, such dipole-dipole interaction would be orders of magnitude smaller between one-electron Rydberg ions since the wavefunction is concentrated more tightly around a highly charged nucleus.

Experiments with cold, one-electron Rydberg ions may be possible for a wide range of nuclear charge Z and angular momentum ℓ. A considerable array of tools and techniques have emerged in the last two decades which enable 'engineered atoms' to be formed in traps and tailored in specific states of experimental interest. Cooling techniques [62] routinely used with singly- or doubly-ionized atoms, for instance, can be exploited to cool highly charged ions. Man-made quantum systems—for example, antihydrogen [63] or a single electron in a Penning trap [15]—have the potential to extend the range of precision measurements that probe nature. In this section, we describe an experimental effort at NIST to produce one-electron ions in Rydberg states by isolating bare nuclei in a trap for recombination via electron capture from laser-excited alkali atoms.

11.6.1 Source of Fully-Stripped Ions

Various methods have been used to produce Rydberg states [20]. Generally, one (or a combination) of the following mechanisms is used: (1) electron impact excitation, (2) photo-excitation, or (3) charge exchange. High angular-momentum states present some challenges when dealing with highly charged ions. Electron impact excitation from the ground state tends to produce low-ℓ Rydberg states since it changes angular momentum by only one unit. Even for low-Z one-electron ions, lasers with the proper frequency are not readily available for photo-excitation from the ground state. On the other hand, recombination by electron capture from a highly-excited atom produces Rydberg states with high ℓ; it is noteworthy that, following charge exchange, the recombined Rydberg ion tends to evolve via radiative cascade towards a circular state (maximum angular momentum) [66].

To use charge exchange for producing hydrogen-like ions, fully-stripped ions (bare nuclei) must be isolated in a trap. Fully-stripped ions can be obtained from an electron beam ion trap (EBIT). An EBIT has been in operation at NIST since 1993 [67]. The EBIT at NIST has an existing beamline for ion extraction [65]. Figure 11.7 provides a simplified schematic diagram of the experimental set-up. As discussed in the next section, we recently demonstrated that a compact Penning trap [68] can be used for capturing ions extracted from the EBIT at NIST.

Highly charged ions (HCI) are produced in an EBIT by repeated electron impact ionization of atoms injected as a neutral gas target via a nozzle (alternatively, low charge-state metal ions can be injected from a metal-vapor vacuum arc [MeVVA] source [69]). In an EBIT, a nearly-monoenergetic electron beam is accelerated through high voltage U_e to reach high energies $e\,U_e$ as it travels through the axis of a stack of three electrodes called drift tubes; for the EBIT at NIST, an electron beam current as high as 150 mA can be used, accelerated to energies up to 30 keV. Coaxial with the drift tubes is a strong magnetic field ($\approx$3 T) which compresses the radial extent of the electron beam, yielding very high current density. Confinement of positively-charged ions is similar to that in a Penning trap, with the drift tubes biased to form an electrostatic potential well along the trap symmetry axis; however, radial confinement of ions is dominated by attraction to the tightly-compressed, axial electron beam. Confined ions collide with the intense electron beam repeatedly and

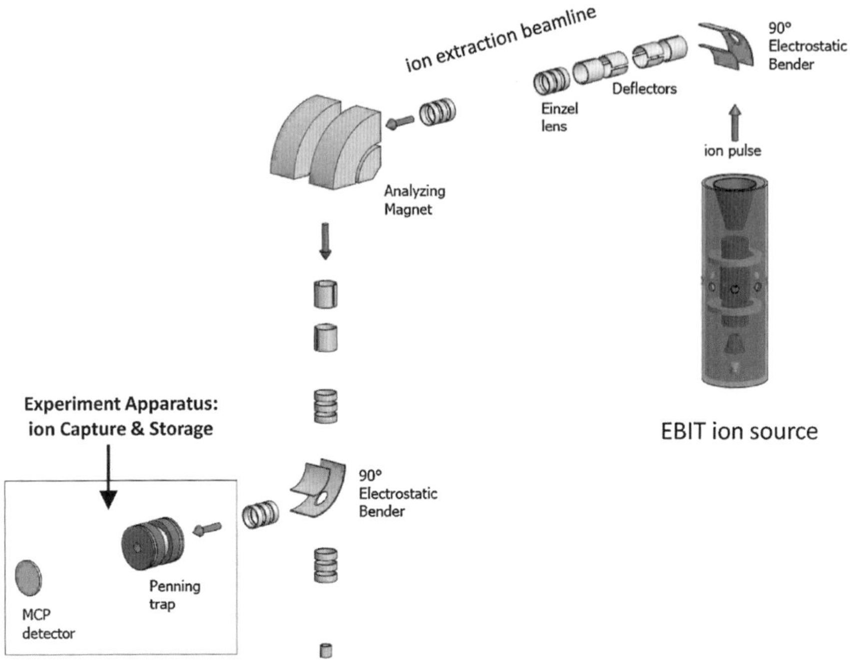

Fig. 11.7 Simplified schematic of the experimental set-up (not to scale). Highly stripped ions are produced in an EBIT (*right*). Extracted ion pulses are transported to an analysing magnet in the beamline. Detailed descriptions of the extraction beamline have been published in [64, 65]. Ions of a selected charge state are deflected by the analysing magnet into the vertical beamline, where a 90° bender deflects them into the experimental apparatus (*left*)

are ionized to successively higher charge states limited by the kinetic energy of the electron beam.

A mixture of several charge states is produced in an EBIT, bounded by the ionizing energy of the electron beam, with a relative distribution determined by various tuning parameters. For the purpose of capturing and isolating the ions in an experimental apparatus, the HCI mixture is extracted in bunches by quickly ramping the EBIT middle drift tube up in voltage. The extracted ion pulse accelerates out of the drift tube region, with each ion of charge number Q acquiring substantial kinetic energy ($\approx Q\,\mathrm{e}\,U_\mathrm{e}$) as it departs from the potential 'hill' that accelerates the electron beam. Typical ion pulse energies for the experiments of interest here are between 2,000 Q eV and 4,000 Q eV. The process of slowing and capturing the extracted ions in a compact Penning trap is discussed in detail in a forthcoming publication, and is briefly described below in Sect. 11.6.2.

During ion extraction, an ion pulse from the EBIT is injected vertically into the extraction beamline (described elsewhere [65]). As shown in Fig. 11.7, a 90° electrostatic bender turns the ion pulse into the horizontal beamline, which transports the ion bunch to an analysing magnet located ≈3.8 m from the EBIT. The analysing magnet allows selection of one specific ion charge state by adjusting its magnetic

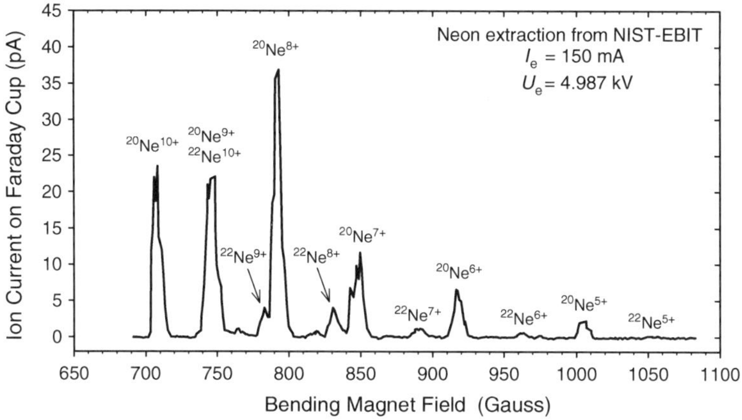

Fig. 11.8 Charge states of neon resolved by the analysing magnet, detected on a Faraday cup near the *bottom* of the vertical beamline. The Faraday cup signal is plotted as a function of the magnetic field generated by the analysing magnet. The leftmost peak corresponds to fully-stripped $^{20}Ne^{10+}$ (bare nuclei)

field to the proper value necessary for deflecting an ion of a chosen mass/charge ratio into the vertical section of the beamline. Figure 11.8 illustrates the selection of a specific charge state by controlling the magnetic field of the analysing magnet. Downstream in the vertical beamline, another 90-degree electrostatic bender is used to divert the ion pulse to the experimental apparatus as illustrated in Fig. 11.9, where ions can be detected on a Faraday cup or a microchannel plate. The total path length from the EBIT to the compact Penning trap in the experimental apparatus is $\approx$7.2 m.

11.6.2 Capturing Bare Nuclei in a Compact Penning Trap

We have recently demonstrated that a compact Penning trap can be useful for capturing ions extracted from an EBIT [68]. The experimental apparatus shown in Fig. 11.9 was designed to utilize the remaining space ($\approx 1\,m^3$) at the end of the ion extraction beamline. It incorporates a compact Penning trap for capturing ion pulses extracted from the EBIT. A Penning trap can be made very compact by embedding annular rare-earth (NdFeB) magnets within its electrode structure. The electrode structure uses electrical-iron to close the magnetic circuit. The unitary architecture of this two-magnet, compact Penning trap is represented in Fig. 11.10, with a three-dimensional rendering shown in the inset of Fig. 11.9; a more detailed description has been presented in Ref. [68]. The endcap electrodes have holes to allow passage of ions along the trap axis. Not shown in Fig. 11.9 are the vacuum pumps (non-evaporable getter, turbo-molecular and ion pumps) which evacuate the room-temperature chamber to a residual gas pressure of $\approx 1.2 \times 10^{-7}$ Pascal (or 9×10^{-10} Torr).

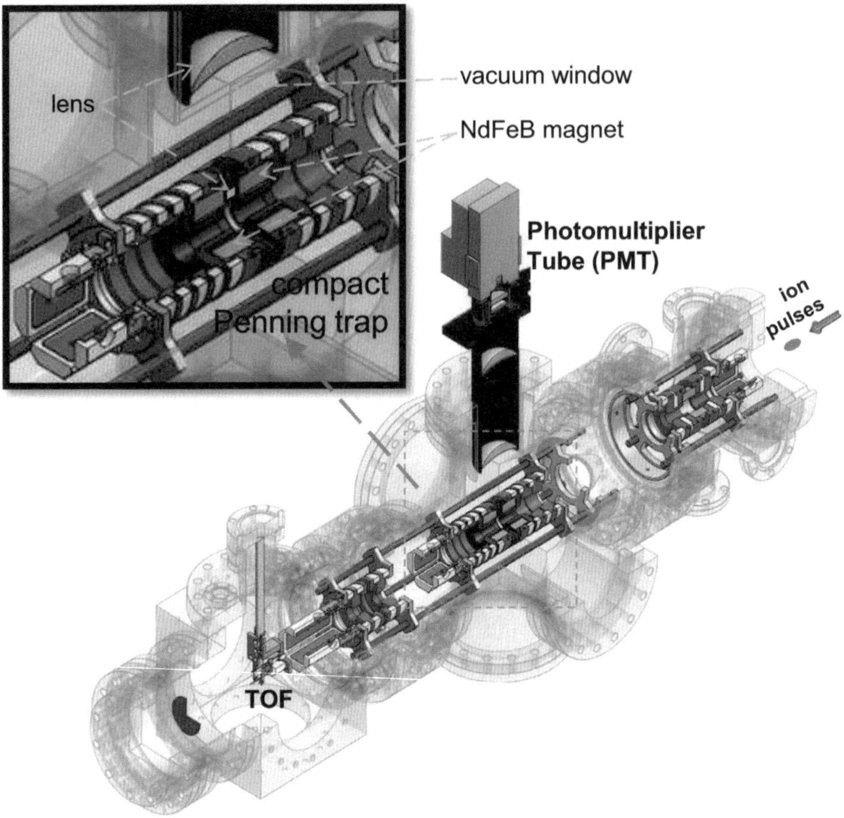

Fig. 11.9 Experimental apparatus for capturing ions extracted from an EBIT using a compact Penning trap. Quarter-cut three-dimensional representation of major components with vacuum chamber in lighter shades. *Inset* shows a zoomed view of the unitary Penning trap and the light collection optics. Ions enter the apparatus from the right hand side

An ion beam or pulse entering the apparatus is steered by two orthogonal pairs of deflection plates, and focussed by an Einzel lens into the compact Penning trap. The electrodes in the trapping region have been carefully designed to optimize ion capture. Details on capturing ions with a compact Penning trap are presented in a forthcoming publication [70]. Here we give a brief summary.

The axial confinement well of the compact Penning trap is formed by symmetrically biasing the endcaps to higher electrical potential than on the ring electrode, typically between 10 V and 40 V. In preparation for injecting an ion pulse into the Penning trap, the potential of the 'front' endcap facing the incoming ion pulse is momentarily lowered below the ring electrode potential. The trap is kept open during the transit of the ions pulse from the EBIT to the compact Penning trap. Ions steered and focussed into the Penning trap are captured by switching the front endcap rapidly back to its high potential setting, ideally when the ions are turning around from the 'back' endcap. Simulations aid the design of trap components to attain

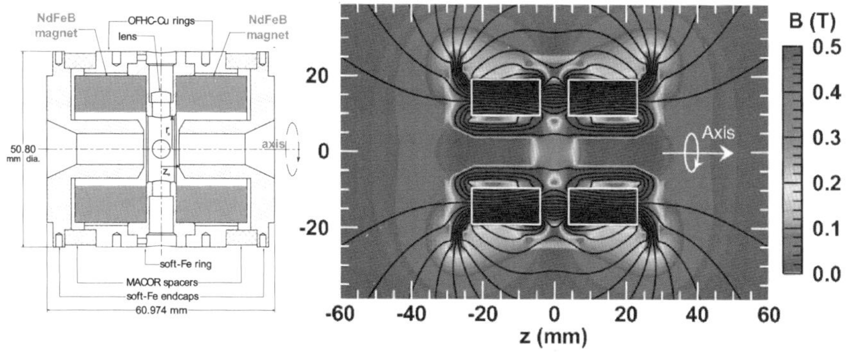

Fig. 11.10 A Penning trap made extremely compact with a unitary architecture. (*Left*) Cross-sectional view showing 2 annular NdFeB magnets embedded within the electrode structure. Re-entrant, soft-iron endcap electrodes are used to close the magnetic circuit to strengthen the field at the centre. (*Right*) Calculated magnetic field generated by the two rare-earth magnets is represented in false colour. More details are provided in [68]

good capture efficiency. Various operating parameters are also important; foremost are: (1) proper timing for closing the trap; and (2) matching the electrical potential of the compact Penning trap to the extraction energy in order to slow the ions as much as possible in the trapping region. Various ion species extracted from an EBIT (including: Ne^{10+}, Ne^{9+}, Ne^{8+}, Ar^{16+}, Ar^{15+}, Ar^{14+}, Ar^{13+}, and Kr^{17+}) have been captured and isolated in the two-magnet Penning trap [71].

The apparatus provides relatively easy access to the stored ions along several directions for various experiments. Occupying less than $150\,\mathrm{cm}^3$ of volume, the compact Penning trap is readily configured with several detectors. In particular, the ring electrode has four equidistant holes to provide midplane line-of-sight access to the stored ions. The trap is centred on a six-way cross with one of the ring electrode holes aligned with the vertical axis of the vacuum cross, and an orthogonal hole aligned with the horizontal window. As illustrated in the inset of Fig. 11.9, the top hole in the ring electrode has an embedded aspheric lens which, together with the lens system mounted outside the top window, collects and focusses light emitted by the trapped ions into a photomultiplier. Very recently, highly charged ions isolated in this unitary Penning trap have been used in experiments to observe forbidden radiative decay [72].

Stored ions can be detected also by ejection to a retractable, time-of-flight (TOF) micro-channel plate (MCP) detector, with $\approx$1 ns response time to resolve different charge states. Alternatively, if the TOF detector is retracted, a position-sensitive MCP detector can be used. Details of the ion detection scheme are in [68]. These capabilities have been useful in fine-tuning the ion beam and in measuring the storage lifetime of captured ions [71, 73].

Fully-stripped neon ions (bare Ne^{10+} nuclei) have been extracted from an EBIT and captured in the unitary Penning trap. An ion cloud in a Penning trap attains dynamical equilibrium by rotation (spin) about the trap axis to generate a $\mathbf{v} \times \mathbf{B}$

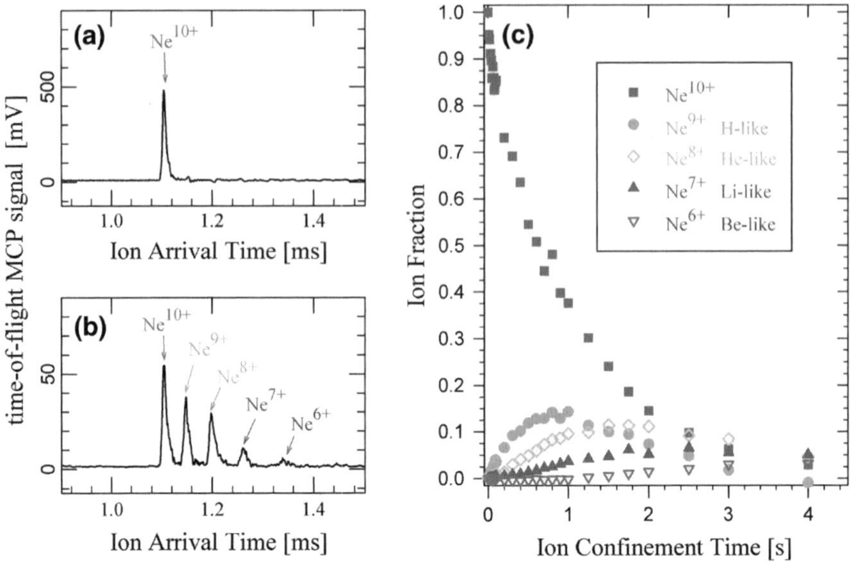

Fig. 11.11 Time-of-flight detection of highly charged ions ejected from the two-magnet Penning trap after various storage times [68, 71], showing charge-state evolution of captured bare nuclei due to electron capture from residual background gas. Neon ions are stored in the Penning trap for times ranging from 1 ms to 4 s, with applied potential difference $\Delta V = 10\,\text{V}$ between the ring and endcap electrodes. In particular, output of the TOF detector is plotted versus arrival time (relative to the ejection trigger) for storage times of (**a**) 1 ms and (**b**) 2 s. The detector signal scale is magnified by a factor of 10 from (**a**) to (**b**). The TOF signal peak for each charge state is converted to ion counts in order to show (**c**) the evolution of charge states due to charge exchange, normalized to the initial population of Ne^{10+} ions observed at the 1 ms storage time. *Error bars* represent 1σ uncertainty

compressive force which counter-balances the Coulomb repulsion (or space-charge repulsion) between ions. For most practical cases, an ion cloud expands radially if the rotation slows down due to some torque acting on the cloud (which may come from trap imperfections or misalignment). Radial expansion of the ion cloud can lead to number loss when some ions strike the wall of the ring electrode. Even in a well-constructed trap wherein cloud expansion is very slow, an ion can change charge state due to electron capture from residual gas atoms.

The charge-state composition of an ion cloud can be analysed from the time-of-flight of ejected ions to the fast MCP detector. The endcap nearest the TOF detector is switched to 400 V below the ring voltage in about 50 ns, ramping out the ions in a pulse. Figure 11.11a shows the TOF signal of bare Ne^{10+} nuclei ejected to the detector after 1 ms of confinement in the two-magnet Penning trap. If there are multiple charge states, several peaks will be observed. Lower charge stages have later arrival times due to their smaller charge-to-mass ratio. Figure 11.11b illustrates the mixture of charge states produced from the charge exchange between a cloud of bare nuclei (initial state) and the background gas atoms, detected after 2 s confinement. The TOF is sufficiently fast to fully resolve the lower charge states produced by

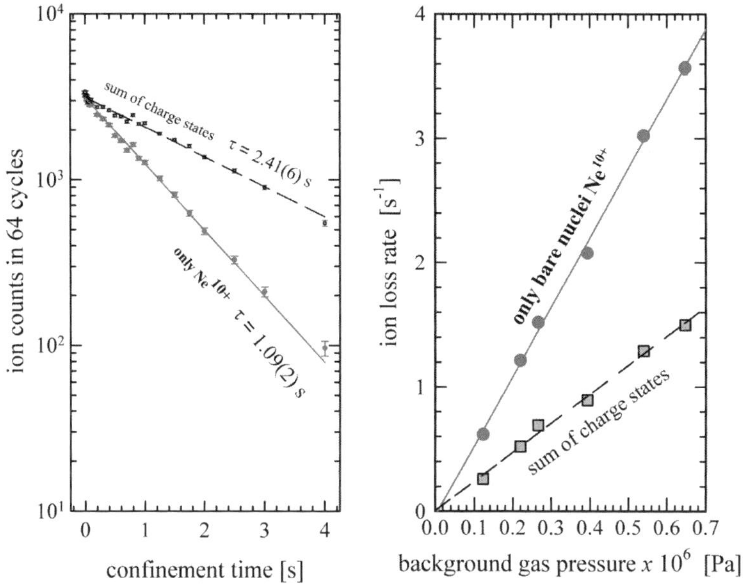

Fig. 11.12 Loss rate of ions stored in the compact Penning trap. (*Left*) Exponential decay of detected ions as a function of confinement time. The vacuum chamber pressure for these measurements is $\approx 1.7 \times 10^{-7}$ Pa (1.3×10^{-9} Torr). (*Right*) The ion loss rate varies linearly with the residual gas pressure in the vacuum chamber. A voltage difference of $\Delta V = 10$ V was applied between the ring and endcap electrodes of the Penning trap. *Error bars* represent 1σ uncertainty. See Ref. [68] for more details

charge exchange. Figure 11.11c shows the charge-state evolution as a function of confinement time.

Collisions with the residual gas atoms or molecules in the vacuum chamber deplete the initial ion species captured in the Penning trap. In Fig. 11.11c, the initial charge state (Ne^{10+} bare nuclei; red squares) decays exponentially with time constant 1.09(2)s, as lower charge states grow; the sum of all observed charge states decays with a longer time constant 2.41(6)s. The variation of these decay rates with the amount of residual gas is illustrated in Fig. 11.12, showing their linear proportionality to the background gas pressure. Loss of stored bare nuclei is largely due to charge exchange. Ion cloud expansion can explain the decay of the charge-state sum [68]. It is noteworthy that the decay rates extrapolate to nearly zero, indicating that the compact Penning trap imperfections do not contribute significantly to the torque driving the ion cloud expansion. If the background gas pressure is reduced, for example by cooling the vacuum envelope with a cryo-cooler, then the lifetime of the charge state of interest can be lengthened (and the ion cloud expansion rate reduced) considerably.

11.6.3 Charge-Exchange Recombination: One-Electron Rydberg Ions

One-electron ions in high-ℓ Rydberg states have attractive features for testing theory. Undoubtedly challenging to realize, such elegantly-simple quantum system can be engineered using an array of tools and techniques that have emerged in atomic-molecular-optical physics within the last two decades. For example, production of cold antihydrogen at the European Laboratory for Particle Physics (CERN) demonstrated two ways for an antiproton to capture a positron in high-ℓ Rydberg states: (1) three-body recombination during positron cooling of antiprotons; or (2) positron capture from positronium in a two-stage charge-exchange mechanism [63]. In addition, Doppler-tuned laser spectroscopy of Rydberg states has been used to study the atomic cores of highly charged ions in crossed-beam experiments [74].

The effort under way at NIST will develop a unitary, compact Penning trap (Fig. 11.10) to facilitate the interaction of stored bare nuclei with an excited atomic beam for the formation of one-electron ions in Rydberg states via charge exchange. In the planned experiment, a beam of alkali atoms enters the compact Penning trap through one of the holes in the ring electrode and they are excited by lasers to Rydberg states. The charge-exchange reaction between a bare nucleus A of charge Z colliding a target atom B prepared in an excited state with a principal quantum number n is represented by the formula

$$A^{+Z} + B(n) \rightarrow A^{+Z-1}(\tilde{n}) + B^{+}\,, \tag{11.23}$$

where $\tilde{n}$ is the principal quantum number of the one-electron ion A^{+Z-1} immediately after recombination. A classical-trajectory Monte Carlo (CTMC) method has been used to estimate the electron capture cross-section, with the low-velocity limit given by [75]

$$\sigma_{cx} \approx 5.5Z\,\pi\,n^4 a_o^2\,, \tag{11.24}$$

which indicates that the large geometrical cross-section for a Rydberg target atom with radius $r_n = n^2 a_o$ is further enhanced by a factor 5.5 Z.

CTMC simulations also showed that the excited electron in the target atom attempts to preserve its original orbital dimension and energy during the electron capture process, leading to an initial post-collisional quantum number distribution given by

$$n\sqrt{Z} \leq \tilde{n} \leq nZ\,, \tag{11.25}$$

with the most probable state being roughly $\tilde{n} \approx [n\,Z + n\,Z^{1/2}]/2$ [75]. Results of crossed-beam experiments which studied ion-Rydberg atom collisions are consistent with this model [20, 76]. As an illustration, the CTMC model predicts that Rydberg states in the range $63 < \tilde{n} < 200$ are obtainable for Ne^{10+} nucleus colliding with a target atom prepared in $n = 20$ excited state, with the peak near $\tilde{n} \approx 130$.

11.7 Summary

With the assumption that the theory is correct, QED is used to determine values of the relevant fundamental constants by adjusting their values to give the best agreement with experiments [77]. The Rydberg constant, for example, is currently determined to 5.0×10^{-12} uncertainty by adjusting its value to give predictions of atomic transition frequencies that agree best with twenty-four spectroscopic measurements in hydrogen and deuterium. The accuracy of the Rydberg constant determination using hydrogen spectroscopy is limited by theoretical uncertainties associated with the nuclear-size effects and, to a lesser extent, the two-loop QED contributions. In such a comparison between theory and experiments, the proton radius can also be adjusted to give the best agreement.

Surprisingly, recent measurements of the Lamb shift in muonic hydrogen, when compared with QED calculations, require a proton radius that is almost seven standard deviations smaller than the radius inferred from hydrogen spectroscopy and electron scattering experiments. The discrepant proton-radius value obtained from muonic hydrogen Lamb shift measurements has a significant impact on the determination of the Rydberg constant if used together with spectroscopic measurements of hydrogen and deuterium. This has generated renewed interest in alternative measurements that are independent of the proton radius. Although transitions between circular states of neutral hydrogen have been measured in the microwave regime and yield such a value of the Rydberg constant, the uncertainty is larger than the aforementioned discrepancy, and may be dominated by large dipole-dipole interactions.

In view of the availability of optical frequency combs, earlier theoretical work at NIST has addressed the possibility of comparing theory with measurements of optical transitions between Rydberg states in one-electron ions isolated in a trap. Attractive features of using one-electron ions include the robustness of Rydberg states when the energy spacing between neighbouring states is magnified by some power of a larger nuclear charge, making the Rydberg states more immune to external perturbations. Moreover, in contrast to neutral Rydberg atoms, the dipole-dipole interaction (which may have limited the MIT Rydberg hydrogen experiment [61]) would be orders of magnitude weaker due to the smaller orbitals around a stronger nuclear attraction. Finally, as reviewed here, recent calculations at NIST have shown that the theory of Rydberg states in one-electron ions is remarkably simplified for high angular momentum ($\ell > 2$).

Acknowledgments Experimental work by S.M. Brewer and N.D. Guise has been supported by the Chemical Physics Program of the University of Maryland, and the Research Associateship Program of the U.S. National Research Council, respectively. Valuable contributions to the theory of Rydberg states have been made by U.D. Jentschura and B.J. Wundt.

References

1. P.J. Mohr, B.N. Taylor, D.B. Newell, CODATA Recommended values of the fundamental physical constants: 2010. Rev. Mod. Phys. **84**(4), 1527 (2012)
2. G.W. Series (ed.), *The Spectrum of Atomic Hydrogen: Advances* (World Scientific, Singapore, 1988)
3. J.S. Rigden, *Hydrogen: The Essential Element* (Harvard University Press, Cambridge, 2002)
4. J.C. Bernauer et al., High-Precision determination of the electric and magnetic form factors of the proton. Phys. Rev. Lett. **105**, 242001 (2010)
5. G. Ron et al., Low-Q^2 measurements of the proton form factor ratio $\mu_P G_E/G_M$. Phys. Rev. C **84**, 055204 (2011)
6. I. Sick, Troubles with the Proton rms-Radius. Few-Body Syst. **50**(1–4), 367 (2011)
7. I. Sick, Problems with proton radii. Prog. Part. Nucl. Phys. **67**(2), 473 (2012)
8. C. Adamuscin, S. Dubnicka, A.Z. Dubnickova, New value of the proton charge root mean square radius. Prog. Part. Nucl. Phys. **67**(2), 479 (2012)
9. R. Pohl et al., The size of the proton. Nature **466**(7303), 213 (2010)
10. I. Sick, On the rms-radius of the proton. Phys. Lett. B **576**(1–2), 62 (2003)
11. P. Kusch, H.M. Foley, Precision measurement of the ratio of the atomic 'g values' in the $2P_{3/2}$ and $2P_{1/2}$ states of Gallium. Phys. Rev. **72**, 1256 (1947)
12. W.E. Lamb J, R.C. Retherford, Fine structure of the Hydrogen atom by a microwave method. Phys. Rev. **72**(3), 241 (1947)
13. W.E. Lamb J, Anomalous fine structure of Hydrogen and Singly Ionized Helium. Rep. Prog. Phys. **14**, 19 (1951)
14. R.S. Van Dyck Jr., P.B. Schwinberg, H.G. Dehmelt, New high-precision comparison of electron and positron g factors. Phys. Rev. Lett. **59**(1), 26 (1987)
15. L.S. Brown, G. Gabrielse, Geonium theory: single electrons and ions in a penning trap. Rev. Mod. Phys. **58**(1), 233 (1986)
16. D. Hanneke, S. Fogwell, G. Gabrielse, New measurement of the electron magnetic moment and the fine structure constant. Phys. Rev. Lett. **100**, 120801 (2008)
17. T. Hänsch, J. Alnis, P. Fendel, M. Fischer, C. Gohle, M. Herrmann, R. Holzwarth, N. Kolachevsky, T. Udem, M. Zimmermann, Precision spectroscopy of hydrogen and femtosecond laser frequency combs. Philos. Trans. R. Soc. London Ser. A **363**(1834), 2155 (2005)
18. F. Dyson, quoted in G. Gabrielse, D. Hanneke, T. Kinoshita, M. Nio, B. Odom, New determination of the fine structure constant from the electron g value and QED. Phys. Rev. Lett. **97**, 030802 (2006)
19. U.D. Jentschura, P.J. Mohr, J.N. Tan, B.J. Wundt, Fundamental constants and tests of theory in Rydberg states of hydrogenlike ions. Phys. Rev. Lett. **100**, 160404 (2008)
20. T.F. Gallagher, *Rydberg Atoms*, 1st edn. (Cambridge University Press, New York, 1994)
21. I. Sick, Precise proton radii from electron scattering. Can. J. Phys. **85**(5), 409 (2007)
22. I. Sick, in *Precision Physics of Simple Atoms and Molecules*, ed. by S.G. Karshenboim, Lecture Notes in Physics 745 (Springer, Berlin 2008), pp. 57–77
23. M. Fischer, N. Kolachevsky, M. Zimmermann, R. Holzwarth, T. Udem, T.W. Hänsch, M. Abgrall, J. Grünert, I. Maksimovic, S. Bize, H. Marion, F.P. Dos Santos, P. Lemonde, G. Santarelli, P. Laurent, A. Clairon, C. Salomon, M. Haas, U.D. Jentschura, C.H. Keitel, New limits on the drift of fundamental constants from laboratory measurements. Phys. Rev. Lett. **92**, 230802 (2004)
24. M. Weitz, A. Huber, F. Schmidt-Kaler, D. Leibfried, W. Vassen, C. Zimmermann, K. Pachucki, T.W. Hänsch, L. Julien, F. Biraben, Precision measurement of the 1S ground-state Lamb shift in atomic hydrogen and deuterium by frequency comparison. Phys. Rev. A **52**(4), 2664 (1995)
25. C.G. Parthey, A. Matveev, J. Alnis, R. Pohl, T. Udem, U.D. Jentschura, N. Kolachevsky, T.W. Hänsch, Precision measurement of the Hydrogen-Deuterium 1S–2S isotope shift. Phys. Rev. Lett. **104**, 233001 (2010)
26. C. Schwob, L. Jozefowski, B. de Beauvoir, L. Hilico, F. Nez, L. Julien, F. Biraben, O. Acef, A. Clairon, Optical frequency measurement of the 2S–12D transitions in hydrogen and deuterium: rydberg constant and lamb shift determinations. Phys. Rev. Lett. **82**(25), 4960 (1999)

27. O. Arnoult, F. Nez, L. Julien, F. Biraben, Optical frequency measurement of the 1S-3S two-photon transition in hydrogen. Eur. Phys. J. D **60**(2), 243 (2010)
28. S. Bourzeix, B. de Beauvoir, F. Nez, M.D. Plimmer, F. de Tomasi, L. Julien, F. Biraben, D.N. Stacey, High resolution spectroscopy of the hydrogen atom: determination of the 1S Lamb shift. Phys. Rev. Lett. **76**(3), 384 (1996)
29. D.J. Berkeland, E.A. Hinds, M.G. Boshier, Precise optical measurement of lamb shifts in atomic hydrogen. Phys. Rev. Lett. **75**(13), 2470 (1995)
30. E.W. Hagley, F.M. Pipkin, Separated oscillatory field measurement of hydrogen $2S_{1/2}$-$2P_{3/2}$ fine structure interval. Phys. Rev. Lett. **72**(8), 1172 (1994)
31. S.R. Lundeen, F.M. Pipkin, Separated sscillatory field measurement of the Lamb shift in H, n=2. Metrologia **22**(1), 9 (1986)
32. G. Newton, D.A. Andrews, P.J. Unsworth, A precision determination of the Lamb shift in hydrogen. Philos. Trans. R. Soc. London Ser. A **290**(1373), 373 (1979)
33. Z. Zhan et al., High-precision measurement of the proton elastic form factor ratio $\mu_P G_E/G_M$ at low Q^2. Phys. Lett. B **705**(1–2), 59 (2011)
34. R. Pohl, R. Gilman, G.A. Miller, K. Pachucki, Muonic Hydrogen and the Proton Radius Puzzle. Annu. Rev. Nucl. Part. Sci. **63**, 175 (2013)
35. R. Pohl et al., The size of the proton and the deuteron. J. Phys. Conf. Ser. **264**, 012008 (2011)
36. A. Antognini et al., Proton structure from the measurement of 2S–2P transition frequencies of Muonic Hydrogen. Science **339**(6118), 417 (2013)
37. A. Antognini, F. Kottmann, F. Biraben, P. Indelicato, F. Nez, R. Pohl, Theory of the 2S-2P Lamb shift and 2S hyperfine splitting in muonic hydrogen. Ann. Phys. (N.Y.) **331**, 127 (2013)
38. P. Indelicato, Nonperturbative evaluation of some QED contributions to the muonic hydrogen n = 2 Lamb shift and hyperfine structure. Phys. Rev. A **87**, 022501 (2013)
39. E. Borie, Lamb shift in light muonic atoms–Revisited. Ann. Phys. (N.Y.) **327**(3), 733 (2012). ArXiv:1103.1772-v6
40. S.G. Karshenboim, V.G. Ivanov, E.Y. Korzinin, Relativistic recoil corrections to the electron-vacuum-polarization contribution in light muonic atoms. Phys. Rev. A **85**, 032509 (2012)
41. U.D. Jentschura, Lamb shift in muonic hydrogen–I. Verification and update of theoretical predictions. Ann. Phys. (N.Y.) **326**(2), 500 (2011)
42. U.D. Jentschura, Lamb shift in muonic hydrogen–II. Analysis of the discrepancy of theory and experiment. Ann. Phys. (N.Y.) **326**(2), 516 (2011)
43. J.D. Carroll, A.W. Thomas, J. Rafelski, G.A. Miller, Nonperturbative relativistic calculation of the muonic hydrogen spectrum. Phys. Rev. A **84**, 012506 (2011)
44. J.C. De Vries, A precision millimeter-wave measurement of the Rydberg frequency. Ph.D. Thesis, MIT 2001
45. T.W. Hänsch, Nobel lecture: passion for precision. Rev. Mod. Phys. **78**(4), 1297 (2006)
46. M. Hori, A. Dax, J. Eades, K. Gomikawa, R. Hayano, N. Ono, W. Pirkl, E. Widmann, H.A. Torii, B. Juhász, D. Barna, D. Horváth, Determination of the Antiproton-to-Electron mass ratio by precision laser spectroscopy of $\bar{p}He^+$. Phys. Rev. Lett. **96**, 243401 (2006)
47. L.S. Ma, Z. Bi, A. Bartels, L. Robertsson, M. Zucco, R.S. Windeler, G. Wilpers, C. Oates, L. Hollberg, S.A. Diddams, Optical frequency synthesis and comparison with uncertainty at the 10^{-19} level. Science **303**(5665), 1843 (2004)
48. A. Marian, M.C. Stowe, J.R. Lawall, D. Felinto, J. Ye, United time-frequency spectroscopy for dynamics and global structure. Science **306**(5704), 2063 (2004)
49. J.R. Sapirstein, D.R. Yennie, in *Quantum Electrodynamics*, ed. by T. Kinoshita, (World Scientific, Singapore, 1990), Chap. 12, pp. 560–672
50. M.I. Eides, H. Grotch, V.A. Shelyuto, Theory of light hydrogenlike atoms. Phys. Rep. **342**(2–3), 63 (2001)
51. G. Gabrielse, P. Larochelle, D. Le Sage, B. Levitt, W.S. Kolthammer, R. McConnell, P. Richerme, J. Wrubel, A. Speck, M.C. George, D. Grzonka, W. Oelert, T. Sefzick, Z. Zhang, A. Carew, D. Comeau, E.A. Hessels, C.H. Storry, M. Weel, J. Walz, Antihydrogen production within a penning-ioffe trap. Phys. Rev. Lett. **100**, 113001 (2008)

52. U.D. Jentschura, P.J. Mohr, J.N. Tan, B.J. Wundt, Fundamental constants and tests of theory in Rydberg states of hydrogenlike ions. Can. J. Phys. **87**(7), 757 (2009)
53. U.D. Jentschura, P.J. Mohr, J.N. Tan, Fundamental constants and tests of theory in Rydberg states of one-electron ions. J. Phys. B **43**, 074002 (2010)
54. E.H. Wichmann, N.M. Kroll, Vacuum polarization in a strong Coulomb field. Phys. Rev. **101**(2), 843 (1956)
55. U.D. Jentschura, Two-loop Bethe logarithms for non-S levels. Phys. Rev. A **74**, 062517 (2006)
56. E.O. Le Bigot, U.D. Jentschura, P.J. Mohr, P. Indelicato, G. Soff, Perturbation approach to the self-energy of non-S hydrogenic states. Phys. Rev. A **68**, 042101 (2003)
57. G. Audi, A.H. Wapstra, C. Thibault, The AME2003 atomic mass evaluation (II). Nucl. Phys. A **729**(1), 337 (2003)
58. H.A. Bethe, E.E. Salpeter, *Quantum Mechanics of One- and Two-electron Atoms* (Academic Press, New York, 1957)
59. F. Low, Natural Line Shape. Phys. Rev. **88**(1), 53 (1952)
60. U.D. Jentschura, P.J. Mohr, Nonresonant effects in one- and two-photon transitions. Can. J. Phys. **80**(6), 633 (2002)
61. D. Kleppner, Private communication (2012)
62. W.M. Itano et al., Cooling methods in ion traps. Phys. Scripta **T59**, 106 (1995)
63. G. Gabrielse, Atoms made entirely of antimatter: Two methods produce slow antihydrogen in *Advances in Atomic Molecular, and Optical Physics*, ed. by B. Bederson, H. Walthers (Elsevier, San Diego, 2005), **50**, pp. 155–217
64. L.P. Ratliff, E.W. Bell, D.C. Parks, A.I. Pikin, J.D. Gillaspy, Continuous highly charged ion beams from the NIST electron-beam ion trap. Rev. Sci. Instrum. **68**, 1998 (1997)
65. L. Ratliff, J. Roberts, Highly charged ion studies at the NIST EBIT, in *Trapping Highly Charged Ions: Fundamentals & Applications*, ed. by J. Gillaspy (Nova Science Publishers, Hauppauge, 2001), p. 257
66. M.R. Flannery, D. Vrınceanu, Quantal and classical radiative cascade in Rydberg plasmas. Phys. Rev. A **68**, 030502(R) (2003)
67. J.D. Gillaspy, L.P. Ratliff, J.R. Roberts, E. Takács, Highly charged ions: Publications of the ebit project, 1993–2001. Special Publication 972, NIST (2001).
68. J.N. Tan, S.M. Brewer, N.D. Guise, Penning traps with unitary architecture for storage of highly charged ions. Rev. Sci. Instrum. **83**(2), 023103 (2012). doi:10.1063/1.3685246. http://link.aip.org/link/?RSI/83/023103/1
69. G.E. Holland, C.N. Boyer, J.F. Seely, J.N. Tan, J.M. Pomeroy, J.D. Gillaspy, Low jitter metal vapor vacuum arc ion source for electron beam ion trap injections. Rev. Sci. Instrum. **76**(7), 073304 (2005)
70. S.M. Brewer, N.D. Guise, J.N. Tan, Capture and isolation of highly charged ions in a unitary Penning trap. Phys. Rev. A **88**, 063403 (2013). doi:10.1103/PhysRevA.88.063403
71. N.D. Guise, S.M. Brewer, J.N. Tan, Highly charged ions in rare-earth permanent-magnet Penning traps. arXiv:1207.3333v1 e-prints (2012)
72. S.M. Brewer, N.D. Guise, J.N. Tan, Observing forbidden radiative decay of highly charged ions in a compact Penning trap. BAPS.2012.DAMOP.U3.8 (2012)
73. N.D. Guise, S.M. Brewer, J.N. Tan, Charge exchange and spectroscopy with isolated highly charged ions. BAPS.2012.DAMOP.U3.6 (2012)
74. S.R. Lundeen, C.W. Fehrenbach, Polarizability of Kr^{6+} from high-L Kr^{5+} fine-structure measurements. Phys. Rev. A **75**(3), 032523 (2007). doi:10.1103/PhysRevA.75.032523
75. R. Olson, Ion-Rydberg atom collision cross sections. J. Phys. B: Atom. Molec. Phys. **13**, 483 (1980)
76. B.D. DePaola, M.T. Huang, S. Winecki, M.P. Stockli, Y. Kanai, S.R. Lundeen, C.W. Fehrenbach, S.A. Arko, Absolute cross-sections for charge capture from Rydberg targets by slow highly charged ions. Phys. Rev. A **52**, 2136 (1995)
77. P.J. Mohr, B.N. Taylor, CODATA recommended values of the fundamental physical constants: 2002. Rev. Mod. Phys. **77**(1), 1 (2005)

Index

W. Quint and M. Vogel (eds.), *Fundamental Physics in Particle Traps*,
Springer Tracts in Modern Physics 256, DOI: 10.1007/978-3-642-45201-7,

F

G

H

I

Printed by Printforce, the Netherlands